Structural Seismic Design and Evaluation

Structural Seismic Design and Evaluation

Guest Editors

Alfredo Reyes-Salazar
Federico Valenzuela-Beltran
Mario D. Llanes-Tizoc

MDPI

Basel • Beijing • Wuhan • Barcelona • Belgrade • Novi Sad • Cluj • Manchester

Guest Editors
Alfredo Reyes-Salazar
Facultad de Ingeniería
Universidad Autónoma de Sinaloa
Culiacan
Mexico

Federico Valenzuela-Beltran
Facultad de Ingeniería
Universidad Autónoma de Sinaloa
Culiacan
Mexico

Mario D. Llanes-Tizoc
Facultad de Ingeniería
Universidad Autónoma de Sinaloa
Culiacan
Mexico

Editorial Office
MDPI AG
Grosspeteranlage 5
4052 Basel, Switzerland

This is a reprint of the Special Issue, published open access by the journal *Applied Sciences* (ISSN 2076-3417), freely accessible at: https://www.mdpi.com/journal/applsci/special_issues/INNILVU08G.

For citation purposes, cite each article independently as indicated on the article page online and as indicated below:

Lastname, A.A.; Lastname, B.B. Article Title. *Journal Name* **Year**, *Volume Number*, Page Range.

ISBN 978-3-7258-5163-8 (Hbk)
ISBN 978-3-7258-5164-5 (PDF)
https://doi.org/10.3390/books978-3-7258-5164-5

Contents

About the Editors

Alfredo Reyes-Salazar

Alfredo Reyes-Salazar is a faculty member at the Universidad Autónoma de Sinaloa. His main interests lie in the seismic behavior of steel buildings, on which he has completed various projects. He is a Level III SECIHTI Scholar and has also been granted the title of Emeritus Professor. He has authored more than 105 papers published in indexed journals and more than 130 conference papers, and has received over 2440 citations. Furthermore, he has received numerous awards, including the "Civil Engineer of the Year" award from the College of Civil Engineers of the state of Sinaloa, and the "Night of the Suns Award" from the government of the state of Sinaloa.

Federico Valenzuela-Beltran

Federico Valenzuela-Beltran is a faculty member at the Universidad Autónoma de Sinaloa and a Level I SECIHTI Scholar. His primary research interests include the seismic analysis and design of structures, performance-based earthquake engineering, structural reliability, and the design of systems incorporating energy dissipation devices. He has authored more than 50 papers published in indexed journals and international conference proceedings. Additionally, he has collaborated on several research projects funded by Mexico's Secretariat of Science, Humanities, Technology, and Innovation (SECIHTI).

Mario D. Llanes-Tizoc

Mario D. Llanes-Tizoc is a faculty member at the Universidad Autónoma de Sinaloa and a Level I SECIHTI Scholar. His main interests lie in the seismic behavior and performance-based design of reinforced concrete and steel buildings, on which he has completed various projects. He has authored more than 20 papers published in indexed journals and received more than 90 citations. Furthermore, he has received numerous awards, including being granted the title of "Emeritus Researcher" by the government of the state of Sinaloa.

applied sciences

MDPI

Editorial

Structural Seismic Design and Evaluation

Federico Valenzuela-Beltran, Mario D. Llanes-Tizoc and Alfredo Reyes-Salazar *

Faculty of Engineering, Autonomous University of Sinaloa, Ciudad Universitaria, Culiacán 80000, Mexico; fvalenzuelab@uas.edu.mx (F.V.-B.); mariollanes@uas.edu.mx (M.D.L.-T.)
* Correspondence: reyes@uas.edu.mx

1. Introduction

Significant damage to structures, induced by the occurrence of strong earthquakes, has been observed in recent decades. This has led researchers worldwide to address several issues concerning structural seismic design and evaluation (SSDE) for different types of structures, including steel and concrete buildings. In fact, the scope of the term SSDE is extremely extensive, meaning that the following discussion will cover only a small part of this large issue. Topics of interest include (a) the evaluation of the reliability of concrete and steel structures with energy-dissipating devices, (b) seismic performance and risk analyses of structures, (c) deterministic and probabilistic methods of dynamic analysis in structural engineering, (d) the seismic behavior and modeling of nonstructural elements, and (e) the seismic performance of new structural systems.

During the last four decades, based on the results of investigations, important steps have been taken within the field of *SSDE*, including the emergence of new analysis and design procedures. A change associated with this is the considerable shift from traditional elastic to inelastic analyses in the profession. There is no doubt that the most accurate analysis method is the step-by-step nonlinear seismic procedure, provided that the dynamic characteristics of the structure, as well as the loading, unloading, and reloading processes onto the structural members, are adequately represented.

Among the many important efforts relating to the analysis and seismic design of structures, we can mention new smart materials, innovative connections and structural members, energy dissipation devices, and structural systems. The aim of the remainder of this paper is to discuss the results of specific investigations focusing on the improvement of SSDE.

2. Recent Advancements in *SSED*

The use of innovative structural systems to improve the seismic performance of buildings is one of the topics that have been addressed recently. Various studies have evaluated the advantages of incorporating different devices into structures, such as buckling-restrained braces (BRBs), viscous dampers, and base isolators. For instance, Della Corte et al. [1] analyzed experimental and theoretical advancements in BRB design, highlighting their energy dissipation capacity and greater ductility compared with traditional systems. The study addressed key aspects such as deformation capacity, connection design, and the effects of non-adherent material layers. It also compared different design approaches and examined the performance of BRBs using real-world tests and numerical models. The authors concluded that BRBs offer an efficient and cost-effective solution for seismic design, although challenges such as residual displacements remain. Bai and Ou [2] developed a performance-based plastic design method for dual systems of reinforced concrete frames with BRBs. The method was validated through static and dynamic analyses on

5- and 10-story structures, demonstrating that they meet the desired seismic performance levels. This method can be applied to other structural systems with braces to enhance their seismic resistance.

Some studies have evaluated the economic benefits of incorporating innovative structural systems to improve the seismic resilience of communities. For example, Hu et al. [3] compared the seismic performance of mid-rise steel buildings with three lateral force-resisting systems: moment-resisting frames (MRF), buckling-restrained braced frames, and re-centering non-buckling braced frames. Using incremental dynamic analyses (IDA), vulnerability functions were derived, and economic loss performance pyramids were established. The results highlighted the advantages and limitations of each system in reducing economic losses due to seismic events. Laguardia et al. [4] analyzed 12,016 reinforced concrete buildings to develop vulnerability, fragility, and loss curves, along with an Expected Annual Loss (EAL) assessment. Using data from damage observations from the 2009 L'Aquila earthquake, the methodology incorporated undamaged and non-surveyed buildings based on typological distributions in reference municipalities. The results showed that the completed database improves the reliability of seismic risk assessments, although EAL assessments are minimally affected by the choice of reference municipality. Important contributions in the same area can be found in other studies [5–7].

Another current topic of great interest is the use of alternative methodologies for seismic design. One such relatively recent approach is performance-based seismic design (PBSD), which has been the focus of numerous recent studies and incorporated into several building codes around the world. A significant number of studies have focused on the implementation of this methodology in the seismic design and evaluation of various types of structures, highlighting the advantages of these alternative seismic design approaches [8,9]. Steneker et al. [10] proposed a framework to optimize seismic upgrade strategies for buildings, considering both structural and nonstructural components. It introduces the Median Shift Probability method, a modified version of the Performance-Based Earthquake Engineering (PBEE) methodology, to quickly assess the impact of structural upgrades on nonstructural components by analyzing how structural changes affect floor-level hazards. This approach allows for the disaggregation of expected losses and identifies combined upgrade strategies that are tailored to a building's specific conditions, optimizing the use of resources in the early design stages.

Bakhshivand et al. [11] assessed the seismic performance of buildings with dual lateral force-resisting systems combining special moment frames and special concentrically braced frames (SMF-SCBF) using the FEMA P695 methodology. Six building archetypes (2, 4, 8, 12, 16, and 20 stories) were designed according to ASCE 7–16 and AISC 341–16 and analyzed using nonlinear models in OpenSees. The buildings were evaluated through nonlinear static and dynamic analyses under 44 far-field ground motions, examining peak and residual drift ratios, hysteresis behavior, collapse fragility curves, and seismic performance factors (R, Ω, and C_d). The results revealed that long-period dual SMF-SCBF systems (16 stories or higher) that were compliant with current codes performed poorly and were prone to severe damage during strong earthquakes. Ciurlanti et al. [12] highlighted the gap between societal expectations and the seismic performance of modern buildings in the aftermath of an earthquake. They claimed that current life safety codes are insufficient for new structures, necessitating higher performance standards to minimize business disruption and economic losses. This study explored strategies to achieve (a) advanced design methods, (b) improved seismic design performance, and (c) low-damage technologies. A parametric cost–benefit analysis compared reinforced concrete buildings that are designed for varying seismic intensities using traditional vs. advanced approaches (force-based vs. displacement-based) and technologies. The results, enhanced by machine learning, showed that displacement-based designs and low-damage technologies significantly improve performance, reducing economic losses by over 50%, with only a 5–10% increase in initial costs.

Many other studies have also been conducted with different specific objectives regarding the evaluation of the seismic behavior of structures in terms of risk assessment, the effects of irregularity, and response modification factors. For example, Parizat et al. [13] aimed to develop equations for moment-resisting reinforced concrete frames to estimate the total floor acceleration based on the fact that failures in nonstructural components (NSCs) due to earthquakes can result in casualties, significant economic losses, disabled critical infrastructures, and loss of building functionality, as well as the fact that planning under the peak floor response or peak floor acceleration is receiving significantly less attention. The data used were obtained from the results of 984 inelastic response simulations and processed to create an idealized equation based on the earthquake characteristics. The proposed equation offers engineers a quantitative approach to understanding the inertial forces that are applied to NSCs in buildings during earthquakes, allowing them to assess the potential risks. Blasi et al. [14] analyzed the seismic response of regular and irregular reinforced concrete frames. Their main results showed an important influence of irregularity, especially on floor accelerations and displacements caused by the change in mass and stiffness over height. Hussain et al. [15] aimed to overcome the limitation that recommended seismic design coefficients in design codes are based on previous research by conducting a study, focusing primarily on regular low- to medium-rise building configurations and using two-dimensional analysis, to validate overstrength factors, as well as amplification and response modification coefficients, using three-dimensional (3D) dynamic response simulation and bidirectional earthquake records for high-rise buildings with different structural systems and irregularities. They suggested increasing the response modification coefficients by 5% and 10% for the structural systems studied. Ricci et al. [16] evaluated the behavior and failure probabilities for reinforced concrete MRF using nonlinear static analysis. Based on their results, they proposed a simplified equation to evaluate the behavioral factors.

In another study, Yousefabadi and Kazemi [17] investigated the performance of a structural system based on a reinforced concrete column and steel beam (RCS) system, with a focus on evaluating the seismic design coefficients using the FEMA-P695 methodology. The RCS system offers a more efficient and cost-effective solution compared with conventional steel and RC moment-resisting frames, with higher damping and lateral stiffness of RC columns and a greater energy dissipation capacity of the steel beams. To reach their objectives, 32 archetypes were designed with varying building heights, span lengths, concrete strengths, gravity load levels, seismic load levels, and column–beam strength ratios. Nonlinear analytical models were developed for the selected archetypes, and the modeling assumptions were validated through five distinct experimental tests. The authors concluded that the design requirements of the RCS system are efficient, providing a high safety margin. However, the level of conservatism was found to be excessively high. Thus, it is possible to use a larger R-factor in the design process or implement some relaxations in the design requirements relating to this structural system. Nanclares et al. [18] presented the results of a nonlinear dynamic analysis of a highway bridge, idealized as a 3D finite element model according to the design codes from different periods. Based on the observed results, they proposed to provide higher shear strength, higher ductility, and buckling restraint to structures that are designed based on outdated requirements.

3. Contribution of This Special Issue

The central objective of this Special Issue was to provide a publishing platform for the global community of researchers in traditional and emerging subdisciplines of the field relating to the seismic behavior of structures and structural design in order to present and discuss recent advancements in this broad field. Although several topics were considered in the Special Issue, most of the papers published fall within the following categories: (a) evaluations of the reliability of concrete and steel structures with energy-dissipating devices, (b) seismic performance and risk analyses of structures, (c) experimental and computational simulations of dynamic effects on structures, (d) the seismic behavior and modeling of nonstructural elements, and (e) quantification of earthquake demands and structural capacity. The contributions of a few of the included papers are presented below.

The protection of existing structures against strong earthquakes has been the subject of some interest. For example, Di Egidio et al. [19] studied the effectiveness of intermediate discontinuities for the seismic protection of new and existing framed structures under the action of harmonic and seismic excitations. The intermediate elasto-plastic discontinuity used in the study was described using the Bouc–Wen model, which in turn was represented with a 3-DOF reduced model. This technique has received an increasing amount of interest, because it is cheaper and technically easier to implement than base isolation. In another study, Chonratana and Chatpattananan [20] proposed a damage index to assess the seismic-resistant design of masonry wall buildings that are reinforced with double x-bracing concrete frames. This damage index was calculated by using the analytic results obtained from four-story concrete structures with masonry walls under cyclical lateral forces, along with the Park–Ang model. The damage index can be used in the seismic-resistant design of masonry walls that are reinforced with x-bracing concrete frames. Kuria and Kegyes-Brassai [21] analyzed the evolution, utilization, and challenges of typical pushover analyses, such as the capacity spectrum and displacement coefficient methods, for evaluating the seismic performance of structures. These methods, along with the advanced versions (multi-mode, modal, adaptive, and energy-based pushover analysis) help determine seismic demands with great accuracy. The authors stated that such advanced versions improved accuracy by considering major vibration modes; however, it sometimes failed to address specific aspects, such as bidirectional responses and torsional effects. The paper highlighted the continuing refinement of the pushover methods, reflecting the sustained attention that they receive from both industry practitioners and researchers.

The above discussion and that presented in Section 2 of the paper clearly illustrate the considerable advancements in the field of SSDE in recent years. However, as discussed in the Conclusion, it is also clear that much more research is needed to make structures more resilient to the action of strong earthquakes.

4. Conclusions

The considerable damage to structures after the occurrence of strong earthquakes in recent decades has led researchers worldwide to address several issues concerning structural seismic design and evaluation (*SSDE*). It is worth mentioning that the term *SSDE* covers an extremely broad field, so in this paper, only a few aspects of this large issue are covered.

During the last four decades, based on the results of a number of investigations, important steps have been taken within the field of *SSDE*, including the emergence of new analysis and design procedures. Among these important efforts, we can mention new smart materials, innovative connections and structural members, energy dissipation devices, and structural systems. In this paper, a few contributions to the field were briefly mentioned. Nevertheless, it is important to emphasize that further research is needed to make structures more resilient to the action of strong earthquakes.

In general, the limitations of the investigations presented in Chapters 2 and 3 can be described in one or more of the following ways: (a) In the case of buildings, the structural models are based on framed members. In addition, the effects of connections are neglected. (b) When experimental investigations are considered, structural subassemblies are used, thus ignoring important parts of the structure. (c) Cyclic loads are assumed, neglecting the effects of the dynamic characteristics of the structures. (d) The concentrated plasticity model is considered. (e) The Rayleigh model is adopted to simulate viscous damping, (f) In some countries, the adoption of emerging seismic response control systems has been slower compared with some developed nations, mainly due to the lack of experimental studies and the absence of local regulations that would enable large-scale implementation. (g) The estimation of economic losses is limited in many studies to the use of international databases, such as FEMA P-58. To achieve a more accurate estimation when applied to other regions, a database reflecting local construction costs and practices should be developed. (h) Seismic loss analyses have been limited to only a few specific cases. Regional studies should be conducted to estimate losses and enhance community resilience against this phenomenon.

Author Contributions: F.V.-B.: writing—original draft preparation; M.D.L.-T.: writing—original draft preparation; A.R.-S. writing—review and editing. All authors have read and agreed to the published version of the manuscript.

Funding: This research received no external funding.

Conflicts of Interest: The authors declare no conflicts of interest regarding the publication of this paper.

References

1. Della Corte, G.; D'Aniello, M.; Landolfo, R.; Mazzolani, F.M. Review of Steel Buckling-restrained Braces. *Steel Constr.* **2011**, *4*, 85–93. [CrossRef]
2. Liu, S.-W.; Bai, R.; Chan, S.-L. Dynamic Time-History Elastic Analysis of Steel Frames Using One Element per Member. *Structures* **2016**, *8*, 300–309. [CrossRef]
3. Hu, S.; Wang, W.; Qu, B. Seismic Economic Losses in Mid-Rise Steel Buildings with Conventional and Emerging Lateral Force Resisting Systems. *Eng. Struct.* **2020**, *204*, 110021. [CrossRef]
4. Laguardia, R.; D'Amato, M.; Coltellacci, M.; Di Trocchio, G.; Gigliotti, R. Fragility Curves and Economic Loss Assessment of RC Buildings after L'Aquila 2009 Earthquake. *J. Earthq. Eng.* **2023**, *27*, 1126–1150. [CrossRef]
5. Santos-Santiago, M.A.; Ruiz, S.E.; Cruz-Reyes, L. Optimal Design of Buildings under Wind and Earthquake, Considering Cumulative Damage. *J. Build. Eng.* **2022**, *56*, 104760. [CrossRef]
6. Cando, M.A.; Hube, M.A.; Parra, P.F.; Arteta, C.A. Effect of Lateral Stiffness on Expected Economic Losses in Reinforced Concrete Shear Wall Buildings. *Eng. Struct.* **2023**, *297*, 116894. [CrossRef]
7. Chalarca, B.; Filiatrault, A.; Perrone, D. Expected Seismic Response and Annual Seismic Loss of Viscously Damped Braced Steel Frames. *Eng. Struct.* **2024**, *303*, 117569. [CrossRef]
8. González, C.; Niño, M.; Ayala, G. Functionality Loss and Recovery Time Models for Structural Elements, Non-Structural Components, and Delay Times to Estimate the Seismic Resilience of Mexican School Buildings. *Buildings* **2023**, *13*, 1498. [CrossRef]
9. Banihashemi, M.; Miliziano, A.; Zsarnóczay, Á.; Wiebe, L.; Filiatrault, A. Consequences of Consequence Models: The Impact of Economies of Scale on Seismic Loss Estimates. *Earthq. Spectra* **2024**, *40*, 1396–1424. [CrossRef]
10. Steneker, P.; Wiebe, L.; Filiatrault, A.; Konstantinidis, D. A Framework for the Rapid Assessment of Seismic Upgrade Viability Using Performance-Based Earthquake Engineering. *Earthq. Spectra* **2022**, *38*, 1761–1787. [CrossRef]
11. Bakhshivand, E.; Ahmadie Amiri, H.; Maleki, S. Evaluation of Seismic Performance Factors for Dual Steel SMF-SCBF Systems Using FEMA P695 Methodology. *Soil Dyn. Earthq. Eng.* **2022**, *163*, 107506. [CrossRef]
12. Ciurlanti, J.; Bianchi, S.; Pampanin, S. Raising the Bar in Seismic Design: Cost–Benefit Analysis of Alternative Design Methodologies and Earthquake-Resistant Technologies. *Bull. Earthq. Eng.* **2023**, *21*, 2723–2757. [CrossRef]
13. Parizat, S.; Kamai, R.; Shmerling, A. Correlating the Seismic Acceleration of Reinforced Concrete Moment-Resisting-Frames with Structural and Earthquake Characteristics. *Bull. Earthq. Eng.* **2024**, *22*, 5059–5081. [CrossRef]
14. Blasi, G.; Scarlino, A.S.; Chirivì, S.; Perrone, D.; Aiello, M.A. Seismic Response of Irregular RC Buildings Designed for Gravity and Seismic Loads. *Bull. Earthq. Eng.* **2024**, *22*, 5231–5257. [CrossRef]

15. Hussain, N.; Alam, M.S.; Mwafy, A. Seismic Design Coefficients Verification of Regular and Vertically Irregular High-Rise Shear Wall Buildings Using Bidirectional Horizontal Ground Motions and 3D Modeling. *Structures* **2024**, *70*, 107835. [CrossRef]
16. Ricci, P.; Di Domenico, M.; Verderame, G.M. Behaviour Factor and Seismic Safety of Reinforced Concrete Structures Designed According to Eurocodes. *Structures* **2023**, *55*, 677–689. [CrossRef]
17. Tavasoli Yousefabadi, E.; Kazemi, M.T. Quantification of the Seismic Design Coefficients for Composite Special Moment Frames with Reinforced Concrete Columns and Steel Beams. *Adv. Struct. Eng.* **2024**, *27*, 487–505. [CrossRef]
18. Nanclares, G.; Ambrosini, D.; Curadelli, O. Comparative Nonlinear Seismic Analysis of an Existing RC Bridge Designed with Obsolete Codes. *Adv. Struct. Eng.* **2022**, *25*, 1386–1401. [CrossRef]
19. Di Egidio, A.; Pagliaro, S.; Contento, A. Seismic Performance of Frame Structure with Hysteretic Intermediate Discontinuity. *Appl. Sci.* **2023**, *13*, 5373. [CrossRef]
20. Chonratana, Y.; Chatpattananan, V. A Damage Index for Assessing Seismic-Resistant Designs of Masonry Wall Buildings Reinforced with X-Bracing Concrete Frames. *Appl. Sci.* **2023**, *13*, 12566. [CrossRef]
21. Kuria, K.K.; Kegyes-Brassai, O.K. Pushover Analysis in Seismic Engineering: A Detailed Chronology and Review of Techniques for Structural Assessment. *Appl. Sci.* **2023**, *14*, 151. [CrossRef]

applied sciences

MDPI

Article

Optimal Seismic Retrofit Alternative for Shear Deficient RC Beams: A Multiple Criteria Decision-Making Approach

Paola Villalba [1,2], Byron Guaygua [1,2] and Víctor Yepes [1,*]

1 Institute of Concrete Science and Technology (ICITECH), Universitat Politècnica de València, 46022 Valencia, Spain; pxvillalba@uce.edu.ec (P.V.); bguaygu@doctor.upv.es (B.G.)
2 Faculty of Engineering and Applied Sciences, Civil Engineering Degree Programme, Universidad Central del Ecuador, Quito 170521, Ecuador
* Correspondence: vyepesp@cst.upv.es

Abstract: The vulnerability of existing buildings to recent earthquakes underscores the critical need to explore effective retrofit solutions thoroughly. This study presents a comprehensive methodology for ranking seismic retrofit alternatives for reinforced concrete beams with shear deficiencies. It evaluates five alternatives to ensure a 50-year service life, meeting current seismic standards and incorporating specific preventive maintenance measures for each option. A cradle-to-grave life cycle assessment was used to analyze the impacts associated with the sustainability of each alternative. Hybridization of emerging multi-criteria decision-making methods was applied for criteria weighting and final ranking, and a hierarchical model including economic, environmental, social, and functional criteria was developed. The results highlight carbon fiber reinforcements and steel plates with epoxy adhesives as optimal solutions due to their lower environmental and social impact, along with improvements in execution time and minimal architectural impact. This study underscores the necessity of a comprehensive approach to identifying optimal retrofitting alternatives, demonstrating the imperative to complement the conventional structural engineering objective of ensuring safety while minimizing investment.

Keywords: decision making; life cycle assessments; MCDM; retrofit; seismic

1. Introduction

The past decade's seismic events underscore societies' persistent vulnerability to earthquakes. Despite increased efforts by governments, academia, and international organizations to mitigate seismic risks, recent earthquakes reveal that significant challenges remain [1]. Much of the aging building stock was constructed under outdated seismic design standards and fails to meet current seismic resistance codes [2]. As more buildings near the end of their intended service life, additional risks related to material durability emerge; this creates a substantial societal risk, as evidenced by the susceptibility of existing structures to recent earthquakes, which have caused structural damage, significant economic losses, serious injuries, and loss of life [3].

Informal, self-managed housing construction—typically undertaken by residents or builders without formal training and often without permits or adherence to building codes—constitutes a significant proportion of residential construction in low- and middle-income countries [4]. Post-earthquake assessments have consistently highlighted structural vulnerabilities and a high incidence of collapses in informally constructed reinforced concrete and masonry housing [5]. One of the most devastating recent earthquakes, the

April 2016 Pedernales earthquake on the northwest coast of Ecuador, had a moment magnitude (MW) of 7.8. Official statistics reported 668 fatalities, 12 missing persons, 6274 injuries, 113 rescues, and 28,775 individuals displaced to shelters. Approximately 30,000 homes were severely damaged or destroyed, with urban (13,962) and rural (15,710) areas experiencing nearly equal levels of destruction [6]. The widespread damage to low- and mid-rise reinforced concrete buildings revealed that inadequate seismic detailing and poor-quality control contributed to their partial or total collapse [7]. Limited economic resources further exacerbate these problems, as most construction practices in informal buildings do not comply with seismic codes.

Post-earthquake studies worldwide have highlighted the significant impact of shear failures in beams, columns, and beam-column joints on the seismic performance of existing reinforced concrete buildings [8]. Pre-code reinforced concrete structures are particularly susceptible to shear failures during seismic events, underscoring the critical importance of accurately estimating the shear capacity of these elements. Many existing reinforced concrete buildings were designed using outdated standards that often overestimated the shear strength of beams. Unlike flexural damage, shear damage is brittle and far more detrimental to structural performance. Furthermore, poor reinforcement practices—such as insufficient transverse reinforcement and overlapping splices in the plastic hinge region—heighten vulnerability by promoting brittle failure modes, including shear failures. These deficiencies compromise structural integrity, posing significant risks to property and human life [9].

Increasing the shear capacity of reinforced (RC) concrete beams is crucial, as shear failure is sudden and typically accompanied by poor energy dissipation. Numerous parameters influence the shear strength of reinforced concrete beams, and predicting shear capacity is an ongoing area of research. Current design standards employ either empirical or semi-empirical methods. ACI-318 uses a semi-empirical approach based on truss models for shear reinforcement contribution and an empirical factor for the concrete contribution [10]. In contrast, Eurocode 2 adopts the variable strut inclination method [11]. Additionally, other researchers have proposed models based on the modified compression field theory and the general stress field approach [10].

Localized preventive interventions can help mitigate potential damage. Several methods for enhancing the shear strength of reinforced concrete beams have been developed, with the most common being the addition of reinforced concrete jacketing, steel jacketing, and composite jacketing (fiber-reinforced polymers) [12,13]. RC cladding is particularly effective in improving shear performance and is a relatively easy-to-apply reinforcement method [12]. It is also a well-established technique for retrofitting structural elements. The success of the jacketing process largely depends on achieving a strong bond between the old and new materials, with reinforced concrete being highly effective at the interface, making it a reliable material for U-shaped reinforcement of shear-critical beams [14]. Steel jacketing offers the advantage of requiring less space, being lightweight, and easy to install [13]. On the other hand, externally bonded fiber-reinforced polymer (FRP) composites have emerged as an efficient solution, enhancing bending, shear, torsion, and confinement capabilities. FRP composites are characterized by their low invasiveness, high strength-to-weight ratio, and ease of application [15].

Structural engineering has traditionally focused on achieving maximum safety with minimal investment. However, growing concerns about sustainability have elevated the importance of its three pillars: economic, environmental, and social. A circular economy (CE) presents a promising approach to addressing sustainability challenges by extending the useful life of the built environment through the retrofit and refurbishment of buildings and infrastructure [16]. In the building and construction industry, the economic dimension of sustainability has historically received the most attention, while environmental impacts have gained increasing research interest in recent years. Seismic retrofitting of reinforced concrete buildings typically involves the production of new materials and implementing construction processes, contributing to the industry's already significant environmental impact. Given the large number of buildings with seismic deficiencies, these retrofitting activities may exacerbate negative environmental consequences [17]. Meanwhile, the social dimension of sustainability in the construction sector faces substantial challenges due to its relative development and lack of maturity [18].

Multi-criteria decision-making methods (MCDM) are valuable tools for tackling complex engineering problems, particularly when selecting optimal solutions under conflicting criteria [19]. Multiple attribute decision-making methodologies (MADMs) have received much attention recently. MADMs have evolved into comprehensive operations research methods integrating computational and mathematical tools to identify optimal solutions [20]. MADMs enable a holistic assessment of safety and sustainability by addressing conflicting dimensions and criteria. Recent research has applied decision-making methods to rank seismic retrofitting alternatives, incorporating economic, functional, and environmental criteria. These studies include contributions from references [21–25]. However, many building retrofit studies lack social life cycle analysis, leaving a significant gap. Despite the availability of advanced MADMs, their application in seismic retrofit remains limited [26]. Classical decision-making methods continue to dominate the research landscape. With the proliferation of innovative decision-making approaches in recent years, there is considerable potential to improve accuracy and applicability, offering robust solutions to current challenges. The term "postpone effect" aptly describes the state of the field, reflecting the lag in adopting and applying modern MCDMs in research [27].

The scientific community has made significant strides in utilizing MCDM methods to select optimal seismic retrofitting alternatives. However, there is still a need to thoroughly explore retrofit strategies with a holistic approach that can effectively minimize seismic risk. While established MCDM methods have been widely applied, newer methodologies developed in recent years may yield improved results in the weighting of criteria and the final ranking of alternatives. This study addresses these gaps by introducing a methodology that integrates the three pillars of sustainability—economic, environmental, and social—through life cycle assessments alongside a functional dimension. The primary objective is to propose a comprehensive framework for classifying seismic retrofit strategies for reinforced concrete beams with shear deficiencies at the local level. Additionally, two secondary objectives are pursued: first, to evaluate the sustainability dimensions through life cycle assessments, and second, to employ emerging MCDM methods for criteria weighting and ranking alternatives.

The article is structured to achieve these objectives: Section 2 reviews the literature on life cycle assessments and emerging MCDM approaches. Section 3 contains materials and methods and details the proposed methodology. Sections 4 and 5 analyze and discuss the results, including a sensitivity analysis. Finally, Section 6 concludes with the key findings of the study.

2. Literature Review

2.1. Life Cycle Assessment

The life cycle assessment (LCA) framework is a powerful tool for assessing the environmental impacts of products, systems, or processes throughout their life cycle. Life cycle sustainability assessment is increasingly applied in the building and construction sector, with significant economic, environmental, and social impacts [18]. Recent studies have focused on analyzing the environmental aspects of seismic retrofitting of buildings using LCA. For example, a group of researchers compared the environmental performance of reinforced concrete column jacketing and dissipation devices, concluding that the latter is more environmentally friendly than traditional solutions [28]. Similarly, another study evaluated three seismic retrofitting methods for an old RC building: concrete shear wall reinforcement, RC column jacketing with shear wall reinforcement, and base isolation using high-damping rubber bearings and viscous fluid dampers. Of these, the base isolation method was found to have the least environmental impact [29]. Another group of researchers conducted an LCA of three seismic retrofitting alternatives for a reinforced concrete building: RC column jacketing, beam weakening, and shear walls. Among them, shear walls were found to have the highest environmental impact [17]. Similarly, another research analyzed and compared the environmental performance of four seismic retrofitting strategies: FRP-based strengthening solution, FRP-RC jacketing-based strengthening solution, installation of RC shear walls, and base isolation. Their results indicate that base isolation has the lowest environmental impact [30]. Another study conducted an LCA of seismic retrofitting of RC columns, comparing three traditional methods: concrete jacketing, steel jacketing, and carbon fiber incorporation; steel jacketing proved to be the best alternative from an economic and environmental point of view [26].

The building and construction sector faces significant challenges on the social sustainability front. The criteria for assessing social sustainability in the built environment vary considerably, highlighting the need for a standardized framework [31]. This lack of consistency can be attributed to the ambiguity surrounding the definition of social sustainability criteria. Notably, no relevant studies currently address social life cycle assessments of seismic retrofitting alternatives, underscoring a significant knowledge gap.

2.2. Decision-Making Methods

Selecting the most appropriate multi-criteria decision-making methods from the existing literature is complex. No single MCDM model can address all the multifaceted challenges encountered in the construction industry [32]. Among the approaches employed to select retrofit alternatives, the joint application of the Analytical Hierarchy Process (AHP) for criteria weighting and the Technique for Order of Preference by Similarity to Ideal Solution (TOPSIS) for ranking alternatives has gained widespread acceptance within the scientific community. This combined approach generates a comprehensive ranking of alternatives with minimal parameters required from the decision-maker (DM) [26]. However, research on the application of MCDM in the construction industry remains limited. Researchers and decision-makers must integrate emerging MCDM methods to fill this gap.

The weighting stage of criteria is a critical step in the decision-making process. According to the literature, the Best-Worst Method (BWM) outperforms both the AHP and the Analytical Network Process (ANP) [27]. BWM offers a systematic approach that requires fewer pairwise comparisons than AHP, leading to more consistent judgments [33]. For the ranking of alternatives, several state-of-the-art methods have gained prominence. These include EDAS (Evaluation Based on Distance to Average Solution), MABAC (Multi-Attribute Boundary Approximation Area Comparison), CODAS (Combinatorial Distance-Based Assessment), and MARCOS (Measurement of Alternatives and Ranking According to Compromise Solution). These advanced techniques offer robust frameworks for evaluating and ranking alternatives effectively.

2.2.1. Best-Worst Method

The pairwise comparison method proposed by Rezaei in 2015 [34] is based on identifying the best and worst criteria from a set of evaluation criteria, followed by pairwise comparisons between these criteria and all others. The stages of the Best-Worst Method are briefly outlined below: (a) Determine the set of decision criteria; (b) Determine the best criterion (the most desirable, the most important) and the worst criterion (the least desirable, the least significant); (c) Determine the preference of the best criterion over the other criteria, using a number between 1 and 9; (d) Determine the preference of all criteria over the worst criterion, using a number between 1 and 9; (e) Calculate the optimal weights determined from the maximum absolute differences:

$$\left|\frac{w_B}{w_j} - a_{Bj}\right| \; and \; \left|\frac{w_j}{w_W} - a_{jW}\right| \tag{1}$$

For all, *j is* minimized, which results in the following min-max model:
This model is equivalent to the following model:

$$\begin{gathered}\min \xi \\ \left|\frac{w_B}{w_j} - a_{Bj}\right| \leq \xi, \; for \; all \; j\end{gathered} \tag{2}$$

$$\left|\frac{w_j}{w_W} - a_{jW}\right| \leq \xi, for \; all \; j \tag{3}$$

$$\sum_j w_j = 1; \; w_j \geq 0, \; for \; all \; j \tag{4}$$

The solution of this model could lead to multiple optimal solutions; a new model is proposed that minimizes the maximum between the set of $\{|w_B - a_{Bj}w_j|, |w_j - a_{jW}w_W|\}$. The problem can be formulated as follows:

$$\min max_j\{|w_B - a_{Bj}w_j|, |w_j - a_{jW}w_W|\}, \; \sum_j w_j = 1; \; w_j \geq 0, \; for \; all \; j \tag{5}$$

This problem is transformed into the following linear programming problem:

$$\begin{gathered}\min \xi^L \\ |w_B - a_{Bj}w_j| \leq \xi^L, \; for \; all \; j\end{gathered} \tag{6}$$

$$|w_j - a_{jW}w_W| \leq \xi^L, \; for \; all \; j \tag{7}$$

$$\sum_j w_j = 1; \; w_j \geq 0, \; for \; all \; j \tag{8}$$

This problem has only one solution, and we obtain the optimal weights $(w_1^*, w_2^*, \ldots, w_n^*)$ and ξ^{L*}; ξ^{L*}. It is regarded as an indicator of comparison consistency, with values close to zero reflecting a high degree of consistency.

2.2.2. Evaluation Based on Distance to Average Solution

Proposed by Ghorabaee et al. in 2015 [35], it ranks alternatives based on the distance to the median solution, eliminating the need to calculate both the ideal and nadir solutions. Two measures are used to assess the desirability of alternatives: positive distance to the mean (PDA) and negative distance to the mean (NDA). These measures indicate the difference between each alternative and the mean solution. Higher PDA and lower NDA values mean the alternative is better than the median solution. Their application is summarized in the following steps:

1. Construct the decision matrix $X = [X_{ij}]_{nxm}$, X_{ij} is the value of the performance of alternative i in criterion j.
2. Determine the average solution according to all criteria:

$$AV_J = \frac{\sum_{i=1}^{n} X_{ij}}{n} \tag{9}$$

3. Calculate the distance matrices according to the type of benefit or cost criterion:

Benefit:

$$PDA_{ij} = \frac{\max(0, (X_{ij} - AV_j))}{AV_j} \tag{10}$$

$$NDA_{ij} = \frac{\max(0, (AV_j - X_{ij}))}{AV_j} \tag{11}$$

Cost:

$$PDA_{ij} = \frac{\max(0, (AV_j - X_{ij}))}{AV_j} \tag{12}$$

$$NDA_{ij} = \frac{\max(0, (X_{ij} - AV_j))}{AV_j} \tag{13}$$

4. Determine the weighted sum for all alternatives:

$$SP_i = \sum_{j=1}^{m} w_j PDA_{ij} \tag{14}$$

$$SN_i = \sum_{j=1}^{m} w_j NDA_{ij} \tag{15}$$

w_j is the weight of the criterion j.

5. Normalize SP and SN values:

$$NSP_i = \frac{SP_i}{max_i(SP_i)} \tag{16}$$

$$NSN_i = 1 - \frac{SN_i}{max_i(SN_i)} \tag{17}$$

6. Calculate the evaluation score (AS) of all alternatives:

$$AS_i = \frac{1}{2}(NSP_i + NSN_i);\ 0 \leq AS_i \leq 1 \tag{18}$$

Alternatives are ranked in descending order of values (AS).

2.2.3. Multi-Attribute Boundary Approximation Area Comparison

Proposed by Pamucar and Cirovic in 2015 [36], it is based on the definition of the distance of the criterion function of each alternative to the edge approximation area; the method is applied in six steps:

1. Formation of the decision matrix $X = [X_{ij}]_{mxn}$;
2. Normalization of the elements of the initial matrix X, $N = [n_{ij}]_{mxn}$;

Benefit:

$$n_{ij} = \frac{X_{ij} - X_i^-}{X_i^+ - X_i^-} \tag{19}$$

Cost:

$$n_{ij} = \frac{X_{ij} - X_i^+}{X_i^- - X_i^+} \tag{20}$$

$$X_i^+ = \max(X_1, X_2, \ldots X_m) \text{ y } X_i^- = \min(X_1, X_2, \ldots X_m) \tag{21}$$

3. Determine the weighted matrix $V = [v_{ij}]_{mxn}$
   $v_{ij} = w_i * (n_{ij} + 1)$, wi criteria weighting coefficients;
4. Determine the edge approximation area matrix $G = [g_1 g_2 \ldots g_n]$, formed by edge approximation area (BAA);

$$g_i = \left(\prod_{j=1}^{m} vij\right)^{\frac{1}{m}} \tag{22}$$

5. Determine the matrix $Q = [q_{ij}]_{mxn}$:

$$Q = V - G \tag{23}$$

Alternative A_i could belong to the border approach area G, the upper approach area G^+, or the lower approach area G^-, G^+ contains the ideal alternative A^+, and G^- contains the anti-ideal alternative A^-. Membership is determined according to the following:

$$A_i \in \begin{Bmatrix} G^+ \; if \; q_{ij} > 0 \\ G \; if \; q_{ij} = 0 \\ G^- \; if \; q_{ij} < 0 \end{Bmatrix} \tag{24}$$

For A_i to be selected as the best, it must have as many criteria as possible belonging to the upper approximation zone G^+.

To rank the alternatives, the values of the criterion functions for the other options S_i are determined as the sum of the distance of the alternatives from the boundary approach areas q_i.

$$S_i = \sum_{j=1}^{n} q_{ij}, j = 1, 2, \ldots, n; i = 1, 2, \ldots, m \tag{25}$$

2.2.4. Combinatorial Distance-Based Assessment

Proposed by Ghorabaee et al. in 2016 [37], it evaluates the desirability of alternatives using Euclidean and taxonomic distances. The primary metric is the Euclidean distance between the alternatives and the negative ideal solution, while the secondary metric is the Taxicab distance. The alternative farthest from the negative ideal solution is deemed the most desirable. The method is implemented as follows:

1. Definition of the decision matrix $X = [X_{ij}]_{nxm}$;
2. Normalization of the decision matrix:

$$n_{ij} = \begin{Bmatrix} \dfrac{X_{ij}}{max_i X_{ij}} \; if \; j \in N_b; benefit \\ \dfrac{min_i X_{ij}}{X_{ij}} \; if \; j \in N_c; cost \end{Bmatrix} \tag{26}$$

3. Calculating the weighted normalized decision matrix:

$$r_{ij} = w_j \cdot n_{ij};\ 0 < w_j < 1;\ \sum_{j=1}^{m} w_j = 1 \tag{27}$$

4. Determine the ideal negative solution (point):

$$ns = \left[ns_j\right]_{1xm};\ ns_j = min_i r_{ij} \tag{28}$$

5. Calculate the Euclidean and taxicab distances of the alternatives with respect to the negative ideal solution:

$$E_i = \sqrt{\sum_{j=1}^{m} \left(r_{ij} - ns_j\right)^2} \tag{29}$$

$$T_i = \sum_{j=1}^{m} \left|r_{ij} - ns_j\right| \tag{30}$$

6. Construct the relative evaluation matrix $Ra = [h_{ik}]_{nxn}$:

$$h_{ik} = (E_i - E_k) + \psi(T_i - T_k) \tag{31}$$

ψ is a threshold function to recognize the equality of the Euclidean distances of two alternatives:

$$\psi(x) = \begin{cases} 1\ if\ |x| \geq \tau \\ 0\ if\ |x| < \tau \end{cases} \tag{32}$$

τ is the threshold that can be set by the decision-maker. Its value can be between 0.01 and 0.05; usually τ = 0.02 is used.

7. Calculate the evaluation score for each alternative:

$$H_i = \sum_{k=1}^{n} h_{ik} \tag{33}$$

The alternatives are ranked according to the degressive values of Hi.

2.2.5. Measurement of Alternatives and Ranking According to Compromise Solution

The method proposed by Steviç et al. in 2020 [38] measures and ranks alternatives relative to a compromise solution; utility functions must be determined based on the distances to the ideal and anti-ideal solutions and their aggregations. The best alternative is the one closest to the perfect solution and farthest from the anti-ideal benchmark. The method is applied by following these steps:

1. Determining the decision matrix;
2. Determine the extended matrix defining the ideal (IA) and anti-ideal (AAI) solution:

$$X = \begin{array}{c} \\ AAI \\ A_1 \\ A_2 \\ \cdots \\ A_m \\ AI \end{array} \left| \begin{array}{cccc} C_1 & C_2 & \cdots & C_n \\ X_{aa1} & X_{aa2} & \cdots & X_{aan} \\ X_{11} & X_{12} & \cdots & X_{1n} \\ X_{21} & X_{22} & \cdots & X_{2n} \\ \cdots & \cdots & \cdots & \cdots \\ X_{m1} & X_{m2} & \cdots & X_{mn} \\ X_{ai1} & X_{ai2} & \cdots & X_{ain} \end{array} \right| \tag{34}$$

$$AAI = min_i X_{ij}\ if\ j \in B\ and\ max_i X_{ij}\ if\ j \in C \tag{35}$$

$$AI = max_i X_{ij}\ if\ j \in B\ and\ min_i X_{ij}\ if\ j \in C \tag{36}$$

B group of benefit criteria, and C group of cost criteria;

3. Normalize the extended matrix $N = [n_{ij}]_{mxn}$:

$$n_{ij} = \frac{X_{ai}}{X_{ij}}\ if\ j \in C \tag{37}$$

$$n_{ij} = \frac{X_{ij}}{X_{ai}}\ if\ j \in B \tag{38}$$

4. Determine the weighted matrix $V = [v_{ij}]_{mxn}$; $v_{ij} = n_{ij} * w_j$;
5. Calculate each alternative's utility relative to the ideal and anti-ideal solutions:

$$K_i^- = \frac{S_i}{S_{aai}} \tag{39}$$

$$K_i^+ = \frac{S_i}{S_{ai}} \tag{40}$$

$$S_i = \sum_{i=1}^{n} v_{ij} \tag{41}$$

6. Determine the utility function of the alternatives:

$$f(K_i) = \frac{K_i^+ + K_i^-}{1 + \frac{1-f(K_i^+)}{f(K_i^+)} + \frac{1-f(K_i^-)}{f(K_i^-)}} \tag{42}$$

The utility functions of the ideal and anti-ideal solutions are calculated according to the following:

$$f(K_i^-) = \frac{K_i^+}{K_i^* + K_i^-} \tag{43}$$

$$f(K_i^+) = \frac{K_i^-}{K_i^* + K_i^-} \tag{44}$$

The higher the value of the utility function, the better.

3. Materials and Methods

A comprehensive methodology is proposed for selecting the optimal seismic retrofitting solution for reinforced concrete beams with shear deficiencies, adopting a holistic approach. Figure 1 illustrates the methodology, organized into four clearly defined stages: Stage 1: Define the problem and the dimensions to analyze for the strengthening alternatives: economic, environmental, social, and functional. Stage 2: Present five strengthening alternatives to enhance the shear strength of the beam under study. Stage 3: Quantify the economic, environmental, and social impacts using life cycle assessments—Life Cycle Cost Assessment (LCCA), Environmental Life Cycle Assessment (LCA), and Social Life Cycle Assessment (S-LCA). Additionally, two criteria representing the functionality of the strengthening alternatives will be evaluated. Stage 4: Develop a decision model through expert input, structured hierarchically into four dimensions, eight criteria, and 16 specific indicators. Employ the BWM for subjective criteria weighting and the EDAS, MABAC, CODAS, and MARCOS methods to rank the alternatives. This approach identifies the optimal seismic retrofitting solution for reinforced concrete beams with shear deficiencies.

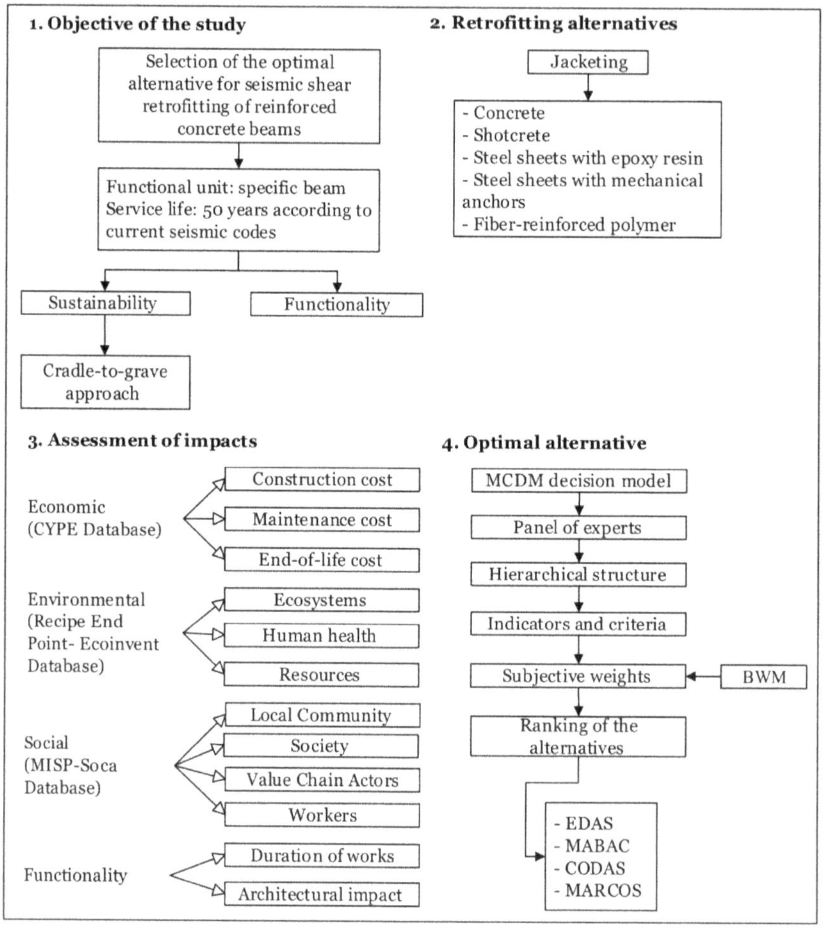

Figure 1. Methodology of the study.

3.1. Description of the Problem and Retrofit Alternatives

This study aims to determine the optimum seismic retrofitting solution for a reinforced concrete beam with shear deficiencies. The analyzed beam has a span of 6.0 m. It presents material properties and longitudinal and transverse reinforcement, typical of beams built according to obsolete standards or non-designed practices. In order to comply with current seismic regulations, the retrofit of the beam includes a 22% increase in shear strength within the intended plastic hinge zone. The analysis assumes a service life of 50 years for the retrofit beam, which ensures compliance with modern seismic requirements and improves the durability of the structural element. The seismic strengthening alternatives are evaluated using a "cradle-to-grave" life cycle analysis, which incorporates economic, environmental, and social dimensions while assessing functionality.

Optimal seismic retrofitting is selected from five alternatives: concrete jacketing (CJ), shotcrete jacketing (SCJ), steel jacketing with structural epoxy adhesive (STE), steel jacketing with anchor bolts (STA), and use of carbon fiber reinforced polymer (CFRP). The methods for the design of retrofitting alternatives are considered to be an additive approach.

The equations in the Ecuadorian standard NEC-15 [39], which are based on ACI-318, were used for the concrete jacketing alternatives. The nominal shear strength considers the contribution of the concrete (V_c) and the shear reinforcement (V_s), according to Equation (45).

$$V_n = V_c + V_s \tag{45}$$

The contribution of concrete to shear was determined using Equation (46).

$$V_c = 0.17\sqrt{f'_c} b_w d \tag{46}$$

where f'_c is the compressive strength of the concrete cylinder in MPa, b_w is the width of the beam, and d is the effective depth of the beam. For the contribution of the vertical stirrup, ACI-318 considers the truss model represented by Equation (47).

$$V_s = \frac{A_v f_y d}{s} \tag{47}$$

where A_v is the effective area of the stirrup legs, f_y is the steel yield strength, and s is the spacing between the stirrups.

For the alternatives incorporating steel plates, equation 48 followed the reference guidelines [40]. This expression considers the shear strength of reinforced concrete beams with continuous horizontal steel plates attached to the web, the addition of the contributions of the concrete, the internal shear reinforcement, and the external steel plates. Equation (49) determines the contribution of shear stress (Vp), considering the sum of the shear stresses along its depth and thickness.

$$V_n = V_c + V_s + V_p \tag{48}$$

$$V_p = \tau_{avg} h_p t_p \tag{49}$$

where τ_{avg} is the average shear stress in the steel plates, h_p is the depth of the plates, and t_p is the thickness of the plates on either side of the beam's web. For the quadratic distribution of shear stress over the depth of the plate, the average stress is calculated using Equation (50):

$$\tau_{avg} = \frac{2}{3}\tau_{max} \tag{50}$$

τ_{max} is the maximum shear stress in a steel plate, which can be determined from Tresca's maximum shear stress theory. This theory considers the yield strength of the material f_{yp}, as indicated in Equation (51):

$$\tau_{max} = \frac{1}{2} f_{yp} \tag{51}$$

Thus, when considering Equations (49)–(51), the contribution of the plates becomes Equation (52):

$$V_p = \frac{1}{3} f_{yp} h_p t_p \tag{52}$$

For the design of the CFRP, the Guide for the Design and Construction of Externally Bonded FRP Systems for Reinforcing Concrete Structures (ACI 440) was used, where the contribution of the FRP shear stress (Vf) in Equation (53) is affected by the coefficient ψ_f with a value of 0.85 for U-shaped wraps.

$$V_n = V_c + V_s + \psi_f V_f \tag{53}$$

The contribution of the shear stress of the FRP is given by Equation (54):

$$V_f = \frac{A_{fv} f_{fe} (sin\alpha + cos\alpha) d_{fv}}{s_f} \tag{54}$$

where for rectangular sections, A_{fv} is determined by Equation (55):

$$A_{fv} = 2nt_f w_f \tag{55}$$

where n is the number of FRP reinforcement layers, t_f is the nominal thickness of a reinforcing sheet, w_f is the nominal width of a reinforcing sheet, f_{fe} is the effective stress in the FRP reinforcement, α is the angle formed by the extreme fiber in traction with the FRP strip, d_{fv} the depth of the FRP shear reinforcement, and sf the distance between the centers of the FRP strips.

Specific preventive maintenance measures based on material properties were incorporated to ensure that the retrofit alternatives achieve the expected service life of 50 years from installation or until structural damage due to an earthquake occurs. Table 1 summarizes the general characteristics of the beam analyzed and the shear strength values of the five retrofit alternatives considered.

Table 1. General information about the original beam and the retrofitting alternatives.

<table>
<tr><td colspan="5">Original beam</td></tr>
<tr><td>A (m)</td><td>B (m)</td><td>E (m)</td><td>DL (mm)</td><td>Number</td></tr>
<tr><td>0.4</td><td>0.6</td><td>0.2</td><td>16</td><td>8</td></tr>
<tr><td>DT (mm)</td><td>S [a] (m)</td><td>f′c (Mpa)</td><td>fy (Mpa)</td><td>V (kN)</td></tr>
<tr><td>10</td><td>0.1</td><td>24</td><td>421</td><td>558.79</td></tr>
<tr><td colspan="5">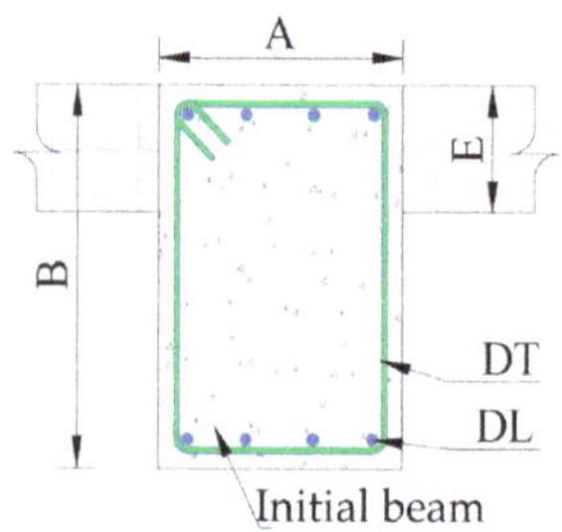
</td></tr>
<tr><td colspan="5">Retrofit beam</td></tr>
<tr><td colspan="5">V (kN)</td></tr>
<tr><td>CJ</td><td>SCJ</td><td>STE</td><td>STA</td><td>CFRP</td></tr>
<tr><td colspan="2">722.57</td><td colspan="2">742.65</td><td>682.77</td></tr>
</table>

[a] Spacing of ties.

The beam requiring retrofit is assumed to be part of a reinforced concrete building in northern Quito, classified as a high seismic hazard zone under Ecuador's current seismic regulations. The following transport distances were considered: 8.0 km from the concrete factory, 42.3 km for steel, 40.2 km for structural adhesives and materials used in preventive maintenance, and 13.2 km to the landfill for waste disposal. For CFRP and epoxy resin, transport distances include 3800 km by air and 37.3 km by land, as these materials are imported into Ecuador. Figure 2 shows the schematics of the retrofitting alternatives. A detailed description of the alternatives to retrofit is provided.

CJ Alternative: A 75 mm thick layer of concrete is applied to the bottom and side faces of the beam. The concrete, with a compressive strength of 24 MPa and a soft consistency, is pre-mixed at the plant and pumped on-site. Before application, a continuous layer of two-component, thixotropic, epoxy resin-based structural adhesive is applied to the hardened concrete surface. The reinforcement consists of four longitudinal bars 12 mm in diameter and stirrups 10 mm in diameter, spaced at 150 mm and anchored to the slab using chemicals. Preventive maintenance includes the application of an anti-carbonation paint.

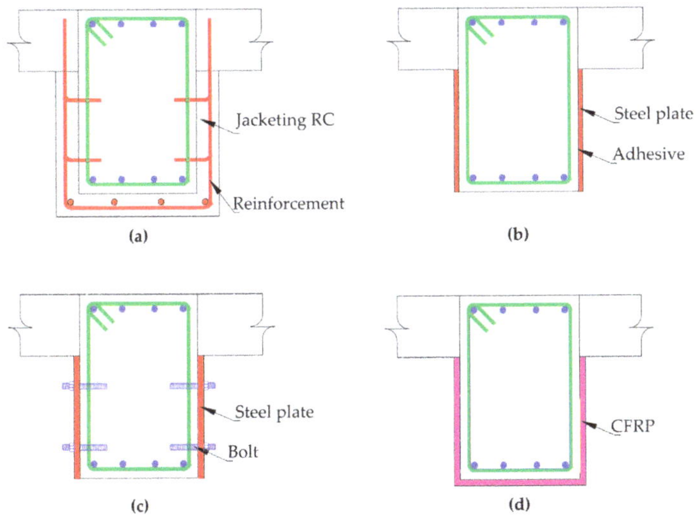

Figure 2. Alternative reinforcement schemes. (**a**) CJ and SCJ; (**b**) STE; (**c**) STA; (**d**) CFRP.

SCJ Alternative: A 75 mm thick shotcrete jacket is applied to the bottom and side faces of the beam. The shotcrete has a compressive strength of 24 MPa and fluid consistency. It is applied using a wet spraying technique. The reinforcement is identical to that used in the CJ alternative. Preventive maintenance consists of the application of anti-carbonation paint.

STE Alternative: 4 mm thick A572 Gr50 hot-rolled steel plates are applied to the lateral faces of the beam and secured using a two-component thixotropic epoxy resin adhesive. Preventive maintenance includes the application of a two-component epoxy resin-based anticorrosive paint and passive fire protection achieved by spraying cement-based fireproof mortar.

STA Alternative: 4 mm thick A572 Gr50 hot-rolled steel plates are secured to the beam's lateral faces using mechanical anchorage with bolts. Preventive maintenance includes the application of two-component epoxy resin-based anticorrosive paint and passive fire protection provided by the projection of cement-based fireproof mortar.

CFRP Alternative: The CFRP is applied using a U-wrap method with two layers of unidirectional carbon fabric, each 0.165 mm thick. The fabric has a tensile strength of 4900 MPa, an elasticity modulus of 230,000 MPa, an ultimate elongation of 2.10%, and a 300 g/m^2 weight. Preventive maintenance includes passive fire protection through the projection of cement-based fireproof mortar.

3.2. Implementation of Life Cycle Assessment

The LCCA includes construction, preventive maintenance, and end-of-life (EoL) costs. Construction costs cover the supply of materials, equipment and machinery usage, labor, and minor tools. For the CJ alternative, this also includes the cost of the recoverable formwork system. Preventive maintenance costs encompass anti-carbonation paint for CJ and SCJ, anti-corrosion paint for STE and STA, fireproof mortar for STE, STA, and CFRP, and decennial maintenance for the first 10 years. EoL costs involve demolition, sorting, and transportation to recycling plants or landfills as appropriate. Future costs are converted to present value using a 2% social discount rate, which aligns with long-term sustainability objectives [41]. The LCCA used unit price analyses from the CYPE construction cost database, which included specific costs for Ecuador with a cutoff date of January 2025.

The environmental and social dimensions are evaluated across four distinct phases. First, the objective and scope of the study are to assess five seismic retrofitting alternatives for reinforced concrete beams with shear deficiencies. The functional unit is the seismic retrofitting of a specific reinforced concrete beam, ensuring structural safety over a 50-year service life. A cradle-to-grave approach is adopted, covering material fabrication, transportation to the site, construction, preventive maintenance, and end-of-life. The end-of-life stage includes pre-treatment of concrete, transporting steel and concrete to recycling plants, and final disposal of CFRP to a landfill.

The second phase consisted of conducting an inventory analysis for each alternative. Environmental impacts were assessed using the Ecoinvent database, while social impacts were assessed using SOCA v2, with both analyses modeled in OpenLCA version 2.3 software. By using SOCA, which integrates data from the PSILCA social database, the same processes applied in the environmental impact assessment were used, ensuring consistency in the results. As the global SOCA data are complete but region-specific data may be missing, the global average was used for process modeling, following the reference recommendation [42]. The energy required for machinery operations during construction was obtained from the BEDEC database. For the SCJ alternative, the additional energy required for shotcrete application was considered and extracted from BEDEC. For the STA alternative, the energy values for machinery operations during construction were adapted from reference [43]. For carbon fiber, the energy results for virgin carbon fiber production from reference [44] were adapted to Ecoinvent's carbon fiber reinforced plastic process used in aircraft manufacturing. Table 2 presents the inventory by functional unit and the associated processes from the Ecoinvent database, which were also used in SOCA. A zero discount rate was applied for future end-of-life quantifications.

Table 2. Life cycle inventory.

Process	Quantity	Unit	Ecoinvent Process
Concrete Jacketing (CJ)			
Concrete	0.27	m^3	Concrete, 24–30 MPa
Reinforcement steel	32.83	kg	Reinforcing steel
Structural adhesive	5.69	kg	Epoxy resin
Column jacketing	462.82	MJ	Diesel, burned in building machine [a]
Anti-carbonation paint	1.44	kg	Epoxy resin
Retrofit demolition	64.96	MJ	Diesel, burned in building machine [a]
Concrete crushing	545.13	kg	Rock crushing
Shotcrete Jacketing (SCJ)			
Concrete	0.27	m^3	Concrete, 24–30 MPa
Shotcrete	386.63	MJ	Diesel, burned in building machine [a]
Reinforcement steel	32.83	kg	Reinforcing steel
Column jacketing	462.82	MJ	Diesel, burned in building machine [a]
Anti-carbonation paint	1.44	kg	Epoxy resin
Retrofit demolition	64.96	MJ	Diesel, burned in building machine [a]
Concrete crushing	545.13	kg	Rock crushing

Table 2. *Cont.*

Process	Quantity	Unit	Ecoinvent Process
Steel Jacketing with epoxy resin (STE)			
Structural steel	63.30	kg	Hot rolling, steel
Structural adhesive	3.84	kg	Epoxy resin
Column jacketing	190.85	MJ	Diesel, burned in building machine [a]
Anti-corrosion primer	0.96	kg	Epoxy resin
Fireproof mortar	38.76	kg	Cement mortar
Steel Jacketing with mechanical anchorages (STA)			
Structural steel	65.82	kg	Hot rolling, steel
Column jacketing	131.21	MJ	Diesel, burned in building machine [a]
Anti-corrosion primer	0.96	kg	Epoxy resin
Fireproof mortar	38.76	kg	Cement mortar
Fiber-Reinforced Polymers (CFRP)			
Carbon Fibers	1.90	kg	Acrylonitrile; Chemical, organic; Electricity, low voltage; Heat, district or industrial; injection moulding
Epoxy Resins	4.12	kg	Epoxy resin
CFRP installation	132.02	MJ	Diesel, burned in building machine [a]
Fireproof mortar	58.15	kg	Cement mortar

[a] BEDEC Database.

In the third phase, the Endpoint Recipe method was employed to assess environmental impacts. At the same time, the Social Impact Weighting Methodology (MISP) was used to evaluate social impacts, and both models were modeled using OpenLCA software. Pedigree matrix approaches were incorporated to account for uncertainty based on five indicators: reliability, completeness, temporal correlation, geographical correlation, and technological correlation.

In the final phase, the environmental results were interpreted by grouping the impacts into three damage categories: ecosystems (measured in species per year), human health (measured in disability-adjusted life years), and resource availability (measured in US dollars). A hierarchical approach was used for a long term scenario (H), with the three damage categories normalized using the WORLD ReCiPe methodology, following H/H hierarchies (person/year). The Final Recipe method was then applied to calculate each retrofit alternative's full life-cycle environmental impact score. For social impacts, the MISP methodology translates the 55 indicators into four stakeholder categories: Workers, Value Chain Actors, Society, and the Local Community. These impacts are measured using a unit average mean risk per hour (MRH).

3.3. Decision Model

The decision model incorporated four dimensions represented by eight criteria and 16 specific indicators, as shown in Table 3. The criteria reflect the impacts identified through life cycle analysis using a cradle-to-grave approach for the economic, environmental, and social dimensions. The functionality dimension includes two criteria commonly employed by researchers in selecting optimal building retrofitting solutions using MCDM: the duration of construction works and the architectural impact of the retrofit alternative [26].

The economic dimension is represented by two criteria: C1: Construction costs, which include construction costs, materials, equipment, machinery, labor, and minor tools. C2: Maintenance and EoL, which include preventive maintenance, such as carbonation protection for CJ and SCJ, anti-corrosion paint for STE and STA, and fireproof mortar for STE, STA, and CFRP. This criterion also includes decennial maintenance and EoL costs, such as crushing for CJ, transportation to material plants for CJ, STE, and STA for recycling, and transportation to landfills for CFRP. According to the LCCA, all costs are measured in dollars. Future costs consider a social discount rate of 2% to reflect long-term sustainability issues.

Table 3. Hierarchical tree of the decision model.

Dimension	Criteria		Indicators	
Economic	C1	Construction cost	I1	Construction cost (USD)
	C2	Maintenance and EoL costs	I2	Preventive maintenance (USD)
			I3	Maintenance (USD first 10 years)
			I4	Demolition, classification, and transport of waste (USD)
Environmental	C3	Manufacturing and construction	I5	Ecosystem (Points)
			I6	Human health (Points)
			I7	Resources (Points)
	C4	Maintenance and EoL	I8	Ecosystem (Points)
			I9	Human health (Points)
			I10	Resources (Points)
Social	C5	Local Community and Society	I11	Local Community (MRH)
			I12	Society (MRH)
	C6	Value Chain Actors and Workers	I13	Value Chain Actors (MRH)
			I14	Workers (MRH)
Functionality	C7	Duration of works	I15	Duration of works/disruption to occupants (scale)
	C8	Architectural impact	I16	Architectural impact (scale)

Environmental dimensions encompass two criteria reflecting the environmental impacts throughout the life cycle: C3: Production and Construction covers the impacts associated with the production of materials, transportation to the site, and energy consumed by machinery during the installation of the retrofit alternative. C4: Maintenance and EoL consider the impacts of producing raw materials required for preventive maintenance of each alternative and the end-of-life phase, including treatment before recycling CJ and SCJ and transport to recycling facilities or landfills. These criteria are evaluated using indicators that measure impacts within three damage categories: Ecosystems, Human Health, and Resources. Impacts are quantified in points derived from the Endpoint ReCiPe method.

The social dimension includes two criteria grouping four stakeholder categories: C5: Local community and society and C6: Stakeholders and value chain workers. Social impacts are measured in Mean Risk per Hour (MRH).

The Functionality dimension is represented by two criteria: C7: Duration of works, evaluated using a scale based on CYPE cost analysis performances. C8: Architectural impact, evaluated by a scaling approach translating the qualitative judgments of the decision-makers using the AHP and pairwise comparisons of the five alternatives. Six decision-makers with professional experience in civil engineering and architecture were consulted to ensure a diversity of perspectives and approaches. These experts have between 10 and 35 years of academic and professional experience in structural engineering, construction, and sustainability.

4. Analysis of Results

4.1. Results of Life Cycle Assessments

Figure 3 illustrates the results of the LCCA for alternatives to retrofit the beam with shear deficiencies. The costs are expressed as present values as of January 2025. Construction costs are the most significant component of the total costs, accounting for an average of 82% across all alternatives. The CFRP alternative incurs the highest construction costs due to the expensive raw materials. CJ ranks second, mainly due to the high cost of the formwork system required. The STE and STA alternatives have the highest costs in the maintenance stage. This is attributed to the inclusion of two types of preventive maintenance: anti-corrosion paint with fireproofing mortar, which represents a 2.3 times increase compared to the preventive maintenance cost of the CJ alternative. In the EOL phase, the costs are especially significant for the CJ and SCJ alternatives due to the requirements for the pretreatment of concrete waste. The percentages shown in Figure 3 represent the total LCCA values, with the CJ alternative as the benchmark. The SCJ alternative has the lowest total costs, with a 45% reduction, while the CFRP alternative has an increase of 1%.

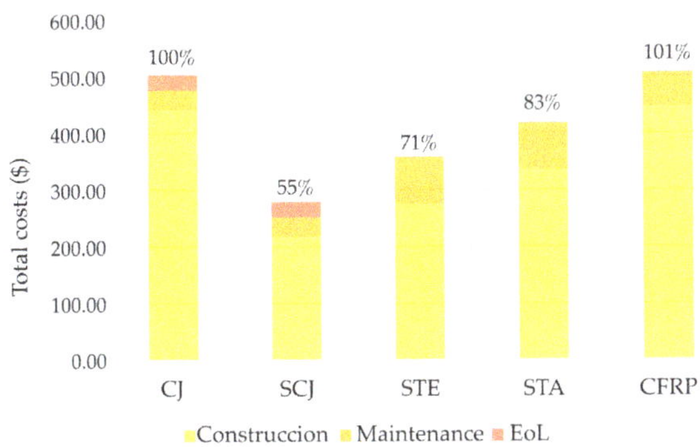

Figure 3. Economic life cycle assessment.

The ReCiPe endpoint approach provides straightforward LCA results for environmental impacts, offering a clear overview of the effects associated with each alternative. Figure 4 highlights the impacts across damage categories: Ecosystems, Human Health, and Resources. The manufacturing phase has the highest impact in all damage categories, with the exception of the STE alternative, which has the highest values for the preventive maintenance stage in the Ecosystems category. In the EoL phase, impacts are substantial for CJ and SCJ. Regarding total impact values measured in points, the Resources category has the highest representation, followed by Human Health and Ecosystems. Among all alternatives, CJ exhibits the highest environmental impact. SCJ ranks second but achieves the most significant reduction in the Resources damage category, with a 21% decrease compared to CJ. Despite requiring higher energy consumption for machinery operations during the manufacturing and application of shotcrete, SCJ benefits from the reduction of structural adhesives, which helps lower its overall environmental impact. STE and STA alternatives offer substantial ecological advantages, particularly in the Human Health category, with reductions of 69% for STE and 79% for STA compared to CJ. Similarly, CFRP presents significant benefits with reductions of 54% in the Ecosystems category, 70% in Human Health, and 56% in Resources compared to CJ. Similarly, the Figure 4d) summarizes the total environmental impacts of the retrofit alternatives. CJ has the highest overall impact, while SCJ achieves a 19% reduction. CFRP shows a 59% decrease, and the lowest environmental impacts are observed for the STE and STA alternatives, with decreases of 62% and 77%, respectively, compared to CJ.

The impacts associated with the S-LCA are represented in Figure 5, where the highest values are for the Workers and Society stakeholders of the alternatives. The most significant social impacts are found in the CJ alternative. SCJ presents a total decrease in social impacts of 21% regarding CJ, and the STE alternative is 61%. The most significant reductions correspond to the CFRP alternative, with a reduction of 68%, and STA, with the most significant reduction of 75%.

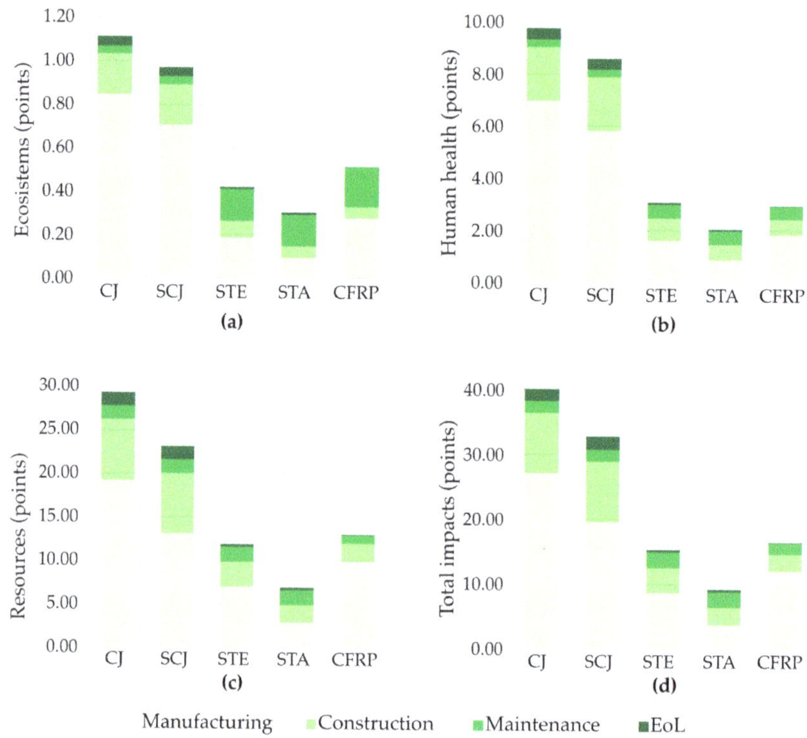

Figure 4. Environmental life cycle assessment. (**a**) ecosystems; (**b**) human health; (**c**) resources; (**d**) total impacts.

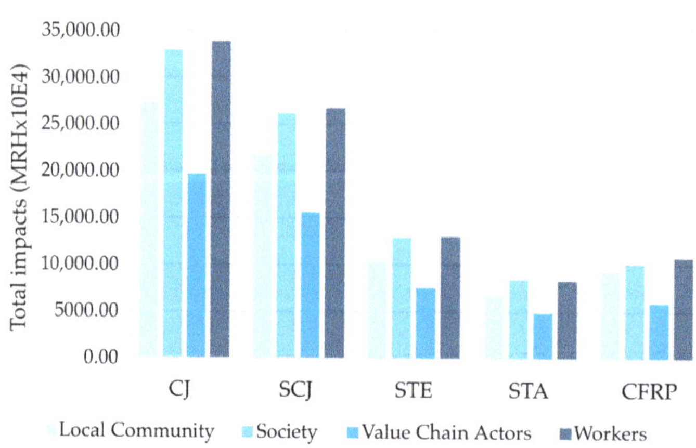

Figure 5. Social life cycle assessment.

4.2. Selecting the Optimal Retrofit Alternative

Table 4 presents the values of indicators representing the economic, environmental, social, and functional dimensions for the five retrofit alternatives considered. Costs are expressed in dollars, environmental indicators in points, social indicators in average mean risk per hour, and functionality on a standardized scale. Based on these values, the eight criteria were determined for use in the decision-making process. Figure 6 illustrates the normalized values for each criterion across all alternatives. The CJ alternative shows the highest values for environmental, social, and functional criteria, while the CFRP alternative scores highest for criterion C1. STE and STA exhibit the highest values for criterion C2, corresponding to preventive maintenance costs and end-of-life impacts. SCJ has the lowest construction cost, STE achieves the lowest value for criteria C7, and STA for criteria C3, C5, and C6. Lastly, the CFRP alternative demonstrates the weakest values for criteria C2, C4, and C8.

Table 4. Values of the indicators for each alternative.

Indicators		CJ	SCJ	STE	STA	CFRP
I1	Construction cost USD	439.80	215.35	274.96	334.46	445.16
I2	Preventive maintenance (USD)	22.18	22.18	51.46	51.46	33.14
I3	Maintenance (USD first 10 years)	14.06	14.06	31.64	31.64	28.49
I4	Demolition, classification, and transport waste (USD)	28.15	28.15	1.26	1.26	0.37
I5	Ecosystem (Points)	1.03	0.89	0.26	0.15	0.33
I6	Human health (Points)	9.04	7.89	2.46	1.44	2.42
I7	Resources (Points)	26.23	20.08	9.85	4.89	11.95
I8	Ecosystem (Points)	0.08	0.08	0.16	0.16	0.19
I9	Human health (Points)	0.73	0.73	0.61	0.61	0.53
I10	Resources (Points)	3.12	3.12	2.01	2.02	1.09
I11	Local Community (MRHx10E4)	2.73	2.17	1.05	0.68	0.94
I12	Society (MRHx10E4)	3.29	2.62	1.30	0.85	1.01
I13	Value Chain Actors (MRHx10E4)	1.97	1.56	0.76	0.49	0.60
I14	Workers (MRHx10E4)	3.38	2.67	1.31	0.83	1.09
I15	Duration of works (scale)	0.37	0.24	0.07	0.23	0.09
I16	Architectural impact (scale)	0.35	0.35	0.09	0.15	0.06

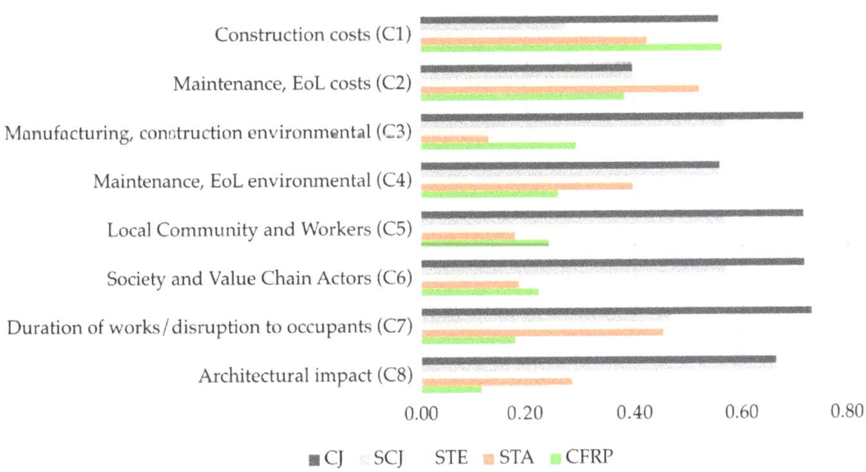

Figure 6. Normalized values of the criteria.

Table 5 presents the geometric mean values of the weights derived using the BWM method for both the criteria and dimensions analyzed. The criterion with the lowest weighting was the environmental impact of the maintenance and EoL phases, followed by the costs associated with maintenance and EoL and the social impacts on the local community and workers. In contrast, architectural impact and construction cost were identified as the most important criteria. Table 6 shows the results of the ranking methods applied in this study. The CJ alternative was the worst-rated in all cases, followed by the SCJ alternative. The STA alternative ranked third in all methods. The EDAS and MABAC methods identified the STE alternative as the best option for seismic retrofitting of shear-deficient beams, with CFRP ranking second. CODAS and MARCOS methods ranked CFRP as the best option, followed by the STA alternative.

Table 5. Weights derived from BWM.

Dimension	Weight	Criteria	Global Weight
Economic (D1)	0.27	C1 C2	0.19 0.08
Environment (D2)	0.23	C3 C4	0.16 0.07
Social (D3)	0.17	C5 C6	0.08 0.09
Funcionality (D4)	0.33	C7 C8	0.13 0.20

Table 6. Classification of alternatives with MCDM.

Alternatives	EDAS		MABAC		CODAS		MARCOS	
	Score	Ranking	Score	Ranking	Score	Ranking	Score	Ranking
CJ	0.01	5	−0.40	5	−1.83	5	0.31	5
SCJ	0.34	4	−0.06	4	−0.75	4	0.43	4
STE	0.99	1	0.28	1	0.76	2	0.66	2
STA	0.85	3	0.19	3	0.66	3	0.63	3
CFRP	0.98	2	0.27	2	1.17	1	0.70	1

5. Discussion

In the search for optimal seismic retrofit strategies, the scientific community has predominantly relied on traditional MCDM methods, often overlooking innovative approaches that could provide better solutions. This study presents a comprehensive methodology for ranking seismic retrofit alternatives using emerging MCDM techniques. The methodology integrates functionality with the three pillars of sustainability-economic, environmental, and social life cycle analysis. The proposed approach is applied to evaluate and rank five seismic retrofit alternatives for a reinforced concrete beam with shear deficiencies. These retrofit alternatives were designed to achieve the same increase in shear strength to meet seismic regulations. The analysis considers a time frame beginning from the implementation of the retrofit and extending up to a useful life of 50 years or until the beam sustains damage from an earthquake. The inclusion of specific preventive maintenance, tailored to the material characteristics of each alternative, ensures the validity of this time frame. This approach eliminates the possibility of one retrofit alternative having a longer lifespan, allowing for a fair comparison [17].

Determining the shear capacity of reinforced concrete beams remains an evolving area of research. Most shear design methods use an additive approach, combining the contributions of concrete and steel. However, Eurocode 2 considers only the steel contribution to shear strength [10], which can lead to an underestimation of the shear capacity in lightly shear-strengthened beams where the concrete contribution is more significant [11]. This manuscript's methods for designing the retrofitting alternatives are based on an additive approach that considers the contributions of concrete, steel, and composites. This approach resulted in comparable increases in shear strength capacity across the different retrofit alternatives, assuming that all resistance mechanisms reach their maximum values simultaneously without any interaction between them. Although this additive approach is widely accepted, recent research has indicated potential interactions between the experimental strength contributions of FRP shear reinforcement [45].

The relatively high increase in shear force observed in the concrete-based alternatives is mainly due to construction limitations, as additional concrete is required even when it is not structurally necessary [43]. For alternatives involving steel plates, the contribution to shear stress is considered a function of the depth and thickness of the continuous plates. However, in the case of STE, where the plates are not fully anchored to the concrete, they may not reach maximum strength at the elastic limit. To take this into account, previous studies [40,46] have introduced modification factors that consider the tensile capacity of the concrete at the concrete-adhesive interface, with a reduction factor of 0.76 being applicable in this case. For the STA, the contribution of the bolts to the shear capacity could also be considered [46]; however, theoretical models to accurately predict shear capacity are still under development [47].

The depth of the beam below the slab limits the bond length or connection length of the steel and CFRP coatings. As a result, the bond length can be shorter than the required development length, leading to premature failure of the rehabilitation system and limiting the effective utilization of the high-strength properties of the laminates. For the STE alternative, steel plates contribute more significantly to the shear resistance of the beam when the ratio of plate depth to beam depth reaches its maximum feasible value. Studies have shown that, for ratios greater than 0.75, the contribution of the plate remains unchanged, which is particularly advantageous for T beams and beams located under slabs [40]. In this study, the ratio was 0.70. On the other hand, shear reinforcement using CFRP laminates with anchors can increase shear capacity by approximately 20% or more compared to CFRP laminates without anchors [48,49]. However, the ACI 440 design guidelines provide conservative and reliable predictions, guaranteeing safe retrofitting designs for beams with a shear deficiency with CFRP laminates [50]; there are currently no standardized guidelines for anchor design.

By analyzing the cost-effectiveness of the retrofitting alternatives, their economic viability can be assessed by determining the cost required to achieve a 1% increase in shear strength [51]. Table 7 presents the retrofitting costs obtained through LCCA for a 1% increase in shear strength in the different alternatives. Among the alternatives, the SCJ method demonstrates the lowest cost per 1% increase in shear strength, making it the most cost-effective option. In contrast, the CFRP alternative incurs the highest costs. However, despite its cost-effectiveness, concrete jacketing has certain drawbacks, especially in labor-intensive construction processes, long execution times, and reduced usable space inside the building [52]. The viability of the retrofitting strategies was evaluated using two criteria: C7–Duration of the work and C8–Architectural impact. The results indicate that CJ has the highest value for C7, while both CJ and SCJ have the highest values for C8.

Table 7. Cost-effectiveness of the alternatives.

Alternatives	CJ	SCJ	STE	STA	CFRP
Shear force increase ratio (%)	29%	29%	33%	33%	22%
Cost LCCA [a]	504.19	279.74	359.32	418.82	507.16
Cost to increase the strength by 1% [a]	17.39	9.65	10.89	12.69	23.05

[a] Costs are expressed in US dollars.

Retrofitting alternatives involving the use of concrete generate significant environmental, social, and functional impacts. These high environmental impacts and associated social consequences are primarily due to concrete production, further exacerbated by the large quantities required for these retrofit methods, even when not structurally necessary. The environmental and social impacts of SCJ are lower than CJ due to the reduction of structural adhesive in the retrofit process. Although the shotcrete and equipment required for SCJ are more expensive than conventional concrete, the cost of SCJ is reduced by avoiding extensive formwork systems and the reduction of structural adhesives required for CJ. Alternatives that include concrete have notable architectural impacts, but CJ has the highest execution time, contributing to its lower ranking. Consequently, CJ and SCJ are ranked fifth and fourth among the retrofit alternatives.

Steel jacketing alternatives significantly reduce environmental and social impacts and lower construction costs; this is attributed to the efficient use of materials by placing continuous steel plates on both sides of the beam web. This retrofitting technique effectively enhances shear strength and ductility, achieving higher shear capacity by increasing the height of the plates rather than their thickness [12]. However, the highest preventive maintenance costs for steel jacketing alternatives are due to the need for anti-corrosion paint and fireproof mortar. The STA alternative shows reduced environmental and social impacts compared to STE, primarily because it eliminates using epoxy resin-based thixotropic adhesives. Nevertheless, STA involves higher construction costs, longer execution times, and more significant architectural impacts due to the anchor bolts required, ranking it third among the alternatives. The incorporation of carbon fibers presents reductions in environmental and social impacts, shorter execution times, and minimal architectural impacts, although its construction costs are high. The STE and CFRP retrofit alternatives lead the alternatives in the ranking.

In construction engineering, decision-making often involves conflicting criteria. In the search for optimal seismic retrofit solutions, economic, environmental, and social impacts, lead times, and architectural considerations can result in conflicting criteria for different retrofit alternatives. Various MCDM methods are available in the literature to address such complexities. Among them, the widely used AHP and TOPSIS combine antiquity and ease of application. However, little attention has been given to exploring newer, more advanced decision-making methods. AHP remains the most commonly applied method for criteria weighting. However, the BWM has emerged as a robust alternative in the MCDM field, offering reliable and relevant results. Unlike the arbitrary pairwise comparisons in AHP, BWM employs a systematic approach [53]. It requires fewer pairwise comparisons, reducing inconsistencies and producing more reliable outcomes than AHP [54]. One of the concerns in applying different MCDM methods is the rank reversal problem. This issue occurs when the ranking of alternatives changes depending on the method used, leading to inconsistencies in decision-making [27].

The STE alternative ranks first in the EDAS and MABAC methods, followed closely by CFRP. In the MARCOS method, CFRP ranks first, narrowly edging out STE. In contrast, the CODAS method places CFRP in first place, with a significant advantage over STE in second place. This study has considered input from diverse experts to capture different professional, academic, and research perspectives. However, the weighting of criteria often introduces uncertainty in the application of MCDM methods, which can significantly influence decision-making results. The results indicate that STE is the highest-ranked alternative in the two methods, while CFRP leads in the remaining two. An important step in the application of MCDM is the sensitivity analysis of aspects that can significantly alter the results, thus ensuring consistency in the final decision [55]. In order to support the robustness of these results, a sensitivity analysis was performed to examine how variations in the weighting of the criteria might affect the rankings. The analysis included positive and negative weighting variations of 10% and 15%, resulting in 16 scenarios. An additional scenario was considered in which all criteria had the exact weighting. Rankings were then generated for each scenario across all MCDM methods, providing valuable information on the consistency and reliability of the decision-making process.

Figure 7 presents the results for a ±15% variation in the criteria weights, as no significant changes were observed with a 10% variation. Across the 68 scenarios analyzed, CJ consistently ranks last, while SCJ occupies fourth place in all MCDM rankings. In the EDAS method, STE ranks first in 14 scenarios. CFRP leads the ranking when the weight of the construction cost criterion is reduced when the architectural impact criterion is increased and in the scenario with equal weights for all criteria. Similarly, in the MABAC method, STE ranks first in 15 scenarios. At the same time, CFRP emerges as the best option when construction costs are lower and in the equal-weight scenario. CFRP is consistently ranked as the best option across all scenarios analyzed for the CODAS and MARCOS methods. Notably, STA ranks second when the environmental impact weight during the manufacturing stage is increased in the CODAS method and when equal weights are applied in both the CODAS and MARCOS methods. Based on the results from the four MCDM methods and the weight variations applied in the sensitivity analysis, CFRP emerges as the best option in 57% of cases. At the same time, STE is the top choice in 43% of cases.

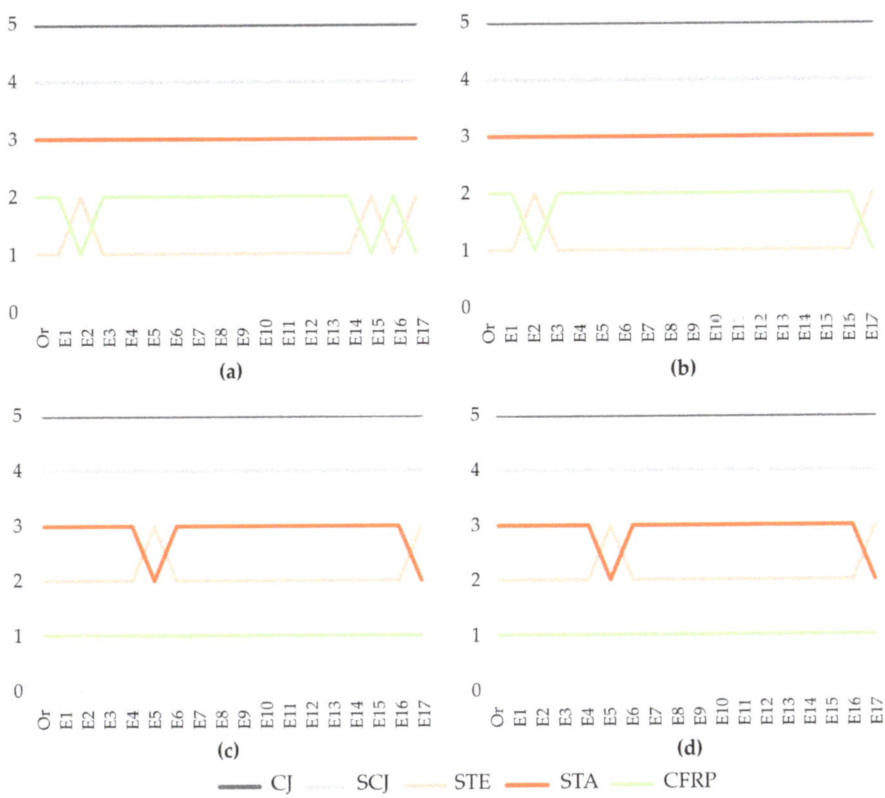

Figure 7. Sensitivity analysis. (**a**) EDAS; (**b**) MABAC; (**c**) CODAS; (**d**) MARCOS.

Traditional retrofit techniques, which consist of increasing the concrete section and using steel profiles, are widely used worldwide [15]. Applying U-jacketing reinforced concrete sleeves to retrofit beams with shear deficiencies is an ongoing development within the scientific community to address key challenges, such as limited section increase. A group of researchers conducted an experimental study using thin layers of U-shaped reinforced concrete to retrofit shear-critical beams, evaluating the performance of these jackets [14]. Similarly, another group explored the effectiveness of U-shaped reinforced cementitious mortar liners for repairing damaged shear-critical beams [52]. In another study, they analyzed beams representing older construction practices with inadequate stirrup spacing, which leads to premature shear failure. They applied a three-sided coating method using thin sleeves of self-compacting reinforced concrete [56].

The retrofitting alternatives in this study focus on improving shear strength in reinforced concrete beams. However, concrete jackets allow the sections to behave similarly to conventional sections of the exact dimensions. A group of researchers analyzed the load-displacement curves and the corresponding ductilities, concluding that retrofitting reinforced concrete beams exhibits mechanical behavior comparable to and slightly better than ordinary reinforced concrete beams of the same size [57]. As for CFRP, significant improvements in ductility have been determined up to the point of lamina detachment [58]. Similarly, another group studied the retrofitting of shear-critical reinforced concrete beams using various FRP systems. It concluded that retrofitting improved shear capacity and transformed shear failure into more ductile flexural failure [59]. In addition, an experimental study examined the effects of steel plates attached to the web of the beam, either with adhesives or bolts, showing significant improvements in beam stiffness—by 29% and 42%, respectively—as well as more excellent ductility when more bottomless plates are used [46].

This study classified five retrofit alternatives based on coating techniques for jacketing beams with shear deficiencies. However, the literature highlights additional techniques that achieve the same retrofit objective. One such method is shear beam retrofit through external prestressing, where a restoring force is applied via externally placed, non-adhered tendons. This approach is known for its ease of application, cost-effectiveness, and straightforward maintenance [51,60]. In recent years, ultra-high-performance concrete (UHPC) has been proposed to retrofit reinforced concrete structural elements. In beams, UHPC has been shown to prevent shear failure and enhance flexural performance, increasing strength by 40% to 125%, making it a promising emerging alternative [61].

The durability of the retrofitting alternatives is considered within the time frame of this study, starting from implementation and extending up to 50 years or until the retrofitting beam sustains damage due to an earthquake. While seismically retrofit structures typically perform well under expected earthquake scenarios, this does not guarantee their continued operability after a seismic event [62]. For the CJ and SCJ alternatives, the brittle shear failure mode is effectively mitigated. In shallow beam jacketing with a span-to-depth ratio greater than 2.5, the retrofitted section can achieve ductile flexural failure modes [63], improving flexural strength, energy dissipation, and ductility [64]. The STE alternative effectively controls cracking under the steel plate; however, its anchoring capacity is limited by the tensile strength of the concrete under the adhesive. This limitation can lead to a risk of explosive detachment of the steel plate. In the case of the STA alternative, stresses at the fixing points can cause localized creep of the steel plate and crushing of the adjacent concrete. However, with adequate anchoring, more excellent ductility can be achieved at the point of failure [65]. Finally, the CFRP alternative improves crack resistance, promoting a shift to a ductile failure mode. However, two main failure modes are observed: detachment from the concrete surface and rupture of the CFRP. In the first scenario, the beams may fail prematurely under the initial load before the CFRP reaches its maximum capacity. In the second scenario, rupture governs the shear capacity [66,67]. Future research could focus on the need for resilient buildings that account for the challenges posed by seismic damage, including restoring functionality amidst physical, social, and economic disruptions over time [68].

6. Conclusions

This paper presents a comprehensive methodology for selecting seismic retrofitting solutions for reinforced concrete beams with shear deficiencies. Five retrofit alternatives designed to improve the shear capacity in the expected plastic hinge formation zone were evaluated to ensure compliance with current seismic regulations. A service life of 50 years was considered for each alternative, with preventive maintenance adapted to the specific properties of the material. The retrofit alternatives evaluated include concrete jacket (CJ), shotcrete jacket (SCJ), steel jacket with structural epoxy adhesive (STE), steel jacket with mechanical anchors (STA), and incorporation of carbon fiber reinforced polymer (CFRP). A hierarchical decision-making model encompassed four dimensions-functionality, economic, environmental, and social- and was analyzed through eight criteria and measured by 16 specific indicators. The BWM method was applied to weigh the criteria and rank the reinforcement alternatives. At the same time, four new MCDM methods were used for the final ranking: EDAS, MABAC, CODAS, and MARCOS.

Life cycle assessments of the economic, environmental, and social dimensions were conducted using a "cradle to grave" approach. The SCJ and STE alternatives demonstrate cost reductions compared to the traditional CJ option. The STE, STA, and CFRP alternatives offer substantial environmental and social advantages over the concrete-based CJ and SCJ alternatives. Compared to the CJ alternative, the STA alternative achieves the most significant reductions in environmental and social impacts, with decreases of 76% and 73%, respectively. STE shows reductions of 61% and 59%, while CFRP achieves reductions of 58% and 66%, respectively, in environmental and social impacts. In addition, STE and CFRP show the shortest execution times and lowest architectural impacts among the alternatives evaluated.

Based on the assumptions of this study, the CFRP and STE alternatives emerge as the preferred options when economic, environmental, social, and functional criteria are considered together. Retrofitting with carbon fiber significantly reduces environmental and social impacts while enhancing functionality despite the high costs associated with the raw materials. Steel jacketing with structural epoxy adhesive is also viable, providing balanced performance across all criteria and outperforming the alternative incorporating structural steel with mechanical anchors. In contrast, alternatives involving concrete are the least favorable due to their high environmental and social impacts coupled with lower functionality.

This study evaluated five commonly used alternatives for seismic retrofitting shear-deficient beams. Future research could explore additional options, including fiber-reinforced polymeric materials and innovative composite materials, such as concrete composite linings and fiber-reinforced cement mortars. Further studies could also incorporate criteria that address the final capacity characteristics of the retrofit element and address the need for resilient buildings capable of recovering quickly after an earthquake to maintain functionality in a shorter time frame. Additionally, the application of MCDM methods could be improved by integrating approaches to manage the uncertainty inherent in expert opinion and considering interactions between criteria, ultimately improving the robustness of solutions.

Author Contributions: This study represents a result of teamwork Conceptualization, P.V., B.G. and V.Y.; methodology, P.V., B.G. and V.Y.; validation, P.V. and B.G.; writing—original draft, P.V. and B.G.; writing—review and editing, V.Y.; resources, V.Y.; supervision, V.Y.; project administration, V.Y.; funding acquisition, V.Y. All authors have read and agreed to the published version of the manuscript.

Funding: Grant PID2023-150003OB-I00 funded by MCIN/AEI/10.13039/501100011033 and by "ERDF A way of making Europe".

Institutional Review Board Statement: Not applicable.

Informed Consent Statement: Not applicable.

Data Availability Statement: Data will be made available on request.

Acknowledgments: The first two authors are grateful to the Universidad Central del Ecuador for funding to pursue a doctoral program at the Universitat Politècnica de València.

Conflicts of Interest: The authors declare no conflicts of interest.

Abbreviations

The following abbreviations are used in this manuscript:

RC	Reinforced concrete
FRP	Fiber-reinforced polymer
CE	Circular economy
MCDM	Multi-criteria decision-making
MADM	Multiple attribute decision-making
LCA	Life cycle assessment
AHP	Analytical Hierarchy Process
TOPSIS	Technique for Order of Preference by Similarity to Ideal Solution
DM	Decision-maker
ANP	Analytical Network Process
BWM	Best-Worst Method
EDAS	Evaluation Based on Distance to Average Solution
MABAC	Multi-Attribute Boundary Approximation Area Comparison
CODAS	Combinatorial Distance-Based Assessment
MARCOS	Measurement of Alternatives and Ranking According to Compromise Solution
LCCA	Life Cycle Cost Assessment
S-LCA	Social Life Cycle Assessment
CJ	Concrete jacketing
SCJ	Shotcrete jacketing
STE	Steel jacketing with structural epoxy adhesive
STA	Steel jacketing with anchor bolts
CFRP	Carbon fiber-reinforced polymer
EoL	End-of-life
MISP	Social Impact Weighting Methodology
MRH	Mean risk per hour

References

1. Hoyos, M.C.; Silva, V. Exploring benefit cost analysis to support earthquake risk mitigation in Central America. *Int. J. Disaster Risk Reduct.* **2022**, *80*, 103162. [CrossRef]
2. Harirchian, E.; Jadhav, K.; Mohammad, K.; Hosseini, S.E.A.; Lahmer, T. A Comparative Study of MCDM Methods Integrated with Rapid Visual Seismic Vulnerability Assessment of Existing RC Structures. *Appl. Sci.* **2020**, *10*, 6411. [CrossRef]
3. Pohoryles, D.A.; Bournas, D.A.; Da Porto, F.; Caprino, A.; Santarsiero, G.; Triantafillou, T. Integrated seismic and energy retrofitting of existing buildings: A state-of-the-art review. *J. Build. Eng.* **2022**, *61*, 105274. [CrossRef]
4. Murray, P.B.; Feliciano, D.; Goldwyn, B.H.; Liel, A.B.; Arroyo, O.; Javernick-Will, A. Seismic safety of informally constructed reinforced concrete houses in Puerto Rico. *Earthq. Spectra* **2023**, *39*, 5–33. [CrossRef]
5. Goldwyn, B.; Vega, Y.G.; Javernick-Will, A.; Liel, A.B. Identifying misalignments between the informal construction sector's perceptions and engineering assessments of housing safety in future disasters for capacity development. *Int. J. Disaster Risk Reduct.* **2022**, *77*, 103105. [CrossRef]
6. Carrillo, J.; Arteta, C.A.; Vera, X. Post-earthquake safety assessment of schools after the 2016 Ecuador M7.8 earthquake. *Soil Dyn. Earthq. Eng.* **2024**, *179*, 108561. [CrossRef]

7. Goretti, A.; Hutt, C.M.; Hedelund, L. Post-earthquake safety evaluation of buildings in Portoviejo, Manabi province, following the M_w7.8 Ecuador earthquake of April 16, 2016. *Int. J. Disaster Risk Reduct.* **2017**, *24*, 271–283. [CrossRef]
8. Scala, S.A.; De Risi, M.T.; Verderame, G.M. Code-based brittle capacity models for seismic assessment of pre-code RC buildings: Comparison and consequences on retrofit. *Bull. Earthq. Eng.* **2024**, *22*, 6643–6674. [CrossRef]
9. Liu, X.; Thermou, G.E. A review on the shear performance of reinforced concrete (RC) beams strengthened with externally bonded mortar-based composites. *Structures* **2023**, *58*, 105474. [CrossRef]
10. Olalusi, O.B.; Viljoen, C. Assessment of simplified and advanced models for shear resistance prediction of stirrup-reinforced concrete beams. *Eng. Struct.* **2019**, *186*, 96–109. [CrossRef]
11. De Domenico, D.; Ricciardi, G. Shear strength of RC beams with stirrups using an improved Eurocode 2 truss model with two variable-inclination compression struts. *Eng. Struct.* **2019**, *198*, 109359. [CrossRef]
12. Chung, P.-T.; Chou, C.-C. One-sided shear retrofit of reinforced concrete beams in existing high-rise buildings. *Eng. Struct.* **2022**, *252*, 113634. [CrossRef]
13. Liu, X.; Lu, Z.-D.; Li, L.-Z. The Use of Bolted Side Plates for Shear Strengthening of RC Beams: A Review. *Sustainability* **2018**, *10*, 4658. [CrossRef]
14. Naoum, M.C.; Papadopoulos, N.A.; Sapidis, G.M.; Chalioris, C.E. Advanced Structural Monitoring Technologies in Assessing the Performance of Retrofitted Reinforced Concrete Elements. *Appl. Sci.* **2024**, *14*, 9282. [CrossRef]
15. Gonzalez-Libreros, J.H.; Sneed, L.H.; D'Antino, T.; Pellegrino, C. Behavior of RC beams strengthened in shear with FRP and FRCM composites. *Eng. Struct.* **2017**, *150*, 830–842. [CrossRef]
16. Lei, H.; Li, L.; Yang, W.; Bian, Y.; Li, C.-Q. An analytical review on application of life cycle assessment in circular economy for built environment. *J. Build. Eng.* **2021**, *44*, 103374. [CrossRef]
17. Salgado, R.A.; Apul, D.; Guner, S. Life cycle assessment of seismic retrofit alternatives for reinforced concrete frame buildings. *J. Build. Eng.* **2020**, *28*, 101064. [CrossRef]
18. Dong, Y.; Ng, S.T.; Liu, P. Towards the principles of life cycle sustainability assessment: An integrative review for the construction and building industry. *Sustain. Cities Soc.* **2023**, *95*, 104604. [CrossRef]
19. Munoz-Medina, B.; Ordonez, J.; Romana, M.G.; Lara-Galera, A. Typology Selection of Retaining Walls Based on Multicriteria Decision-Making Methods. *Appl. Sci.* **2021**, *11*, 1457. [CrossRef]
20. Zavadskas, E.K.; Govindan, K.; Antucheviciene, J.; Turskis, Z. Hybrid multiple criteria decision-making methods: A review of applications for sustainability issues. *Econ. Res.-Ekon. Istraz.* **2016**, *29*, 857–887. [CrossRef]
21. Formisano, A.; Castaldo, C.; Chiumiento, G. Optimal seismic upgrading of a reinforced concrete school building with metal-based devices using an efficient multi-criteria decision-making method. *Struct. Infrastruct. Eng.* **2017**, *13*, 1373–1389. [CrossRef]
22. Fiore, P.; Donnarumma, G.; Falce, C.; D'Andria, E.; Sicignano, C. An AHP-Based Methodology for Decision Support in Integrated Interventions in School Buildings. *Sustainability* **2020**, *12*, 181. [CrossRef]
23. Santa-Cruz, S.; Cordova-Arias, C.; Brioso, X.; Vazquez-Rowe, I. Transparency-based protocol for decision-making regarding seismic rehabilitation projects of public buildings. *Int. J. Disaster Risk Reduct.* **2021**, *55*, 102116. [CrossRef]
24. Clemett, N.; Gallo, W.W.C.; O'Reilly, G.J.; Gabbianelli, G.; Monteiro, R. Optimal seismic retrofitting of existing buildings considering environmental impact. *Eng. Struct.* **2022**, *250*, 113391. [CrossRef]
25. Villalba, P.; Sánchez-Garrido, A.J.; Yepes, V. Life cycle evaluation of seismic retrofit alternatives for reinforced concrete columns. *J. Clean. Prod.* **2024**, *455*, 142290. [CrossRef]
26. Villalba, P.; Sánchez-Garrido, A.J.; Yepes, V. A review of multi-criteria decision-making methods for building assessment, selection, and retrofit. *J. Civ. Eng. Manag.* **2024**, *30*, 465–480. [CrossRef]
27. Zhu, X.; Meng, X.; Zhang, M. Application of multiple criteria decision making methods in construction: A systematic literature review. *J. Civ. Eng. Manag.* **2021**, *27*, 372–403. [CrossRef]
28. Cavalieri, F.; Bellotti, D.; Caruso, M.; Nascimbene, R. Comparative evaluation of seismic performance and environmental impact of traditional and dissipation-based retrofitting solutions for precast structures. *J. Build. Eng.* **2023**, *79*, 107918. [CrossRef]
29. Pushkar, S.; Halperin, I.; Ribakov, Y. Life-Cycle Assessment of Contemporary and Classical Seismic Retrofitting Approaches Applied to a Reinforced Concrete Building in Israel. *Buildings* **2022**, *12*, 1854. [CrossRef]
30. Vitiello, U.; Salzano, A.; Asprone, D.; Di Ludovico, M.; Prota, A. Life-Cycle Assessment of Seismic Retrofit Strategies Applied to Existing Building Structures. *Sustainability* **2016**, *8*, 1275. [CrossRef]
31. Passos Neto, G.d.M.; Alencar, L.H.; Valdes-Vasquez, R. Multiple-Criteria Methods for Assessing Social Sustainability in the Built Environment: A Systematic Review. *Sustainability* **2023**, *15*, 16231. [CrossRef]
32. Sánchez-Garrido, A.J.; Navarro, I.J.; Yepes, V. Multi-criteria decision-making applied to the sustainability of building structures based on Modern Methods of Construction. *J. Clean. Prod.* **2022**, *330*, 129724. [CrossRef]
33. Goldani, N.; Ishizaka, A. A hybrid fuzzy multi-criteria group decision-making method and its application to healthcare waste treatment technology selection. *Ann. Oper. Res.* **2024**, 1–26. [CrossRef]
34. Rezaei, J. Best-worst multi-criteria decision-making method. *Omega-Int. J. Manag. Sci.* **2015**, *53*, 49–57. [CrossRef]

35. Ghorabaee, M.K.; Zavadskas, E.K.; Olfat, L.; Turskis, Z. Multi-Criteria Inventory Classification Using a New Method of Evaluation Based on Distance from Average Solution (EDAS). *Informatica* **2015**, *26*, 435–451. [CrossRef]
36. Pamucar, D.; Cirovic, G. The selection of transport and handling resources in logistics centers using Multi-Attributive Border Approximation area Comparison (MABAC). *Expert Syst. Appl.* **2015**, *42*, 3016–3028. [CrossRef]
37. Ghorabaee, M.K.; Zavadskas, E.K.; Turskis, Z.; Antucheviciene, J. A new combinative distance-based assessment (codas) method for multi-criteria decision-making. *Econ. Comput. Econ. Cybern. Stud. Res.* **2016**, *50*, 25–44.
38. Stevic, Z.; Pamucar, D.; Puska, A.; Chatterjee, P. Sustainable supplier selection in healthcare industries using a new MCDM method: Measurement of alternatives and ranking according to COmpromise solution (MARCOS). *Comput. Ind. Eng.* **2020**, *140*, 106231. [CrossRef]
39. *MIDUVI, Norma Ecuatoriana de la Construcción, 2015, NEC*; Ministerio de Desarrollo Urbano y Vivienda: Quito, Ecuador, 2015.
40. Adhikary, B.B.; Mutsuyoshi, H. Shear strengthening of RC beams with web-bonded continuous steel plates. *Constr. Build. Mater.* **2006**, *20*, 296–307. [CrossRef]
41. Allacker, K. Environmental and economic optimisation of the floor on grade in residential buildings. *Int. J. Life Cycle Assess.* **2012**, *17*, 813–827. [CrossRef]
42. Jiang, T.; Yin, P.; Jin, Q. Performances of typical photovoltaic module production from the perspective of life cycle sustainability assessment. *Sustain. Energy Technol. Assess.* **2024**, *64*, 103703. [CrossRef]
43. Palacios-Munoz, B.; Gracia-Villa, L.; Zabalza-Bribian, I.; Lopez-Mesa, B. Simplified structural design and LCA of reinforced concrete beams strengthening techniques. *Eng. Struct.* **2018**, *174*, 418–432. [CrossRef]
44. Akbar, A.; Liew, K.M. Multicriteria performance evaluation of fiber-reinforced cement composites: An environmental perspective. *Compos. Part B-Eng.* **2021**, *218*, 108937. [CrossRef]
45. Bertolli, V.; Sneed, L.H.; Focacci, F.; D'Antino, T. Shear Strengthening of RC Beams with U-Wrapped FRCM Composites: State of the Art and Assessment of Available Analytical Models. *J. Compos. Constr.* **2025**, *29*, 4024091. [CrossRef]
46. Rakib, M.A.; Morshed, A.Z. Shear strengthening design of pre-cracked reinforced concrete beams using bonded and bolted steel plates. *Aust. J. Struct. Eng.* **2023**, *24*, 159–172. [CrossRef]
47. Li, L.Z.; Cai, Z.W.; Lu, Z.D.; Zhang, X.L.; Wang, L. Shear performance of bolted side-plated reinforced concrete beams. *Eng. Struct.* **2017**, *144*, 73–87. [CrossRef]
48. Mhanna, H.H.; Hawileh, R.A.; Abdalla, J.A. Shear strengthening of reinforced concrete T-beams using CFRP laminates anchored with bent CFRP splay anchors. *Procedia Struct. Integr.* **2020**, *28*, 811–819. [CrossRef]
49. Kalfat, R.; Jumaah, R.; Al-Mahaidi, R.; Abdouka, K.; Hashemi, J. Post-Tensioned Concrete Beams Strengthened in Shear Using Fiber-Reinforced Polymer Laminates and Patch Anchors. *J. Compos. Constr.* **2020**, *24*, 4019065. [CrossRef]
50. Mhanna, H.H.; Hawileh, R.A.; Abdalla, J.A.; Salama, A.S.D.; Alkhrdaji, T. Shear Strengthening of Reinforced Concrete T-Beams with Anchored CFRP Laminates. *J. Compos. Constr.* **2021**, *25*, 4021030. [CrossRef]
51. Lee, S.-H.; Lee, H.-D. Shear strengthening effect by external post-tensioning steel rods in damaged reinforced concrete beams. *Structures* **2024**, *62*, 106327. [CrossRef]
52. Chalioris, C.E.; Kytinou, V.K.; Voutetaki, M.E.; Papadopoulos, N.A. Repair of Heavily Damaged RC Beams Failing in Shear Using U-Shaped Mortar Jackets. *Buildings* **2019**, *9*, 146. [CrossRef]
53. Balali, A.; Valipour, A. Prioritization of passive measures for energy optimization designing of sustainable hospitals and health centres. *J. Build. Eng.* **2021**, *35*, 101992. [CrossRef]
54. Pamucar, D.; Ecer, F.; Cirovic, G.; Arlasheedi, M.A. Application of Improved Best Worst Method (BWM) in Real-World Problems. *Mathematics* **2020**, *8*, 1342. [CrossRef]
55. Navarro, I.J.; Yepes, V.; Marti, J.V. A Review of Multicriteria Assessment Techniques Applied to Sustainable Infrastructure Design. *Adv. Civ. Eng.* **2019**, *2019*, 6134803. [CrossRef]
56. Chalioris, C.E.; Thermou, G.E.; Pantazopoulou, S.J. Behaviour of rehabilitated RC beams with self-compacting concrete jacketing—Analytical model and test results. *Constr. Build. Mater.* **2014**, *55*, 257–273. [CrossRef]
57. Altun, F. An experimental study of the jacketed reinforced-concrete beams under bending. *Constr. Build. Mater.* **2004**, *18*, 611–618. [CrossRef]
58. Waqas, R.M.; Elahi, A.; Kırgız, M.S. Experimental and finite element analysis of shear deficient of reinforced concrete beam retrofitted externally with carbon fiber reinforced polymer sheet. *Structures* **2025**, *72*, 108232. [CrossRef]
59. Baggio, D.; Soudki, K.; Noël, M. Strengthening of shear critical RC beams with various FRP systems. *Constr. Build. Mater.* **2014**, *66*, 634–644. [CrossRef]
60. Giuseppe Santarsiero, V.P.; Masi, A. Structural rehabilitation of half-joints in RC bridges: A state-of-the-art review. *Struct. Infrastruct. Eng.* **2025**, *21*, 227–250. [CrossRef]
61. Li, C.; Aoude, H. Blast retrofit of shear-deficient high-strength concrete beams with ultra-high performance concrete. *Eng. Struct.* **2024**, *304*, 117619. [CrossRef]

62. Gallo, W.W.C.; Clemett, N.; Gabbianelli, G.; O'Reilly, G.; Monteiro, R. Seismic Resilience Assessment in Optimally Integrated Retrofitting of Existing School Buildings in Italy. *Buildings* **2022**, *12*, 845. [CrossRef]
63. Chalioris, C.E.; Pourzitidis, C.N. Rehabilitation of Shear-Damaged Reinforced Concrete Beams Using Self-Compacting Concrete Jacketing. *Int. Sch. Res. Not.* **2012**, *2012*, 816107. [CrossRef]
64. Tahsiri, H.; Sedehi, O.; Khaloo, A.; Raisi, E.M. Experimental study of RC jacketed and CFRP strengthened RC beams. *Constr. Build. Mater.* **2015**, *95*, 476–485. [CrossRef]
65. Barnes, R.A.; Baglin, P.S.; Mays, G.C.; Subedi, N.K. External steel plate systems for the shear strengthening of reinforced concrete beams. *Eng. Struct.* **2001**, *23*, 1162–1176. [CrossRef]
66. Askar, M.K.; Hassan, A.F.; Al-Kamaki, Y.S.S. Flexural and shear strengthening of reinforced concrete beams using FRP composites: A state of the art. *Case Stud. Constr. Mater.* **2022**, *17*, e01189. [CrossRef]
67. Chen, J.F.; Teng, J.G. Shear capacity of FRP-strengthened RC beams: FRP debonding. *Constr. Build. Mater.* **2003**, *17*, 27–41. [CrossRef]
68. Hao, H.; Bi, K.; Chen, W.; Pham, T.M.; Li, J. Towards next generation design of sustainable, durable, multi-hazard resistant, resilient, and smart civil engineering structures. *Eng. Struct.* **2023**, *277*, 115477. [CrossRef]

applied sciences

MDPI

Article

Assessing the Suitability of Damage Indexes for Digital Twin Applications in RC Buildings Considering Masonry Infills

Luca Danesi, Andrea Belleri *, Michelle Gualdi and Simone Labò

Department of Engineering and Applied Sciences, University of Bergamo, Viale Marconi 5, 24044 Dalmine, Italy; luca.danesi@unibg.it (L.D.); michelle.gualdi@unibg.it (M.G.); simone.labo@unibg.it (S.L.)
* Correspondence: andrea.belleri@unibg.it

Abstract: Given the significant damage caused by earthquakes over the years, accurate prediction and assessment of the extent of structural damage is critical to ensure safety and guide post-disaster recovery efforts. This study examines the effectiveness and reliability of various damage indexes for reinforced concrete buildings, particularly in the context of seismic events. It highlights the potential of these indexes for future use in digital twin applications or for direct analysis of sensor data recorded during an earthquake, with the ultimate goal of improving real-time damage assessment and decision making. A comprehensive literature review was carried out looking at the damage indexes developed over the last decades. These indexes were applied to a case study involving an RC building with three different structural configurations: a pre-code moment-resisting frame, a code-compliant moment-resisting frame, and a code-compliant shear wall system, both bare and infilled with masonry. The seismic performance of these configurations was evaluated using Multi-Stripe Analyses (MSA) to account for the variability of the seismic input. The results of applying the damage indexes highlight the versatility of these indexes in detecting damage, although some limitations were noted, particularly with cycle-related indicators and their application to infilled structures. The study emphasizes the importance of refining these tools to improve their accuracy and reliability in different structural contexts, ultimately contributing to more accurate seismic damage assessment and damage prediction for specific seismic scenarios.

Keywords: damage index; masonry infills; RC buildings; seismic damage

1. Introduction

Earthquakes pose a significant threat to human life, infrastructure and the economy worldwide. Past earthquakes have shown how vulnerable the existing building stock is to both direct and indirect losses, with the latter being more difficult to quantify and often overlooked in damage assessments. For example, the economic losses caused by the disruption of services or the interruption of industrial production are particularly difficult to estimate.

Nowadays, there is a growing interest in quantifying and assessing potential building damage using specialized methods. Accurate prediction of expected damage prior to an earthquake based on factors such as building design, building materials, seismic retrofitting and historical earthquake data is crucial for urban planning, risk mitigation strategies and insurance. Similarly, effective damage assessment that evaluates the actual damage after the earthquake plays an important role in guiding emergency response, repair efforts and reconstruction, as well as refining future predictive models.

In recent decades, numerous formulations for the calculation of damage indexes have emerged that provide clear and standardized definitions for expected and actual damage. These methods cover both the prediction of damage before an event and the assessment of damage after an event. In this study, the damage indexes developed in recent decades to assess the seismic damage condition of buildings are examined in a comprehensive literature review. The indexes assessment is performed through Multi-Stripe Analyses (MSA), which take into account the variability of the seismic input for a more robust understanding of the potential damage scenarios. Specifically, the study considers six selected damage indexes and a reinforced concrete (RC) reference building with different structural configurations: a pre-code frame system, a code-compliant moment-resisting frame system and a code-compliant shear wall system. Both the bare frame and the masonry-infilled configurations are analyzed.

In this context, the potential application of damage indexes in digital twins and sensor-based systems is emphasized. Through a non-linear dynamic analysis performed on digital twins, damage indexes can be calculated in near real time, allowing for a fast and accurate post-event assessment. This approach provides a robust tool for continuous structural health monitoring and facilitates the simulation of different earthquake scenarios to better understand and prepare for potential risks. Similarly, the use of the data recorded by the sensors and their elaboration, such as floor accelerations and inter-storey displacements, may enable the calculation of damage indexes without requiring a finite element model. This enables rapid quantification of damage to buildings, including structural and non-structural components, and provides information for timely decision making for emergency response and repair strategies.

In this study, we provide a comprehensive overview of selected damage indexes for the assessment of seismic damage to buildings, validate their effectiveness and highlight their strengths and limitations. The main novelty of this work lies in the comprehensive evaluation of the effectiveness and reliability of damage indexes thorough Multi-Stripe Analyses (MSA) and in the proposal of specific formulations for the consideration of masonry infills. The use of MSA is particularly important to understand how each index performs in different structural configurations—such as existing buildings, code-compliant framing systems and code-compliant wall systems—and how accurately they quantify damage at different limit states.

The effectiveness of the damage indexes in the analyses carried out on finite element models suggests that these indicators can also be widely applied to digital twin models. Damage indexes can be implemented in digital twin models to assess potential damage due to extreme events such as earthquakes in near real time. The knowledge gained from this study will help to select the most appropriate existing index to assess the predicted and expected damage in different scenarios, considering both serviceability and collapse conditions. The same procedure could also be used to evaluate the effectiveness of new damage indexes.

2. Seismic Damage Indexes for RC Buildings

The growing interest of the scientific community in damage detection and control in buildings has led to the definition of a variety of damage indexes that can express the condition of structural elements and components or the entire building after a seismic event by a scalar measure [1]. Since the 1970s, systematic attempts have been made to quantitatively estimate the local or global extent of seismic damage to structural systems. In the past, the damage indexes for reinforced concrete (RC) buildings proposed in the literature have been extensively discussed and studied [2,3]. This section presents the main damage indexes that have been developed to assess local damage to RC structural members, as well as a global index suitable for assessing the overall damage of the whole building.

As for the local damage in a single element, Lybas and Sozen (1977) [4] introduced the following index:

$$DI_{\mathrm{L-S}} = \frac{k_0}{k_m} \tag{1}$$

where k_0 and k_m are the elastic and the secant stiffness at the maximum displacement, respectively.

Assuming a monotonically increasing load path, this index starts at a value of 1.0 in the undamaged state and gradually increases until it reaches a maximum value of 3.5 in the final state.

In the same year, a flexural damage index was introduced by Banon et al. (1977) [5]:

$$DI_{\mathrm{FDR}} = \frac{\frac{\phi_m}{M_m} - \frac{\phi_y}{M_y}}{\frac{\phi_u}{M_u} - \frac{\phi_y}{M_y}} \tag{2}$$

where ϕ_m, ϕ_y and ϕ_u are the maximum, yield and ultimate curvature, respectively, and M_m, M_y and M_u are the maximum, yield and ultimate bending moment, respectively. The index is designed to provide a value of zero in the absence of damage and to gradually increase to a value of 1.0 when the structure reaches its final state.

Banon and Veneziano (1982) [6] introduced the first damage index based on the dissipated energy, which is particularly relevant for cyclic loads, such as those experienced during seismic events. This index is calculated for each loading cycle:

$$DI_{\mathrm{B-V}} = \sqrt{\left(\frac{d_m}{d_y - 1}\right)^2 + \left[\left(\frac{2E_h}{F_y d_y}\right)^{0.38}\right]^2} \tag{3}$$

where d_m and d_y are the maximum and the yield displacement, respetively, F_y is the yield force and E_h is the dissipated energy. This damage index is calculated for each cycle and is defined as the maximum value among the indexes determined. It assumes values greater than zero even before the yield point and exceeds 9.0 under failure conditions.

Hwang and Scribner (1984) [7] defined the following index based on the dissipated energy:

$$DI_{\mathrm{H-W}} = \sum\nolimits_{i=1}^{n} \frac{k_i {d_i}^2}{k_e {d_y}^2} \tag{4}$$

where k_e and k_i are the element's initial elastic flexural stiffness and the element's flexural stiffness at the i-th cycle, respectively, d_y and d_i are the element's yield displacement and the element's displacement at the i-th cycle, respectively, and n is the number of cycles. As can be seen from its mathematical formulation, the index represents the cumulative damage across all cycles. This results in values that are already greater than zero before the yield point and exceed 4.0 in the final state of the structural element.

The index proposed by Park and Ang (1985) [8] is

$$DI_{\mathrm{P-A}} = \frac{d_m}{d_u} + \frac{\beta}{F_y d_u} \int dE \tag{5}$$

in which

$$\beta = \left(-0.447 + 0.073\frac{L}{d} + 0.24n + 0.314\rho_t\right) \cdot 0.7^{\rho_\varpi} \tag{6}$$

where d_u is the ultimate displacement under monotonic loading, d_m is the maximum displacement experienced during the earthquake, dE is the differential dissipated energy, L is the structural element length, d is the effective depth of the cross-section, n is the normalized axial load, ρ_t is the percentage of longitudinal reinforcement and ρ_ω is the confinement ratio. This index assumes values greater than zero when the element is undamaged and approaches a value of 2.0 when the element reaches failure.

Stephensen and Yao (1987) [9] introduced an index based on the cumulative displacement ductility:

$$DI_{S-Y} = \sum_{i=1}^{n} \left(\frac{\Delta d^+}{\Delta d_f} \right)^{1-b \cdot r} \tag{7}$$

where Δd^+ and Δd^- are the incremental increase in the displacements of the element in the positive and negative direction, respectively, Δd_f is the collapse displacement, b is a coefficient that assumes a value of 0.77 and $r = \frac{\Delta d^+}{\Delta d^-}$ is the ratio between the two incremental increases. The sum of the damage over all cycles leads to a value greater than zero before the yielding point and greater than 1.0 when the structural element is approaching failure.

Wang and Shah (1987) proposed an index based solely on the forces and bending moment value experienced by the element [10]. However, this characteristic makes the index difficult to apply for real-time structural monitoring, as it requires specialized measurement equipment. The index can be computed as

$$DI_{W-S-1} = 1 - \frac{F_y}{F_m} \tag{8}$$

where F_y and F_m are, the yield and the maximum shear force or bending moment in the element during the cyclic loading, respectively.

This index assumes negative values under loading conditions below the yielding point, reaches 0.0 at the yielding point and generally approaches values close to 0.1 when the element reaches its maximum force during cyclic loading.

A second formulation of the index assumes that the damage to an RC member subjected to cyclic loading depends on the maximum displacements experienced by the element during each cycle [10]. A cyclic-loading parameter is defined as

$$\beta = C \cdot \sum_{i=1}^{n} \frac{\vartheta_i}{\vartheta_u} \tag{9}$$

where ϑ_i and ϑ_u are the rotation at the i-th cycle and the ultimate rotation of the section of the analyzed element, respectively. The index is computed as follows:

$$DI_{W-S-2} = \frac{e^{\eta\beta} - 1}{e^{\eta} - 1} \tag{10}$$

where η can be -3 and -1 while C is a constant equal to 0.15.

This index assumes values greater than zero even before the element reaches the yield point and remains below 0.2 as the structural element approaches failure.

Powell and Allahabadi (1988) [11] developed an index for RC elements based on plastic deformation and displacement ductility:

$$DI_{P-Al} = \frac{\mu_m - 1}{\mu_u - 1} \tag{11}$$

where μ_m is the maximum ductility experienced by the element and μ_u is the ultimate ductility of the element. The mathematical formulation of the index defines values between 0.0 and 1.0, corresponding to no-damage and failure conditions, respectively.

As for the global damage of a structure, Di Pasquale and Cakmak (1988) [12] presented an index based on the variation of the fundamental period of vibration of the structure following an earthquake:

$$DI_{\mathrm{DP-C}} = 1 - \frac{T_a}{T_m} \tag{12}$$

where T_a and T_d are the initial and final fundamental period of vibration of the structure, respectively.

Bracci et al. (1989) [13] suggested a damage index based on the residual curvature of the structural element:

$$DI_{\mathrm{Br-al}} = \frac{\varphi_m - \frac{M_u}{k_m}}{\varphi_u - \frac{M_u}{k_u}} \tag{13}$$

where φ_m and φ_u are the maximum and ultimate curvature, respectively, M_u is the ultimate bending moment, and k_m and k_u are the maximum and the ultimate stiffness, respectively. The index is formulated to start at zero in the absence of damage and to gradually increase to a maximum value of 1.0 as the structure approaches its ultimate condition.

Colombo and Negro (2005) [14] defined a general damage index as the ratio between the initial and the ultimate resistance capacity of RC elements.

$$DI_{\mathrm{C-N}} = 1 - \left[\left(1 - \frac{\vartheta_m}{\vartheta_u}\right)^{\frac{1}{\beta_1}} \cdot 0.5\left(1 - \tanh\left(\beta_2 \frac{\int dE}{E_u^*} - \pi\right)\right) \cdot e^{-\beta_3 \frac{\int dE}{E_u^*}}\right] \tag{14}$$

where $\beta_1 = \beta_3 = 0.1$, $\beta_2 = 2.4$, ϑ_m is the maximum rotation, ϑ_u is the ultimate rotation and E_u^* is a parameter that considers the dissipated energy.

Rodriguez and Padilla (2009) [15] proposed the following index considering using earthquake records which did not result in building collapse:

$$DI_{\mathrm{R-P}} = \frac{E_h}{E_\lambda} \tag{15}$$

in which the energy absorbed in elastic system is

$$E_\lambda = k_\theta \theta_c^2 \tag{16}$$

where, for a single degree of freedom (SDOF), E_h is the hysteretic energy, k_θ is the elastic energy absorbed by SDOF system while it achieves $\theta = \pm 1$, θ_c is the maximum drift ratio.

Ladinović et al. (2011) [16] suggested a damage index based on displacement history, hysteretic energy and plastic deformation, modifying the existing Park and Ang model [7] to enhance damage assessment accuracy.

$$DI_{\mathrm{L-al}} = \frac{d_m - d_y}{d_u - d_y} + \alpha\beta \frac{E_h}{F_y(d_u - d_y)} \tag{17}$$

in which

$$\alpha = 1 - \frac{\mu_c}{\mu_{ac}} \tag{18}$$

where d_u is the ultimate displacement under monotonic loading, d_m is the maximum displacement experienced during the earthquake, d_y is the yield displacement, μ_c ($\mu_c = d_{c,m}/d_y$) is the cyclic ductility depending on the maximum cyclic displacement request $d_{c,m}$ pending a ground movement [17], μ_{ac} is the accumulative ductility depending on the total inelastic displacements, β is the coefficient introduced by Park and Ang [7] defined in Equation (6), E_h is the hysteretic energy.

Guo et al. (2016) [18] developed another implementation of the existing Park and Ang index [7], proposing a three-dimensional damage index to account for the biaxial effect due to earthquake events.

$$DI_{\text{G-al}} = (1-\beta)\,\mu_{\theta,m}^{-} + \beta \int \alpha_m \frac{dE_p}{E_{mon}} \quad (19)$$

where β is the weight factor of energy dissipation, $\mu_{\theta,m}^{-}$ is the maximum normalized rotation ductility factor, dE_p is the plastic energy dissipation increment, E_{mon} is the value of plastic energy dissipation due to the effect of monotonic loading at the failure limit of the structure member, α_m is the correction of the energy dissipation term.

The aforementioned damage indexes can be classified based on the parameters involved in their formulation, as listed in Table 1, or based on the required knowledge of the structure as represented in Table 2.

Table 1. Classification of the damage indexes based on the parameters involved in the formulation.

Parameter	Indexes
Stiffness	Lybas and Sozen [4]; Banon et al. [5]; Hwang and Scribner [7]; Wang and Shah-1st [10]; Di Pasquale and Cakmak [12]; Bracci [13]
Ductility	Banon et al. [5]; Banon and Veneziano [6]; Park and Ang [8]; Wang and Shah-2nd [10]; Powell and Allahabadi [11]; Bracci [13], Colombo and Negro [14]; Ladinović et al. [16]; Guo et al. [18]
Number of cycles	Hwang and Scribner [7]; Stephensen and Yao [9]; Wang and Shah [10]
Energy	Banon and Veneziano [6]; Park and Ang [8]; Colombo and Negro [14]; Rodriguez and Padilla [15]; Ladinović et al. [16]; Guo et al. [18]
Combined	Banon and Veneziano [6]; Hwang and Scribner [7], Park and Ang [8]; Wang and Shah-2nd [10]; Bracci [13]; Colombo and Negro [14]; Ladinović et al. [16]; Guo et al. [18]

Table 2. Classification of the damage indexes based on the level of knowledge required.

Level of Knowledge	Indexes
Dynamic properties	Lybas and Sozen [4]; Hwang and Scribner [7]; Stephensen and Yao [9]; Di Pasquale and Cakmak [12]
Structural details	Banon et al. [5]; Banon and Veneziano [6]; Park and Ang [8]; Wang and Shah-1st [10]; Wang and Shah-2nd [10]; Powell and Allahabadi [11]; Bracci [13], Colombo and Negro [14]; Rodriguez and Padilla [15]; Ladinović et al. [16]; Guo et al. [18]

More precisely, the level of knowledge in Table 2 refers to parameters obtained from dynamic identification methods or on construction details, materials and dimensions required to define ductility, displacements and ultimate rotations.

To ensure a meaningful comparison, integration into decision-making processes and consistent model calibration, it is important that all index values are within the same range. A review of the literature shows that each index typically operates within its own range, so rescaling is necessary for consistency. In this context, the previous indexes have been rescaled so that they lie between 0, for no damage, and 1, for collapse. The rescaled damage indexes in this section are denoted as DI_*. The Lybas and Sozen index is rescaled as follows:

$$DI_{L-S*} = \frac{C}{ID_0} - C \quad (20)$$

with

$$C = \frac{k_e}{k_u - k_e} \tag{21}$$

where k_e and k_u are the secant stiffness at yield and at collapse of the element, respectively.

The indexes proposed by Banon and Veneziano, Hwang and Scribner, Park and Ang, Stephensen and Yao and Wang and Shah (first formulation) can be rescaled as follows:

$$DI_* = \frac{DI_0}{DI_u - DI_y} - \frac{DI_y}{DI_u - DI_y} \tag{22}$$

where DI_0 is the damage index calculated according to the original definition, DI_y is the damage index calculated at yield and DI_u is the damage index calculated at ultimate conditions.

The Wang and Shah index (second formulation) can be rescaled through the cyclic-loading parameter β as follows:

$$\beta_n = \frac{\beta_0}{\beta_u - \beta_y} - \frac{\beta_y}{\beta_u - \beta_y} \tag{23}$$

where β_0, β_y and β_u are the coefficients calculated according to the original definition, at yield and at ultimate conditions, respectively.

A graphical example of the scaling process is shown in Figure 1, where EDP stands for Engineering Demand Parameter and represents general structural response metrics used to estimate the extent of damage to structural components and systems, e.g., inter-story drift and element rotation. The indexes not mentioned do not need to be rescaled because they are already in the range from 0 to 1.

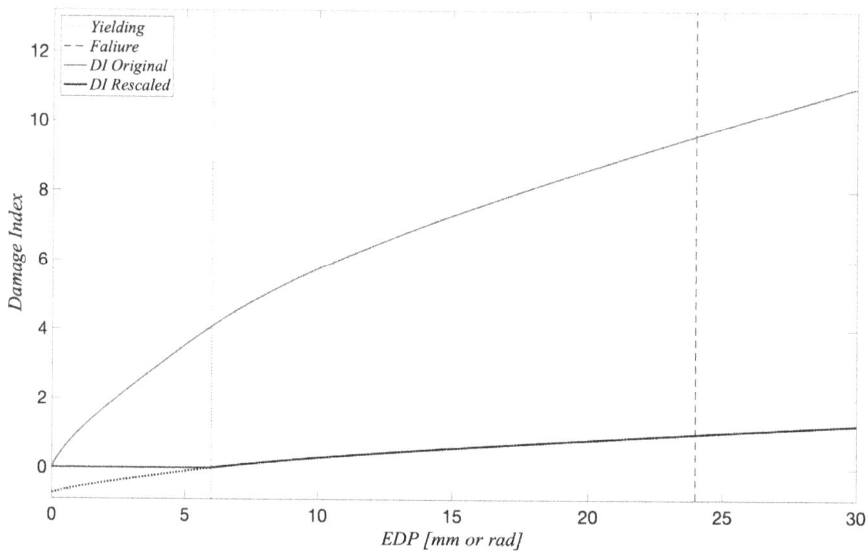

Figure 1. Example of rescaling for the Banon and Veneziano [6] index.

Since only Di Pasquale and Cakmak refer to the whole structure, the other indexes, referring to single structural elements, should be combined to obtain a global index. For this purpose, a method based on the weighted average of the indexes assessed for each element can be used. This approach considers the dissipated energy of each element: first, an index is computed to reflect the global damage state at each building level (Dlevel), then, such indexes are combined to obtain a global damage index (Dbuilding). The formulation of the indexes is

$$D_{level} = \frac{\sum_{i=1}^{n} D_i E_i}{\sum_{i=1}^{n} E_i} \quad (24)$$

$$D_{building} = \frac{\sum_{j=1}^{N} D_{level,j} E_{level,j}}{\sum_{j=1}^{N} E_{level,j}} \quad (25)$$

where D_i is the local damage index of the i-th structural element, n is the number of structural elements per building level, N is the total number of building levels, E_i is the energy dissipated by the i-th element and Elevel, j is the dissipated energy of the j-th level, the latter assessed as the sum of the dissipated energy of the structural elements belonging to the j-th level.

3. Damage Indexes Effectiveness

The efficiency of damage indexes in predicting and assessing the level of damage in an RC structure is evaluated using a reference case study in three different structural configurations, in particular:

1. An existing RC-frame building representative of a typical post-World War II construction (EX_MRF) designed for gravity load only;
2. A seismic code-conforming RC Moment-Resisting Frame structure (CC_MRF);
3. A seismic code-conforming RC wall building (CC_W).

For each configuration, the bare frame (BF) and the infilled frame (IF) configurations are considered to highlight the strengths and weaknesses of the damage indexes both with and without the interaction of infill panels.

3.1. Considered Case Study

The reference case study is a four-storey RC residential building in L'Aquila (AQ) (an area of high seismicity in Italy). It has a rectangular floor plan (21.05 m × 11.45 m). The first floor is 3.40 m high, while the inter-floor height for the rest of the building is 3.05 m (Figure 2). The one-way beam and block flooring system consists of RC beams and hollow bricks with a 5 cm thick RC overlay. The floor spans along the X-axis. The longitudinal and internal transverse RC beams are 25 cm deep; the depth of the transverse beams and edge beams varies depending on the configuration under consideration. The infill panels of the perimetral frame consist of two layers of hollow bricks (12 cm + 12 cm).

Concrete and steel materials with the following mechanical properties are considered as concrete C25/30 (f_{ck} = 25 N/mm^2, f_{cd} = 14.2 N/mm^2) and steel B450C (f_{yk} = 450 N/mm^2, f_{yd} = 391 N/mm^2). The existing building was designed for gravity loads only, whereas for the code-compliant cases, seismic actions were also considered. The applied loads are listed in Table 3.

Table 3. Permanent and live load applied to the structure (in kN/m^2). The infill load takes into account 15% of opening. Note: G refers to dead loads while Q for live loads.

Load	G_1-Floor and Roof	G_2-Floor	Q-Floor	G2-Roof	Q-Roof (Snow)	G2-Staircase	Q-Staircase	G2-Infill
Value	3.20	3.10	2.00	1.60	2.18	0.80	4.00	3.50

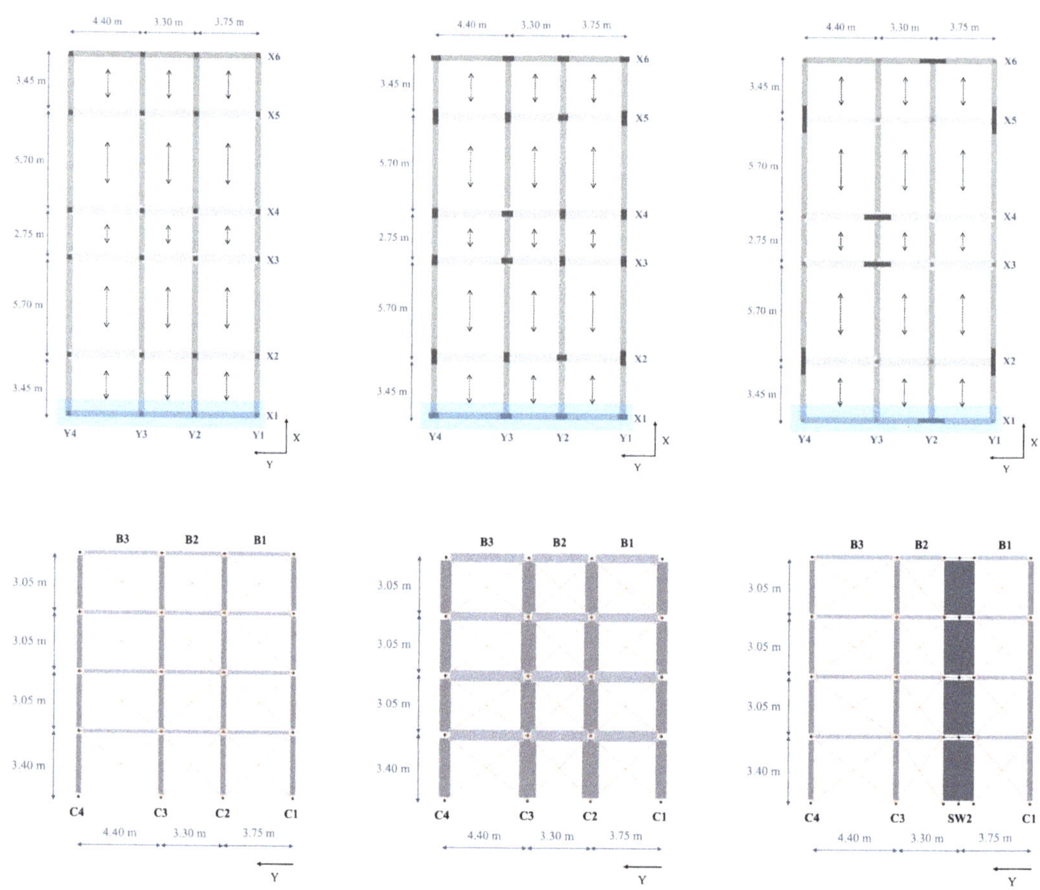

(a) (b) (c)

Figure 2. Reference plan and frontal view of the perimetral frame for (**a**) the existing structure (EX_MRF), (**b**) code-conforming moment-resisting frame (CC_MRF) and (**c**) code-conforming wall (CC_W).

For an initial assessment and development of methods and indicators that can accurately represent the state of the building after a seismic event, a 2D frame was extracted and considered for performing non-linear dynamic analysis: frame X1 highlighted in Figure 2. A planar frame analysis helps to avoid the influence of torsional phenomena on the interpretation of the damage indexes.

3.2. Numerical Modeling

For the numerical analyses, two finite element models have been developed. For the design phase, elastic models were developed with the software Midas GEN v.1.1 [19] and response spectrum analyses were conducted. For the performance assessment phase, non-linear models were developed with the software Opensees [20] and non-linear response history analyses were conducted.

For the response spectrum analyses, RC elements were modeled as *beam*-type elements with flexural stiffness scale factors equal to 0.7 and 0.5 for columns and beams, respectively. As for the shear walls, a stiffness scale factor equal to 0.7 was considered and the beam-to-shear wall connection was simulated by rigid elements linking the wall centroid to adjacent beams. Columns and shear walls were fixed at the base. Only the bare frame configuration (BF) was considered and masonry infill panels were accounted for only in terms of mass. The details of the frame cross-sections resulting from the design are listed in Tables 4–7.

Table 4. Geometrical properties and reinforcement details of beams and columns for the EX_MRF building.

ID	Element Type	Storey	Cross-Section (mm)	Longitudinal Reinforcement	Stirrups
C1-C2-C3-C4	Column	All	300 × 300	8ϕ14	ϕ10/200
B1-B2-B3 (i and j)	Beam	All	400 × 250	4ϕ14 top 4ϕ14 bot	ϕ10/200

Table 5. Geometrical properties and reinforcement details of beams and columns for the CC_MRF building.

ID	Element Type	Storey	Cross-Section (mm)	Longitudinal Reinforcement	Stirrups
C1-C4	Column	1	700 × 350	20ϕ24	3bϕ10/100
C2-C3	Column	1	900 × 350	26ϕ24	3bϕ10/75
C1-C4	Column	2, 3	600 × 350	18ϕ24	3bϕ10/100
C2-C3	Column	2, 3	800 × 350	22ϕ24	3bϕ10/75
C1-C4	Column	4	600 × 350	18ϕ24	3bϕ10/100
C2-C3	Column	4	700 × 350	20ϕ24	3bϕ10/100
B1-B2-B3 (i and j)	Beam	1, 2	350 × 600	6ϕ20 top 5ϕ20 bot	3bϕ10/100
B1-B2-B3 (i and j)	Beam	3, 4	350 × 500	6ϕ16 top 6ϕ16 bot	3bϕ10/100

Table 6. Geometrical properties and reinforcement details of beams and columns for the CC_W building.

ID	Element Type	Storey	Cross-Section (mm)	Longitudinal reinforcement	Stirrups
C1-C3-C4	Column	All	250 × 250	8ϕ16	ϕ8/90
B1-B2-B3 (i and j)	Beam	All	400 × 250	4ϕ16 top 4ϕ16 bot	4bϕ8/90

Table 7. Geometrical properties and reinforcement details of shear walls for the CC_W building.

ID	Element Type	Storey	Cross-Section [mm]	L_c [mm]	Reinforcement in L_c	Reinforcement in L_w	Stirrups
SW2	Shear wall	1	1600 × 300	600	20ϕ18	8ϕ10	ϕ10/100
SW2	Shear wall	2	1600 × 300	600	20ϕ16	8ϕ10	ϕ10/150
SW2	Shear wall	3	1600 × 300	600	20ϕ14	8ϕ10	ϕ10/150
SW2	Shear wall	4	1600 × 300	600	20ϕ12	8ϕ10	ϕ10/150

After completing the design phase, the final cross-section, the amount of longitudinal and transverse reinforcement, and the axial load in seismic condition were used to define the non-linear behavior of the structure. Two non-linear models were defined for each building configuration: the bare frame (BF) model, where only the infill mass is considered, and the infill frame (IF) model, which considers the contribution of the infills with an equivalent compression-only strut. Lumped plastic hinges were introduced to capture the non-linear behavior with the PeakOriented hysteretic model developed by Ibarra–Medina–Krawinkler (IMK) [21] (Figure 3a). Such hinges were applied by means of *zeroLength* elements in the end sections of the elements, while the inner part of the member was modeled with *ElasticBeamColumn* elastic frame elements.

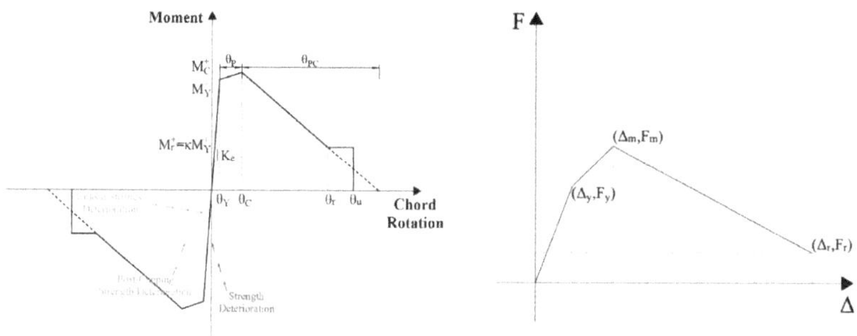

Figure 3. (**a**) Modified Ibarra–Medina–Krawinkler deterioration model (image from [20]) and (**b**) masonry infills backbone curve.

The infill panels were modeled as compression-only diagonal struts whose backbone curve values were derived from Decanini et al. [22] and adapted according to available experimental tests [23,24] and on the masonry properties defined by Hak et al. [25]. An example of the backbone curve is represented in Figure 3b whose values for a typical 30 cm thick infill panel with 2.65 m × 4.15 m dimensions are Δ_y, F_y (0.30%, 388 kN), Δ_m, F_m (0.50%, 485 kN) and Δ_m, F_m (1.86%, 0 kN).

For the non-linear assessment, Multi-Stripe Analyses (MSA) were conducted. MSA is a method that enables the evaluation of the seismic behavior of structures under earthquake scenarios with increasing intensity measures (IM) of ground motion. This method stems from the work of Cornell et al. [26] and aims to calculate the distributions of one or more Engineering Demand Parameter for several scenarios of ground motion records represented by an intensity measure, herein represented by the elastic spectral acceleration, $S_a(T_j)$, at the fundamental period of vibration of the building, T_j.

Ten intensity measure levels (IMLs) were considered for each building type and configuration (i.e., BF, IF) corresponding to return periods of 10, 50, 100, 250, 500, 1000, 2500, 5000, 10,000 and 100,000 years for the site of L'Aquila (high seismicity area in Italy), with a soil category C [27]. For each intensity level, 20 different ground motions were selected from the DPC-ReLUIS project and were used to evaluate the Engineering Demand Parameters (EDP) [28,29] adopting the conditional mean spectrum approach [30]. The conditioning periods at which the various spectra of the chosen ground motion converge to the spectral acceleration at each intensity level are reported in Table 8.

Table 8. Conditioning periods chosen for each configuration of the building.

Case Study	Building Fundamental Period (T_1) [s]	Selected Conditioning Period (T_j) [s]
Existing building (BF)	1.55	1.50
Existing building (IF)	0.27	0.15
Code-conforming wall building (BF)	0.62	0.50
Code-conforming wall building (IF)	0.17	0.15
Code-conforming MRF building (BF)	0.56	0.50
Code-conforming MRF building (IF)	0.18	0.15

3.3. MSA Results

In this study, the results of the MSA are used to calculate the values of the considered damage indexes and to assess their effectiveness and reliability compared to various damage scenarios and performance levels, selected according to the DPC-ReLUIS project [28,29] and Eurocode 8 [27]: Usability Preventing Damage (UPD), Life Safety (LS), Collapse Prevention (CP) and Global Collapse (GC). Table 9 lists the limit states and the corresponding structural conditions.

Table 9. Considered limit states and performance levels.

Limit State/Performance Level	Description
Usability Preventing Damage (UPD)	Defined according to Eurocode 8: it aims to withstand a seismic action having a larger probability of occurrence than the design seismic action, without the occurrence of damage and the associated limitations of use, the costs of which would be disproportionately high in comparison with the costs of the structure itself.
Life Safety (LS)	Defined according to Eurocode 8: it aims to protect occupants from serious injury and to allow safe evacuation. Although significant structural and non-structural damage may occur, the critical load-bearing elements of the building must remain intact to prevent complete structural failure. Herein considered when the inter-story drift ratio reaches a value of 2%.
Collapse Prevention (CP)	Defined according to Eurocode 8: it represents the stage at which the structure is severely damaged, with large permanent drifts, but retains its vertical load-bearing capacity and most of the ancillary components, if any, have collapsed. Herein considered when the top displacement reaches the value corresponding to a 20% degradation of the maximum base shear obtained from a non-linear static, pushover, analysis.
Global Collapse (GC)	Defined according to Eurocode 8: it represents the stage at which the structure could be considered to have collapsed. Herein considered when the top displacement reaches the value corresponding to a 50% degradation of the maximum base shear obtained from a non-linear static, pushover, analysis.

Figure 4a provides an example of the results obtained from the analysis as diagrams showing the damage index values as a function of the spectral acceleration. The complete graphical representation is reported in Appendix A for different building configurations.

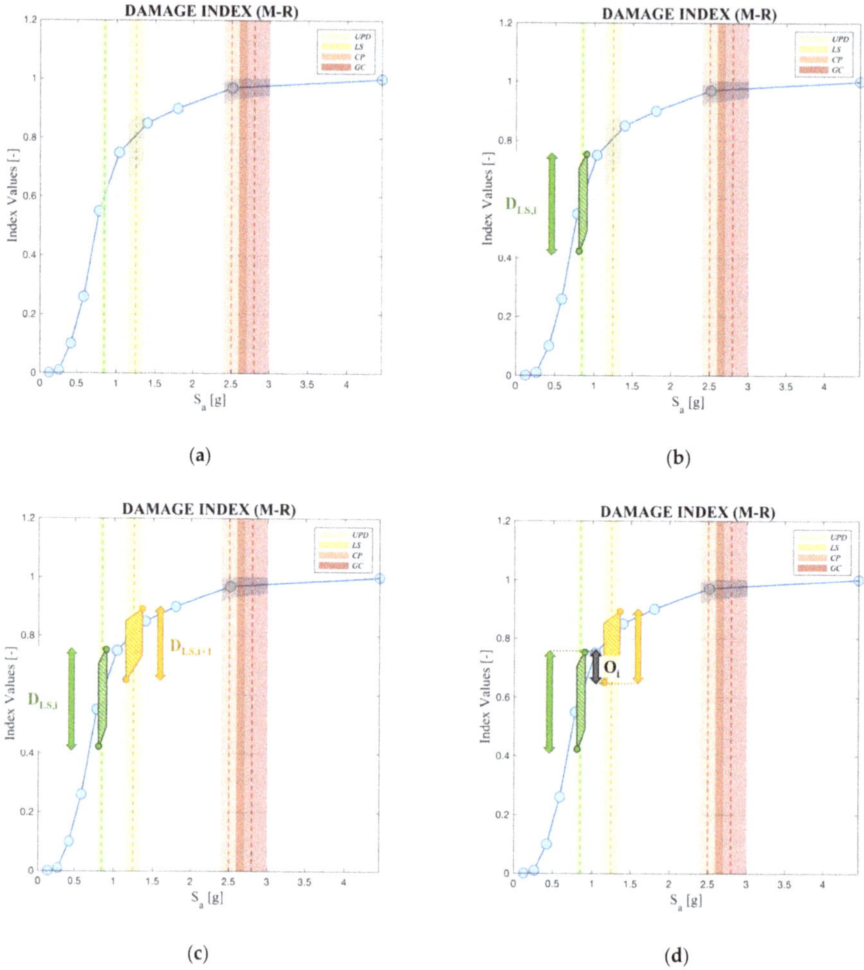

Figure 4. (**a**) Example of the computation of the damage index using the M-R approach. (**b**) Graphical representation of the dispersion (D) for a generic limit state i. (**c**) Visualization of the dispersion $D_{LS,i}$ and $D_{LS,i+1}$ for consecutive limit states. (**d**) Representation of the overlap (O_i) between two consecutive limit states-performance levels.

In Figure 4a, the solid blue line and the blue circles show the median value of the damage indexes calculated over the 10 IM levels. This provides a representation of the central tendency of the damage indexes. The light blue shaded area shows the interquartile range (25th to 75th percentile), which illustrates the variability of the index, with wider ranges reflecting greater uncertainty. The vertically shaded areas correspond to the defined performance levels: UPD (green), LS (orange), CP (red) and GC (dark red). The width of these areas represents the interquartile range of the performance levels while the dashed vertical line represents the median value. This representation highlights the damage index value dispersion within the same intensity level and offers a clear view of the variability and robustness of the indicators. Herein, only the damage indexes calculated starting from the moment-rotation values of the RC elements are represented. Similar results are obtained from the damage indexes calculated starting from the floor shear-interstory drift values.

In order to better assess the efficiency of each damage index, specific parameters are introduced to highlight their accuracy in detecting damage while minimizing the overlap between the limit states and reducing the dispersion of the data. The dispersion (D) measures the variability of the values within a given limit state-performance level by calculating the difference between the maximum and minimum values in the interquartile range and provides an indication of the accuracy of the data. This is shown graphically in Figure 4b,c. The overlap parameter (O) evaluates the potential overlap between consecutive limit states and indicates the risk of misinterpretation of the damage levels and the uncertainty in distinguishing between them. Figure 4d graphically represents this parameter, which reflects the precision and reliability of the damage indexes in the assessment of seismic performance.

Tables 10 and 11 summarize the results of damage index curves plotted in Appendix A, where the damage index values are plotted as a function of the spectral acceleration. The comparison of the individual indexes aims to provide a clear evaluation of the efficiency of the damage indicators for the different structural systems in both bare and infilled frame configurations. The results for each building configuration are reported in Appendix B. In these tables, indexes are cited as L-S for Lybas and Sozen (1977) [4], FDR for Banon et al. (1977) [5], B-V for Banon and Veneziano (1982) [6], P-A for Park and Ang (1985) [8], W-S-1 for Wang and Shah (first formulation (1987)) [10] and P-Al for Powel and Allahabadi (1988) [11]. When analyzing the damage indexes for different building typologies (as also reported graphically in Appendix A), especially in the bare frame configuration, the indexes show strong capabilities in detecting and quantifying damage. In this configuration, the indexes exhibit remarkable uniformity in damage quantification, with overlaps O1 and O2 equal to zero for all damage indexes (except for Lybas and Sozen) and low dispersion values (D). This underlines their potential for effective "real-time" damage detection in BF configurations for all building types. Conversely, the performance of these indexes decreases when evaluating infilled structures. While all indexes successfully track incremental damage with increasing intensity, they exhibit greater variability, with larger dispersion values (D) and increased sensitivity to building typology. This greater heterogeneity leads to a significant overlap between the different damage states, with higher values already from O_2 for LS-CP. These overlaps reduce the effectiveness of the damage indexes and limit their applicability for the reliable prediction of damage in infilled structures.

In particular, in the bare-frame configuration for existing MRF and SW buildings, the L-S and P-A indexes show high efficiency for the UPD and LS limit states. They exhibit no overlap O1 and O2 and a low to medium dispersion, ranging from less than 5% for UPD to 25% for LS. However, for the CP and GC limit states, their efficiency decreases significantly due to a significant overlap (about 95%) and a larger dispersion of about 10% to 80%. Similarly, the B-V, W-S-1 and P-Al indexes show high efficiency for UPD for all building types, with a very good LS performance for frame systems. They show no overlap O1 and O2 and a low dispersion, ranging from 0% to 15% for existing and MRF buildings and up to 40% for LS in the shear wall configuration. On the other hand, their efficiency decreases for CP and GC due to overlaps of more than 95% and higher dispersion values of 10% to 40%, reflecting the trends observed for other indexes. Conversely, the FDR Banon index performs well for UPD as it has no overlap and very low dispersion. However, its efficiency decreases for LS, where the overlap O2 is between 5% and 20% with increasing dispersion. For CP and GC, the FDR index mirrors the trend of the other indexes, with lower efficiency due to significant overlap O3 and higher dispersion. Therefore, to improve the damage index performance, it is suggested to consider just an ultimate performance level, condensing CP and GC.

Table 10. Overview of the dispersion (D) of all damage indexes for each limit state for all the building configurations.

	DISPERSION (D)																							
	EX_MRF_BF				EX_MRF_IF				CC_MRF_BF				CC_MRF_IF				SW_MRF_BF				SW_MRF_IF			
Index	UPD [%]	LS [%]	CP [%]	GC [%]	UPD [%]	LS [%]	CP [%]	GC [%]	UPD [%]	LS [%]	CP [%]	GC [%]	UPD [%]	LS [%]	CP [%]	GC [%]	UPD [%]	LS [%]	CP [%]	GC [%]	UPD [%]	LS [%]	CP [%]	GC [%]
L-S	4.2	9.0	6.9	7.0	27.6	71.0	8.4	5.4	0.0	23.4	12.5	12.6	23.4	41.8	37.0	44.9	9.9	12.8	5.7	5.1	58.6	29.5	8.8	9.2
FDR	20.6	4.4	7.3	7.3	6.4	75.3	19.3	12.5	0.0	14.0	23.4	23.4	19.2	32.9	56.9	62.7	6.6	36.1	20.7	18.3	21.9	52.5	44.0	44.5
B-V	0.9	14.4	11.3	9.6	12.3	68.4	2.0	1.3	0.0	7.2	40.7	32.8	1.5	23.7	91.2	94.4	0.2	40.4	43.0	37.9	1.5	56.9	71.4	71.8
P-A	0.0	25.1	43.0	44.8	13.2	73.3	14.1	9.1	0.0	4.7	66.2	81.9	1.4	18.1	93.2	95.1	0.3	24.1	49.0	45.8	7.3	60.2	75.5	75.8
W-S-1	1.9	11.0	20.4	17.1	3.9	125	48.0	31.0	0.0	7.4	23.9	20.7	21.0	27.3	63.9	68.6	1.4	39.3	28.2	25.2	8.4	55.5	50.1	50.7
P-Al	1.7	13.3	8.7	7.4	6.4	75.3	19.3	12.5	0.0	11.7	30.3	23.8	4.8	33.2	78.7	85.0	1.0	40.1	28.4	25.1	8.0	57.4	51.1	51.7

Table 11. Overview of the overlap (O) of all damage indexes for each limit state for all the building configurations.

	OVERLAP (O)																	
	EX_MRF_BF			EX_MRF_IF			CC_MRF_BF			CC_MRF_IF			SW_MRF_BF			SW_MRF_IF		
Index	UPD LS [%]	LS CP [%]	CP GC [%]	UPD LS [%]	LS CP [%]	CP GC [%]	UPD LS [%]	LS CP [%]	CP GC [%]	UPD LS [%]	LS CP [%]	CP GC [%]	UPD LS [%]	LS CP [%]	CP GC [%]	UPD LS [%]	LS CP [%]	CP GC [%]
L-S	0.0	0.0	96.9	0.0	11.8	64.7	0.0	0.0	91.4	0.0	48.3	100	0.0	0.5	87.1	0.0	19.2	100
FDR	0.0	18.1	94.7	0.0	25.6	64.7	0.0	0.0	86.1	0.0	53.1	100	0.0	5.8	87.1	0.0	54.7	100
B-V	0.0	0.0	79.8	0.0	2.9	64.7	0.0	0.0	66.7	0.0	73.5	100	0.0	0.0	87.6	0.0	65.4	100
P-A	0.0	0.0	94.4	0.0	19.2	64.7	0.0	0.0	93.6	0.0	78.3	100	0.0	0.0	88.7	0.0	78.0	100
W-S-1	0.0	0.0	80.5	0.0	38.5	64.7	0.0	0.0	70.4	0.0	53.4	100	0.0	0.0	87.7	0.0	47.4	100
P-Al	0.0	0.0	80.1	0.0	25.6	64.7	0.0	0.0	65.5	0.0	61.0	100	0.0	0.0	87.6	0.0	47.7	100

In the case of infilled configurations, the FDR and W-S-1 indexes show a high efficiency in distinguishing damage related to UPD and LS for all building types. These indexes show no overlap O1 and a low to medium dispersion between 5% and 20%. However, their efficiency decreases for LS and CP due to significant overlap O2 between 25% and 55%, accompanied by an increased dispersion over 20%. The B-V and P-A indexes maintain their high efficiency for UPD, as they have no overlap O1 and scatter values between 1% and 15%. However, their reliability decreases for LS, CP and GC, with overlap O2 increasing from 5% to 80% and dispersion values increasing from 20% to 65%. Similarly, the P-Al index for UPD has a minimum dispersion from 5% to 10% and no overlap O1, but its efficiency decreases for LS, CP and GC due to a larger overlap O_2 ranging from 25% to 60% and a larger dispersion. Conversely, the L-S index shows reasonable efficiency but lower precision for UPD, with no overlap O_1 and dispersion values between 20% and 60%. For LS, CP and GC, its behavior mirrors the trend observed for the other indexes, with reliability decreasing at higher damage states due to greater overlap and scatter.

After analyzing the behavior of the damage indexes in the BF and IF configurations, the most appropriate index for each building configuration can be identified. For BF configurations in all analyzed building types, most of the indexes effectively detect damage without misinterpreting the LS and CP/GC states, except for the FDR-Banon index. Among these indexes, Lybas and Sozen's index proves to be the optimal choice for all BF configurations, as it consistently has lower dispersion values for the CP and GC limit states. As far as IF configurations are concerned, the FDR-Banon index is the most efficient index for the pre-code moment-resisting frame. This index has the lowest overlap between the LS and CP/GC limit states and one of the lowest dispersion values for the CP and GC limit states. In contrast, Lybas and Sozen's index proves to be the most effective choice for both the moment-resisting frame and the shear wall system for code-compliant buildings. It minimizes overlap and dispersion while optimizing calculation time, making it the preferred index for these configurations.

Although originally developed for structural RC elements, the evaluation of the efficiency and reliability of damage indexes in this study is extended to non-structural elements, such as masonry infills, which significantly affect repair costs after a seismic event. To achieve this, specific indexes are defined and tested to quantify the damage level of masonry infills. Among these, the index of Lybas and Sozen and the index of Powell and Allahabadi are selected as the most efficient and computationally effective options for calculating the damage level of masonry infills. In addition, a new dissipated energy-based index is introduced, based on the ratio between the area under the trilinear curve of the infill panel up to the point reached during the analysis and the total area under the trilinear curve.

The effectiveness and adaptability of the analyzed infill indexes are evaluated by comparing the median values that each index assumes in each infill during the analyzed intensity measures (IMs) in the Multiple Stripe Analysis. Three threshold values are determined, corresponding respectively to the elastic limit of the infill, the maximum capacity of the infill and the failure of the infill. These three thresholds define four states of damage for the infill: the elastic and uncracked phase, the cracked phase, the softening phase and, finally, failure. These are shown respectively with green, yellow, light red and dark red color codes. The stage of each infill is defined based on the median value that each index assumes for the infill under investigation during the analyzed strip.

The results showed that all three indexes were able to capture the growth and increase of infill damage in all building configurations studied, making these indexes an effective tool for detecting damage to closure panels regardless of building typology. In particular, the P-Al index and the energy-based index are particularly effective in progressively assessing damage to these elements. Conversely, the L-S index could potentially lead to an underestimation of damage, especially if it is close to the elastic limit. An example of the damage progression in infills of the shear wall building is represented in Figure 5.

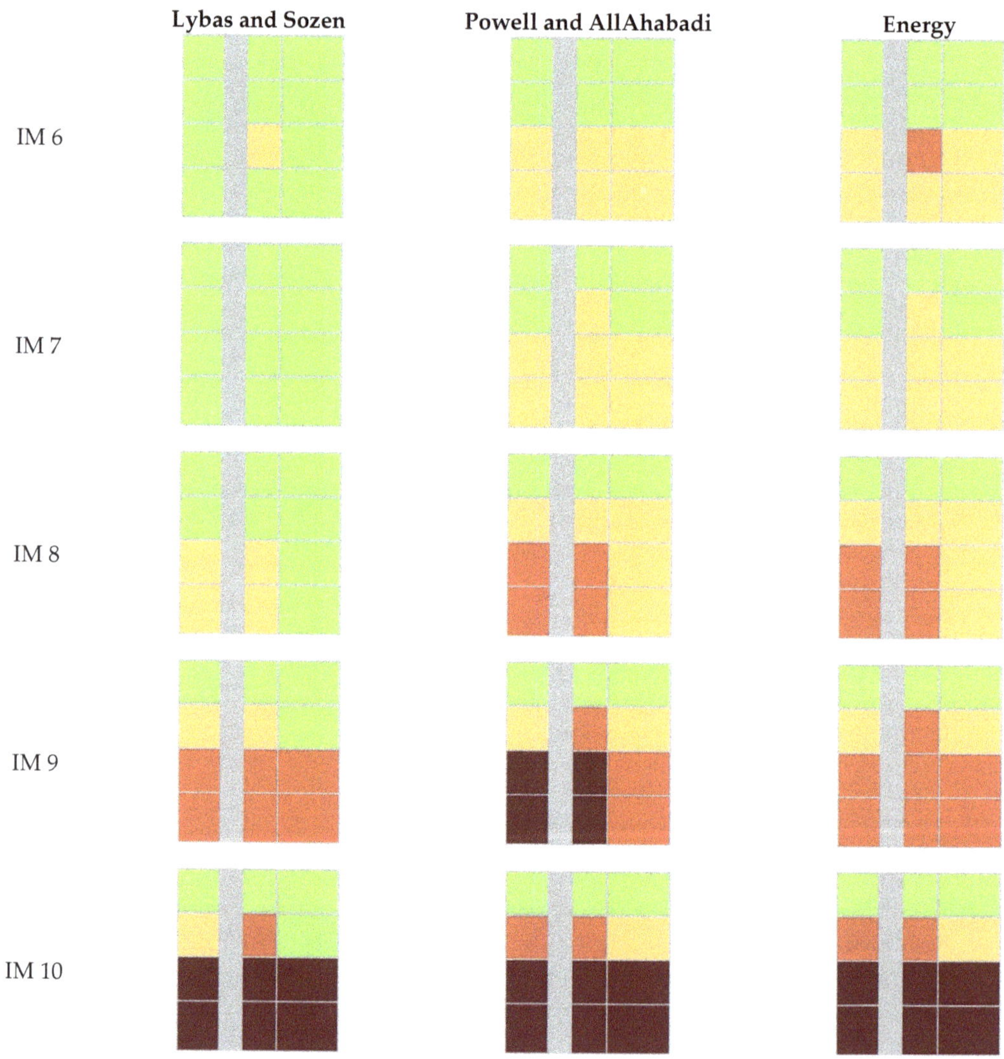

Figure 5. Visualization of the progression of damage through Lybas and Sozen, Powell and Allahabadi and energy-based indexes for the CC_W_IF building. Note: the green, yellow, orange and dark red refer to the UPD, LS, CP and GC performance levels, respectively.

4. Conclusions

The detection and quantification of damage to reinforced concrete (RC) buildings is an important area of research driven by the recurrent and significant damage caused by earthquakes over the years and by the possibility of creating digital twins of existing buildings to directly assess damage after earthquakes. The aim of this work was to evaluate the effectiveness and reliability of different damage indexes to ultimately identify tools that can improve seismic risk assessment and demonstrate their potential applicability for digital twins of RC buildings, which is crucial for post-earthquake assessment.

First, a comprehensive literature review was conducted to examine damage indexes that have been developed and used in the past. Notable examples include the indexes proposed by Lybas and Sozen, Park and Ang, and Banon and Veneziano. Following the literature review, the study focused on evaluating the efficiency of selected damage indexes for an RC reference building. This building was analyzed under three different structural configurations: an existing gravity-only moment-resisting frame, a code-compliant moment-resisting frame, and a code-compliant shear wall system that accounts for seismic loading. In addition, both the bare frame and the masonry infilled frame conditions were analyzed for each building configuration. The seismic performance of the building was thoroughly evaluated using Multi-Stripe Analyses, which allowed for a detailed assessment across different intensity levels.

The study shows which indexes are most suitable for certain situations and which are less effective. It is worth noting that cycle-related indexes were not considered as they cannot be directly compared to other indexes unless a rescaling approach is provided. This limitation results from their mathematical formulation. In particular, by summing the damage from each hysteretic cycle, the index value can exceed 1, so a direct comparison with other indexes—which are typically defined between 0 (no damage) and 1 (collapse)—is not possible. In the case of cycle-related indicators, specific performance values must be defined for each index.

In BF configurations, most indexes effectively detect damage without misinterpreting the LS and CP/GC states, with the exception of the FDR-Banon index. Among these, Lybas and Sozen's index and Powell and Allahabadi's index have been found to be the optimal choice for all BF configurations, as they consistently have lower scatter values for the CP and GC limit states.

For the IF configurations, the FDR-Banon index proved to be the most efficient index for the pre-coded moment-resisting frame. This index has the lowest overlap between the LS and CP/GC limit states and one of the lowest scatter values for the CP and GC limit states. In contrast, Lybas and Sozen's index is the most effective choice for both moment-resisting frames and shear wall systems in code-compliant buildings. It minimizes overlap and scatter while optimizing calculation time, making it the preferred index for these configurations.

For non-structural elements, such as masonry infill, both the Lybas and Sozen index and the Powell and Allahabadi index proved to be the most efficient options for assessing the extent of damage. In addition, a new dissipated energy-based index was developed based on the ratio between the area under the trilinear curve of the infill panel up to the point reached during the analysis and the total area under the curve. The results show that all three indexes can capture the growth and increase in infill panel damage in all building configurations studied, making them effective tools for detecting damage to closure panels regardless of building typology. In particular, the Powell and Allahabadi index and the energy-based index are especially effective for the progressive assessment of damage to these elements, while the Lybas and Sozen index can lead to an underestimation of damage.

In general, the damage indexes proved to be very effective in detecting not only the presence of damage, but also its extent. Importantly, the results of this study highlight the potential benefits of applying damage indexes in advanced monitoring and assessment systems, particularly in digital twin systems and sensor-based applications.

Damage indexes can improve real-time assessment of seismic damage and decision making after seismic events, which is an important aspect of engineering practice. In particular, seismic design should consider structural safety together with optimization of available resources and prioritization of actions. The use of damage indexes to quickly identify damage and vulnerabilities enables more informed decisions leading to a more efficient use of resources in post-earthquake rehabilitation. In addition, the use of damage indexes could promote a more integrated and systematic approach to resource allocation for seismic rehabilitation measures by moving from assessing the vulnerability of individual buildings to a damage index ranking of the different buildings in the urban context.

Author Contributions: Conceptualization, L.D., M.G., S.L. and A.B.; methodology, L.D. and A.B.; software, L.D.; validation, L.D., M.G., S.L. and A.B.; formal analysis, L.D.; writing—original draft preparation, M.G.; writing—review and editing, L.D., M.G., S.L. and A.B.; visualization, L.D., M.G., S.L. and A.B.; supervision, A.B. All authors have read and agreed to the published version of the manuscript.

Funding: This research received no external funding.

Institutional Review Board Statement: Not applicable.

Informed Consent Statement: Not applicable.

Data Availability Statement: The raw data supporting the conclusions of this article will be made available by the authors on request.

Acknowledgments: The studies presented here were partially carried out as part of the activities envisaged by the Agreement between the Italian Department of Civil Protection (DPC) and the ReLUIS Consortium. The contents of this paper represent the authors' ideas and do not necessarily correspond to the official opinion and policies of DPC.

Conflicts of Interest: The authors declare no conflicts of interest.

Appendix A

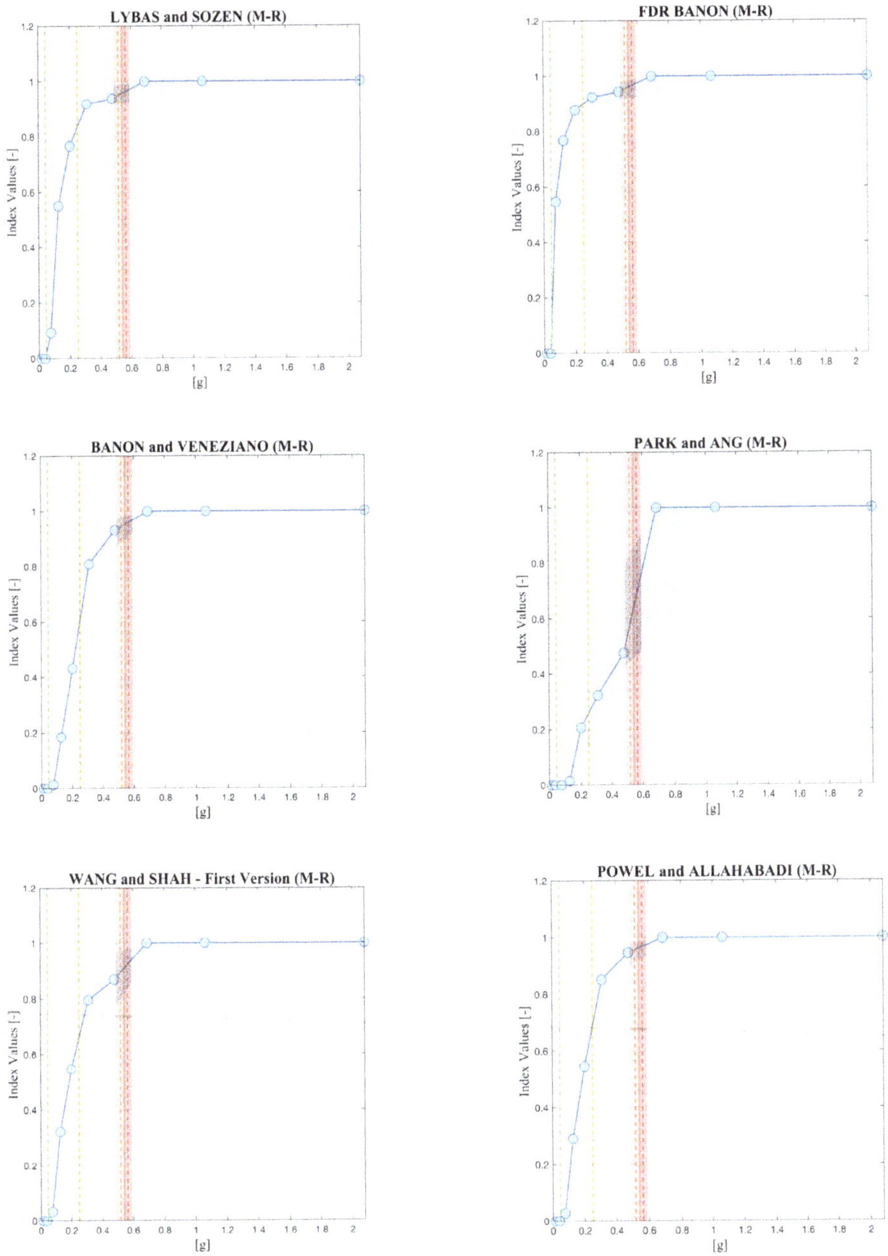

Figure A1. Damage indexes for the existing RC frame building in the bare frame configuration (EX_MRF_BF). Note: the green, yellow, orange and red refer to the UPD, LS, CP and GC performance levels, respectively.

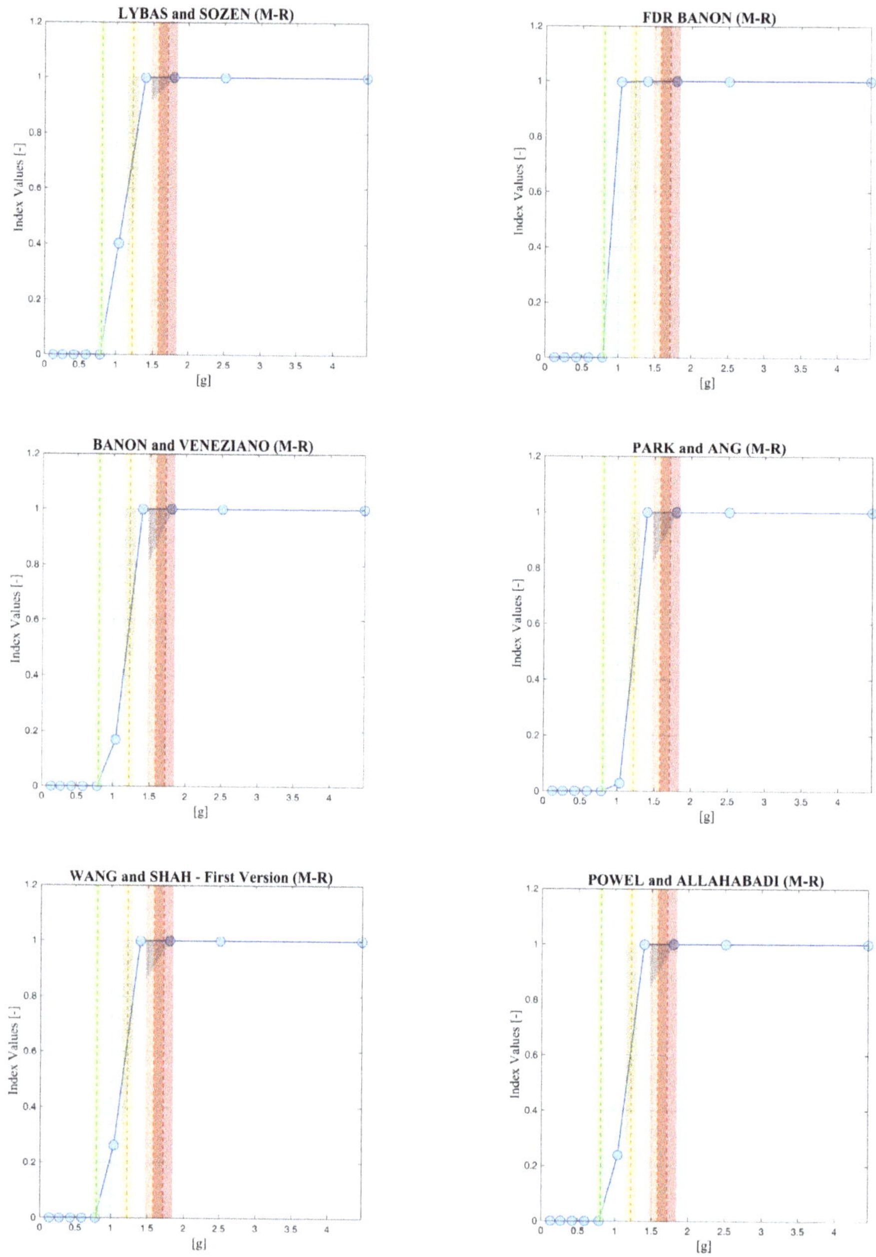

Figure A2. Damage indexes for the existing RC frame building in the infilled frame configuration (EX_MRF_IF). Note: the green, yellow, orange and red refer to the UPD, LS, CP and GC performance levels, respectively.

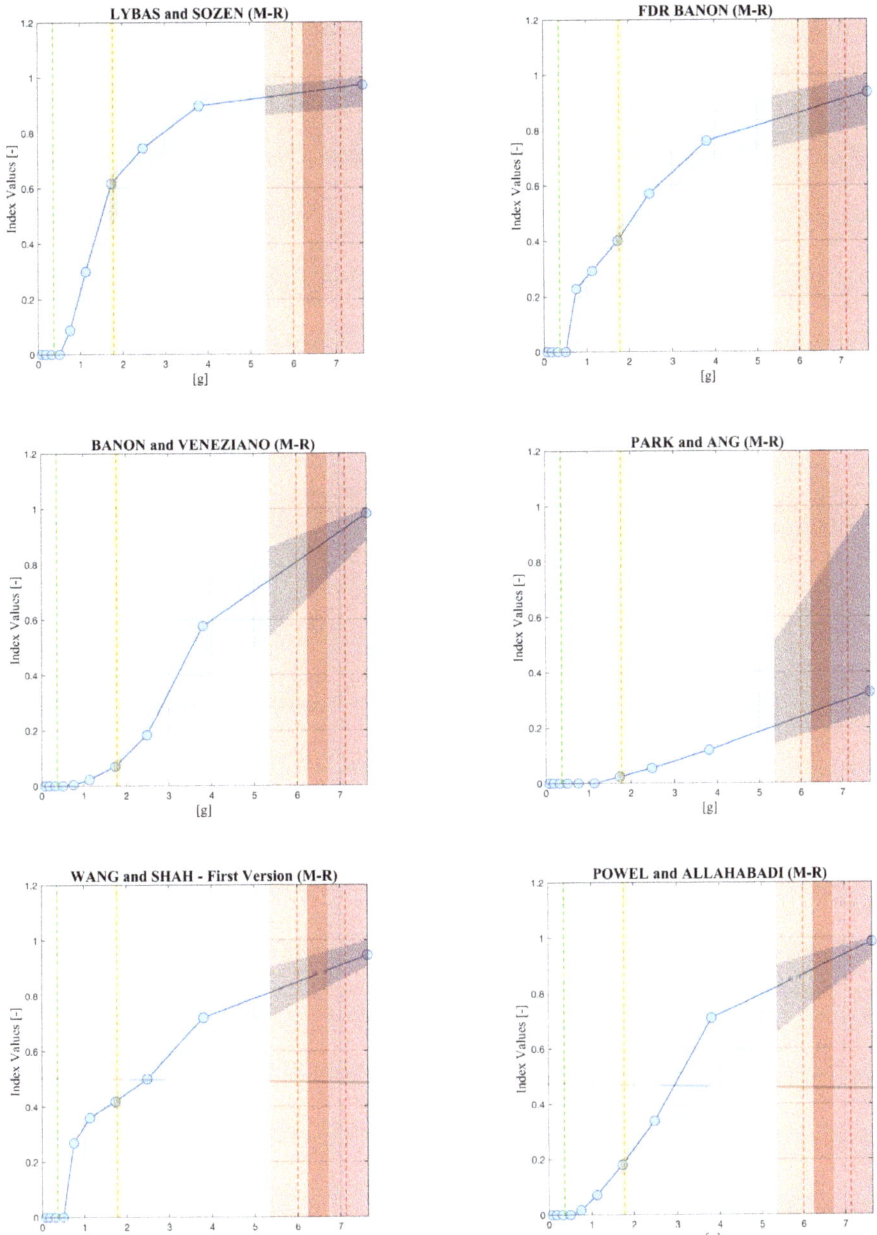

Figure A3. Damage indexes for the code-conforming MRF building in the bare frame configuration (CC_MRF_BF). Note: the green, yellow, orange and red refer to the UPD, LS, CP and GC performance levels, respectively.

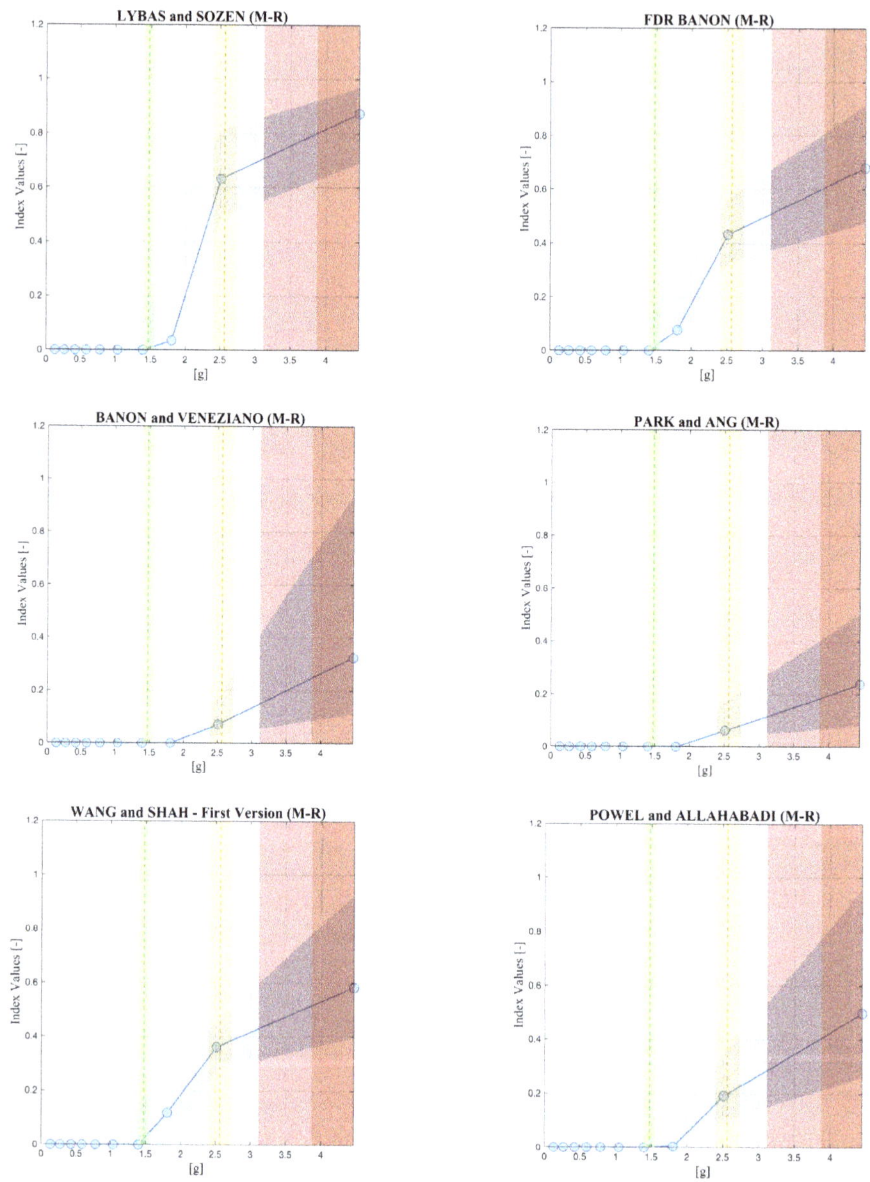

Figure A4. Damage indexes for the code-conforming MRF building in the infilled frame configuration (CC_MRF_IF). Note: the green, yellow, orange and red refer to the UPD, LS, CP and GC performance levels, respectively.

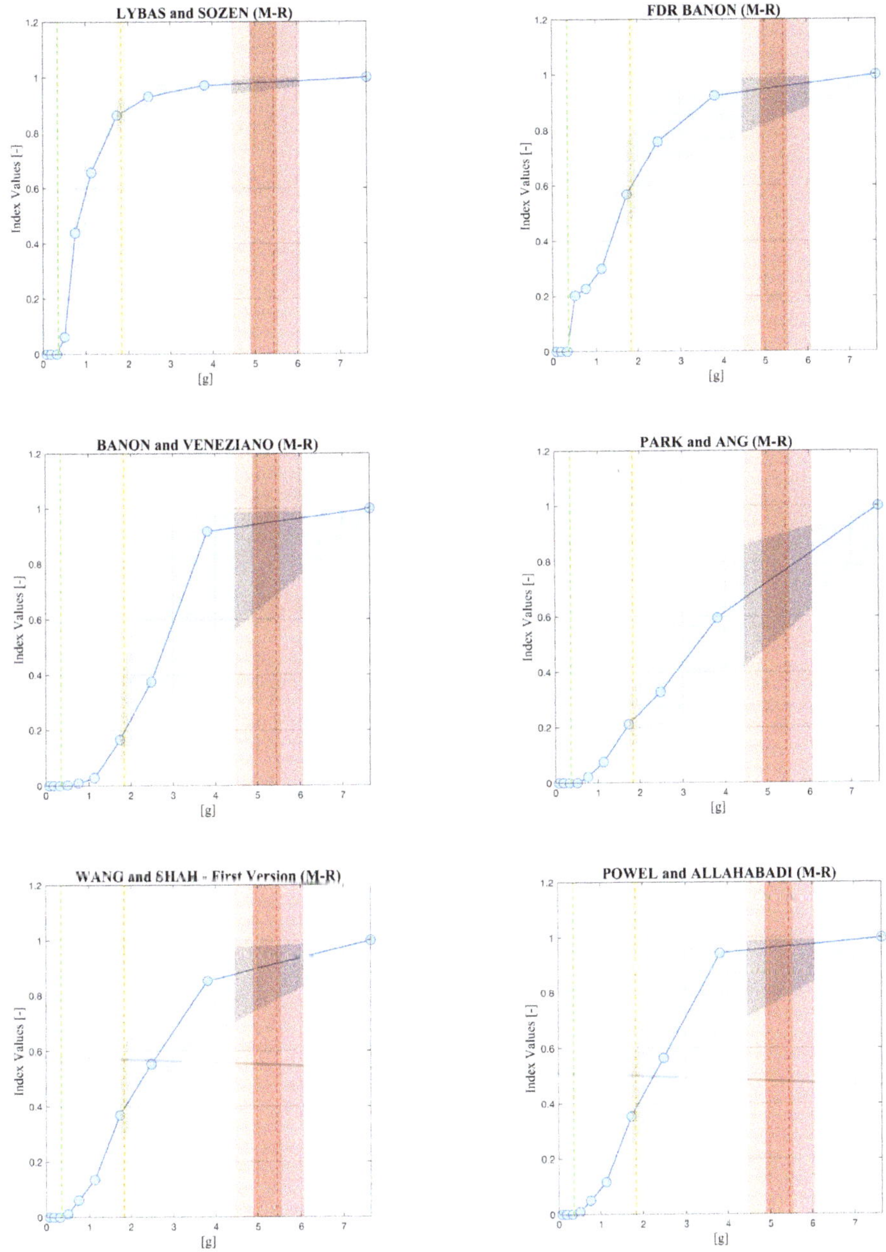

Figure A5. Damage indexes for the code-conforming SW building in the bare frame configuration (CC_W_BF). Note: the green, yellow, orange and red refer to the UPD, LS, CP and GC performance levels, respectively.

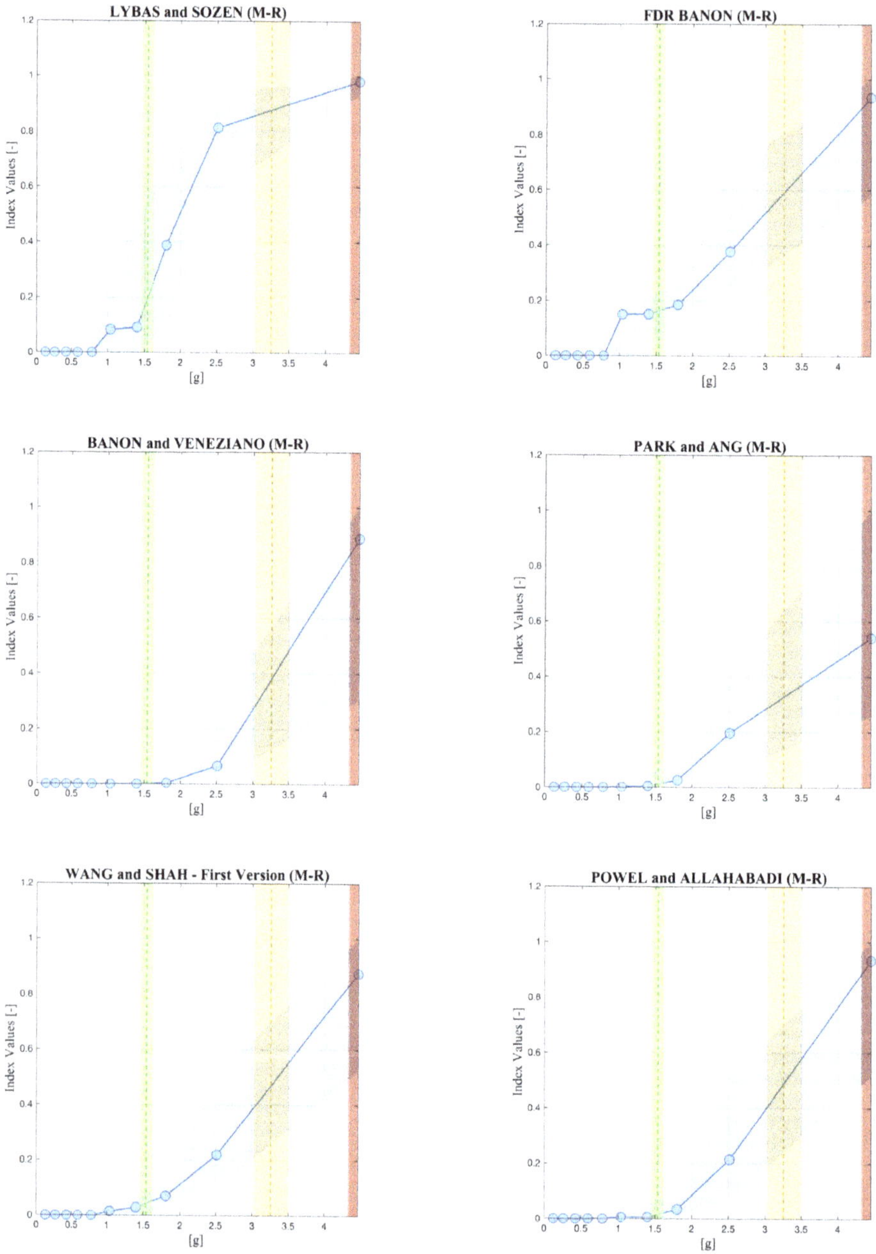

Figure A6. Damage indexes for the code-conforming SW building in the infilled frame configuration (CC_W_IF). Note: the green, yellow, orange and red refer to the UPD, LS, CP and GC performance levels, respectively.

Appendix B

Table A1. Overview of the range, dispersion (D) and overlap (O) of all damage indexes for each limit state for the EX_MRF_BF building based on MR values.

	UPD				LS				CP				GC		
Index	Min	Max	D [%]	O_1 [%]	Min	Max	D [%]	O_2 [%]	Min	Max	D [%]	O_3 [%]	Min	Max	D [%]
L-S	0.00	0.05	4.2	0.0	0.79	0.88	9.0	0.0	0.92	0.99	6.9	96.9	0.92	0.99	7.0
FDR	0.02	0.22	20.6	0.0	0.88	0.92	4.4	18.1	0.91	0.99	7.3	94.7	0.92	0.99	7.3
B-V	0.00	0.01	0.9	0.0	0.54	0.69	14.4	0.0	0.87	0.98	11.3	79.8	0.89	0.99	9.6
P-A	0.00	0.00	0.0	0.0	0.12	0.37	25.1	0.0	0.43	0.86	43.0	94.4	0.45	0.90	44.8
W-S-1	0.00	0.02	1.9	0.0	0.62	0.73	11.0	0.0	0.77	0.98	20.4	80.5	0.81	0.98	17.1
P-Al	0.00	0.02	1.7	0.0	0.62	0.76	13.3	0.0	0.90	0.99	8.7	80.1	0.92	0.99	7.4

Table A2. Overview of the range, dispersion (D) and overlap (O) of all damage indexes for each limit state for the EX_MRF_IF building based on MR values.

	UPD				LS				CP				GC		
Index	Min	Max	D [%]	O_1 [%]	Min	Max	D [%]	O_2 [%]	Min	Max	D [%]	O_3 [%]	Min	Max	D [%]
L-S	0.00	0.28	27.6	0.0	0.29	1.00	71.0	11.8	0.92	1.00	8.4	64.7	0.95	1.00	5.4
FDR	0.00	0.06	6.4	0.0	0.25	1.00	75.3	25.6	0.81	1.00	19.3	64.7	0.88	1.00	12.5
B-V	0.00	0.12	12.3	0.0	0.32	1.00	68.4	2.9	0.98	1.00	2.0	64.7	0.99	1.00	1.3
P-A	0.00	0.13	13.2	0.0	0.27	1.00	73.3	19.2	0.86	1.00	14.1	64.7	0.91	1.00	9.1
W-S-1	0.00	0.04	3.9	0.0	0.31	1.56	125	38.5	1.08	1.56	48.0	64.7	1.25	1.56	31.0
P-Al	0.00	0.06	6.4	0.0	0.25	1.00	75.3	25.6	0.81	1.00	19.3	64.7	0.88	1.00	12.5

Table A3. Overview of the range, dispersion (D) and overlap (O) of all damage indexes for each limit state for the CC_MRF_BF building based on MR values.

	UPD				LS				CP				GC		
Index	Min	Max	D [%]	O_1 [%]	Min	Max	D [%]	O_2 [%]	Min	Max	D [%]	O_3 [%]	Min	Max	D [%]
L-S	0.00	0.00	0.0	0.0	0.45	0.68	23.4	0.0	0.86	0.99	12.5	91.4	0.87	1.00	12.6
FDR	0.00	0.00	0.0	0.0	0.35	0.19	11.0	0.0	0.73	0.97	23.4	86.1	0.77	1.00	23.4
B-V	0.00	0.00	0.0	0.0	0.05	0.12	7.2	0.0	0.54	0.94	40.7	66.7	0.67	1.00	32.8
P-A	0.00	0.00	0.0	0.0	0.01	0.05	4.7	0.0	0.14	0.80	66.2	93.6	0.18	1.00	81.9
W-S-1	0.00	0.00	0.0	0.0	0.39	0.46	7.4	0.0	0.72	0.96	23.9	70.4	0.79	1.00	20.7
P-Al	0.00	0.00	0.0	0.0	0.13	0.25	11.7	0.0	0.66	0.96	30.3	65.5	0.76	1.00	23.8

Table A4. Overview of the range, dispersion (D) and overlap (O) of all damage indexes for each limit state for the CC_MRF_IF building based on MR values.

	UPD				LS				CP				GC		
Index	Min	Max	D [%]	O_1 [%]	Min	Max	D [%]	O_2 [%]	Min	Max	D [%]	O_3 [%]	Min	Max	D [%]
L-S	0.00	0.23	23.4	0.0	0.41	0.83	41.8	48.3	0.63	1.00	37.0	100	0.55	1.00	44.9
FDR	0.00	0.19	19.2	0.0	0.28	0.61	32.9	53.1	0.43	1.00	56.9	100	0.37	1.00	62.7
B-V	0.00	0.01	1.5	0.0	0.03	0.26	23.7	73.5	0.09	1.00	91.2	100	0.06	1.00	94.4
P-A	0.00	0.01	1.4	0.0	0.03	0.21	18.1	78.3	0.07	1.00	93.2	100	0.05	1.00	95.1
W-S-1	0.00	0.21	21.0	0.0	0.23	0.51	27.3	53.4	0.36	1.00	63.9	100	0.31	1.00	68.6
P-Al	0.00	0.05	4.8	0.0	0.08	0.42	33.2	61.0	0.21	1.00	78.7	100	0.15	1.00	85.0

Table A5. Overview of the range, dispersion (D) and overlap (O) of all damage indexes for each limit state for the CC_W_BF building based on MR values.

	UPD				LS				CP				GC		
Index	Min	Max	D [%]	O_1 [%]	Min	Max	D [%]	O_2 [%]	Min	Max	D [%]	O_3 [%]	Min	Max	D [%]
L-S	0.00	0.10	9.9	0.0	0.81	0.94	12.8	0.5	0.94	1.00	5.7	87.1	0.95	1.00	5.1
FDR	0.00	0.07	6.6	0.0	0.44	0.81	36.1	5.8	0.78	0.99	20.7	87.1	0.81	0.99	18.3
B-V	0.00	0.00	0.2	0.0	0.09	0.49	40.4	0.0	0.56	0.99	43.0	87.6	0.61	0.99	37.9
P-A	0.00	0.00	0.3	0.0	0.12	0.36	24.1	0.0	0.42	0.91	49.0	88.7	0.47	0.93	45.8
W-S-1	0.00	0.01	1.4	0.0	0.26	0.65	39.3	0.0	0.70	0.98	28.2	87.7	0.74	0.99	25.2
P-Al	0.00	0.01	1.0	0.0	0.25	0.65	40.1	0.0	0.71	0.99	28.4	87.6	0.74	0.99	25.1

Table A6. Overview of the range, dispersion (D) and overlap (O) of all damage indexes for each limit state for the CC_W_IF building based on MR values.

	UPD				LS				CP				GC		
Index	Min	Max	D [%]	O_1 [%]	Min	Max	D [%]	O_2 [%]	Min	Max	D [%]	O_3 [%]	Min	Max	D [%]
L-S	0.01	0.60	58.6	0.0	0.67	0.97	29.5	19.2	0.91	1.00	8.8	100	0.91	1.00	9.2
FDR	0.02	0.24	21.9	0.0	0.32	0.85	52.5	54.7	0.56	1.00	44.0	100	0.56	1.00	44.5
B-V	0.00	0.01	1.5	0.0	0.09	0.66	56.9	65.4	0.29	1.00	71.4	100	0.28	1.00	71.8
P-A	0.00	0.07	7.3	0.0	0.11	0.71	60.2	78.0	0.24	1.00	75.5	100	0.24	1.00	75.8
W-S-1	0.00	0.09	8.4	0.0	0.21	0.76	55.5	47.4	0.50	1.00	50.1	100	0.49	1.00	50.7
P-Al	0.00	0.08	8.0	0.0	0.19	0.76	57.4	47.7	0.49	1.00	51.1	100	0.48	1.00	51.7

References

1. Angiolilli, M.; Eteme Minkada, M.; Di Domenico, M.; Cattari, S.; Belleri, A.; Verderame, G.M. Comparing the Observed and Numerically Simulated Seismic Damage: A Unified Procedure for Unreinforced Masonry and Reinforced Concrete Buildings. *J. Earthq. Eng.* **2022**, *28*, 1157–1193. [CrossRef]
2. Williams, M.S.; Sexsmith, R.G. Seismic damage indeces for concrete structures: A state-of-the-art review. *Earthq. Spectra* **1995**, *11*, 319–349. [CrossRef]
3. Chung, Y.S.; Meyer, C.; Shinozuka, M. *Seismic Damage Assessment of Reinforced Concrete Members*; Technical Report NCEER 97-0022; State University of New York at Buffalo: Buffalo, NY, USA, 1987.
4. Lybas, J.; Sozen, M. *Effect of Beam Strength and Stiffness on the Dynamic Behavior of Reinforced Concrete Coupled Walls*; Civil Engineering Studies, Structural Research Series No. 444; University of Urbana: Champaign, IL, USA, 1977.
5. Banon, H.; Irvine, H.M.; Biggs, J.M. Seismic damage in reinforced concrete frames. *J. Struct. Div.* **1981**, *107*, 1713–1729. [CrossRef]
6. Banon, H.; Veneziano, D. Seismic safety of reinforced concrete members and structures. *Earthq. Eng. Struct. Dyn.* **1982**, *10*, 179–193. [CrossRef]
7. Hwang, T.H.; Scribner, C.F. R/C member cyclic response during various loadings. *J. Struct. Eng.* **1984**, *110*, 477–489. [CrossRef]
8. Park, Y.J.; Ang, A.H.-S. Mechanistic seismic damage model for reinforced concrete. *J. Struct. Eng.* **1985**, *111*, 722–739. [CrossRef]
9. Stephens, J.E.; Yao, J.T.P. Damage assessment using response measurements. *J. Struct. Eng.* **1987**, *113*, 787–801. [CrossRef]
10. Wang, M.-L.; Shah, S.P. Reinforced concrete hysteresis model based on the damage concept. *Earthq. Eng. Struct. Dyn.* **1987**, *15*, 993–1003. [CrossRef]
11. Powell, G.H.; Allahabadi, R. Seismic damage prediction by deterministic methods: Concept and procedures. *Earthq. Eng. Struct. Dyn.* **1998**, *16*, 719–734. [CrossRef]
12. DiPasquale, E.; Cakmak, A.S. *Identification of the Serviceability Limit State and Detection of Seismic Structural Damage*; Technical Report NCEER-88-0022; State University of New York, Buffalo: Buffalo, NY, USA, 1989.
13. Bracci, J.M.; Reinhorn, A.M.; Mander, J.B.; Kunnath, S.K. *Deterministic Model for Seismic Damage Evaluation of Reinforced Concrete Structures*; Technical Report NCEER-89-0033; State University of New York, Buffalo: Buffalo, NY, USA, 1989.
14. Colombo, A.; Negro, P. A damage index of generalised applicability. *Eng. Struct.* **2005**, *27*, 1164–1174. [CrossRef]
15. Rodriguez, M.E.; Padilla, D. A damage index for the seismic analysis of reinforced concrete members. *J. Earthq. Eng.* **2009**, *13*, 364–383. [CrossRef]
16. Ladinović, D.; Radujković, A.; Rašeta, A. Seismic performance assessment based on damage of structures-part 1: Theory. *Archit. Civ. Eng.* **2011**, *9*, 77–88.
17. Mahin, S.A.; Bertero, V.V. An evaluation of inelastic seismic design spectra. *J. Struct. Eng.* **1981**, *107*, 1775–1795. [CrossRef]

18. Guo, J.; Wang, J.J.; Li, Y.; Zha, W.G.; Du, Y.L. Three-dimensional extension for Park and Ang damage model. *Structures* **2016**, *7*, 184–194. [CrossRef]
19. Midas GEN. Analysis Manual for Midas GEN. 2022. Available online: https://manual.midasuser.com/KR/Gen/935/whnjs.htm (accessed on 9 April 2024).
20. McKenna, F.; Fenves, G.L. OpenSees Manual. 2013. Available online: https://opensees.berkeley.edu/wiki/index.php/Command_Manual (accessed on 8 May 2024).
21. Ibarra, L.F.; Medina, R.A.; Krawinkler, H. Hysteretic models that incorporate strength and stiffness deterioration. *Earthq. Eng. Struct. Dyn.* **2005**, *34*, 1489–1511. [CrossRef]
22. Decanini, L.D.; Gavarini, C.; Bertoldi, S.H. Telai tamponati soggetti ad azioni sismiche, un modello semplificato: Confronto sperimentale e numerico. In Proceedings of the VI Italian National Seismic Engineering Conference, Perugia, Italy, 13–15 October 1993.
23. da Porto, F.; Donà, M.; Verlato, N.; Guidi, G. Experimental Testing and Numerical Modelling of Robust Unreinforced and Reinforced Clay Masonry Infill Walls, With and Without Openings. *Front. Built Environ.* **2020**, *6*, 591985. [CrossRef]
24. Sassun, K.; Sullivan, T.J.; Morandi, P.; Cardone, D. Characterising the in-plane seismic performance of infill masonry. *Bull. N. Z. Soc. Earthq. Eng.* **2016**, *49*, 100–117. [CrossRef]
25. Hak, S.; Morandi, P.; Magenes, G. *Damage Control of Masonry Infills in Seismic Design*; Research EUCENTRE, 2013/1; IUSS Press: Pavia, Italy, 2013.
26. Cornell, C.A.; Jalayer, F.; Hamburger, R.O.; Foutch, D.A. Probabilistic basis for 2000 SAC federal emergency management agency steel moment frame guidelines. *J. Struct. Eng.* **2002**, *128*, 526–533. [CrossRef]
27. *EN 1998-1:2004*; Eurocode 8: Design of Structures for Earthquake Resistance—Part 1: General Rules, Seismic Actions and Rules for Buildings. European Committee for Standardization: Brussels, Belgium, 2004.
28. Iervolino, I. *The Implicit Seismic Risk of Code-Conforming Structures in Italy—A Joint ReLUIS-EUCENTRE Project*; Results of the 2015–2017; RINTC Workgroup: Naples, Italy, 2018.
29. Iervolino, I.; Baraschino, R.; Belleri, A.; Cardone, D.; Della Corte, G.; Franchin, P.; Lagomarsino, S.; Magliulo, G.; Andrea Marchi Penna, A.; Viggiani, L.R.S.; et al. Seismic Fragility of Italian Code- Conforming Buildings by Multi-Stripe Dynamic Analysis of Three-Dimensional Structural Models. *J. Earthq. Eng.* **2023**, *27*, 4415–4448. [CrossRef]
30. Baker, J.W. Conditional Mean Spectrum: Tool for ground motion selection. *J. Struct. Eng.* **2011**, *137*, 322–331. [CrossRef]

applied sciences

MDPI

Article

Data-Driven Machine-Learning-Based Seismic Response Prediction and Damage Classification for an Unreinforced Masonry Building

Nagavinothini Ravichandran [1,*], Butsawan Bidorn [1,2,*], Oya Mercan [3] and Balamurugan Paneerselvam [1]

1 Center of Excellence in Interdisciplinary Research for Sustainable Development, Faculty of Engineering, Chulalongkorn University, Bangkok 10330, Thailand; balamurugan.phd10@gmail.com
2 Department of Water Resources Engineering, Chulalongkorn University, Bangkok 10330, Thailand
3 Department of Civil Engineering, University of Toronto, Toronto, ON M5S 1A4, Canada; oya.mercan@utoronto.ca
* Correspondence: vino.civil35@gmail.com (N.R.); butsawan.p@chula.ac.th (B.B.)

Abstract: Unreinforced masonry buildings are highly vulnerable to earthquake damage due to their limited ability to withstand lateral loads, compared to other structures. Therefore, a detailed assessment of the seismic response and resultant damage associated with such buildings becomes necessary. The present study employs machine learning models to effectively predict the seismic response and classify the damage level for a benchmark unreinforced masonry building. In this regard, eight regression-based models, namely, Linear Regression (LR), Stepwise Linear Regression (SLR), Ridge Regression (RR), Support Vector Machine (SVM), Gaussian Process Regression (GPR), Decision Tree (DT), Random Forest (RF), and Neural Networks (NN), were used to predict the building's responses. Additionally, eight classification-based models, namely, Naïve Bayes (NB), Discriminant Analysis (DA), K-Nearest Neighbours (KNN), Adaptive Boosting (AB), DT, RF, SVM, and NN, were explored for the purpose of categorizing the damage states of the building. The material properties of the masonry and the earthquake intensity were considered as the input parameters. The results from the regression models indicate that the GPR model efficiently predicts the seismic response with larger coefficients of determination and smaller root mean square error values than other models. Among the classification-based models, the RF, AB, and NN models effectively classify the damage states with accuracy levels of 92.9%, 91.1%, and 92.6%, respectively. In conclusion, the overall performance of the non-parametric models, such as GPR, NN, and RF, was found to be better than that of the parametric models.

Keywords: damage state; machine learning; numerical analysis; seismic response; unreinforced masonry

1. Introduction

Unreinforced masonry (URM) buildings constitute a significant portion of residential and heritage structures worldwide. Notably, such buildings represent a considerable number of the occupied buildings in developing nations such as India, Nepal, Pakistan, and Thailand [1,2]. These structures undergo extensive damage or complete collapse during seismic events due to their poor lateral load-bearing capacity. Hence, seismic response prediction and damage classification in URM buildings is critical due to the existence of non-engineered URM buildings in several earthquake-prone regions of the world [3].

Furthermore, modern computational methods involving machine learning algorithms will suitably aid in the effective seismic risk assessment of URM buildings.

The investigation of the seismic response of URM buildings holds significant importance for enhancing the structural safety of the buildings. It is also essential to formulate effective strategies to safeguard these vulnerable structures during seismic events. Modern computational tools, specifically machine learning algorithms, present a promising alternative to time-consuming numerical and analytical simulations, addressing the critical need for seismic risk assessment in URM buildings more effectively. The significance of the research lies in its innovative approach to the identification of damage in URM buildings, paving the way for the development of machine learning models for seismic damage assessment in such structures.

The application of machine learning in structural engineering has increased considerably in recent years. Machine learning (ML) algorithms utilize statistics to make inferences from the data to train developed models by using past experiences and behaviour patterns [4]. Hence, developing the data usually presented as predictors and response variables is crucial for implementing ML models for a particular application [5]. A comprehensive dataset will enable the ML algorithms to predict or to classify possible future samples. Since the numerical and analytical simulations of complex structural systems are time-consuming, ML could be a suitable alternative for various applications in structural engineering and earthquake engineering [6,7].

Several studies have been conducted on the structural response prediction and damage classification associated with a wide range of structures, using ML-based models [8–12]. Various ML models were developed to assess the seismic response, failure mode classification, and strength estimation of reinforced concrete structures at the structural element and building levels (including the reinforced concrete shear walls, frames with infills, and beam–column joints) [9–12]. In addition, ML applications in the optimization and fragility analysis of masonry-infilled reinforced concrete and steel moment frames have also been reported recently [13–15]. Studies also reported the application of ML algorithms in post-earthquake structural safety assessment, risk assessment, and fatigue analysis [16–18]. The literature clearly indicates the effectiveness of the machine learning approach in its application to complex structural engineering problems.

In the case of masonry structures, recent notable research on damage assessment using ML approaches can be found in the literature. Chomacki et al. [19] implemented ML algorithms to assess the damage intensity of existing masonry buildings subjected to the adverse impacts of the industrial environment of mines, and better performance was obtained from the Bayesian belief network. Loverdos and Sarhosis [20] utilized deep learning networks to improve crack detection in masonry walls and automation in brick segmentation, and the results indicated a better outcome from the ML models, compared to the typical image-processing techniques. Mishra et al. [21] analyzed the applicability of three ML algorithms for predicting the compressive strength of masonry constructions, establishing a database using 44 masonry specimens. The results revealed that the ML models can successfully predict the compressive strength of historical constructions.

Regarding the damage assessment of masonry, Rezaie et al. [22] implemented simple ML algorithms for the damage assessment of rubble stone masonry piers and reported the better performance of the K-Nearest Neighbour model. Siam et al. [23] developed an ML-based framework for the performance prediction and damage classification of structural components. They demonstrated the use of the developed framework to classify the reinforced masonry shear walls and to evaluate several key parameters that affect the ultimate drift of walls. Recently, Loverdos and Sarhosis [24] developed a workflow for crack detection in masonry structures and developed a geometric digital twin of the masonry structures using ML. Harichian et al. [25] utilized four machine learning techniques to develop fragility curves that assessed the probability of collapse associated with masonry walls in Malawi. They reported that the Random Forest (RF) technique exhibited efficient performance, with appropriate levels of accuracy. Zain et al. [26] conducted a seismic vulnerability assessment of school buildings in high-intensity seismic zones, employing machine learning methodologies including RF, NN, XGBoost, and the Extremely Randomized Tree Regressor (ERTR). This investigation, which focused on reinforced concrete (RC) and URM school buildings, demonstrated the superior performance of the RF and XGBoost algorithms. Additionally, Coskun and Aldemir [27] devised a straightforward visual screening method aimed at predicting damage to masonry structures using various machine learning algorithms, including Logistic Regression, DT, RF, KNN, SVM, and an Ensemble Learning Method (ELM).

Many existing studies have focused on damage assessment through crack detection and capacity estimation in masonry walls rather than entire buildings [28,29]. Further, studies focusing on seismic response prediction in masonry buildings were found to be scarce in the literature, and the existing studies never attempted the application of a wide range of ML algorithms. More specifically, a study involving both ML-based seismic response and damage prediction for a non-engineered URM building has not been attempted so far. This can be attributed to the need for experimental results and time-consuming numerical analysis procedures required to develop the database. Also, shaking-table tests involving real-scale prototype buildings, particularly URM buildings, would demand large facilities, making the experimental data not readily available for different masonry structures. Hence, the research gap pertains to the ML-based seismic response prediction associated with non-engineered URM buildings and the need for a comprehensive database in order to develop ML models for complete seismic risk assessments of URM buildings.

In this context, the present paper aims to examine the effectiveness of various ML techniques used to predict the seismic response and to classify the damage state of a non-engineered URM building, based on the methodology shown in Figure 1. A finite-element (FE) model of a benchmark building based on the nonlinear layered shell elements approach was used to carry out a nonlinear time history analysis in SAP2000 v.22. With the FE model, a database of 1080 numerical results of the analysis of the benchmark URM building was developed, for which the damage levels were identified using the definitions from the literature. The material properties of the URM building and the intensity of ground motion were used as the input variables. ML models for regression and classification were then implemented to identify the structural response and damage state of the URM building. The performance assessments of the prediction and classification models were then presented using standard performance metrics. Finally, an analysis of the relative importance of the predictor variables was also presented. The results indicated the effectiveness of the ML models in predicting the seismic response and classifying the damage state of URM buildings. The findings from the study also demonstrated the capability of ML models for future relevant studies in the field of structural engineering.

The novelty of the present study lies in creating a comprehensive database and implementing various ML models to analyse the seismic response of URM buildings for the first time. The database developed specifically for this study focuses on the seismic analyses of entire buildings, as opposed to individual masonry structural components, an approach which has not, to this point, been documented in the existing literature. Further, a complete investigation of the performance of ML models in the seismic response prediction and damage state classification of URM buildings is still being explored. All of these features represent the main highlights of the present study, which is focused mainly on developing a wide range of ML models for the seismic damage assessment of URM buildings.

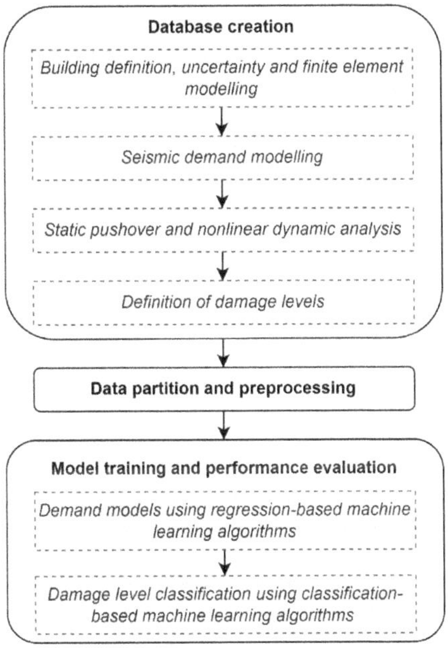

Figure 1. Methodology of the study.

The paper is organized as follows: Section 2 describes the methodology of the study. The development of the database through the selection of building archetype, followed by FE modelling, is discussed in Section 3. Then, the optimized parameters of the ML algorithms used for regression and classification are detailed in Section 4. Finally, Section 5 presents the results of the developed ML framework and the evaluation of the performance of the various algorithms in predicting the structural response and identifying the damage level associated with the URM building.

2. Methodology of the Study

The study aims to identify a suitable predictive model which could be used to assess the seismic response of the URM building archetype, representing non-engineered masonry buildings generally, followed by the identification of building damage using ML techniques. First, regression-based ML techniques were used to develop demand models incorporating the input variables (ground-motion intensity measure and material properties) and an output variable (seismic response of the building). The maximum inter-story drift ratio (IDR), which is the ratio of the relative displacement (measured at the roof of the building) to the height of the building, was utilized as the building response. Next, the classification-based ML techniques were explored to classify the masonry building damage, using the material properties, ground-motion intensity, and seismic response as input parameters. The study involved three major steps: (i) database creation, (ii) data partition and preprocessing, and (iii) model training and performance evaluation.

2.1. Database Creation

To create the database, a benchmark URM building was selected; this was followed by the development of a numeric model of the building using a simplified finite-element modelling approach, using layered shell elements to model the masonry. This methodology was chosen to minimize the computational time required for database development. Compared to other detailed finite-element modelling techniques that employ solid elements, this shell-based three-dimensional modelling approach demonstrates enhanced computational efficiency while maintaining an acceptable level of accuracy. Then, the material properties of unreinforced brick masonry reported in the literature were collected. To consider the uncertainties associated with material properties, 'N' random samples were assembled using the Latin Hypercube Sampling (LHS) technique. This methodology was employed to minimize the number of simulations, while encompassing a broader spectrum of material properties. The random samples were created by partitioning each material property into an equal number of sub-intervals according to their probability distribution and selecting a sample from each interval. To incorporate the uncertainties associated with seismic events, 'N' compatible ground-motion records were identified and randomly paired with the material property samples.

For every random sample, static push-over analyses were carried out to record the lateral load-carrying capacity of the URM building. Incremental dynamic analyses were then carried out to identify the seismic response of the URM building for every input sample. The maximum IDR values were then identified through numerical analyses. Finally, the database for the regression-based ML models was developed, with the predictor variables being material properties and ground-motion intensity and the response variable being IDR. The definitions of damage levels associated with URM buildings, obtained from the literature, were then used to classify the building damage based on the values obtained for the engineering demand parameter (IDR). Using the damage class identification, the database for the classification-based ML models was developed with the predictor variables (material properties, ground-motion intensity, and IDR) and categorical response variables (damage states).

2.2. Data Partition and Pre-Processing

The data partition is necessary to avoid overfitting the model, a situation in which the model performs well with the training data but may not be accurate with the new data. Hence, the data were divided among a training set (70%) and a test set (30%) to check the performance of the ML models. In data pre-processing, the data were transformed and scaled to improve the performance of the ML models. The data transformation can be performed based on the identified distribution of the input features, and it can be implemented to improve the model performance. Furthermore, feature scaling helps to avoid the dominance of high-range features over lower-range features. Particularly, feature scaling will improve classification techniques which are based on the similarities and distances between data, such as K-Nearest Neighbours, Support Vector Machine, and Neural Networks [6].

2.3. Model Training and Performance Evaluation

In the model training stage, the ML models were trained with the database by tuning the model parameters to minimize the specific cost function. Regression-based models such as LR, SLR, RR, DT, RF, SVM, GPR, and NN were implemented to identify the seismic response of the URM building. The performance of the models was evaluated through the coefficient of determination (R^2) and root mean square error (RMSE), and an optimum demand model was identified. Classification-based ML models such as NB, DA, KNN, DR, RF, AB, SVM, and NN were trained to classify the building damage states. The performance of the classification models was evaluated through metrics such as accuracy, precision, recall, and F1 score, along with the confusion matrix.

3. Database Development

A well-developed database is mandatory for the successful development of prediction and classification models. The selected building archetype, the uncertainty modelling considering the random material properties and ground-motion records, the numerical modelling of the URM building, and the nonlinear dynamic analyses carried out to procure data are explained in this section.

3.1. Finite-Element Modelling of the Benchmark URM Building

3.1.1. Benchmark Building

Numerous URM buildings exist in various regions worldwide, encompassing residential dwellings, government facilities, educational institutions, and historic monumental structures. These structures, situated in seismically active areas, were built using traditional methodologies devoid of engineering oversight. Additionally, the literature indicates that many older URM buildings remain inhabited, owing to their cultural significance and traditional ties [3]. Notably, contemporary URM constructions built in the last decade often showcase the fusion of masonry walls and reinforced concrete slabs, as depicted in Figure 2.

Figure 2. Typical URM building with reinforced concrete slab in India.

A single-storey, single-bay masonry building was considered the benchmark building archetype for database development [30]. This box-type URM building geometry is commonly found in the rural areas of several countries worldwide. The plan and elevation of the benchmark URM building are shown in Figure 3. This particular building is considered the benchmark for the present study due to the availability of experimental and numerical investigations in the literature [30,31]. The plan area of the building is 14.42 m^2 with 0.23 m thick walls, and 0.15 m thick reinforced concrete lintel beams and roof slab. The building has a 0.914 m height parapet wall, and a 0.25 m thick sand layer was placed on the roof slab. The building is made up of brick units with cement mortar. The wall density values of the building along the X and Y directions are 13.1% and 9.7%, respectively. The walls were assumed to be sufficiently anchored to the diaphragm.

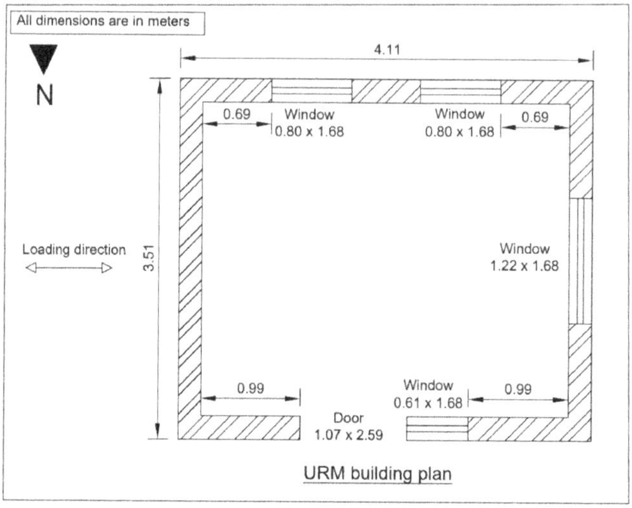

Figure 3. Plan of the benchmark URM building [30,31].

3.1.2. Development of the Simplified Numerical Model

A simplified 3D model of the benchmark URM building developed in SAP2000 v.22 with nonlinear layered shell elements was used in the present study to obtain the seismic response of the building. This is a macro-modelling technique, in which the masonry is modelled as a homogeneous material with equivalent mechanical properties. This technique is suitable for studying the overall behaviour of the masonry with reduced computational time. In this layered-element approach, various layers in the thickness direction with different associated constitutive laws can be used [32].

In the present study, masonry was considered as a nonlinear isotropic material and the material angle in each layer was taken as zero. The orthotropic behaviour of the masonry was not considered due to the lack of availability of experimental data for suitable calibration of the required parameters. Two stress–strain curves, corresponding to normal stress (compression and tension) and shear stress, were used in the model to represent the nonlinear behaviour of the masonry, using two layers of quadrilateral shell elements in the model [33]. The bottom nodes of the URM building were constrained in all three degrees of freedom. The RC lintels and roof were modelled using linear elastic elements, and the total weight of the 0.25 m thick sand layer was applied on the roof of the building as distributed mass. A rigid diaphragm constraint was applied at the roof level to constrain the nodes. The optimal mesh size (equal to 0.114 m) for the model was identified through detailed sensitivity analysis, for which 5184 shell elements were ultimately used. The FE model of the URM building is shown in Figure 4a. Readers can refer to [2,31,34] for additional insights as to the advantages and reproducibility of this simplified numerical model.

The basic motivation behind the use of this simplified numerical modelling approach is to reduce the computational time. Since a considerable amount of data is required to develop ML models, this numerical model can aid in the procurement of data satisfactorily. Further, the application of this simplified numerical modelling approach for the seismic response and fragility analyses of base-isolated URM buildings in the Himalayan regions highlights the suitability of the model for the real-world applications [2,33]. Additionally, it is also important to note that experimental data on non-engineered masonry buildings are scarce in the literature, and insufficient for the development of ML models. Hence, this attempt by the authors to develop the seismic damage prediction models along with data development is unique.

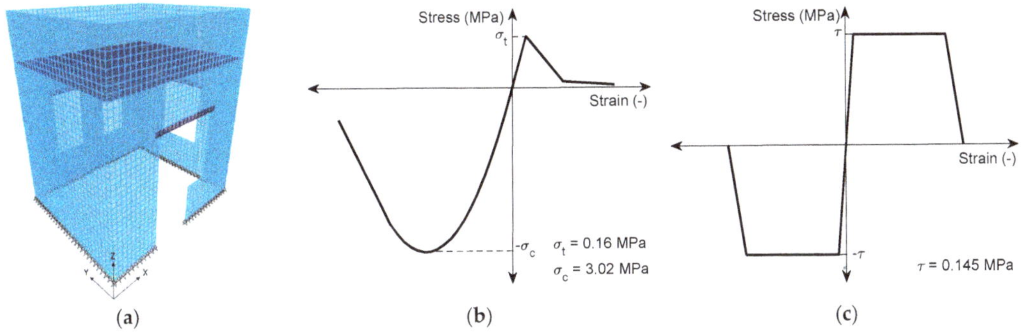

Figure 4. (**a**) FE model of the benchmark URM building; (**b**) normal stress–strain curve; and (**c**) shear stress–strain curve.

In this simplified modelling approach, the model parameters were defined through the direct implementation of stress–strain points. Hence, existing models in the literature were used to develop the stress–strain curves, using the available masonry strength parameters from the experiments. The constitutive models used in the study are shown in Figure 4b,c, with the material properties obtained from Shahzada et al. [30]. The analytical model developed by Kaushik et al. [35] was used to define the uniaxial compressive behaviour of the masonry. The uniaxial tensile behaviour of the masonry was defined by the relationship proposed by Akhaveissy and Milani [36]. The trilinear stress–strain curve was used to reduce the convergence issues that arise during the early stages of the analysis. The FE model was calibrated through a series of push-over analyses by applying incremental horizontal displacement at the control node in the roof level of the building. According to the experimental study, the masonry building reached its peak lateral resistance of 99.5 kN at a lateral displacement equal to 7.89 mm (corresponding to an inter-storey drift of 0.23%). The maximum load-carrying capacity of the URM building model developed in SAP2000 was 98.7 kN at 6.9 mm lateral displacement, which differs from the experimental value by less than 1%.

3.1.3. Model Validation

Modal analysis and nonlinear time history analysis were performed to validate the developed simplified FE model. The comparisons of the fundamental frequencies of the first three vibration modes of the simplified model with the detailed 3D model in the literature is listed in Table 1. The directions obtained for the first three modes were along the X direction, along the Y direction, and in the XY plane (Torsion); these were found to be similar to the results reported in the literature.

Table 1. Comparisons of the fundamental frequencies of the URM building.

Mode	Frequency (Hz)	
	Detailed 3D Solid Element Model [31]	Layered Shell Element Model [Present Study]
1	8.60	8.61
2	9.43	9.68
3	14.04	15.04

In addition, the developed model was subjected to the Bhuj earthquake record (with moment magnitude = 7.9). The maximum inter-storey drift of the URM building obtained from the simplified model was 0.0135% at 47.2 s, whereas the IDR from the detailed ANSYS model reported in the literature was 0.017 [31]. The results indicated the effectiveness of the developed numerical model with reduced computational time, an approach which is suitable for the development of the data required for training ML models.

3.2. Ground-Motion Selection

Ground-motion records tend to produce higher uncertainty than the levels associated with capacity and numerical modelling in seismic response analysis. Hence, the uncertainties associated with the seismic demand must be considered by accounting for various ground-motion records to represent the exact seismicity of the considered region [34]. Therefore, ground-motion records associated with previous earthquakes in various parts of the world (Armenia, Chile, Costa Rica, India, Iran, Mexico, Nepal, Peru, and Turkey) were collected, and 54 records were used for the nonlinear response analysis. The ground-motion time histories of the earthquake records were obtained from the Strong Motion Virtual Data Centre, developed at the University of California, Santa Barbara [37]. The variations associated with the moment magnitude and epicentral distance in the selected records are shown in Figure 5. The moment magnitude of the ground-motion records varies from 5.2 to 7.9. The epicentral distance varies from 2.2 to 250 km, considering both near- and far-field earthquakes. This indicates the consideration of a wide range of earthquakes of varying intensity and frequency intervals. Further, the site soil condition of the selected ground-motion records varies from soft soil to rock. The analysis was carried out by scaling the Peak Ground Acceleration (PGA) of the ground-motion records from 0.01 g to 2.0 g at an increment of 0.01 g.

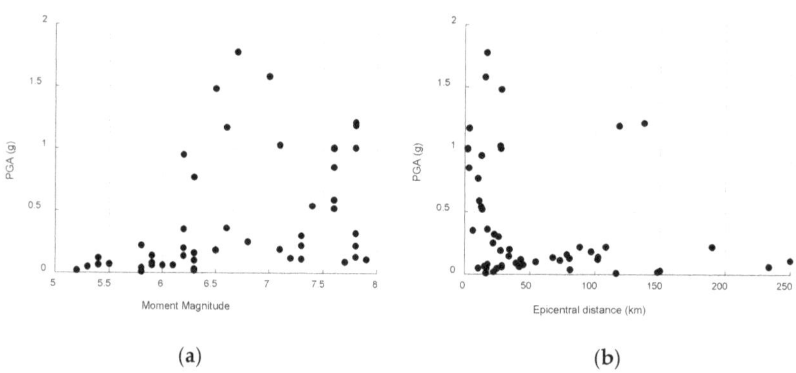

Figure 5. Variation of PGA for the selected ground-motion records with (**a**) moment magnitude; and (**b**) epicentral distance.

3.3. Material Properties

Using locally available materials and construction methods for URM buildings increases the uncertainty associated with the material properties of masonry. Hence, the material properties of URM buildings in different countries were collected and used for the probabilistic modelling. Masonry elastic modulus (E), density (ρ), compressive strength (σ_c), tensile strength (σ_t), and shear strength (τ) were identified as the major material properties that are sources of uncertainty in the seismic capacity of URM buildings. The statistical distributions (mean and coefficient of variation (CoV)) of the material properties were calculated based on the properties of URM buildings reported in India, Nepal, Pakistan, and Iran, and assuming a lognormal random distribution [38–46]. The values are listed in Table 2. The consideration of uncertainties associated with the masonry material properties indicates the consideration of a building class rather than a single building. Based on the statistical distributions, 54 sets of masonry material properties were generated to carry out the seismic analysis, and the ranges of the selected properties are also listed in Table 2.

Table 2. Masonry: material properties [38–46].

Material Properties	Mean	CoV	Range
Mass density (kg/m^3)	1778	0.07	[1272, 2013]
Compressive strength (MPa)	4.79	0.39	[1.11, 8.67]
Tensile strength (MPa)	0.25	0.61	[0.03, 0.43]
Shear strength (MPa)	0.20	0.58	[0.02, 0.63]
Young's modulus (MPa)	1669	0.62	[620, 2982]

3.4. Damage States

One of the critical components in the risk assessment of buildings is the classification of building damage levels, which is helpful in the probabilistic estimation of seismic losses. The damage levels in the URM building were defined in accordance with the definitions established by Park et al. [44], utilizing the identification of IDR threshold values on the push-over curves, as depicted in Figure 6.

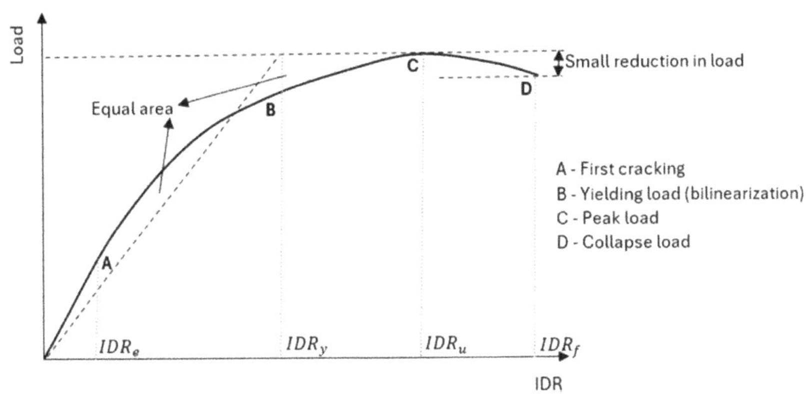

Figure 6. Limit-states definition of the URM building.

Four distinct damage levels were taken into consideration based on the push-over curves: (i) slight damage—DS1, (ii) moderate damage—DS2, (iii) heavy damage—DS3, and (iv) complete damage—DS4. DS1 was defined using IDR_e, corresponding to the initiation of nonlinear behaviour, indicating the onset of macroscopic cracking and nonlinear behaviour of the structure. DS2 corresponds to the yielding resisting force obtained through equal area bilinearization, and was defined by IDR_y. DS3 and DS4 correspond to the extensive and complete damage states, the IDR values for which were defined by the peak load (IDR_u) and the collapse load (IDR_f). The modelling approach considered in the study is not capable of reproducing the exact post-peak softening behaviour of the masonry. Hence, an IDR value corresponding to a 10% reduction in the peak load is considered to be DS4, beyond which the structure is considered inoperable. It is also important to note that the inability of the model to predict the softening behaviour does not limit the aim of the present study, in that the damage classification was conducted based on different levels of damage rather than with the intention of definitively categorizing the structure as safe or unsafe for occupancy.

3.5. Numerical Analyses of Probabilistic Models

The Latin Hypercube Sampling technique was used to combine the uncertainties associated with the material properties and ground-motion records in the analysis [47]. Based on this approach, 54 sets of randomly generated material properties were combined with the ground-motion records, resulting in 54 sets of analysis conditions. In this way, the uncertainties of the masonry material properties associated with the URM building class and the ground-motion record uncertainties were considered together in the seismic demand prediction.

Push-over analyses for the 54 probabilistic modelling cases were carried out by applying horizontal displacement at the control node in the roof level of the URM building, which ensures the global response of the building, ignoring the local failures. Based on the push-over analysis results, the maximum IDR values corresponding to different damage states associated with the 54 sets of material properties were identified.

Each probabilistic model generated by randomly sampling material properties was then paired with the selected ground-motion records. Incremental dynamic analyses for the sets were then carried out by scaling the PGA of the ground-motion records from 0.01 g to 2.0 g, with an increment of 0.01 g, which accounts for 20 analysis cases for each material properties set. The nonlinear dynamic analysis of the URM building was carried out with a time step of 0.01 s for all of the cases. A proportional Rayleigh damping ratio of 4% was set for the first two vibration modes to avoid convergence issues, based on reports in the literature [48]. Finally, 1080 (54 sets of material properties ×20 sets of PGAs of ground-motion records) nonlinear time history analyses were carried out to identify the response of the URM building.

3.6. Database Development and Processing

The response-prediction analyses utilized seven input variables, comprising the five randomly generated material properties (outlined in Table 2), the shear modulus of the masonry, and the intensity of the ground-motion record. The shear modulus for every analysis set was calculated from the corresponding values for Young's modulus and Poisson's ratio (equal to 0.25) [31]. The range of shear modulus values used in the analysis was {248, 1193} MPa. The output variable was the maximum IDR of the URM building. Consequently, a database matrix of size 1080 × 7 was employed in the development of the response-prediction models. The general predictive equation for the structural response prediction is as follows:

$$Y_{MIDR} = f(X_1, X_2, X_3, X_4, X_5, X_6, X_7) \quad (1)$$

where Y_{MIDR} is the structural response, which is the maximum IDR. The predictor variables include the modelling and ground-motion-related uncertainties: X_1 is the mass density of masonry in kg/m^3, X_2 is the compressive strength of masonry in MPa, X_3 is the tensile strength of masonry in MPa, X_4 is the shear strength of masonry in MPa, X_5 is the Young's modulus of the masonry in MPa, X_6 is the shear modulus of masonry in MPa, and X_7 is the intensity measure from the earthquake records. PGA, one of the most commonly used Intensity Measures (IMs) in seismic demand analysis, was used as the IM. The effects of other IMs were not considered in the present study.

In the context of collapse status identification, the maximum IDR of the models was also incorporated as an input variable, resulting in eight input variables. The output variable for the classification models is the damage state (DS1 to DS4). A database matrix with a size of 1080 x 8 was used in developing the classification-based ML models. The general form of the classification models is as follows:

$$Y_{Damage} = f(X_1, X_2, X_3, X_4, X_5, X_6, X_7, X_8) \tag{2}$$

where Y_{Damage} is the damage states DS1 through DS4, the X_1 through X_7 variables are the same as those used in the prediction models, and X_8 is the IDR of the URM building.

In the next step, the original data were pre-processed to improve the performance of the machine learning models. The input variables X_1 through X_7 are transformed into logarithmic scale before fitting the models in order to reduce the nonlinearity in the models. The data obtained from the nonlinear dynamic analysis of the URM building archetype were then divided into training and test sets. The training set, which accounted for 70% of the data (756 data points), was used to train both the regression-based and classification-based prediction models. The remaining 30% of the data (324 data points) was used to test the developed models, which helped to determine their performance on unknown data.

The correlations between the input variables were not considered in the study. The distribution of the damage states in the developed database is shown in Figure 7. Considering the scaling of the ground-motion records to 2.0 g and the comparatively lower load-carrying capacity of the URM building, 55% of the data falls under the collapse damage state (DS4).

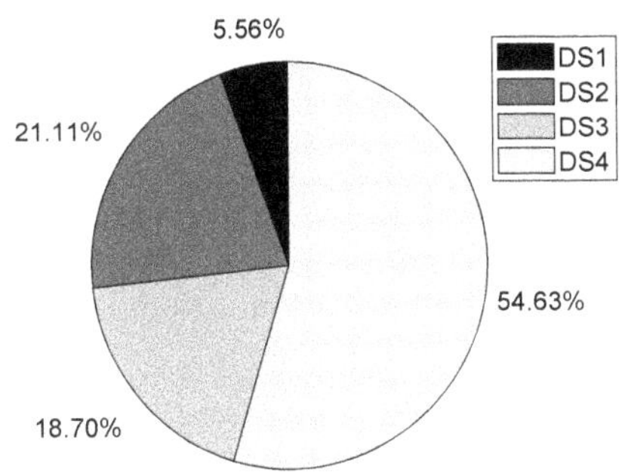

Figure 7. Distribution of the damage states in the developed database.

4. Development of Machine Learning Models

The objective of this data-driven analysis is to explore the applicability of the various ML algorithms for the identification of the structural response and subsequent classification of the damage states. The regression and classification of damage states of URM buildings using ML algorithms are highly advantageous in identifying the underlying relationships between the input (structural parameters, material properties, and exciting force) and output variables (structural response and damage level) with minimal engineering judgement. These approaches are computationally efficient and are highly effective in identifying the hidden patterns in the structural response. Further, the models can be easily updated with additional data from experimental and numerical investigations.

The maximum IDR was considered the continuous quantitative output variable in the regression-based models. On the other hand, the categorical output considered in the classification-based approaches was the damage state, which was classified based on the structural response of the URM building to the earthquake's excitation. Further, both parametric and non-parametric approaches were used in the study, in both prediction and classification, to encompass a wider range of ML algorithms. In each model, the model parameters were optimized by reducing the five-fold cross-validation loss. The models were programmed and developed using MATLAB R2024b.

Linear models such as LR, SLR, and RR were initially employed in prediction modelling due to their straightforward implementation and interpretative capabilities. To address the challenges associated with overfitting in linear models, techniques such as cross-validation and regularization were implemented. Following the development of linear models, various non-linear models were explored to capture the complex relationships between predictor and response variables in seismic response prediction. Notably, DT, RF, and SVM were utilized for prediction and classification tasks, owing to their robustness against outliers and their flexibility in accommodating diverse data types. Additionally, the NN model was incorporated into the prediction and classification processes due to its adaptability and learning capabilities.

As for the classification-based models, NB, DA, and AB were examined for their ease of implementation and interpretability, and may prove beneficial in future applications related to damage classification in URM buildings. The KNN model was selected to classify damage states due to its efficacy with small datasets and its straightforward application, which could be helpful for the present study. Moreover, the literature also suggests that the aforementioned models demonstrate superior performance across various structural-engineering applications [6,9–11]. The model development and parameter selection for every ML model considered in the present study are discussed in this section.

4.1. *Regression-Based Machine Learning Techniques*

Linear Regression, Stepwise Linear Regression, Ridge Regression, Support Vector Machine, Gaussian Process Regression, Decision Tree, Random Forest, and Neural Networks were used to construct a structural response prediction model.

4.1.1. Linear Regression (LR)

The LR technique identifies the linear relationship between the predictor and response variables. The present study used a quadratic model containing linear and squared terms for each predictor and products of pairs of distinct predictors. Maximum Likelihood estimation was used to formulate the robust model equations. It was solved using Iteratively Reweighted Least Squares (IRTS) to improve the fit by minimizing the effect of outliers by including weight as an additional scale factor.

4.1.2. Stepwise Linear Regression (SLR)

In this method, a linear model is developed by adding and removing predictors from the constant model based on a specific criterion [47]. In the present study, the *p*-value for an F-test for the change in the sum of squared error that results from removing or adding the term was used as the criterion. Further, a quadratic model was used for the predictors to improve the model performance.

4.1.3. Ridge Regression (RR)

This model develops the linear relationship by shrinking the regression coefficients by imposing a penalty based on the size of this value. Thus, the input parameters with less influence on the output variable will be removed [49]. The coefficients in Ridge Regression were chosen by minimizing the mean squared error, which is the difference between the observed and predicted values. The penalty term, composed of fit coefficients and a tuning parameter λ, was calculated using the L^2 norm.

4.1.4. Decision Tree (DT)

The LR, SLR, and RR models fit the data by developing a linear relationship, but the relationship between the predictor and response variables might be nonlinear. In such cases, non-parametric approaches, such as Decision Tree and Random Forest, can be used to develop the nonlinear relationship. These models can provide more accurate predictions but are challenging to interpret. In the DT model, a tree-like structure is formed by splitting the regression region into a hierarchy of simple decisions involving one or several input features until a stopping criterion is met. The root node, or the first decision node, is the best predictor [50,51]. The best split is identified through the minimization of impurity or node error. In the present study, 563 splits were used to identify the response variable.

4.1.5. Random Forest (RF)

RF is an ensemble-based model which consists of several trees constructed from a bootstrap sample of the training data [52]. The RF model includes a layer of randomness to improve the bagging. For each spit, only a randomly drawn subset of input features is used, to avoid overfitting by reducing the correlation between the trees. As a result, the Random Forest model usually outperforms other simple ML algorithms. In the present study, the bootstrap method was used, along with 64 learning cycles. To attain better accuracy, the learning rate for shrinkage was assigned a value of 0.3, with a minimum number of leaf node observations of 3.

4.1.6. Support Vector Machine (SVM)

This model, established as a binary classifier, uses an optimal hyperplane to separate the data points by maximizing the margin [53–55]. The nonlinear boundary between the data points was identified through the nonlinear transformation of the features to a higher dimensional space using a kernel function. In the present study, a Gaussian kernel function was used for both regression and classification. Further, the model parameters were optimized through the minimization of the five-fold cross-validation loss.

4.1.7. Gaussian Process Regression (GPR)

GPR is a nonparametric kernel-based probabilistic model [56]. The model explains the response by introducing latent variables from the Gaussian process (GP) and explicit basis function h, projecting the inputs into a p-dimensional feature space. In the present study, the rational quadratic kernel function was used, along with the optimized hyperparameters, by minimizing the five-fold cross-validation loss. In the optimized model, the constant basis function and initial value σ of 0.234 for the noise standard deviation were used.

4.1.8. Neural Networks (NN)

The NN approach is modelled based on the human brain; the responses are identified through the nonlinear functional relationships associated with the input features [57]. The architecture of a neural network consists of input, hidden, and output layers, comprising neurons or nodes. The neurons combine the inputs from the input data with suitable weights, assigned based on the importance of the input. The present study used a sigmoid activation function with three fully connected layers to train the model with the standardized data.

4.2. *Classification-Based Machine Learning Techniques*

Naïve Bayes, Discriminant Analysis, K-Nearest Neighbours, Adaptive Boosting, Decision Tree, Random Forest, Support Vector Machine, and Neural Networks algorithms were used for developing the classification model to identify the damage state associated with the URM building subjected to the earthquake.

4.2.1. Naïve Bayes (NB)

NB is based on the Bayes theorem and assumes that the value of an attribute in a particular class is independent of the values of other attributes [58]. The method classifies the data according to the posterior probability of a sample belonging to each class. In the present study, a multiclass Naïve Bayes model was trained using a kernel smoothing density estimate and an optimized kernel smoothing window width.

4.2.2. Discriminant Analysis (DA)

DA is a classification method which assumes that the data generated by different classes are based on different multivariate Gaussian distributions [59]. The method creates distributions for each class based on the independent variables and calculates the probability of the dependent variable using Bayes' theorem. For this study, linear Discriminant Analysis was used, which assumes that the predictors in each class have equal covariance and uses a linear decision surface.

4.2.3. K-Nearest Neighbours (KNN)

The KNN algorithm incorporates no assumption on the decision boundaries, and it works based on feature similarity [60]. This non-parametric ML algorithm classifies the sample based on the determination of the K most similar samples in the sample space. In the present study, optimized hyperparameters were used to train the model by five-fold cross-validation. The optimized model, with k = 12, an Euclidean distance metric and squared inverse distance weighting function, was used to classify the labels [61].

4.2.4. Adaptive Boosting (AB)

AB implements a simple decision tree as a weak base model and ultimately obtains a stronger model through an adaptive reweighted training of the data at each iteration [62]. This iterative algorithm begins with the uniform weights and updates the error value and weights at every iteration step.

The DT, RF, SVM, and NN classification methods follow the same methodologies explained in the previous section to classify the data. In the case of the DT model, 268 splits with the deviance-based split criterion were used to train the model. Standardized predictor variables were used with a polynomial kernel function to train the SVM model. Finally, a single-layered network with a *tanh* activation function was used in neural network mode.

4.3. *Performance Evaluation Metrics*

The values associated with coefficient of determination (R^2), adjusted R^2, and Root Mean Square error (RMSE) are used for evaluating the performance of the prediction models. Smaller values for the errors and higher values of R^2 indicate the better performance of the models.

In case of classification models, the performance of the models was evaluated using accuracy, precision, recall, and F1 score. These metrics were defined based on the confusion charts, which display the actual damage states and the predicted damage states by the classification models. The diagonal elements of the matrix represent cases accurately classified by the algorithms, while the off-diagonal elements indicate misclassified cases. Additionally, the confusion chart includes row-normalized and column-normalized values, which provides the percentage of accurately classified samples relative to the total number of samples.

Accuracy is defined as the percentage of correctly predicted damage states relative to the total number of damage states or samples analysed. Precision is measured as the ratio of true positives to the total number of positives, either true or false; this value is represented by column-normalized values. This metric reflects the ability of the model to classify damage states while minimizing false positives accurately. Recall, indicated by row-normalized values, is the ratio of correctly classified data to the total number of actual cases. Recall is crucial in seismic damage classification, as it ensures the correct identification of severely damaged structures, a situation in which misclassification may lead to significant consequences. Higher values of accuracy, precision, and recall indicate superior performance of the machine learning algorithm.

In order to fully examine the effectiveness of the classification model, both recall and precision should be assessed. However, tuning the model to increase recall may result in a decrease in precision, and vice versa. As a trade-off, the F1-score is commonly used to examine the accuracy of the model. The F1-score is the weighted harmonic mean of the precision and recall, and it is given by the following:

$$\text{F1} = 2\left(\frac{\text{Precision} * \text{Recall}}{\text{Precision} + \text{Recall}}\right) \quad (3)$$

5. Results and Discussion

5.1. Building Response Prediction Models

The ML algorithms described in the previous section were used to predict the seismic response (IDR) of the URM building, and their performance was assessed through the metrics developed based on the training set and the test set. The values of R^2, adjusted R^2, and RMSE associated with the prediction models are listed in Table 3. The following inferences can be drawn from the results of the prediction models:

1. After examining several models, the RF, GPR, and NN models demonstrated effective performance in predicting the IDR of the URM building. This is evidenced by their higher R^2 values and the lower RMSE values associated with the training data, the least being the RMSE of the RF model. However, the GPR model exhibits comparatively lower RMSE relative to the test data. The RF model showed a lower R^2 value for the test set compared to the training set (3.6% reduction), indicating poor performance or high variance relative to the new dataset. On the other hand, the GPR model performed better in both the training and test sets (less than 0.2% variation), which is again evident from the RMSE values associated with the training and test sets of the model. This highlights the efficiency of GPR model in capturing the complex relationships in data.
2. The performance levels of the DT and SVM were found to be effective while examining the R^2 and RMSE values of the training set. However, the results obtained from the test set of DT and SVM models showed a slightly poor performance, with comparatively smaller R^2 values of 0.923 and 0.958, and larger RMSE values of 0.476 and 0.327, respectively, compared to other non-parametric models (RF, GPR, and NN). This indicates the overfitting of the DT model and the instability of the model relative to changes in the data.
3. The LR, SLR, and RR models are ineffective in accurately predicting structural response due to the nonlinearity in the relationship between the predictor and response variables. The findings also indicate a tendency for these linear models to overfit the data. The relatively poor performance of the linear models in comparison to their nonlinear counterparts may also be a result of their sensitivity to outliers in the seismic response data, despite the implementation of cross-validation and regularization techniques during the model development phase.
4. Even though the variation in the R^2 value of the parametric models (LR, SLR, and RR) exhibits a variation from the best performing GPR model by 10% to 17%, a notable variation in the RMSE of the training set (55% to 65%) can be observed. This fact was also established by comparing the actual and predicted response (in natural logarithmic scale) of the test set obtained through various ML algorithms, as illustrated in Figure 8. It is also clear that non-parametric models, particularly the GPR model, show a much better fit. With its high RMSE values (0.70), and the least fit in the regression graph, the RR exhibits comparatively poor performance among the parametric models.

5. The variations of the R^2 and RMSE values associated with the prediction models are illustrated in Figure 9. The R^2 values of the training and test sets for all the non-parametric models (DT, RF, SVM, GPR, and NN) ranged between 0.9 and 1.0, indicating the best performance, as compared to parametric models (LR, SLR, and RR). Furthermore, the RMSE values for SVM, GPR, and NN were all below 0.4, demonstrating their better performance.
6. Furthermore, the percentage variation in the R^2 and RMSE values of the training and test sets of the ML models were calculated, and the results are compared in Figure 10. The lowest percentage variations for both the R^2 and RMSE values were found in the GPR model, which again substantiates the effective performance of the model in predicting the response. Based on these results, the performance of SVM model was found to be relatively better than the DT, RF and NN models. Even though the percentage variations associated with the results of the parametric models (LR, SLR, and RR) were comparatively lower than those of the non-parametric models, the values were quantitatively higher than those of the values reported by non-parametric models.
7. The performance of the ML models can be influenced by the underlying database, which in turn is impacted by factors such as building type, input variables, dataset size, and output variables in structural-engineering applications. The existing literature contains consistent findings which demonstrate that non-parametric regression models outperform parametric regression models. Specifically, RF, SVM, and DT have demonstrated superior performance in accurately estimating the structural responses of steel moment-resisting frame buildings [6].

Table 3. Performance of the prediction models.

Model	Training			Test		
	R^2	Adjusted R^2	RMSE	R^2	Adjusted R^2	RMSE
Linear Regression (LR)	0.878	0.877	0.556	0.862	0.858	0.596
Stepwise Linear Regression (SLR)	0.877	0.876	0.558	0.860	0.857	0.601
Ridge Regression (RR)	0.806	0.804	0.700	0.818	0.814	0.677
Decision Tree (DT)	0.966	0.966	0.289	0.923	0.921	0.476
Random Forest (RF)	0.978	0.978	0.226	0.945	0.943	0.410
Support Vector Machine (SVM)	0.969	0.968	0.281	0.958	0.957	0.317
Gaussian Process Regression (GPR)	0.976	0.976	0.248	0.974	0.973	0.257
Neural Networks (NN)	0.978	0.977	0.239	0.972	0.972	0.307

5.2. Damage Levels Classification Models

After training the classification models, their performance was evaluated using standard metrics to measure their effectiveness on the test dataset. These metrics include accuracy, precision, recall, and F1 score, as defined based on the confusion charts shown in Figure 11. The diagonal cells, highlighted in various shades of blue, represent the correct predictions for different damage states. In contrast, the off-diagonal cells, highlighted in orange, indicate incorrect predictions. The recall and precision values for each damage state are presented in the final column and row of the chart, respectively, and are marked in grey.

It is also important to note that in the seismic assessment of URM buildings, recall values are crucial, as they show the number of samples classified as having reduced damage. This aspect is essential, as misclassifying a sample with Damage State 4 (DS4) as either Damage State 3 (DS3) or Damage State 2 (DS2) may erroneously categorize a severely damaged building as one with moderate or heavy damage. Such misclassification during post-earthquake evaluations could result in a heavily damaged structure being ascribed a lower damage rating, thus posing a significant risk to repair personnel and occupants following an earthquake. The accuracy, training error, and misclassification rates associated with the test sets and the F1-scores of the classification models are listed in Table 4.

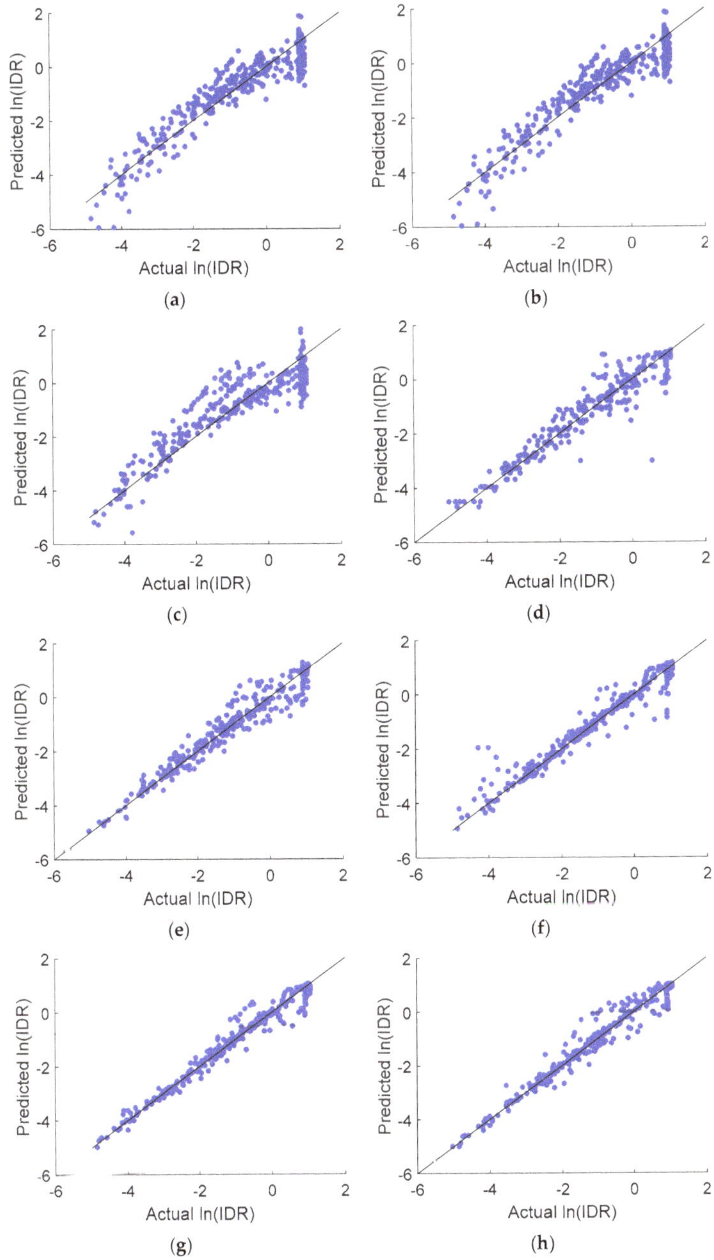

Figure 8. Actual response and predicted response by the regression models: (**a**) LR, (**b**) SLR, (**c**) RR, (**d**) DT, (**e**) RF, (**f**) SVM, (**g**) GPR, and (**h**) NN.

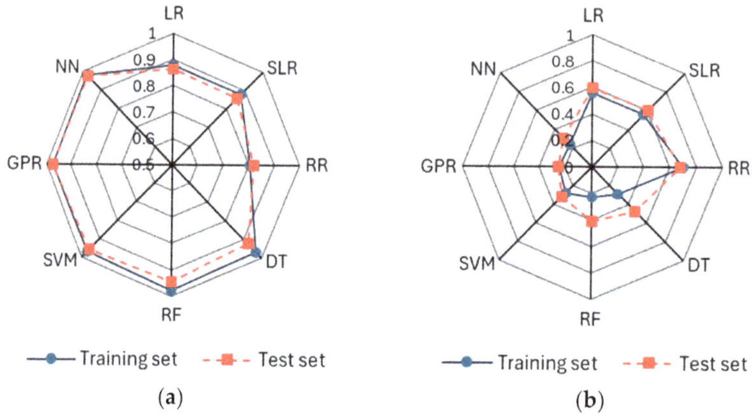

Figure 9. Comparison of results of training and test sets of regression models: (**a**) R^2 values and (**b**) RMSE values.

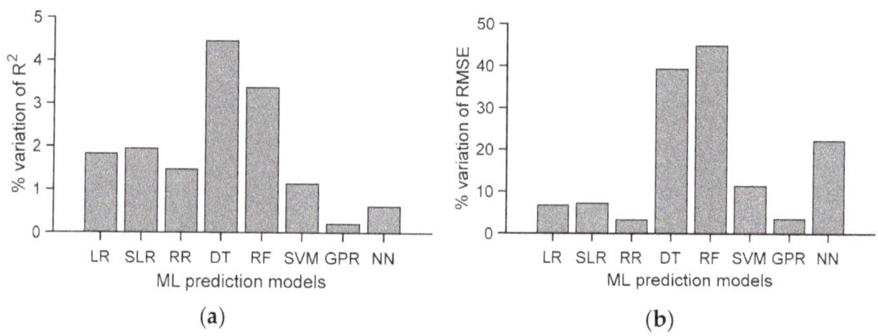

Figure 10. Percentage variations for the (**a**) R^2 and (**b**) RMSE values of the training and test sets of the regression models.

Table 4. Performance of classification models in identifying the damage state of a URM building.

Model	Accuracy (%)	Misclassification Rate	Training Error	F1 Score			
				DS1	DS2	DS3	DS4
NB	85.8	0.1420	0.1098	83.86	78.92	70.03	94.52
DA	83.3	0.1667	0.1032	89.62	83.21	59.68	91.34
KNN	89.8	0.1019	0	89.62	87.06	73.69	96.10
DT	88.5	0.1142	0.0503	90.31	89.48	71.82	93.55
RF	92.9	0.0710	0	93.33	91.66	81.03	96.60
AB	91.1	0.0895	0.0013	93.33	89.05	79.37	95.79
SVM	87.7	0.1235	0.0026	71.96	85.30	73.20	94.85
NN	92.6	0.0741	0.1323	93.33	91.04	81.03	96.90

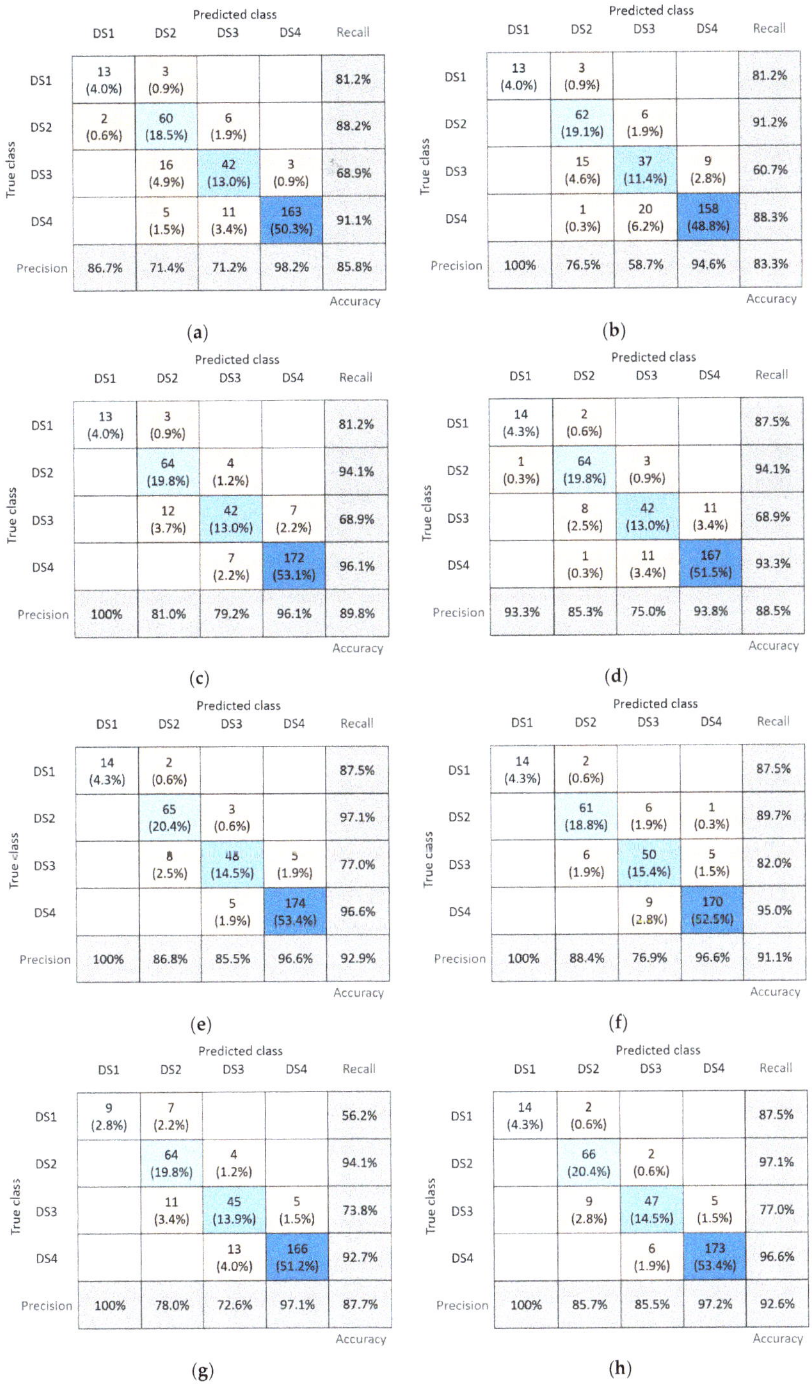

Figure 11. Confusion matrices for the classification models in terms of damage state identification: (**a**) NB, (**b**) DA, (**c**) KNN, (**d**) DT, (**e**) RF, (**f**) AB, (**g**) SVM, and (**h**) NN.

The following inferences were drawn from the results of the trained classification models:

1. The results show that the RF model had the highest accuracy rate, at 92.9%, followed by the NN and AB models. The good performance of the tree-based models suggests that there are non-linear decision boundaries between the failure modes. The results clearly indicate that the non-parametric tree-based models performed better overall than did parametric non-tree-based models like NB.
2. Identifying damage state DS4, which indicates the collapse of the URM building, is critical. The RF model had a precision and recall rate of 96.6% in identifying DS4 in the test set. Similarly, the NN model showed a high recall rate for DS4 and slightly higher precision than the RF model.
3. Compared to other ML models, the RF model had a higher F1 score for all damage states, followed by the NN and AB models. These models also had lower misclassification rates. The F1 score of the DA model for all damage states was comparatively lower than other models, indicating poor performance in classifying the damage states. The AB model had 1.8% lower accuracy and 1.6% lower DS4 recall than the RF model. Based on the results, it can be clearly seen the bagging-based RF model outperforms gradient-based methods like AB in terms of accuracy.
4. Despite having a lower training error, the SVM model was outperformed by the neural network model in terms of accuracy, precision, and recall on the test set. Additionally, the KNN model with zero training error had lower accuracy than the tree-based models. This highlights the importance of splitting the data into training and test sets to evaluate model performance accurately.
5. In Figure 12, the accuracy, F1 score, precision, and recall of the classification models are compared. It is evident that the RF, SVM, and NN models demonstrate higher accuracy levels, ranging from 90% to 100%. Additionally, the F1 score values for these models exceed 75.0% across all damage states. While accurate classification of DS3 and DS4 is crucial, models exhibiting superior performance in classifying all damage states will significantly aid in the post-earthquake risk assessment of URM buildings. Notably, the precision values of the NB, DA, KNN, DT, and SVM models fall below 70% for damage states DS1 and DS3, indicating their poor performance. Furthermore, the recall values also reflect the inferior performance of NB and DA models, with recall values below 70% for the damage state DS3.
6. The results clearly demonstrate the superior performance of RF, AB, and NN models in categorizing the damage states of the URM building. Comparable findings validating the efficacy of these classification models can be found in the existing literature. For instance, the AB model achieved over 80% accuracy in classifying failure modes of reinforced concrete frames with infills [9]. Furthermore, the RF and AB models utilized for the collapse status classification of RC frame buildings exhibited enhanced performance [10]. Similarly, the RF model employed for identifying the flexure–shear failure mode of concrete shear walls attained the highest accuracy, with 70% recall and 84% precision [11].

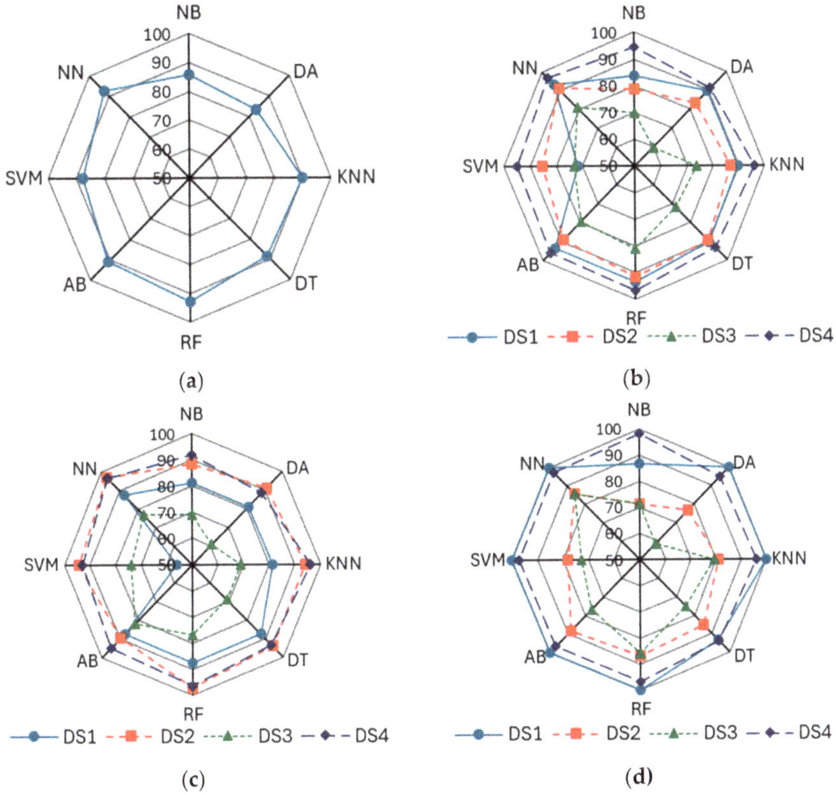

Figure 12. Comparison of results of the classification models: (**a**) accuracy, (**b**) F1 score, (**c**) precision, and (**d**) recall.

Based on the higher values for accuracy and recall, the RF model can be seen as suitable for identifying the damage states in the URM building. In order to identify the importance of the input parameters that affect the performance of the RF model, the relative importance of the predictor variables was evaluated, and this is shown in Figure 13. The importance of the predictors was assessed by permuting out-of-bag observations across the trees. In this study, out-of-bag predictor importance values were initially computed through permutation, reflecting the impact of predictor variables on output estimation. The relative importance of each variable was then determined by dividing its importance value by the sum of the importance values for all predictor variables. The results indicated that the IDR is the most important predictor, with a relative-importance value of 26.84%, followed by PGA (16.81%) and shear strength of masonry (11.5%). The relative-importance values for the masonry material properties range from 7.14% to 11.5%.

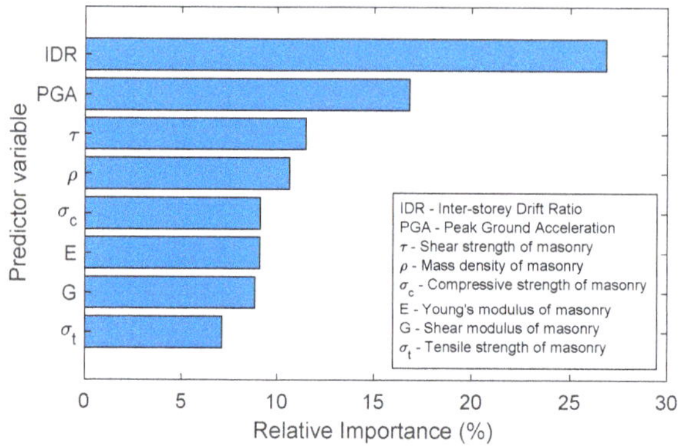

Figure 13. Relative importance of input parameters affecting the Random Forest classification model.

6. Conclusions

URM buildings are a valuable part of cultural heritage and residential architecture worldwide, despite their limited ability to bear lateral loads. Further, these buildings are prevalent in regions highly prone to earthquakes. This study examines the effectiveness of various ML algorithms in identifying the seismic response and damage classification of a benchmark URM building. A comprehensive database was established by performing numerical analysis on a benchmark URM building, due to the absence of experimental data in the literature suitable for training ML models. This study presents a novel application of a diverse range of ML algorithms for the classification of seismic damage, which may improve the seismic risk assessment of non-engineered, unreinforced masonry buildings.

Eight ML algorithms were used to predict the seismic response of the URM building. The results showed that the GPR model exhibited better performance, with smaller R^2 values and larger RMSE values in both training and test sets. Additionally, the comparison of the actual and predicted responses from the GPR model showed better fit than other regression models. The percentage variation in the R^2 and RMSE values of the training and test sets (0.20% and 3.50%, respectively) for the GPR model were also found to be smaller than those of the other prediction models. The lowest RMSE value of the training set was found to be associated with the RF model, but the RMSE value of the test set was higher (44.88% higher), indicating the overfitting of the model. Hence, the results indicated that the overall performance of the non-parametric models such as GPR, RF, and NN was better than that of the linear models, indicating the nonlinear relationship between the predictor and response variables.

In the case of damage classification, eight ML algorithms were explored to identify the suitability of the models for the classification of the damage in a URM building. The results showed that the RF model had a higher accuracy rate, at around 92.90%, and a misclassification rate of 0.07. The precision and recall values associated with DS4 in the RF model were found to be equal to 96.6%, which indicates the accurate classification of completely damaged building cases, which is crucial in the seismic damage assessment of buildings. Followed by the RF model, the performance of the NN and AB models was found to be comparatively better than that of the other classification models used in the study. Additionally, the F1 score of these models was also higher, suggesting that the models performed better in distinguishing different damage states accurately. Finally, the relative-importance assessment of the input parameters in RF model showed that the IDR value was the most important predictor, with a relative importance of about 26.84%, followed by PGA (16.81%) and shear strength of masonry (11.5%).

The results of the study highlight the effectiveness of various ML models in the prediction of seismic response and classifying damage in URM buildings. The ML models, in conjunction with the optimized parameters outlined in the study, can be effectively employed to address similar challenges related to the prediction of seismic response in URM buildings worldwide. Moreover, these results will significantly influence the post-earthquake assessment of URM buildings and aid researchers in designing their experimental and numerical studies on URM buildings.

The important limitation in this demonstrated application of ML models for the damage prediction of a URM building is the requirement of the use of similar features for the input and output variables. Furthermore, accurate prediction of response and damage classification necessitate the use of precise values for masonry material properties and detailed descriptions of damage states. Therefore, ensuring the reliability of the material properties is essential for the appropriate utilization of these developed models in seismic damage assessment. It is essential to recognize that the present study primarily focuses on demonstrating the effectiveness of ML models in classifying the damage states of URM buildings, without considering various failure modes and the reasons for failure.

Future research efforts should aim to develop a comprehensive database that encompasses URM buildings with diverse floor plans and increased numbers of stories, thus encompassing a wider range of buildings globally. Additionally, there is also a need for the development of ML-based damage pattern prediction and fragility assessment for URM buildings. This expansion of scope will contribute to a more comprehensive understanding of URM building behaviour under seismic events.

Author Contributions: Conceptualization, N.R.; methodology, N.R.; software, N.R.; validation, N.R., O.M. and B.P.; formal analysis, N.R.; investigation, N.R. and B.P.; resources, N.R. and B.B.; data curation, N.R.; writing—original draft preparation, N.R. and B.P.; writing—review and editing, O.M. and B.B.; visualization, N.R. and B.P.; supervision, B.B. and O.M.; project administration, N.R. and B.B.; funding acquisition, B.B. All authors have read and agreed to the published version of the manuscript.

Funding: This research is supported by Ratchadapisek Somphot Fund for Postdoctoral Fellowship, Chulalongkorn University.

Institutional Review Board Statement: Not applicable.

Informed Consent Statement: Not applicable.

Data Availability Statement: The raw data supporting the conclusions of this article will be made available by the authors on request.

Conflicts of Interest: The authors declare no conflicts of interest.

Abbreviations

The following abbreviations are used in this manuscript:

AB	Adaptive Boosting
CoV	Coefficient of Variation
DA	Discriminant Analysis
DS	Damage State
DT	Decision Tree
ELM	Ensemble Learning Method
ERTR	Extremely Randomized Tree Regressor
FE	Finite Element
GDI	Gini's Diversity Index
GPR	Gaussian Process Regression
IDR	Inter-story drift ratio

IM	Intensity Measure
IRTS	Iteratively Reweighted Least Squares
KNN	K-Nearest Neighbors
LHS	Latin Hypercube Sampling
LR	Linear Regression
ML	Machine Learning
NB	Naïve Bayes
NN	Neural Networks
PGA	Peak Ground Acceleration
RC	Reinforced Concrete
RF	Random Forest
RMSE	Root Mean Square Error
RR	Ridge Regression
SLR	Stepwise Linear Regression
SVM	Support Vector Machine
URM	Unreinforced Masonry

References

1. Arya, A.S.; Boen, T.; Ishiyama, Y. *Guidelines for Earthquake Resistant Non-Engineered Construction*; UNESCO: Paris, France, 2014.
2. Ravichandran, N.; Losanno, D.; Parisi, F. Low-cost base isolation systems for unreinforced masonry buildings in developing countries. In Proceedings of the 8th International Conference on Computational Methods in Structural Dynamics and Earthquake Engineering Methods in Structural Dynamics and Earthquake Engineering, Eccomas Proceedia, Athens, Greece, 28–30 June 2021.
3. Choudhury, T.; Milani, G.; Kaushik, H.B. Comprehensive numerical approaches for the design and safety assessment of masonry buildings retrofitted with steel bands in developing countries: The case of India. *Constr. Build. Mater.* **2015**, *85*, 227–246. [CrossRef]
4. Michalski, R.S.; Carbonell, J.G.; Mitchell, T.M. *Machine Learning an Artificial Intelligence Approach*; Springer: Berlin/Heidelberg, Germany; New York, NY, USA; Tokyo, Japan, 1984.
5. Sun, H.; Burton, H.V.; Huang, H. Machine learning applications for building structural design and performance assessment: State-of-the-art review. *J. Build. Eng.* **2021**, *33*, 101816. [CrossRef]
6. Kiani, J.; Camp, C.; Pezeshk, S. On the application of machine learning techniques to derive seismic fragility curves. *Comput. Struct.* **2019**, *218*, 108–122. [CrossRef]
7. Thai, H.T. Machine learning for structural engineering: A state-of-the-art review. *Structures* **2022**, *38*, 448–491. [CrossRef]
8. Wang, X.; Mazumder, R.K.; Salarieh, B.; Salman, A.M.; Shafieezadeh, A.; Li, Y. Machine learning for risk and resilience assessment in structural engineering: Progress and future trends. *J. Struct. Eng.* **2022**, *148*, 03122003. [CrossRef]
9. Huang, H.; Burton, H.V. Classification of in-plane failure modes for reinforced concrete frames with infills using machine learning. *J. Build. Eng.* **2019**, *25*, 100767. [CrossRef]
10. Hwang, S.H.; Mangalathu, S.; Shin, J.; Jeon, J.S. Machine learning-based approaches for seismic demand and collapse of ductile reinforced concrete building frames. *J. Build. Eng.* **2021**, *34*, 101905. [CrossRef]
11. Mangalathu, S.; Jang, H.; Hwang, S.H.; Jeon, J.S. Data-driven machine-learning-based seismic failure mode identification of reinforced concrete shear walls. *Eng. Struct.* **2020**, *208*, 110331. [CrossRef]
12. Mangalathu, S.; Jeon, J.S. Classification of failure mode and prediction of shear strength for reinforced concrete beam-column joints using machine learning techniques. *Eng. Struct.* **2018**, *160*, 85–94. [CrossRef]
13. Charalampakis, A.E.; Tsiatas, G.C.; Kotsiantis, S.B. Machine learning and nonlinear models for the estimation of fundamental period of vibration of masonry infilled RC frame structures. *Eng. Struct.* **2020**, *216*, 110765. [CrossRef]
14. Latif, I.; Banerjee, A.; Surana, M. Explainable machine learning aided optimization of masonry infilled reinforced concrete frames. *Structures* **2022**, *44*, 1751–1766. [CrossRef]
15. Wu, J.R.; Di Sarno, L. A machine-learning method for deriving state-dependent fragility curves of existing steel moment frames with masonry infills. *Eng. Struct.* **2023**, *276*, 115345. [CrossRef]
16. Mojaddadi, H.; Pradhan, B.; Nampak, H.; Ahmad, N.; Ghazali, A.H.B. Ensemble machine-learning-based geospatial approach for flood risk assessment using multi-sensor remote-sensing data and GIS. *Geomat. Nat. Hazards Risk* **2017**, *8*, 1080–1102. [CrossRef]
17. Zhang, Y.; Burton, H.V.; Sun, H.; Shokrabadi, M. A machine learning framework for assessing post-earthquake structural safety. *Struct. Saf.* **2018**, *72*, 1–16. [CrossRef]
18. Zhu, J.; Zhang, W. Probabilistic fatigue damage assessment of coastal slender bridges under coupled dynamic loads. *Eng. Struct.* **2018**, *166*, 274–285. [CrossRef]

19. Chomacki, L.; Rusek, J.; Słowi, L. Machine Learning Methods in Damage Prediction of Masonry Development Exposed to the Industrial Environment of Mines. *Energies* **2022**, *15*, 3958. [CrossRef]
20. Loverdos, D.; Sarhosis, V. Automatic image-based brick segmentation and crack detection of masonry walls using machine learning. *Autom. Constr.* **2022**, *140*, 104389. [CrossRef]
21. Mishra, M.; Bhatia, A.S.; Maity, D. Predicting the compressive strength of unreinforced brick masonry using machine learning techniques validated on a case study of a museum through nondestructive testing. *J. Civ. Struct. Health Monit.* **2020**, *10*, 389–403. [CrossRef]
22. Rezaie, A.; Godio, M.; Achanta, R.; Beyer, K. Machine-learning for damage assessment of rubble stone masonry piers based on crack patterns. *Autom. Constr.* **2022**, *140*, 104313. [CrossRef]
23. Siam, A.; Ezzeldin, M.; El-Dakhakhni, W. Machine learning algorithms for structural performance classifications and predictions: Application to reinforced masonry shear walls. *Structures* **2019**, *22*, 252–265. [CrossRef]
24. Loverdos, D.; Sarhosis, V. Geometrical digital twins of masonry structures for documentation and structural assessment using machine learning. *Eng. Struct.* **2023**, *275*, 115256. [CrossRef]
25. Harirchian, E.; Hosseini, S.E.A.; Novelli, V.; Lahmer, T.; Rasulzade, S. Utilizing advanced machine learning approaches to assess the seismic fragility of non-engineered masonry structures. *Results Eng.* **2024**, *21*, 101750. [CrossRef]
26. Zain, M.; Dackermann, U.; Prasittisopin, L. Machine learning (ML) algorithms for seismic vulnerability assessment of school buildings in high-intensity seismic zones. *Structures* **2024**, *70*, 107639. [CrossRef]
27. Coskun, O.; Aldemir, A. Machine learning network suitable for accurate rapid seismic risk estimation of masonry building stocks. *Nat. Hazards* **2023**, *115*, 261–287. [CrossRef]
28. Li, H.; Li, J.; Farhangi, V. Determination of piers shear capacity using numerical analysis and machine learning for generalization to masonry large scale walls. *Structures* **2023**, *49*, 443–466. [CrossRef]
29. Valero, E.; Forster, A.; Bosché, F.; Hyslop, E.; Wilson, L.; Turmel, A. Automated defect detection and classification in ashlar masonry walls using machine learning. *Autom. Constr.* **2019**, *106*, 102846. [CrossRef]
30. Shahzada, K.; Khan, A.N.; Elnashai, A.S.; Ashraf, M.; Javed, M.; Naseer, A.; Alam, B. Experimental seismic performance evaluation of unreinforced brick masonry buildings. *Earthq. Spectra* **2012**, *28*, 1269–1290. [CrossRef]
31. Ravichandran, N.; Losanno, D.; Parisi, F. Comparative assessment of finite element macro-modelling approaches for seismic analysis of non-engineered masonry constructions. *Bull. Earthq. Eng.* **2021**, *19*, 5565–5607. [CrossRef]
32. CSI. *Analysis Reference Manual*; Computers and Structures Inc.: Berkeley, CA, USA, 2014.
33. Thuyet, V.N.; Deb, S.K.; Dutta, A. Mitigation of seismic vulnerability of prototype low-rise masonry building using U-FREIs. *J. Perform. Constr. Facil.* **2018**, *32*, 04017136. [CrossRef]
34. Losanno, D.; Ravichandran, N.; Parisi, F. Seismic fragility of base-isolated single-storey unreinforced masonry buildings equipped with classical and recycled rubber bearings in Himalayan regions. *J. Build. Eng.* **2022**, *45*, 103648. [CrossRef]
35. Kaushik, H.B.; Rai, D.C.; Jain, S.K. Uniaxial compressive stress–strain model for clay brick masonry. *Curr. Sci.* **2017**, *92*, 497–501.
36. Akhaveissy, A.H.; Milani, G. Pushover analysis of large scale unreinforced masonry structures by means of a fully 2D non-linear model. *Constr. Build. Mater.* **2013**, *41*, 276–295. [CrossRef]
37. CESMD. Center for Engineering Strong Motion Data, Strong Motion Virtual Data Center. Available online: www.strongmotioncenter.org (accessed on 10 October 2023).
38. Su, L.; Li, X.L.; Jiang, Y.P. Comparison of methodologies for seismic fragility analysis of unreinforced masonry buildings considering epistemic uncertainty. *Eng. Struct.* **2020**, *205*, 110059. [CrossRef]
39. Bakhshi, A.; Karimi, K. Performance evaluation of masonry buildings using a probabilistic approach. *Sci. Iran.* **2008**, *15*, 295–307.
40. Choudhury, T.; Milani, G.; Kaushik, H.B. Experimental and numerical analyses of unreinforced masonry wall components and building. *Constr. Build. Mater.* **2020**, *257*, 119599. [CrossRef]
41. Giordano, N.; De Luca, F.; Sextos, A. Analytical fragility curves for masonry school building portfolios in Nepal. *Bull. Earthq. Eng.* **2021**, *19*, 1121–1150. [CrossRef]
42. Habieb, A.B.; Milani, G.; Tavio, T. Two-step advanced numerical approach for the design of low-cost unbonded fiber reinforced elastomeric seismic isolation systems in new masonry buildings. *Eng. Fail. Anal.* **2018**, *90*, 380–396. [CrossRef]
43. Kadam, S.B.; Singh, Y.; Bing, L. Seismic fragility reduction of an unreinforced masonry school building through retrofit using ferrocement overlay. *Earthq. Eng. Eng. Vib.* **2020**, *19*, 397–412. [CrossRef]
44. Park, J.; Towashiraporn, P.; Craig, J.I.; Goodno, B.J. Seismic fragility analysis of low-rise unreinforced masonry structures. *Eng. Struct.* **2009**, *31*, 125–137. [CrossRef]
45. Sarkar, A.; Halder, L.; Sharma, R.P. Seismic damage evaluation of unreinforced masonry buildings in high seismic zone using the nonlinear static method. *Adv. Struct. Eng. Dynam* **2015**, *2*, 1039–1053.
46. Cross, T.; De Luca, F.; De Risi, R.; Camata, G.; Petracca, M. Micro-modelling of stone masonry template buildings as a strategy for seismic risk assessment in developing countries. *Eng. Struct.* **2023**, *274*, 114910. [CrossRef]
47. Castaldo, P.; Palazzo, B.; Della Vecchia, P. Seismic reliability of base-isolated structures with friction pendulum bearings. *Eng. Struct.* **2015**, *95*, 80–93. [CrossRef]

48. Betti, M.; Galano, L.; Vignoli, A. Time-history seismic analysis of masonry buildings: A comparison between two non-linear modelling approaches. *Buildings* **2015**, *5*, 597–621. [CrossRef]
49. Friedman, J.; Hastie, T.; Tibshirani, R. *The Elements of Statistical Learning*; Springer Series in Statistics; Springer: Berlin/Heidelberg, Germany, 2001.
50. Leo, B.; Jerome, F.; Charles, J.S.; Olshen, R.A. *Classification and Regression Trees*, 1st ed.; Taylor & Francis: New York, NY, USA; Routledge: Oxfordshire, UK, 2017.
51. Quinlan, J.R. Induction of decision trees. *Mach. Learn.* **1986**, *1*, 81–106. [CrossRef]
52. Ho, T.K. The random subspace method for constructing decision forests. *IEEE Trans. Pattern Anal. Mach. Intell.* **1998**, *20*, 832–844.
53. Bishop, C.M.; Nasrabadi, N.M. *Pattern Recognition and Machine Learning*; Springer: New York, NY, USA, 2006.
54. Cristianini, N.; Shawe-Taylor, J. *An Introduction to Support Vector Machines and Other Kernel-Based Learning Methods*; Cambridge University Press: Cambridge, UK, 2000.
55. Cortes, C.; Vapnik, V. Support-vector networks. *Mach. Learn.* **1995**, *20*, 273–297. [CrossRef]
56. Rasmussen, C.E.; Williams, C.K. *Gaussian Processes for Machine Learning*; MIT Press: Cambridge, UK, 2006.
57. Haykin, S. *Neural Networks—A Comprehensive Foundation*, 2nd ed.; Pearson Prentice Hall: Noida, India, 2004.
58. Friedman, N.; Geiger, D.; Goldszmidt, M. Bayesian network classifiers. *Mach. Learn.* **1997**, *29*, 131–163. [CrossRef]
59. Lachenbruch, P.A.; Goldstein, M. Discriminant analysis. *Biometrics* **1979**, *35*, 69–85. [CrossRef]
60. Cover, T.; Hart, P. Nearest neighbor pattern classification. *IEEE Trans. Inf. Theory* **1967**, *13*, 21–27. [CrossRef]
61. Hu, L.Y.; Huang, M.W.; Ke, S.W.; Tsai, C.F. The distance function effect on k-nearest neighbor classification for medical datasets. *SpringerPlus* **2016**, *5*, 1304. [CrossRef]
62. Hastie, T.; Rosset, S.; Zhu, J.; Zou, H. Multi-class adaboost. *Stat. Interface* **2009**, *2*, 349–360. [CrossRef]

applied sciences

MDPI

Article

Field Investigation and Numerical Modeling for the Seismic Assessment of the Castle of Lanjarón, Spain

Hayden Luger [1], Rafael Ramirez [1], Paloma Pineda [2] and Paulo B. Lourenço [1,*]

1 Department of Civil Engineering, ISISE, ARISE, University of Minho, Campus de Azurém s/n, 4800-085 Guimaraes, Portugal; hayden.luger@alumni.ucalgary.ca (H.L.); rramirez@civil.uminho.pt (R.R.)
2 Department of Building Structures and Geotechnical Engineering, Universidad de Sevilla, Avda. Reina Mercedes 2, 41012 Seville, Spain; palomapineda@us.es
* Correspondence: pbl@civil.uminho.pt

Abstract: The Castle of Lanjarón is a 16th century stronghold located in Andalucía, Spain. After losing its military function, the castle was abandoned, leading to significant decay. Designated a national heritage site in 1985, recent efforts have sought to preserve its historical and cultural value. This study outlines an inspection and diagnosis campaign carried out on the castle. Non-destructive tests (NDTs) were employed to characterize the properties of the masonry, using both mechanical and wave-based methods. Dynamic identification was performed to determine dynamic and modal properties of the structure, which were used to develop and calibrate a three-dimensional (3D) finite element model (FEM) of the west wall, based on homogenized masonry material. Limit analysis and non-linear static (pushover) analysis under various boundary conditions were conducted to determine the maximum relative load factor in the out-of-plane direction. The results were compared to the expected peak ground acceleration (PGA) of the area, showing that the maximum load capacity of the wall exceeds local seismic demands with a safety factor of 1.39. The study highlights the efficacy of pairing a homogenized macro-modeling approach with wave-based and dynamic identification methods, particularly for resource efficiency. Finally, recommendations for future conservation efforts have been provided.

Keywords: historical structure; limestone masonry; inspection; diagnosis; non-destructive testing; dynamic identification; finite element method; finite element analysis; limit analysis; seismic design

1. Introduction

In regions with little recent seismic activity, the effect of a seismic event can often be overlooked. With a low seismic hazard, a return period of, at times, hundreds of years, and the efficacy of modern building materials at mitigating risk, seismic risk can be easy to ignore. However, for assets of cultural heritage and historical buildings that have existed for centuries, these risks are inevitable. The seismic evaluation of heritage structures is of particular interest since they are often unique and irreplaceable, holding significant historical and cultural values as built heritage. These structures are commonly built with traditional building materials such as unreinforced masonry and were not designed to withstand the significant lateral loading often accompanying seismic events. In addition, they can be expensive to maintain and difficult to work on. Many of these buildings are characterized by irregular geometry, massive size and weight, and masonry composing them that is brittle and vulnerable to tensile and shear loading. These characteristics influence the seismic behavior and vulnerability of a historical structure [1].

Numerical methods and modeling have successfully been employed for the analysis of heritage structures by allowing the simulation of an applied loading condition [2–5]. This application is commonly used to simulate the response of a structure to either a static loading condition, such as a lateral load being applied to a wall, or a dynamic loading condition consisting of a time-history seismic input. The methods are also commonly considered in both the linear elastic and non-linear regimes for materials and are therefore entirely dependent on the constitutive and material models used in the modeling approach. Typical methods for the modeling of masonry involve the different treatment of the mortar–brick interface that characterizes the material. Micro-modeling approaches are more computationally demanding and can be used to model individual masonry units and the interfaces between them as discontinuous. Macro-modeling approaches represent the masonry structure as a homogenized approximation of its composition; they are therefore less computationally demanding and require fewer individual properties as inputs [6].

The need for validation and support of the model-based approach is essential; experimentally obtained data can complement a numerical model by obtaining experimental values for the mechanical and physical properties of a specific site [7]. However, for a culturally significant and protected site, it is difficult to properly assess their mechanical and compositional properties using ex situ laboratory setups or other means that would involve the partial or total destruction of an element. The importance of performing non-destructive tests (NDTs) to assess these parameters becomes immediately apparent; being able to understand the behavior of the structure and its materials without negatively affecting the existing elements is paramount to assessing its current state responsibly. Physical properties, such as compressive strengths for individual units and mortars, can be obtained using more analog methods such as those outlined in this report (Schmidt hammer, Schmidt pendulum hammer, penetrometer, and scratch testing instrument). More wholistic parameters that represent more global properties can be assessed using wave-based methods and dynamic identification, in addition to being able to highlight potential damage in the site [7]. Furthermore, these methods can be used to calibrate a developed numerical model according to their modal parameters of vibration frequencies, mode shapes, and damping [8].

Lanjarón is in one of the most seismically vulnerable regions in Spain [9]. Despite local interest in the preservation of its historical castle, no innovative academic studies within the framework of seismic evaluation and historical structural conservation have been performed. Such investigations complement the use of finite element methods to simulate the response of a structure to external loading by providing material properties that can be used directly in a finite element (FE) model. Within this context, this novel study aims to outline the results of extensive investigations and experimental campaigns of NDTs that were performed on the site to characterize structural elements and identify potential damage to the castle. The study further developed and calibrated an FE model based on a variety of modal parameters and using the Modal Assurance Criterion (MAC) for validation. Finally, efficient methods for the safety assessment of historical structures were used to determine the safety of the structure, all within a comprehensive case study. Future studies may consider using dynamic non-linear analysis to further evaluate the rest of the structure.

2. Materials and Methods

The Castle of Lanjarón (Figure 1a) is a 16th century Christian stronghold that occupies a strategic position upon the southern slopes of the Penibaetic Mountains of the Sierra Nevada adjacent to the town of Lanjarón [10]. It sits atop a steep rocky slope and is, today, composed of two main components that have largely fallen into a state of decay: a keep, on the southern side of the compound (Figure 1b), and the outer enclosure that surrounds it to the north (Figure 1c). A barbican flanks the entrance to the south, running parallel to the walls of the keep above. In this way, the castle is very characteristic of a stronghold on the Iberian Peninsula from its time: the angled approach created by the barbican and the keep above the entrance echo the influence of the Muslim cultures that occupied the region before [11,12].

Figure 1. Visual overview of the Castle of Lanjarón: (**a**) aerial view of the castle looking west [13]; (**b**) view of the castle from the south showing the barbican and keep; (**c**) view of the inner enclosure from the entrance under the keep; (**d**) view showing the long west wall.

The castle itself occupies a footprint of around 500 m^2 and does not have a regular polygonal shape. Its construction is primarily of stone rubble masonry, except for a lime-based cementitious cistern in the base of the large keep on the southern side of the structure. The keep rises to a height of 12 m above the approach to the south, flanked by the 1.2 m thick barbican. A barrel-vaulted passage perpendicular to this approach provides access to the outer enclosure. The enclosure is fortified on all sides by 1.7 m thick walls that rise to heights of almost 7.5 m at their tallest point along the long western wall (Figure 1d). The terrain inclines steeply from south to north, rising from the keep to the apex of the hill the castle sits upon. At its northernmost point, a massive watchtower looks out towards the town to the north.

The existence of the castle in its current state has its origins in the early 16th century, shortly after the Christian occupation of Granada. Material analysis suggests that it is a Christian masonry construction built during the annexation of the area to the Kingdom of Castile [10]. The primary function of the castle may have been surveillance and control of the area and its inhabitants during that initial period and beyond, until the eventual expulsion of the Muslim population rendered the castle obsolete. This led to its conclusive abandonment in the centuries following. Archeological surveys further support this conclusion, implying a gradual decline in the structural condition of the castle from its historical state to its current one. The collapsed state of the keep is a product of centuries of abandonment and continuous exposure to the elements [14]. The castle was added to the Spanish Heritage Monument Listing in 1985; it is protected and designated today with the status of Bien de Interés Cultural (BIC), or Cultural Heritage Site.

Concern over the collapse of the structure led to a variety of restorative and preservative efforts during the 1990s and the 2000s, including the consolidation of cracks in the keep. Masonry elements were reconstructed with fallen stones to their original state. Other masonry units were covered with a thick mortar where it had been removed, in addition to deep, medium depth, and superficial repointing being performed. Some deep voids were filled and consolidated; heavily deteriorated stone masonry units were replaced, and fallen masses were relocated or repositioned to avoid further deterioration. The most recent intervention, completed in 2007, also featured the construction of an earlier planned metallic frame in the keep outlining the historical shape of the structure, and the addition of stairs and safety rails [15].

Even with the recency of the intervention in 2007, the conditions of many of the elements continue to decline. Much of the paint on the metal elements has chipped or been removed, exposing these elements to decay. Many of the limestone masonry units have been heavily exposed to alveolization. Two drainage tubes are present to remove water and could easily become blocked. Significant plant growth threatens the state of the masonry slowly. For an asset of significant cultural interest such as the Castle of Lanjarón, the main elements and materials of a site are often fragile and of significant value in their current state. Figure 2 demonstrates some of these damages.

An intensive in situ experimental campaign was carried out to assess and identify properties and conditions of the elements comprising the castle. These works included rock characterization, testing with various NDT equipment (Schmidt hammer, Schmidt pendulum hammer, penetrometer, scratch testing instrument), a comprehensive wave-based campaign based on sonic tests, and concluded with dynamic identification.

(a)

(b)

Figure 2. *Cont.*

(c) (d)

Figure 2. Current conservation state of the Castle of Lanjarón: (**a**) state of metallic elements in the castle; (**b**) alveolized limestone ashlar; (**c**) blocked drainage tube; (**d**) significant plant growth on structural elements hides a masonry arch.

2.1. Rock Test Procedures

The rock tests were performed on visually similar rocks in the area surrounding the castle and used common geological approaches to characterize calcareous rocks in situ through both visual means and an acid test. The acid test characterizes rocks containing calcium carbonate according to the reaction of a weak acid with the compound. The frequency and nature of the gaseous carbon dioxide bubbles formed by the acid-base reaction can serve as an indication of whether limestone or another calcareous rock is present [16]. Units that reacted readily and bubbled aggressively in the presence of acid were labeled as limestone. Units first requiring the surface to be scratched to a powder before reacting were considered dolomitic. Units that did not readily react did not contain significant traces of calcium carbonate. Based on these results, a Schmidt hammer with a rebound energy of 2.207 Nm was also used to evaluate the compressive strengths of the masonry units using 10 consecutive blows according to standard testing procedures, followed using densities and calibration tables provided in the literature [17–19].

2.2. Mortar Test Procedures

2.2.1. Schmidt Hammer

To quantitatively assess the compressive strength of the mortar of the castle, a Schmidt hammer (PASI SRL Type M) with an impact energy of 0.169 Nm was used (Figure 3a). Using this apparatus, a spring-loaded mass is propelled to impact the surface of a mortar three times in succession before measuring the rebound number. The resulting value was then converted to compressive strength using the operation instructions [20]. Each location was tested in accordance with common application methods from the EN Eurocodes for the application of a Schmidt hammer [21].

Figure 3. (**a**) Schmidt hammer; (**b**) Schmidt pendulum hammer; (**c**) penetrometer; (**d**) scratch test instrument.

2.2.2. Schmidt Pendulum Hammer

An alternative to the Schmidt hammer that uses similar principles is the Schmidt pendulum hammer, which instead uses gravity to swing a pendulum and impact a surface to return a rebound number measurement (Figure 3b). This process was performed using the 'measuring' instructions outlined in the device's guidelines for vertical surfaces [22]. The hammer was placed upon the surface of the mortar with the head facing downwards as vertically as possible using light pressure, and the head was raised into the loaded position. When well positioned, this allows the head to fall and impact the surface, rebounding and returning an analog measurement that can then be used to qualitatively assess the mortar from the operation manual.

2.2.3. Penetrometer

With the advantage of not being as restricted by the required available surface area of the Schmidt hammer tests, a penetrometer (Diagnostic Research Company SRL RSM-15) was also used to quantitatively evaluate the compressive strength of mortar (Figure 3c). The device is based on the same principles as the Schmidt hammer, using an impact-driven accelerated mass to drive a small needle into a mortar. It is considered non-destructive. The 'Type-A—Fast' procedure from the user manual was followed, in which ten consecutive blows from the device drove the needle perpendicularly into the mortar. The needle was then removed, and the depth of penetration was measured. The resulting depth can be converted into a compressive strength value [23].

2.2.4. Scratch Test

A scratch test instrument (Enertren Pty. Ltd. MortarCheck II) was used for further onsite evaluation of the mortar condition and strength (Figure 3d) [24]. The device was placed upon a flat surface of mortar, and properly oriented by placing its legs on the surrounding wall. The abrasive probe was then lowered onto the surface with the spacer still attached and zeroed accordingly. The device was locked into place, the spacer was removed, and the force of the spring drove the probe into the mortar as the yoke was turned using two full rotations. Care was taken not to apply additional pressure to the probe while supporting the legs, and the resulting reading (in inches) was used to qualitatively evaluate the mortar scratch index.

2.3. Sonic Test Procedures

Tests based on the propagation of a wave through a medium are of particular interest within the context of historical constructions due to their non-invasive characteristics and ability to gain information about what is below the visible surface. Examining material through the behavior of wave movement can reveal several characteristics of the medium, such as the presence of cracks or separation of wythes within a masonry wall. The speed at which a wave propagates is the basis of these methods, alongside some theoretical assumptions commonly used in civil engineering. Additionally, there is a relation between the elastic modulus, density, and speed of the wave. The equations used, presented later, are based primarily on the homogenous and isotropic behavior of a linear elastic material and how a wave propagates through it. As masonry is not homogeneous itself, it becomes important to note the common homogenization approaches used for heterogeneous materials when studying them at a macro-level [25].

Testing was performed using an impact hammer (PCB Piezotronics Model 086D05) and accelerometer (Model 352B) in combination with a National Instruments (Model NI USB-4431) Data Acquisition (DAQ) system with a maximum input range of ± 10 V. This setup was used to measure the time between the impact of the hammer and the moment the accelerometer received the signal. The hammer had a sensitivity of 0.23 mV/N and a signal input range of $\pm 22{,}240$ N pk. The accelerometer used had a sensitivity of 1 V/g and a signal input range from -5 g to 5 g. The program used in data acquisition was set up to perform continuous sampling with a 100,000 Hz sampling rate, to collect 50,000 samples. A series of six repeated samples were collected from each point to ensure at least three usable tests were gathered.

A 1 m $\times$ 1 m area was selected for testing in two different locations, one for direct (through the wall) and one for indirect (across the surface of the wall) testing. An approximately evenly spaced 5 $\times$ 5 grid of points was used for indirect testing (Figure 4), while a 6 $\times$ 6 grid of points on opposite sides of a wall was selected for direct testing (Figure 5). Points crossing a crack that was repaired in a previous intervention were also investigated using the indirect method. Indirect tests were performed in both a vertical and horizontal direction by striking a point at either extreme of the grid and subsequently progressing through the points along the line and repeating the process.

The collected data were then processed, and the modulus of elasticity for the materials was calculated using the following formula:

$$E = V_p^2 \rho \frac{(1+\nu)(1-2\nu)}{(1-\nu)} \tag{1}$$

where E [N/m^2] is the modulus of elasticity, V_p [m^2/s] is the wave propagation speed, ρ [kg/m^3] is the material density, and ν [–] is the Poisson's ratio [3,26]. Material properties from the literature were used for values of density and Poisson's ratio [6].

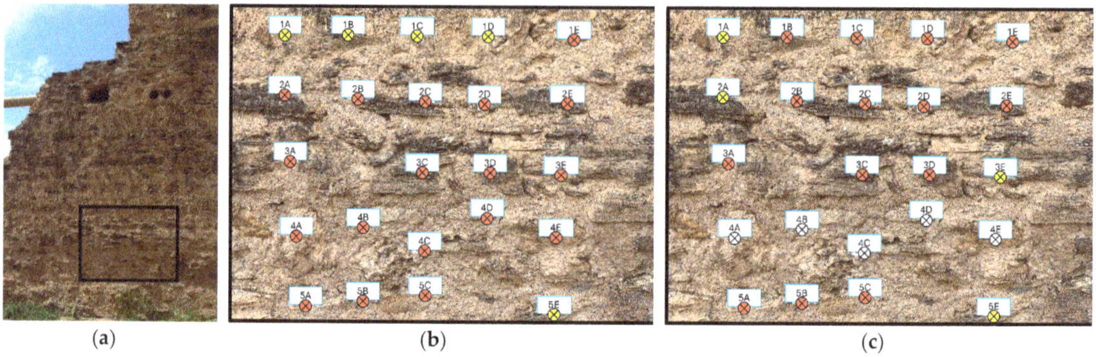

Figure 4. (**a**) Indirect sonic test location (extent shown by black rectangle); (**b**) test grid with locations for vertical indirect sonic tests; (**c**) test grid with locations for horizontal indirect sonic tests. Hammer impact points are highlighted in yellow, accelerometer points are red, and points with null readings are white.

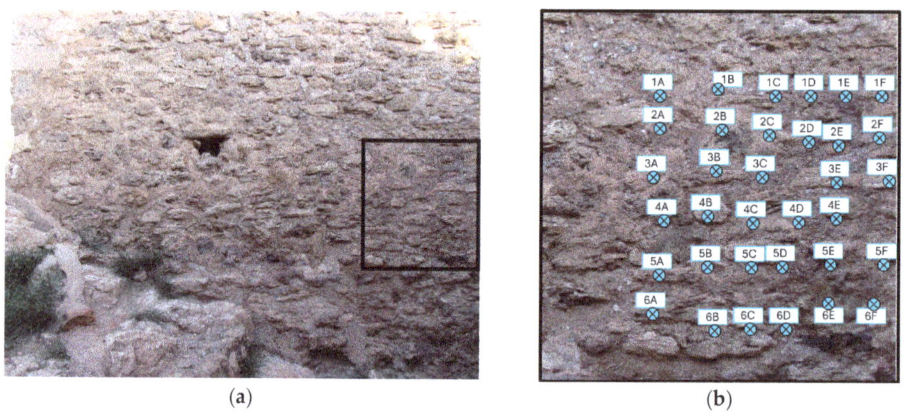

Figure 5. (**a**) Direct sonic test location outside the keep (extent shown by black rectangle); (**b**) test grid with locations for direct sonic tests.

2.4. Dynamic Identification Procedures

Real-time dynamic identification can be used to obtain a variety of dynamic structural properties that describe how a structure oscillates and behaves. The oscillation of a structure is mathematically governed by its mass, stiffness, and capacity to dissipate energy; these parameters are represented generally in dynamic identification through modal parameters. Modal parameters are typically frequencies of vibration, the relative shapes of the modes associated with these frequencies, and damping [8]. The modal parameters can wholistically represent the response of a structure to an input, regardless of any damage or heterogeneity in the structure [4]. They can represent the current state of a structure, together with its material and mechanical properties. Periodically performing dynamic identification on a structure, especially to collect data before and after some damage has occurred, is an effective way to continuously monitor its health non-destructively and without the need for more invasive testing [8].

For this campaign, 8 piezoelectric accelerometers (Model 393B12 PCB Piezotronics) with a sensitivity of 10,000 mV/g, a measurement range of ±0.5 g pk, and a frequency range of 0.15 Hz to 1000 Hz were used with a National Instruments DAQ (Model NI 9174 with two NI 9234) for data collection. A sampling rate of 200 Hz for 30 min was used for each acquisition; this frequency is commonly applied for masonry heritage structures as it can capture frequency content up to 100 Hz [2,7]. Under ambient conditions for excitation, which is considered an Ambient Vibration Test (AVT), the setups are unlikely to capture any bandwidth above 20 Hz. Using a common reference point and accelerometer, multiple acquisitions of data can be combined. For the analysis of the west wall, two instances of 30 min acquisitions with two distinct setups were used to measure the response of the structure to an unexcited input. This method of output-only dynamic identification measures the ambient vibration of the structure, assumed to follow the input of a Gaussian white noise stochastic process for Operational Modal Analysis (OMA) [8]. Acquisitions were then processed using ARTeMIS 6.0 modal identification software to identify modal parameters according to their spectral densities.

3. Results

3.1. Mortar Test Results

A variety of NDT equipment was employed to characterize the mortar onsite, and to identify differences in the types of mortar present from previous interventions. Figure 6 presents the locations of the tests performed and highlights the west wall.

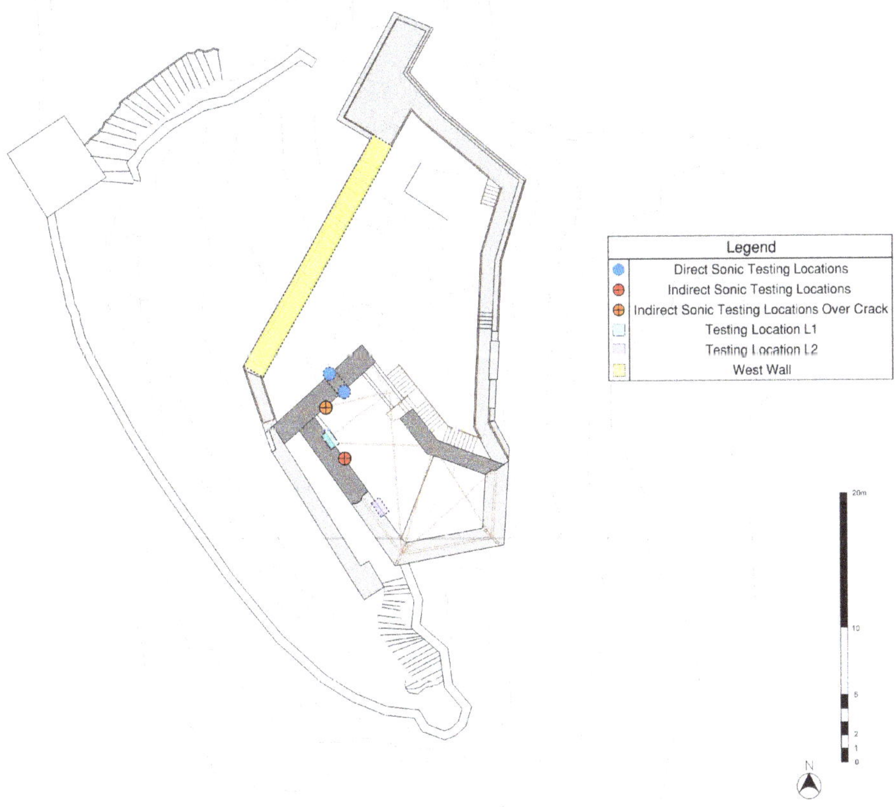

Figure 6. Plan view of site with test locations, highlighting the west wall.

Two locations were identified and investigated using these instruments, each on the south-western wall of the keep. The first was a harder and more densely mortared section of the wall, representative of much of the mortar that has been repointed in previous interventions, called 'L1'. The second was a more eroded variant of the first mortar with more visible aggregates and was expected to be softer; this was called 'L2'. The obtained results are summarized in Table 1.

Table 1. Summary of results from mortar testing.

Test	Result	Unit	Location L1	Location L2
Schmidt Hammer	Compressive strength	MPa	10.8	2.8
Schmidt Pendulum Hammer	Quality	-	Average	Reasonable
Penetrometer	Compressive strength	MPa	1.8	0.6
Scratch Test	Durability grade	-	-	-

3.1.1. Schmidt Hammer

The Schmidt hammer calibrated for mortars was used to evaluate the compressive strength of the two test locations. Calibration tables yielded a compressive strength of 10.8 MPa for L1 and 2.8 MPa for L2. These values are notably high for mortars when compared to values from the literature [6]. This could be explained by the fact that the test is performed on a superficial layer of mortar and additionally representative of the mortar strength from recent interventions on the castle.

3.1.2. Schmidt Pendulum Hammer

Some notable difficulty was observed in the application of the Schmidt pendulum hammer; the apparatus relied mainly on perpendicular application against the wall, often a significant challenge in historical structures due to the irregularity of the masonry. Too significant an inclination in either direction could artificially increase or decrease the measurement. Nevertheless, nine appropriate measurements for each testing location were obtained, returning rebound values between 25 and 36. These numbers were then used to qualitatively assess the mortar condition, observed as "average" for the more intact location (L1) and "reasonable" for the more eroded one (L2).

3.1.3. Penetrometer

Penetrometer measurements are likely to be more representative of the mortar below the surface due to the nature of the tests and returned lower values than the superficial impact hammer tests. The resulting compressive mortar strengths of L1 and L2, respectively, were 1.8 MPa and 0.6 MPa, implying a significant difference in the quality of the mortars.

3.1.4. Scratch Test

The scratch test relies heavily on the proper application of the instrument using three moveable legs to orient the device perpendicular to the surface, and a spring to drive the abrasive probe into the mortar. This process is notably more difficult to do on an irregular surface. Care was taken to not apply additional pressure to the probe as the measurement was being taken, and multiple additional test locations were attempted as a result. Despite best efforts, the practical applicability of the test on a very irregularly surfaced historical structure was very difficult to perform and interpret properly. As such, no reliable readings were obtained from this test.

3.1.5. Mortar Test Conclusions

The overall state of the mortar appears to be reasonable, supported by the results from the Schmidt pendulum hammer. Despite vastly different interpretations of the compressive strength of the mortar from the Schmidt hammer and penetrometer test, the mortar of location L1 is observed to be around three times stronger than that of its more eroded counterpart L2. The discrepancy in the results from the differing tests could be explained by the fact that the results from the Schmidt hammer represent only the strength of the newer mortar added in more recent historical interventions on the castle. By the nature of the application, the penetrometer could be more likely to represent the true strength of the historical mortar used in the castle, which is the working assumption of this paper.

3.2. Rock Test Results

Three unique types of masonry units were identified. Limestone was the most evident, particularly in heavily alveolized units that had degraded significantly. A layered dolomite or dolomitic limestone was also identified, as well as an amount of a harder marble. Identification was possible based on the reactivity of a weak (acetic) acid on the surface and is further supported by a geological survey from the Spanish Geological and Mining Institute [27]. The compressive strengths of the units tested are reported in Table 2, assuming material densities available in the literature [19]. The rock identified as marble presented the highest compressive strength at 40 MPa, while the dolomitic samples reported the weakest at 34 MPa. The similarity of these values, especially compared to the much weaker compressive strength of mortar, further supports the use of a homogenized approach to modeling masonry.

Table 2. Summary of results from rock testing.

Material	Specific Weight (kN/m^3)	Compressive Strength (MPa)
Dolomite	24	34
Limestone	24	37
Marble	26	40

3.3. Sonic Test Results

Sonic test results were divided into direct and indirect methods, totaling 336 results for the direct test and 353 for the indirect test over 79 different testing locations. The data were organized twice using both test type (direct, horizontal indirect, vertical indirect, or over the crack indirect) and approximate distance traveled by the wave to group and remove outliers. The distance category is further subdivided into categories A (20–25 cm), B (40–50 cm), C (60–75 cm), and D (80–100 cm). For the purpose of removing outliers, these grouped datasets were divided into four quartiles with equal populations, and the interquartile range between the first and third quartiles was calculated. Values beyond 1.5 times this range from the median in either direction were defined as outliers and removed, and the results were reported [28]. The results are shown in Figure 7 using boxplots.

Reconstructing the dataset and comparing the resulting elastic moduli after filtering by each method suggests a value between 1.5 and 2.0 GPa. The former is more in line with the results from the indirect testing and the latter with those from the direct testing methods. Boxplots present the reconstructed datasets after each type of filtering; note that the results from the direct test are unchanged since the filtering criteria do not affect them, and that filtering by distance was able to remove one additional data point (Figure 8). The higher value of 2.0 GPa was used as a starting point for initial calibration of the numerical model in Section 4. The results are summarized in Table 3.

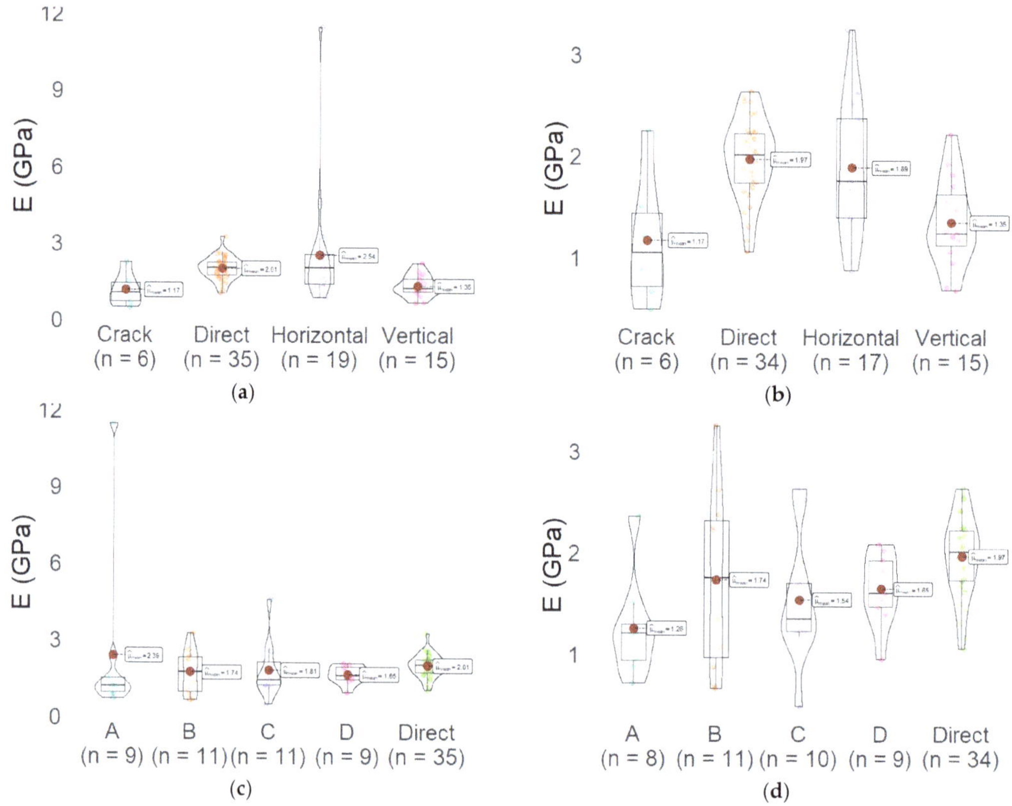

Figure 7. Results from sonic testing: (**a**) unfiltered elastic moduli by test type; (**b**) filtered elastic moduli by test type; (**c**) unfiltered elastic moduli by test distance; (**d**) filtered elastic moduli by test distance.

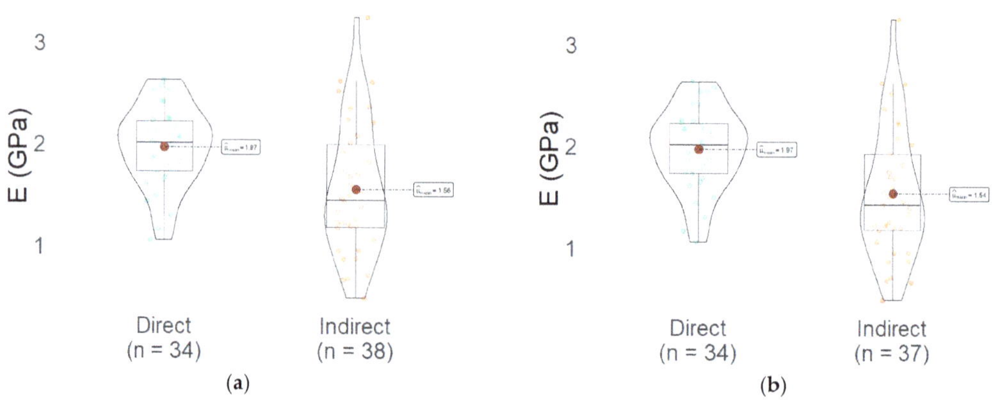

Figure 8. Results from sonic testing: (**a**) reconstructed dataset after test type filtering; (**b**) reconstructed dataset after distance filtering.

Table 3. Mean values from sonic testing before and after outlier filtering based on test type and distance.

Test ID	Filtering	Mean of Elastic Modulus Before Filtering (GPa)	Mean of Elastic Modulus After Filtering (GPa)
Crack	Type	1.17	1.17
Horizontal	Type	2.54	1.89
Vertical	Type	1.35	1.35
A (20–25 cm)	Distance	2.39	1.10
B (40–50 cm)	Distance	1.74	1.74
C (60–75 cm)	Distance	1.81	1.54
D (80–100 cm)	Distance	1.65	1.65
Indirect *	Type	-	1.56
Indirect *	Distance	-	1.54
Direct	Type	2.01	1.97

* Reconstructed datasets after separation and filtering.

3.4. Dynamic Identification Results

Data for the dynamic identification of the west wall were collected with two, 30 min acquisitions. Using multiple acquisitions allowed for multiple recordings of data to be combined, thereby improving the amount of data collected on the wall. Using the same accelerometer and location for multiple acquisitions is essential for this. Fifteen accelerometer locations were selected; with eight accelerometers, this allowed for seven to be moved between setups, while one was kept in place. Five vertical lines across the wall were selected, allowing for three accelerometers to be placed vertically to record the shape and curvature of the wall at each line. An additional point below the lowest accelerometer was also added as a base, with no assumed movement or assigned measurement. The highest point on the midspan, shown as point 12 in the diagram, was selected as the reference point; for each acquisition, three accelerometers were placed along a line to the left of the reference, three accelerometers were placed in a line to the right, and one was placed below the reference point (Figure 9). It is important to note that accelerometers were only placed in the out-of-plane direction, as the stiffness of the wall in the in-plane direction was considered too high to present any descriptive mode shapes.

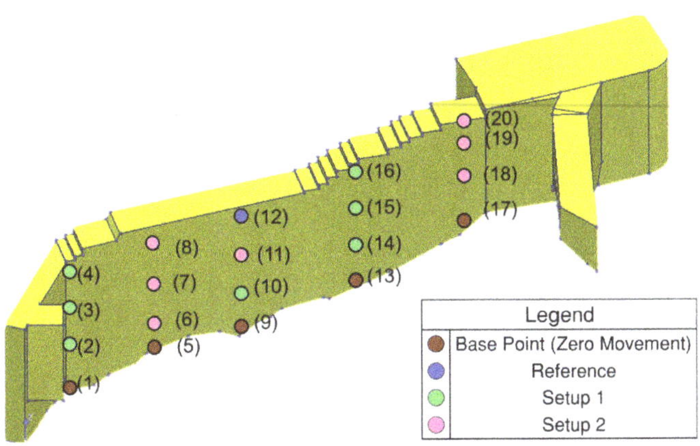

Figure 9. FE model of west wall with nodes for MAC calculations.

The acquired data were processed using ARTeMIS 6.0 modal software for dynamic identification. The data were filtered to consider the frequency content from 0 Hz to 100 Hz for decimation, with a resolution of 1024. Raw signals were converted from a time domain to a frequency domain for processing in this way. Identification of the modes was performed using both a manual peak-selection process and the support of the Enhanced Frequency Domain Decomposition (EFDD) and Stochastic Subspace Identification-Unweighted Principal Components (SSI-UPC) to validate the selected shapes (Figure 10a). Three modes were identified for the west wall setup, consistent with those from a fixed string vibration system due to the length of the wall. The frequencies for the first, second, and third mode were, respectively, 7.58 Hz, 10.84 Hz, and 14.13 Hz. Spatially, the first mode was composed of a single antinode at the midspan where the maximum displacement of the mode occurred, with two nodes at the ends of the wall. The second mode introduced an additional node and antinode, with the maximum displacement occurring at one-quarter of the length of the wall to effectively half the wavelength of the mode. The third mode introduced a final node and antinode pair, returning the maximum displacement to the midspan and introducing another half-wavelength to the system. These shapes are visible in Figure 10b.

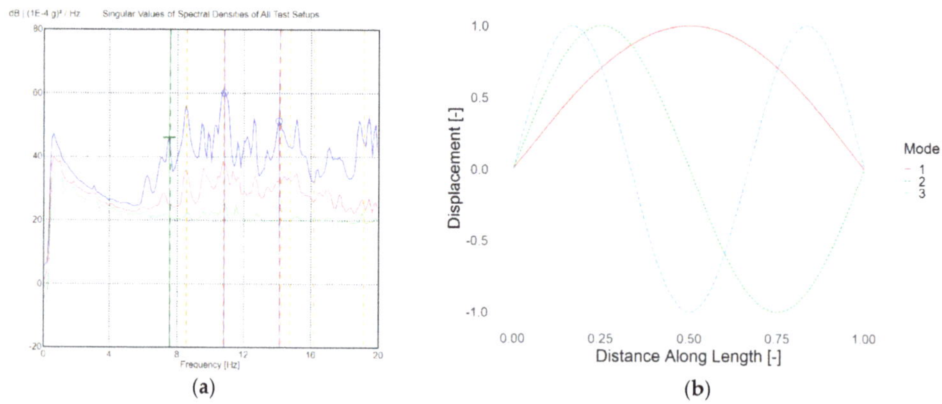

Figure 10. Overview of results from dynamic identification: (**a**) spectral density for west wall setups with estimators; (**b**) first three mode shapes for the fundamental frequencies of a fixed string.

3.5. Validation

Validation for the experimentally obtained properties was explored by their ability to calibrate a numerical model in Section 4. Of particular interest were the values obtained by sonic testing and dynamic identification, which were used to calibrate the elastic modulus and outline the mode shapes used in the modeling and calibration process. The ability of the model to replicate these characteristics will be used to validate experimentally obtained values.

4. Finite Element Modeling Approach

4.1. Modeling Approach

Various methods are commonly applied for the evaluation of masonry structures using numerical models. Masonry as a building material comprises a combination of bricks and mortar, with each element being equally important to the composition. Individual units generally make up the majority of masonry, while the mortar fills in the gaps between the units. Thus, the models used to represent masonry vary in their treatment of the interface and interaction between these elements. Detailed micro-modeling represents all elements individually using continuum elements to model the masonry units and mortar and introducing a discontinuous element as an interface between these units. This allows for the detailed application of mechanical and constitutional models to all of the material types individually. A more simplified approach can be adopted in simplified micro-modeling, in which the masonry units are geometrically expanded to contact one another using continuum elements, and the mortar and unit–mortar interface are modeled together using discontinuous elements [6]. For the application of both these methods, knowledge about the exact geometry and composition of not only a structure, but also the individual masonry elements that comprise it, is required. This is often extremely difficult to apply in a historical context, especially where very irregular and random arrangement and size of units is expected as in the case of the castle. These approaches are also very computationally expensive, requiring much greater computational time and storage capabilities [29]. Macro-modeling was therefore used for the purpose of this study, in which the masonry units and mortar interface are all treated wholistically as a homogenous anisotropic material. This homogenized approach is well supported by the application of non-destructive tests where little additional information about the nature of a site can be gathered, further increasing the efficiency of the methods. Sonic testing and modal analysis, therefore, are complimentary approaches in ascertaining homogenous material properties [6].

A 3D finite element model was prepared using DIANA Finite Element Analysis (DIANA FEA) 10.8 for numerical simulations [30]. The model considered the extent of the masonry elements comprising the castle in three dimensions, based on an existing photogrammetric survey of the site (available in Supplementary Materials). The exterior walls were modeled from the foot of the wall outside the castle to the height of the masonry, disregarding railings or the infill of the soil inside the enclosure. The walls were then extruded inwards. The faces of the underside of the model were considered with translationally fixed boundary conditions. For this study, a separate model considering the large western wall of the enclosure was also prepared. The locations of the accelerometers from dynamic identification were also imprinted on the model to directly compare the models (Figure 9) and calculate the MAC later.

4.2. Mesh Definition

To model the 3D structure, 20-node solid elements (Type CHX60) were used. A maximum element size of 0.34 m was used to ensure that the model would consider five elements across the thickness of the wall to account for possible curvature in a non-linear model.

4.3. Material Properties

The basic material properties used for the FEA are outlined in both a linear elastic and non-linear regime. Linear properties were obtained experimentally and from the literature, namely the Italian code NTC 2018 [31]. Non-linear properties were considered in the model using a total strain-based crack model with rotating crack orientation. The compression regime considered a parabolic compression curve with no residual compressive strength, reduction due to lateral cracking, or stress confinement in the model. The tensile regime considered an exponential stress curve with no residual tensile strength or Poisson's ratio reduction. Compressive strength and fracture energy were evaluated using formulae from the literature, shown in Equations (2) and (3), respectively [6]. Values for tensile strength and fracture energy come from the literature [6]. Table 4 outlines these properties for both regimes.

$$E = 500\ \mathrm{f_c} \tag{2}$$

$$\mathrm{G_c} = \frac{32\ \mathrm{f_c}}{10 + \mathrm{f_c}} \tag{3}$$

Table 4. Linear elastic and non-linear properties used for masonry.

Property	Symbol	Unit	Value
Density	ρ	kg/m^3	1900
Elastic Modulus	E	GPa	2.0
Poisson's Ratio	ν	-	0.2
Compressive Strength	f_c	MPa	3.0
Compressive Fracture Energy	G_c	N/mm	7.40
Tensile Strength	f_t	MPa	0.1
Tensile Fracture Energy	G_f^I	N/mm	0.02

4.4. Model Calibration

Calibration was necessary to tune the initially assumed material properties of the model to better represent those obtained experimentally from dynamic identification. An eigenvalue analysis of the FE model was used to identify the frequencies and mode shapes of the structure in the model for direct comparison with those obtained experimentally. The frequencies determined in this method are directly impacted by both the mass and the stiffness of the structure; therefore, mass was fixed, while stiffness was adjusted for calibration. The relevant frequencies from DIANA are obtained from an eigenvalue analysis, in which the solutions to the free vibration equation (Equation (4)) are represented by the eigenvalue λ (Equation (5)) [32]. K represents the stiffness matrix, M the mass matrix, ϕ the corresponding mode shape vector, ω the angular frequency, and f the frequency:

$$K\phi = \omega^2 M\phi \quad (4)$$

$$\lambda \equiv \omega^2 = 4\pi^2 f^2 \quad (5)$$

For the calibration, the initial elastic modulus was reduced from 2.0 GPa to 1.5 GPa, which reduced the frequency error for similar mode shapes to under 2.0% for the first two frequencies, and under 9% for the third. The average absolute error was therefore 3.8% with a maximum error of 9.0%. Modes were further validated from the eigenvector analysis by selecting the first three that moved above 20% of the mass of the wall each, from an initial analysis that considered up to 70% of the mass moved from 20 eigenfrequencies computed. A visual comparison between the numerical and experimental modes demonstrates their similarity (Figure 11).

4.5. Modal Assurance Criterion

Once the elastic modulus and frequencies had been determined both experimentally and numerically, it was possible to calculate the Modal Assurance Criteria (MAC) to compare the FEA model with the experimental one. The MAC is the most commonly used method in the literature to calculate the correlation between two vectors. In this case, these vectors represent the mode shapes obtained experimentally and numerically. Mathematically, the MAC calculates the scaler product of two normalized vectors [33]. The MAC compares the values of any number, n, entries in two 1 x n matrices, and could therefore directly be used to compare the displacement vectors between the numerical and experimental results for n nodes. Of note is the fact that the formulation of the MAC in this case is based on the vectors used representing only one-dimensional space, since the accelerometers are all placed in an out-of-plane direction. The displacement vector outputs from the numerical model for each of the relevant nodes were therefore expressed only in the out-of-plane direction. One displacement value for each node, at each point that an accelerometer was placed, was then entered into a vector based on the node's ID so that relevant pairs of numerical and experimental displacement vectors could be compared.

The output of the MAC is a single number ranging from 0 to 1 depending on how similarly the methods express the explored mode shape; 1 represents complete correlation, while 0 represents complete disagreement of the values and therefore of the mode shapes. The formulation is written without the need for normalization:

$$\mathrm{MAC}(x_e, x_n) = \frac{x_e^T x_n}{\|x_e\|_2 \, \|x_n\|_2} \tag{6}$$

where x_e is a vector of experimental displacement vectors, and x_n is a vector of numerical displacement vectors [34]. These vectors were obtained from DIANA directly at the nodes specified using the mesh by imprinting and considering only the vectors in the out-of-plane directions.

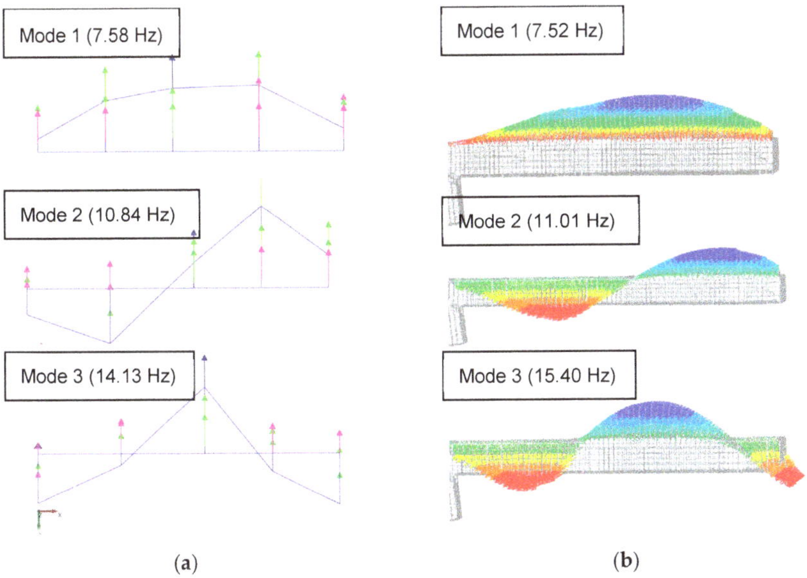

Figure 11. Mode shapes for west wall: (**a**) initial three experimentally determined mode shapes; (**b**) initial three numerically determined mode shapes.

Initial MAC calculations for the system revealed an overall good agreement between the mode shapes obtained, with all values above 0.8. The calculated MAC value was 0.96 for the first mode, 0.84 for the second mode, and 0.94 for the third mode. A graphical comparison between the vectors obtained through both methods revealed that the points of maximum displacement, although similar, occurred in different points between the models. The experimental behavior of the structure did not match the numerical one at the points closest to the north tower. This suggested the need to reduce the stiffness of the tower due to the presence of an arched room which represented a significant void in the structure (Figure 12). The ultimately successful approach was the addition of an approximation of the void room in the structure, which improved the MAC values for the first mode to 0.98 and the second mode to 0.86 while slightly reducing that of the third mode to 0.91. This also slightly increased the frequency error in the first mode to 1.1% while reducing that of the second mode to 0.3% and that of the third mode to 8.8% (Table 5).

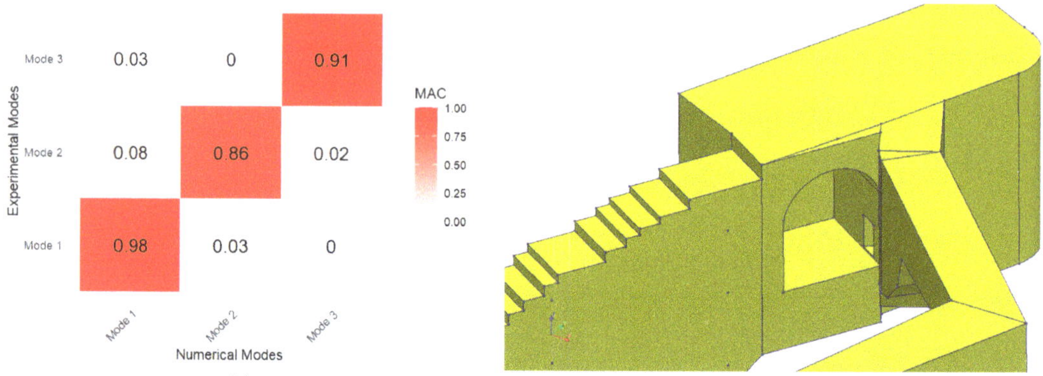

(**a**) (**b**)

Figure 12. (**a**) MAC comparison between numerical and experimental mode shapes for model considering the addition of a void in the north tower; (**b**) model representation of the void in the north tower.

Table 5. Comparison of experimentally obtained frequencies with calibrated and uncalibrated numerically obtained frequencies (percent difference in parentheses).

Mode	Experimental Frequency (Hz)	Uncalibrated Numerical Frequency (Hz)	Initial Calibrated Numerical Frequency (Hz)	Updated Calibrated Numerical Frequency (Hz)
1	7.58	8.68 (14.58%)	7.52 (−0.76%)	7.49 (−1.11%)
2	10.84	12.71 (17.29%)	11.01 (1.60%)	10.87 (0.34%)
3	14.13	17.79 (25.87%)	15.40 (8.99)	15.37 (8.80%)

5. Analysis

The structural performance of the castle was evaluated using two methods of applying an out-of-plane force to overturn the structure. The first of these methods was a simple limit analysis to outline expectations; the second used a non-linear static analysis, also known as a pushover analysis in an FEA environment. This method of evaluating the seismic behavior of an irregular historical masonry structure has been successfully used and is well documented [1]. All methods were compared to the expected peak ground acceleration for the region according to the local design codes [9]. The results of these analyses represent a factor of gravitational acceleration, equal to 9.81 m/s^2, applied to the self-weight of the structure uniformly across its body.

5.1. Limit Analysis

The brief analytical limit analysis was conducted for a 1 m long cross-section of the west wall, considering a thickness in the out-of-plane direction of 1.7 m and the average of the minimum and maximum height of the wall of 6.4 m. The analysis considered a force proportional to the self-weight being applied at exactly the mid-height, using a factor (α) to represent this force. The factor is defined as the ratio of the overturning of the force to the resisting moment of a structure's self-weight at the point of equilibrium. The analysis was performed according to a triangular, elastic distribution of stress at the base of the wall using the following equation:

$$\alpha = \frac{W * \left(\frac{tk}{2} - t\right)}{W * \frac{h}{2}} = \frac{\frac{tk}{2} - \frac{2*W}{3*fc*L}}{\frac{h}{2}} \quad (7)$$

where tk [m] is the thickness of the wall, h [m] is the height of the wall, W [kN] is the self-weight of the wall set in the geometric centroid, fc [kN/m^2] is the compressive strength of the masonry, L [m] is the length of wall considered, and t [m] is the distance of the hinge from the side of the wall. The value of α using this method was 0.25 g, or 25% of the self-weight applied laterally to the structure.

5.2. Non-Linear Static Analysis

The FE model of the west wall was further reduced to represent only the wall itself and not the adjoining structures of the north tower and the southern wall. In this way, it was possible to explore extreme cases of the end conditions of the adjoining masonry. The lowest of these conditions represents a worst-case scenario, in the form of a weak connection to the adjoining masonry. If this minimum condition is considered safe, then any greater state of connection will be as well. The other extreme considered the connecting surfaces as fully translationally fixed with the intent of modeling a failure pattern indicative of arching behavior. The real state of the wall should be somewhere between these two extremes. In each case, the load in the out-of-plane direction was sequentially increased using the secant method for the convergence of the boundary value problem. An energy method with a convergence criterion of 10^{-3} was used, alongside arc-length and line search methods to assist in convergence, following methods outlined in the literature [6]. Each model was first loaded vertically under the self-weight of the wall using five equal load steps before the sequential non-linear static, or pushover iterations were performed.

Ultimately, the walls were able to exhibit non-linear behavior, as evident by the approximately bilinear regimes outlined by the resulting capacity curves for both conditions (Figure 13). The capacity curve modeled by the pushover analysis presents the differences in the two utilized boundary condition models visually. The fully fixed condition immediately exhibits a steeper rise in load factor for a given displacement, attributed to the increased stiffness and capacity provided by the stronger boundary conditions. The slight dip in the load factor that appears just after 2 mm could be indicative of the formation of cracks and the dissipation of energy as the slope changes and the model enters the flatter non-linear regime. The fully free model exhibits a less significant rise in the initial slope, requiring a smaller load factor to achieve a similar amount of displacement. As the non-linear behavior begins at a load factor slightly above 0.2, the lack of energy dissipation capacity due to the absence of boundary constraints results in a much less significant increase in displacement for a given load factor as the masonry continues to deform. The nature of this deformation will be discussed later. The completely free end conditions exhibited a maximum load factor of 0.25 g being applied laterally to the wall, while the fully fixed conditions continued to a value of almost 0.80 g.

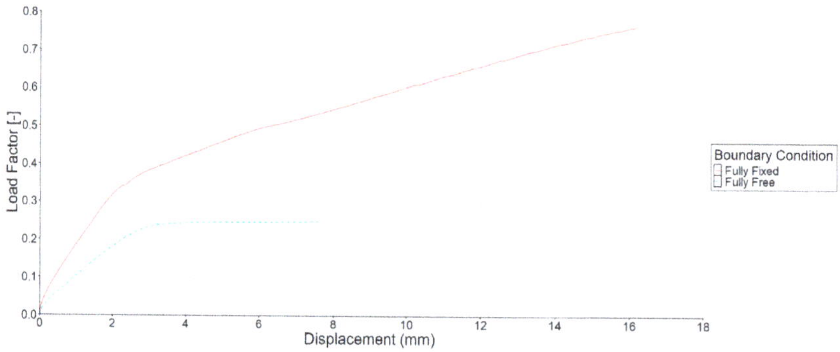

Figure 13. Capacity curve comparing fully fixed and fully free end conditions for pushover analysis of the west wall.

For the version of the model considering fully free end conditions, a total of 37 converged steps were able to outline the deformed shape of the wall as it was sequentially loaded through the linear elastic regime, into non-linear behavior. A maximum displacement of 7.8 mm was reached at the highest portion of the wall, and the structure was able to withstand 0.25 g in the out-of-plane direction. As the structure reached failure, compressive stresses in the outer toe of the wall increased, approaching the crushing strength of the masonry, as outlined by the minimum principal stress in the model (Figure 14a). At the same time, a crack formed on the opposite side of the wall, consistent with the expected behavior of an overturning cantilevered wall (Figure 14b). Cracks began to appear first at the upper base of the wall, indicated by the maximum principal strain in the model. Some nuances were observed as a result of the complexity of the lower portion of the wall used in modeling; the maximum stress did not follow the base across its length and instead jumped over the mid-portion of the base that extends downward.

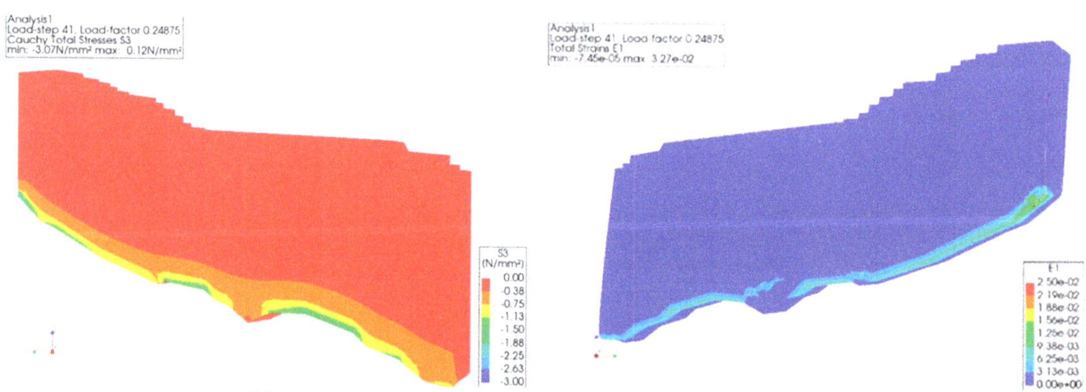

(**a**) (**b**)

Figure 14. Results from the pushover analysis of the west wall with free end condition: (**a**) minimum principal stress (view from the outside of the wall); (**b**) maximum principal strain (view from the in-side of the wall).

The second variation of the analysis, which used fully fixed end conditions, was stopped after 144 successfully converged load steps. The model reached a capacity of 0.76 g, over three times higher than the previous model. The deformed structure demonstrated an arching behavior at the height of the wall as the midspan bulged outwards in the direction the load was applied. From the point of maximum displacement, the displacement at a given point decreased radially until the fixed end conditions were met. Compressive principal stresses indicated crushing at the base of the wall and along the outer edges of the exterior of the wall (Figure 15a). Maximum principal strain outlined the formation of a curved crack on the interior of the wall as it begins to fail, consistent with the arching behavior (Figure 15b). The crack follows the base of the wall at the upper side of the wall, which is both shorter and more elevated than the taller, lower section in which the crack propagates directly through the masonry. The fixed end conditions allow the wall to behave more like a beam and less like a cantilever, allowing it to resolve shear forces and relevant stresses due to horizontal loading at the end conditions.

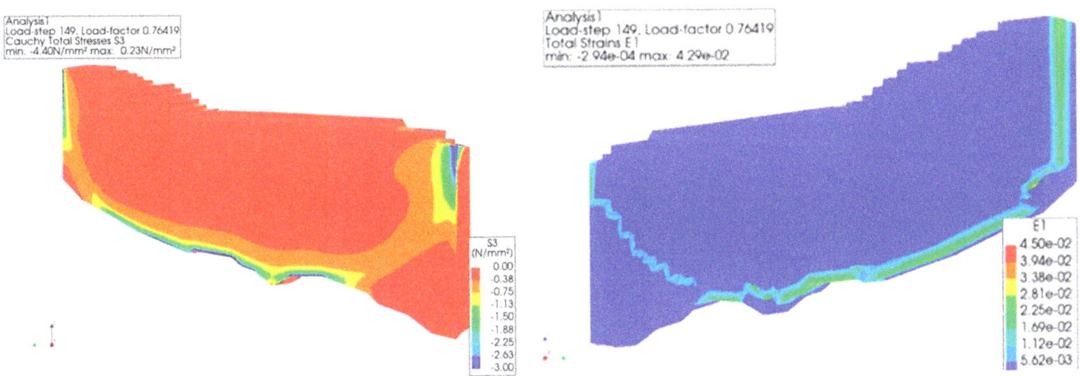

(a) (b)

Figure 15. Results from the pushover analysis of the west wall with fully fixed end condition: (**a**) minimum principal stress (view from the outside of the wall); (**b**) maximum principal strain (view from the inside of the wall).

5.3. Safety Evaluation

The actual conditions of the wall are likely to exhibit behavior with characteristics of both failure patterns. The support conditions suggested by the model calibration are of significantly good quality, though unlikely to be completely translationally fixed. For the purpose of evaluating the safety of the structure under seismic loading, however, the analysis in this case is sufficient. Direct comparisons can be made between the expected peak ground acceleration (PGA) from local design codes and the values resulting from the analyses. Spanish seismic codes use probabilistic seismic hazard analysis to estimate the expected PGA for a given location, integrating data from historical and instrumental records of seismicity, which typically reference magnitude in terms of the moment magnitude scale (Mw). Spanish seismic codes outline Lanjarón as one of the more seismically susceptible regions in Spain, with an expected PGA value of 0.18 g [9]. At minimum, the wall was able to achieve a value of 0.25 g before collapse. This strength can be considered sufficient to withstand local peak ground accelerations according to the code. A safety factor can be obtained from the ratio of the minimum peak value achieved to the PGA of the area, defined as 0.25 g/0.18 g = 1.39. However, the positioning of the castle atop a steep rocky slope and the local aquifers of the surrounding region could destabilize the ground around the structure as suggested in [35].

6. Conclusions

In this work, an extensive array of non-destructive tests for sites of cultural significance were presented and performed for the Castle of Lanjarón, Spain. The works explored the application of a Schmidt hammer, Schmidt pendulum hammer, penetrometer, scratch test device, and a series of direct and indirect sonic tests for wave-based methods to characterize the materials of the structure and their mechanical properties. Dynamic identification was employed to register ambient vibrations and obtain relevant information for the numerical models. Three primary modes were obtained from the dynamic identification process under 15 Hz, being 7.58 Hz, 10.84 Hz, and 14.13 Hz. The fundamental frequencies of vibration were selected manually and further validated using EFDD and SSI-UPC methods. Operational Modal Analysis was used to calibrate and validate a developed homogenized 3D FE model for the west wall of the castle. This was used to assess the seismic behavior of the site using non-linear static analysis in addition to limit analysis. The behavior of the structure was evaluated for safety in accordance with current design codes.

The efficacy of wave-based methods and OMA were highlighted in the study; sonic test results were able to effectively predict the range of the elastic modulus (1.5–2.0 GPa) that resulted from an eigenvector analysis of the model, which considered an initial value of 2.0 GPa but was ultimately successful in calibration with a value of 1.5 GPa. The values obtained from indirect sonic testing were within 3% of the calibrated elastic modulus for the wall, while those from direct testing were around 33% higher. The Modal Assurance Criteria were used to validate the movement of the west wall in the FE model and the experimentally gathered data from OMA to above 85% agreement for all modes presented. The calibrated frequencies for the FE model were 7.52 Hz, 11.01 Hz, and 15.4 Hz, which displayed an average absolute error of 3.8% and maximum absolute error of 9.0%.

Limit analysis and FEA allowed for the evaluation of the seismic behavior of the wall in the out-of-plane direction, the most vulnerable for masonry structures. Common failure patterns were explored, ultimately concluding that the structure could withstand the local peak gravitational accelerations of 0.18 g as defined by the Spanish seismic design codes due to seismic activity. A safety factor of 1.39 was obtained from the ratio of the minimum peak value achieved from non-linear static analysis, 0.25 g, to the PGA of the area, 0.18 g.

The novelty of this case study presents findings that should be observed and considered for all future works on the Castle of Lanjarón. The approach presents a framework that can be adopted for the evaluation of regular and irregular historical masonry structures alike. It is especially relevant for sites that are otherwise difficult to perform common NDTs on. While more traditional NDTs (Schmidt hammer, Schmidt pendulum hammer, penetrometer, and scratch testing instrument) were less successful in their application on the irregularly surfaced masonry, the wave-based tests and modal analysis were extremely successful in their application. These methods were able to wholistically approximate the mechanical behavior of the structure in terms of a homogeneous modulus of elasticity and modal parameters including mode shapes and frequencies of vibration. These wholistic values complemented the use of the macro-modeling approach for masonry and the homogenization assumptions that it entails. These methods can strongly be recommended for the preliminary seismic assessment of a site where resource availability and the applicability of analog methods is difficult.

Further analyses beyond the scope of the paper for the Castle of Lanjarón are recommended. Another non-linear pushover analysis using more accurate boundary conditions for the wall with more computational time could provide a more accurate assessment of the ability of the west wall to resist lateral forces. A non-linear dynamic analysis for the wall or a more general model of the castle could better support the seismic safety conclusions of this paper [5]. Further investigations making use of macro-modeling techniques could utilize deep learning or AI-based imaging models to develop their FE models but may be similarly challenged due to the irregularity of the masonry in the castle [36].

Supplementary Materials: The following supporting information can be downloaded at: https://www.mdpi.com/article/10.3390/app15031518/s1, Appendix S1: Photogrammetric survey and damage mapping of the Castle of Lanjarón.

Author Contributions: H.L.: conceptualization, methodology, validation, formal analysis, investigation, data curation, writing—original draft preparation, and visualization. R.R.: conceptualization, methodology, investigation, data curation, formal analysis, writing—review and editing, and visualization. P.P.: conceptualization, methodology, resources, writing—review and editing, supervision, and project administration. P.B.L.: conceptualization, methodology, resources, writing—review and editing, supervision, project administration, and funding acquisition. All authors have read and agreed to the published version of the manuscript.

Funding: This work was partly financed by FCT/MCTES through national funds (PIDDAC) under the R&D Unit Institute for Sustainability and Innovation in Structural Engineering (ISISE), under reference UIDB/04029/2020 (doi.org/10.54499/UIDB/04029/2020), and under the Associate Laboratory Advanced Production and Intelligent Systems ARISE under reference LA/P/0112/2020.

Institutional Review Board Statement: Not applicable.

Informed Consent Statement: Not applicable.

Data Availability Statement: The original contributions presented in the study are included in the article. Further inquiries can be directed to the corresponding author.

Acknowledgments: This work would not have been possible without the involvement of many people and parties. We would like to thank the Delegación Provincial de Cultura de Granada, and the Lanjarón City Hall for allowing this work to take place; Members of the University of Minho for their technical support in modeling and calibration; and the residents of Lanjarón for their love of the castle and photography, alongside their hospitality and dedication.

Conflicts of Interest: The authors declare no conflicts of interest. The funders had no role in the design of the study; in the collection, analyses, or interpretation of data; in the writing of the manuscript; or in the decision to publish the results.

References

1. Lourenço, P.B.; Mendes, N.; Ramos, L.F.; Oliveira, D.V. Analysis of masonry structures without box behavior. *Int. J. Archit. Herit.* **2011**, *5*, 369–382. [CrossRef]
2. Ramírez, R.; Mendes, N.; Lourenço, P.B. Diagnosis and Seismic Behavior Evaluation of the Church of São Miguel de Refojos (Portugal). *Buildings* **2019**, *9*, 138. [CrossRef]
3. Grazzini, A.; Lacidogna, G. Mechanical Properties of Historic Masonry Stones Obtained by In Situ Non-Destructive Tests on the St. Agostino Church in Amatrice (Italy). *Appl. Sci.* **2021**, *11*, 6352. [CrossRef]
4. Duvnjak, I.; Damjanović, D.; Bartolac, M.; Skender, A. Mode Shape-Based Damage Detection Method (MSDI): Experimental Validation. *Appl. Sci.* **2021**, *11*, 4589. [CrossRef]
5. Pineda, P. Collapse and upgrading mechanisms associated to the structural materials of a deteriorated masonry tower. Nonlinear assessment under different damage and loading levels. *Eng. Fail. Anal.* **2016**, *63*, 72–93. [CrossRef]
6. Lourenço, P.B.; Gaetani, A. *Finite Element Analysis for Building Assessment*; Routledge: New York, NY, USA, 2022. [CrossRef]
7. Vila-Chã, E.; Barontini, A.; Lourenço, P.B. Implementation of a Condition Monitoring Strategy for the Monastery of Salzedas, Portugal: Challenges and Optimisation. *Buildings* **2023**, *13*, 719. [CrossRef]
8. Masciotta, M.; Ramos, L. *Dynamic Identification of Historic Masonry Structures*; Elsevier: Amsterdam, The Netherlands, 2019; pp. 241–264. [CrossRef]
9. *NCSE-02*; Norm for Earthquake-Resistant Construction. Public Works Ministry of Spain: Madrid, Spain, 2002.
10. Navarro, J.D.L. El castillo de Lanjarón (Granada). Un análisis a partir del estudio de la cerámica recogida en la intervención arqueológica de 1995. *Arqueol. Territ.* **2008**, *5*, 141–159.
11. Cuello, A.M. Los castillos en época nazarí. Una primera aproximación. In *Castillos y Territorio en al-Andalus*; Athos-Pérgamos: Granada, Spain, 1998; pp. 246–293.
12. Vílchez, C.V. La Alhambra y su Conservación [1927]. In *La Alhambra de Leopoldo Torres Balbás: Obras de Restauración y Conservación, 1923–1936*; Comares: Granada, Spain, 1998.
13. Khrobostov, A. *Lanjaron Castle or Castillo de los Moros Aerial Panoramic View*; Alamy Stock Images: Oxfordshire, UK, 2021.

14. Porras, A.G.; Forns-Samsó, J.J.B. El Castillo de Lanjarón. Primeros resultados de la última intervención arqueológica. In *Anuario Arqueológico de Andalucía 1996;* Junta de Andalucía: Sevilla, Spain, 2001; pp. 190–200.
15. Ministry of Culture of the Regional Government of Andalusia, Provincial Delegation of Granada. *Resolucion del Ilmo. Sr. Delegado Provincial de la Consejeria de Cultura en Granada de Fecha 26 de Marzo de 2008;* Reference No. BC-JCM/LMM; Regional Government of Andalusia, Ministry of Culture, Provincial Delegation of Granada: Granada, Spain, 2008; Available in the archives of the Ministry of Culture, Provincial Delegation of Granada.
16. Sicree, A.A. An Acidic Reaction. In *Hands-On Demonstrations of Mineral Properties;* Mineral Information Institute: Golden, CO, USA, 2008.
17. *ASTM D5873-14;* Standard Test Method for Determination of Rock Hardness by Rebound Hammer Method. ASTM International: West Conshohocken, PA, USA, 2014. [CrossRef]
18. Schön, J.H. Chapter 4—Density. In *Developments in Petroleum Science;* Schön, J.H., Ed.; Elsevier: Amsterdam, The Netherlands, 2015; Volume 65, pp. 109–118. [CrossRef]
19. Deere, D.U.; Miller, R.P. *Engineering Classification and Index Properties for Intact Rock;* Air Force Weapon Laboratory: Albuquerque, NM, USA, 1966. [CrossRef]
20. PASI. PASI Manuale d'uso SCLEROMETRO TIPO M. Available online: https://www.pasisrl.it/ (accessed on 10 May 2024).
21. *EN 10025-1:2004;* Hot Rolled Products of Structural Steels—Part 1: General Technical Delivery Conditions. European Committee for Standardization: Brussels, Belgium, 2004.
22. *SCHMIDT OS-120 Operating Instructions;* Promat (HK) Ltd.: Hong Kong, China, 2017.
23. DRC. Penetrometro Malta SRM 1.0 User Manual. Available online: https://www.drcitalia.it/ (accessed on 10 May 2024).
24. MORTARCHECK II. Available online: http://enertren.com.au/links.html (accessed on 10 May 2024).
25. Marazzani, J.; Cavalagli, N.; Gusella, V. Elastic properties estimation of masonry walls through the propagation of elastic waves. An experimental investigation. *Appl. Sci.* **2021**, *11*, 9091. [CrossRef]
26. Bourgeois, I.; Di Gregorio, L.T.; Rodrigues, H.; Fonseca, J.; Tavares, A.; Costa, A. Application of Sonic Tests and Modal Identification in the Characterization of Masonry Walls. *Appl. Sci.* **2023**, *13*, 3762. [CrossRef]
27. Instituto Geológico y Minero de España. *Mapa Geológico de España, Scale 1:50000—Sheet 1042—Lanjarón;* Instituto Geológico y Minero de España: Madrid, Spain, 1974.
28. Navadi, W. *Statistics for Engineers and Scientists*, 5th ed.; McGraw-Hill Education: New York, NY, USA, 2020.
29. Truong-Hong, L.; Laefer, D. Micro vs. macro models for predicting building damage underground movements. In Proceedings of the International Conference on Computational Solid Mechanics—CSM2008, Ho Chi Minh City, Vietnam, 27–30 November 2008.
30. Diana, F. *Displacement Method ANAlyser, Release 10.8;* DIANA FEA BV: Delft, The Netherlands, 2024.
31. *NTC 2018;* Italian Building Code for Constructions. Ministry of Infrastructure and Transport: Roma, Italy, 2018.
32. Pástor, M.; Binda, M.; Harčarik, T. Modal Assurance Criterion. *Procedia Eng.* **2012**, *48*, 543–548. [CrossRef]
33. Nguyen, M.; Filippatos, A.; Langkamp, A.; Gude, M. Modal Identification of Output-Only Systems of Composite Discs Using Zernike Modes and MAC. *Sensors* **2019**, *19*, 660. [CrossRef] [PubMed]
34. Lee, E.H.; Bindel, D. Improving Upon Mode to Mode Comparisons in MDOF Systems: The Modal Assurance Fit (MAF). 2018. Available online: https://www.cs.cornell.edu/~erichanslee/maf.pdf (accessed on 17 June 2024).
35. Benito, M.B.; Navarro, M.; Vidal, F.; Gaspar-Escribano, J.; García-Rodríguez, M.J.; Martínez-Solares, J.M. A new seismic hazard assessment in the region of Andalusia (Southern Spain). *Bull. Earthq. Eng.* **2010**, *8*, 739–766. [CrossRef]
36. Ibrahim, Y.; Nagy, B.; Benedek, C. Deep Learning-Based Masonry Wall Image Analysis. *Remote Sens.* **2020**, *12*, 3918. [CrossRef]

applied sciences

MDPI

Article

Understanding the Seismic Resilience of Metallic Cylindrical Tanks Through Parametric Analysis

Emanuele Brunesi [1], Roberto Nascimbene [2,*] and Dumitru Beilic [3]

1 EUCENTRE, European Centre for Training and Research in Earthquake Engineering, Via Ferrata 1, 27100 Pavia, Italy; emanuele.brunesi@eucentre.it
2 Istituto Universitario di Studi Superiori IUSS—Department STS, Piazza della Vittoria n.15, 27100 Pavia, Italy
3 Rose School, Istituto Universitario di Studi Superiori IUSS, Piazza della Vittoria n.15, 27100 Pavia, Italy; dbeilic@roseschool.it
* Correspondence: roberto.nascimbene@iusspavia.it

Abstract: This research investigates the seismic behavior of rigid and flexible cylindrical steel tanks, focusing on tanks with an open top and fully anchored at the base. The primary objective is to evaluate the hydrodynamic pressures exerted by the fluid on the tank walls during seismic excitation. Three widely recognized design approaches—New Zealand NZSEE recommendations, European code UNI EN 1998-4:2006 (CEN, 2006), and American Water Works Association AWWA D100-05 standard (ANSI/AWWA, 2005)—were implemented and compared with high-definition finite element models and then validated against the experimental results. Nonlinear fluid–structure interaction (FSI) was modeled using an Arbitrary Lagrangian–Eulerian (ALE) formulation with the Navier–Stokes equations governing the fluid motion and material and geometric nonlinearities considered in the tank walls. Parametric analyses were conducted to investigate the impact of tank geometry, specifically height-to-radius and radius-to-thickness ratios, on seismic response, identifying a transition between rigid and flexible behavior. The study also examined the influence of seismic input using a set of ten displacement spectrum-compatible ground motions. The findings contribute to a better understanding of the seismic resilience of cylindrical steel tanks, offering valuable insights for improving design standards and safety in earthquake-prone regions where these systems may abound.

Keywords: seismic response; cylindrical steel tanks; fluid–structure interaction (FSI); finite element modeling; hydrodynamic pressure; sloshing

1. Introduction

The vulnerability of steel storage tanks and containment structures in industrial plants to earthquake-induced damage has been well documented across numerous seismic events as well as a wide variety of experimental and/or analytical/numerical studies available in the literature [1–28]. Significant incidents include the 1960 earthquake in Chile [29], the 1964 Alaska earthquake, the 1964 Niigata earthquake, and several major earthquakes in California, such as the 1971 San Fernando, 1979 Imperial Valley, 1989 Loma Prieta, and 1994 Northridge events. Other noteworthy earthquakes include the 1995 Kobe earthquake and the 2006 Silakhor earthquake in western Iran [30]. In Italy, the 2009 L'Aquila earthquake and the 2012 Emilia earthquake [31] highlighted the vulnerabilities of industrial facilities, particularly those that house critical containment structures. The 2013 Sedan and Lake Grassmere earthquakes in New Zealand (with magnitudes of 6.5) and the 2016 Kaikōura earthquake (magnitude 7.8) [32] further demonstrated the severe impact of seismic events on industrial infrastructure, especially on storage tanks and containment facilities that store hazardous materials, chemicals, and fuel. These events underscored the necessity for improved seismic resilience in the design, construction, and retrofitting of such facilities to prevent catastrophic failures that could result in the loss of containment, fires, hazardous spills, and disruption of essential services critical for public safety and environmental protection.

The seismic performance of these tanks is critical not only due to the economic value of the structures and their contents but also because they often play a key role in the operation of emergency services following a seismic event. Disruptions in the supply of essential products such as oil and water can significantly hinder the effectiveness of critical facilities, such as hospitals and fire stations, which rely on these resources. For instance, a lack of water following an earthquake could trigger major fires, potentially causing more damage than the earthquake itself. The failure of tanks holding highly flammable petroleum products can lead to catastrophic, uncontrolled fires, which is similar to those observed after the Niigata and Alaska earthquakes in 1964. Furthermore, the release of hazardous materials, such as liquefied gases or toxic chemicals from containment structures damaged during an earthquake, can pose significant risks to both human health and the environment [26,33].

Past seismic events have revealed various forms of damage and failure in steel tanks [1]. The most common forms of structural damage include buckling of the tank shell, such as elastic–plastic buckling (often referred to as "elephant foot" buckling), which is typically seen in large oil containment structures, and elastic buckling (or "diamond buckling"), which occurs more frequently in thin-shell tanks, like those used in wine storage. Beyond these buckling issues, other common forms of damage have been identified, such as failures in the tank roofs, substantial settlements caused by soil liquefaction beneath the tank foundation, and anchor failures. Additionally, damage to the connecting piping systems is one of the leading causes of product loss during seismic events.

Design and analysis guidelines aimed at addressing these failure modes in steel tanks can be found in documents like Eurocode 8 Part 4 [34] and the recommendations from the NZSEE study group [33]. However, there are ongoing debates about the adequacy of current guidelines, particularly regarding the seismic loads acting on the tank walls, which can affect the necessary design parameters for earthquake-resistant structural systems. The design methods found in current international codes trace their origins back to early studies by Jacobsen [35] and Housner [36] on rigid tanks, which were later expanded by Haroun and Housner [37], Veletsos [38], and Malhotra [39] to include the flexibility of tanks. These methods use simplified mechanical analogies to represent the complex phenomena of fluid–structure interaction, based on solutions to the Laplace equation under specific boundary conditions, allowing for the calculation of hydrodynamic pressure distributions induced by seismic forces.

This research focuses on assessing the seismic response of both rigid and flexible cylindrical steel tanks, specifically above-ground tanks that are fully anchored at their base with an open top, which is commonly used for water or other liquid storage. The primary goal of the study is to evaluate the hydrodynamic pressures exerted by the fluid on the tank walls due to seismic excitation, as these pressures are crucial for determining the tank's seismic response. The circumferential stresses in the tank shell are highly influenced by these hydrodynamic forces, and their accurate evaluation is essential. While mechanical models provided by international standards can calculate global response parameters such as shear and overturning moments at the base of the tank, they do not account for the circumferential stresses, which are vital for a comprehensive analysis of the tank's structural integrity. Cylindrical tanks are typically preferred over rectangular ones because of their ability to withstand large circumferential stresses due to their geometry.

This study examines the effects of various parameters, such as the tank's aspect ratio H/R and wall flexibility, on seismic performance. To assess the accuracy of current code provisions for the seismic analysis and design of liquid-containing structures, the research compares numerical results with those proposed by existing design standards. Specifically, three international codes are analyzed and compared: the New Zealand NZSEE recommendations [33], the European code UNI EN 1998-4:2006 [34], and the American Water Works Association AWWA D100-05 standard [40]. However, before applying numerical techniques for design purposes, the accuracy of these methods must be validated against experimental data. For this purpose, a fully nonlinear fluid–structure interaction (FSI) model based on the Arbitrary Lagrangian–Eulerian (ALE) formulation of the liquid-structure finite element method is used. Two existing experimental studies of anchored tanks are utilized for model validation. Fluid motion is governed by the compressible Navier–Stokes equations, and both material and geometric nonlinearities are considered to accurately determine stress, strain, and pressure distributions throughout the tank. The study proposes a rigorous finite element model to capture the three-dimensional behavior of cylindrical tanks, including the sloshing behavior of the contained liquid.

Given that the methods used by the three standards to assess hydrodynamic pressures vary, this research aims to evaluate their reliability by comparing the results obtained from advanced finite element analyses. The findings from this study will contribute valuable insights into the actual behavior of above-ground cylindrical tanks under seismic motions and offer recommendations for improving the accuracy of seismic analysis and design for these structures.

2. Finite Element Modeling, Analysis and Verification

The analysis of the seismic behavior of steel tanks using the finite element method (FEM) involves addressing an FSI problem, which presents several challenges. One of the main difficulties in solving this problem is the need for specialized finite element (FE) formulations capable of handling the large distortions in computational meshes that arise during simulations of dynamic fluid–structure interactions. To address this issue, several FE formulations for FSI problems are discussed and experimentally validated for use in this research. Particular emphasis is placed on the ALE method, as it is widely regarded as the most effective approach for solving FSI problems in fluid–structure simulations. The ALE method, which allows for the handling of mesh deformations, was specifically chosen for the analyses conducted in this study due to its ability to accurately model the behavior of both the fluid and the tank structure under seismic loads. This method ensures that the evolving geometry of the system during seismic excitation is properly captured, allowing for more accurate predictions of the tank's response to seismic events.

In this research, the general-purpose FE code LS-DYNA is employed to study the FSI problem of deformable steel tanks, which are completely filled with fluid and subjected to seismic loading at the base. The interactions between the fluid and the structure are modeled using the ALE method, which is well suited for handling large deformations in both the fluid and the structure. This method enables an accurate representation of the fluid's behavior and the tank's deformation under seismic excitation. The use of the LS-DYNA FE code allows for the incorporation of both material and geometric nonlinearities, providing a more realistic simulation of the tank's dynamic response. Additionally, LS-DYNA performs dynamic analyses using an explicit time integration scheme based on the central difference method, which is particularly effective for simulating high-speed dynamic events like seismic loading. This combination of advanced modeling techniques and robust numerical methods ensures a comprehensive and accurate assessment of the seismic response of steel tanks under extreme conditions.

Numerical techniques for simulating tank–fluid interaction problems are highly effective tools for evaluating the seismic response of these systems, as they can significantly reduce the need for time-consuming, costly experimental tests, which are typically conducted under specific boundary and excitation conditions. Once validated against experimental data, numerical simulations can serve as reliable design tools to assess various system configurations subjected to different earthquake ground motions. In this study, experimental tests are used as reference solutions for validating the tank models then used in the parametric evaluation of the resilience.

Before developing the FEMs for the case study tanks, the chosen modeling approach is verified by comparing the results of FE analyses with experimental data obtained from two separate tests. The first test is the pioneering work by Clough [41], which involved a small-scale tank model. As a continuation of this research, Manos and Clough [42] conducted a series of experiments on the same model subjected to three different scaled earthquake motions with varying boundary and soil conditions. This study focuses solely on a model tank with an open top, anchored support, and a rigid foundation. The second experimental test considered is presented in Haroun [43], which describes a shake table test performed on an open-top cylindrical tank. Haroun's work also includes numerical results, allowing for a direct comparison between the experimentally observed behavior and the predictions obtained from an LS-DYNA model developed according to the previously described modeling approach.

The comparison is based on several key parameters, including the maximum radial displacement at the top of the tank shell, the maximum meridional force observed in the shell, the maximum base shear, pressure distributions, and the sloshing wave height. By comparing these parameters, the study aims to assess the accuracy and reliability of the numerical model in replicating the observed behavior from the experimental tests, thus validating the chosen modeling approach for simulating the seismic response of steel tanks then used in the parametric evaluation.

2.1. FEM of Manos and Clough Experimental Tank [42]

The tank model used in the experiments was a 1/3 scale model of a steel prototype with a radius of 1.83 m and a total height of 1.83 m. The tank was filled with water (density of 1000 kg/m^3) to a height of 1.53 m. The tank was constructed from aluminum with a density of 2700 kg/m^3, an elastic modulus of 71,000 MPa, and a yield stress of 100 MPa. The thickness of the aluminum bottom plate and the shell section near the bottom was 0.002 m, while the second shell course had a thickness of 0.0013 m. A steel wind girder, shaped in an L configuration, was placed on top of the second shell course. The geometry of the tank is shown in Figure 1a.

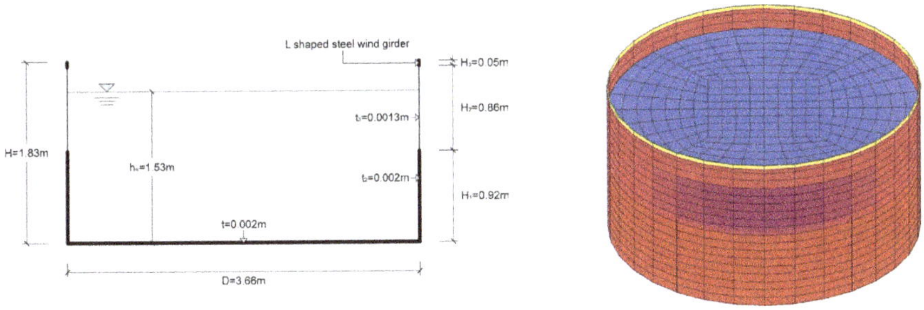

(a) (b)

Figure 1. Manos and Clough experimental tank dimensions (**a**); and finite element model (**b**).

In the experiment, the shaking table motion was derived from the horizontal component of the El Centro 1940 earthquake with a peak acceleration of 0.50 g. The motion was scaled with respect to time by a factor of $1/\sqrt{3}$ to meet similitude requirements. For the numerical verifications, the same tank model with an open top, anchored support, and resting on a rigid foundation was used. The numerical model retained the same dimensions and material properties as the experimental tank. During model development, the effect of element size and formulation, for both the tank and the fluid, was carefully investigated and adjusted to accurately capture the fluid–structure interaction. The tank was discretized using four-noded, fully integrated shell elements with two integration points through the thickness. The resulting finite element model consisted of 5793 nodes, 890 shell elements, and 4320 solid elements, as shown in Figure 1b. An explicit nonlinear transient analysis was performed to replicate the experimental test. The time histories of pressure (excluding hydrostatic pressure) at three specific locations (at R = 1.83 m with heights of 0.05 m, 0.45 m, and 0.86 m above the tank base) during the experiment were presented in Figure 2. These results were used to verify the numerical model and compare the simulated pressures with the experimentally observed values. Pressure is a key parameter in the study of storage tanks because it plays a crucial role in evaluating the dynamic response of the tank to seismic loading. Figure 2 presents a comparison of the pressure distribution at three different levels along the height of the tank: at the base, one-third of the height, and approximately half the height. This comparison is crucial, as it captures the pressure variation along the tank wall during seismic loading, offering valuable insights into the fluid–structure interaction over time. By showing the overlap between the experimental results and the numerically evaluated data, the figure clearly demonstrates that the numerical model successfully replicates the actual behavior of the fluid inside the tank. The pressure profiles at these three heights highlight the nonlinear behavior of the fluid's displacement. The significant variation in pressure over time, especially at higher fluid levels, emphasizes the complex dynamics involved in fluid motion during seismic events. The agreement between the experimental and numerical results confirms that the model effectively captures the real-world fluid behavior. This validation is key to understanding the dynamic response of the tank under seismic loading and supports the accuracy of the proposed numerical methods. In summary, Figure 2 summarizes these findings, illustrating the alignment between experimental and numerical data across different fluid heights, and reinforces the robustness of the methodology used in this research.

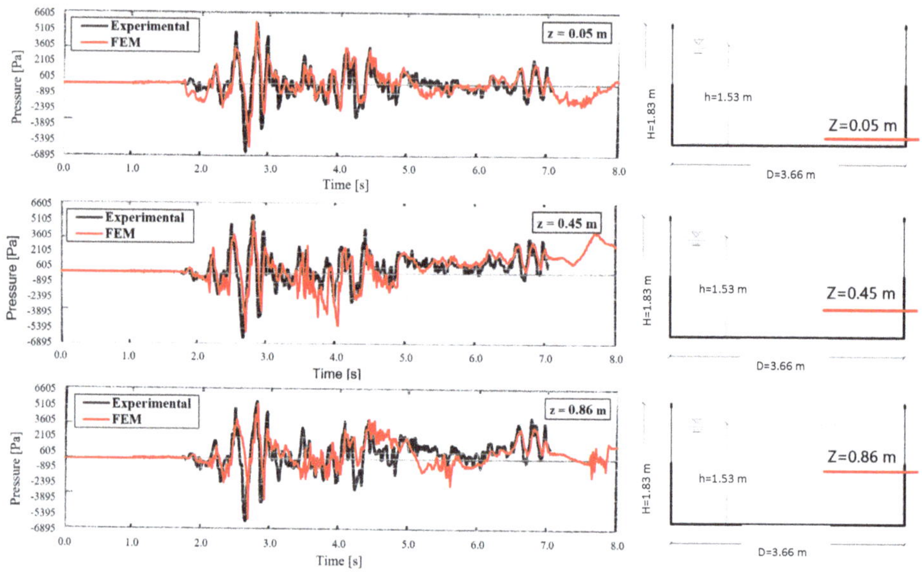

Figure 2. Comparison of the pressure time histories for the experimental and numerical method results.

2.2. *FEM of Haroun Experimental Tank [43]*

The cylindrical tank tested by Haroun [43] was an open-top, anchored support tank filled with water. The shell was made of aluminum with a Young's modulus of E = 68,940 MPa and a density of 2600 kg/m^3. The tank had a radius of 1.18 m and the height of the aluminum shell was 4.57 m, while the height of the water's free surface was 3.96 m. The shell thickness varied with height: up to 3 m, the thickness was 2.3 mm, and above this, the thickness was reduced to 1.6 mm. The shake table test was conducted using a record from the 1994 El Centro earthquake. To account for the scaling of the model, the record was sped up by a factor of 1.73 and scaled by 1.43 to achieve a peak ground acceleration (PGA) of 0.5 g. The FEM was developed using the previously outlined approach. The tank was discretized with 1236 shell elements for the base plate and the wall, while the fluid was modeled using 6084 solid elements with the ALE formulation. The final model, depicted in Figure 3a, consists of 8009 nodes. The discretization was chosen to balance the accuracy of the results with computational efficiency.

An explicit nonlinear transient analysis was performed to replicate the experimental test. The results from the LS-DYNA FE model were compared with the experimental data from Haroun [43]. The comparison of the results is presented in Table 1, and the deformed shape of the model at different times during the transient analysis is shown in Figure 3b. The maximum meridional compression, measured 20 cm above the tank's base plate, was found to be nearly identical in both the experimental test and the FE model with a difference of less than 2%. The maximum radial displacement and base shear were also closely matched with differences between the numerical and experimental results consistently below 3%. These findings suggest that the modeling approach used in this study is sufficiently reliable for simulating the seismic behavior of cylindrical tanks under similar conditions.

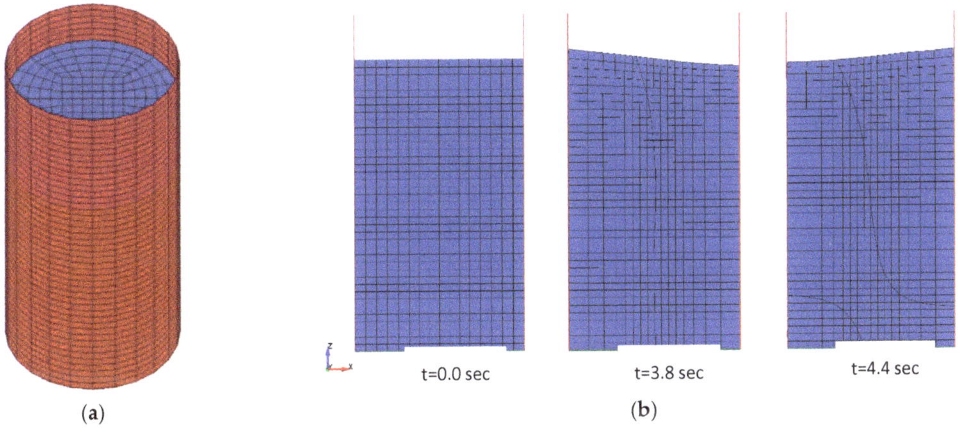

Figure 3. Haroun experimental tank FEM (**a**); and deformed shapes at different time steps (**b**).

Table 1. Comparison of experimental and numerical results.

Parameter	Experimental	FEM	[%]
Max radial disp at shell top [mm]	3.32	3.40	2.41
Max meridional compression [N/m]	63,396	62,205	1.88

3. Case Studies, Material, Geometry and FEM

The case study structures analyzed in this research are vertical cylindrical tanks with both flexible and rigid walls, which were all perfectly anchored at their base. To ensure a comprehensive study, ten different geometric configurations have been considered. The slenderness ratios (H/R) for the case study steel tanks were selected to represent both squat tanks (with $0.5 < H/R < 1.5$) and slender tanks (with $1.5 < H/R < 6$), allowing for an analysis of a wide range of tank geometries. The radius-to-thickness (R/t) ratio was also varied to investigate the effects of both rigid walls (with R/t close to 100) and flexible walls (with R/t close to 2000). Table 2 provides a summary of all the geometric configurations in terms of the H/R and R/t ratios considered for the FE analysis

Table 2. Case studies: geometrical configurations in terms of H/R and R/t ratios.

H/R	0.50	0.75	1.00	1.50	2.00	2.50	3.00	4.00	5.00	6.00
R/t	100	125	150	250	400	600	800	1000	1500	2000

The ten case study tanks were designed to contain water with a fixed radius (R) of 1.83 m. The height (H) of the tanks ranged from 0.91 m to 10.97 m, covering the full range of slenderness ratios ($0.5 < H/R < 6.0$). The tanks were made of aluminum with a density of 2700 kg/m^3, an elastic modulus of 71,000 MPa, and a yield stress of 100 MPa. The shell thickness (t) of the tanks varied between 0.91 mm and 18.29 mm and was kept constant along the entire height of the tank. To avoid unrealistic geometrical configurations in terms of the R/t ratio, preliminary designs were performed, and only significant R/t ratios were selected for each H/R case study tank (see Table 3). Figure 4 illustrates the FE models used for the geometrically and materially nonlinear analyses, which include fluid–structure interaction capabilities in LS-DYNA. These models were designed to simulate the full range of seismic behaviors and provide insight into the dynamic response of the tanks under earthquake loading. In the proposed FE approach, the motion of the fluid was described using compressible Navier–Stokes equations, which were implemented through a mesh of eight-noded, one-point integration ALE solid elements. For modeling the tank wall, four-noded, two-point integration shell elements were selected, as described in [44]. This type of shell element provides a computationally efficient solution, as it is based on a combined co-rotational and velocity–strain formulation. To model the wall's material behavior, a bilinear elastic–plastic constitutive law (MAT_003) was employed, incorporating both isotropic and kinematic hardening. This approach was used to capture the cyclic permanent deformations exhibited by the tank wall under seismic excitation. In this study, null material was used to simulate the fluid, which is conservatively assumed to be water. Consequently, the bulk modulus of the fluid was assumed to be 2.25×10^9 N/m^2. The resulting FE models, based on different tank geometries, vary in complexity, with the number of nodes ranging from 6162 to 25,034, depending on the specific tank configuration being analyzed. This range reflects the model's ability to handle different tank sizes and geometries while maintaining computational efficiency.

Table 3. Main design properties and modelling dimensions of the case studies tanks.

		Thickness Wall, *t* [m]										
	H/R	**0.50**	**0.75**	**1.00**	**1.50**	**2.00**	**2.50**	**3.00**	**4.00**	**5.00**	**6.00**	
R/t	**100**	-	-	-	-	-	-	-	-	-	0.018288	**Rigid**
	125	-	-	-	-	-	-	-	-	0.014630	0.014630	↓
	150	-	-	-	-	-	-	-	0.012192	0.012192	0.012192	
	250	-	-	-	-	-	0.007315	0.007315	0.007315	0.007315	0.007315	
	400	-	-	0.004572	0.004572	0.004572	0.004572	0.004572	0.004572	0.004572	0.004572	
	600	0.003048	0.003048	-	0.003048	0.003048	0.003048	0.003048	0.003048	0.003048	-	
	800	0.002286	0.002286	0.002286	0.002286	0.002286	0.002286	0.002286	0.002286	-	-	
	1000	0.001829	0.001829	0.001829	0.001829	0.001829	0.001829	0.001829	-	-	-	
	1500	0.001219	0.001219	0.001219	0.001219	0.001219	0.001219	0.001219	-	-	-	
	2000	0.000914	0.000914	0.000914	0.000914	0.000914	-	-	-	-	-	**Flex.**
Water height, *H* [m]		0.9144	1.3716	1.8288	2.7432	3.6576	4.5720	5.4864	7.3152	9.1440	10.9728	
N° of shells		16	18	20	24	30	36	40	50	60	72	
Height of shell, *l* [m]		0.057150	0.076200	0.091440	0.114300	0.121920	0.127000	0.137160	0.146304	0.152400	0.152400	
Wall length, *L* [m]		1.20015	1.75260	2.28600	3.31470	4.26720	5.20700	6.17220	8.04672	9.90600	11.73480	
Modulus of stiffening ring, *Z* [cm^3]		0.7196	1.0794	1.4392	2.1587	2.8783	3.5979	4.3175	5.7566	7.1958	8.6350	
Thickness ring, *s* [m]		0.009	0.009	0.010	0.011	0.012	0.013	0.014	0.015	0.017	0.018	

Before conducting a series of case study analyses, the acceleration history of the input motion used in this research was the same as that employed by Manos and Clough [42] for their experimental test. The acceleration history and response spectrum of the input motion, used in both the experimental tests and the numerical simulations of the case studies, are provided in Figures 5a and 5b, respectively. These figures illustrate the time history of the seismic input and its corresponding response spectrum, which serve as the foundation for analyzing the seismic behavior of the tanks in the subsequent case study simulations.

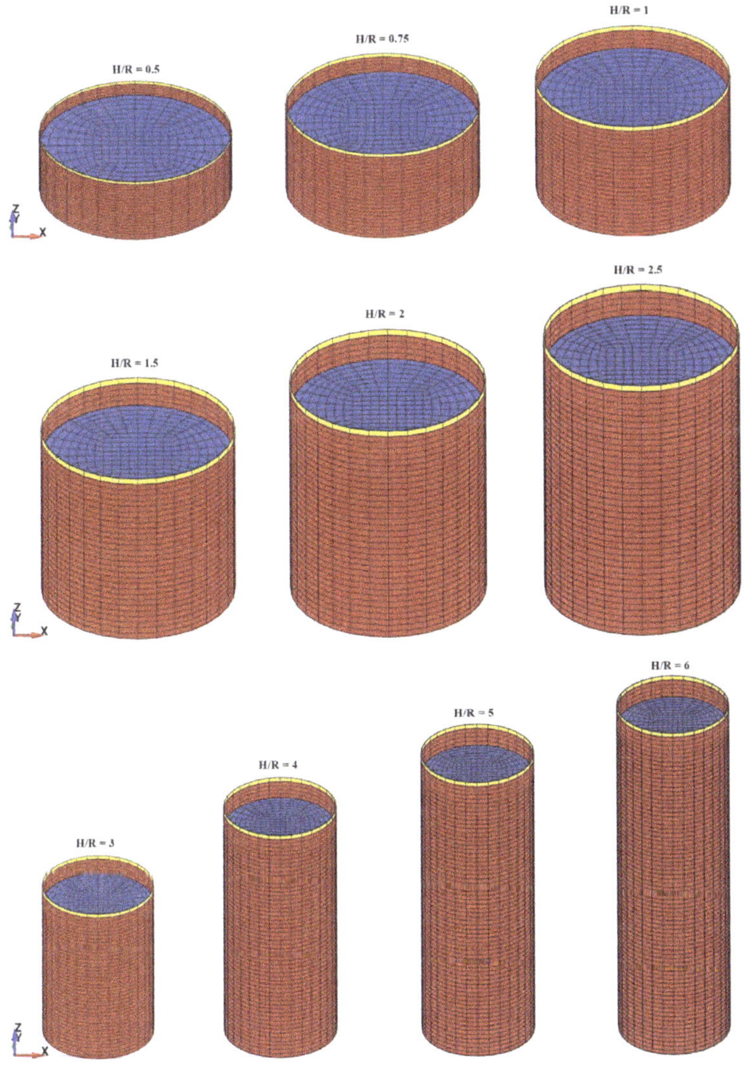

Figure 4. Case studies tanks with slenderness ratio (H/R) from 0.5 to 6.

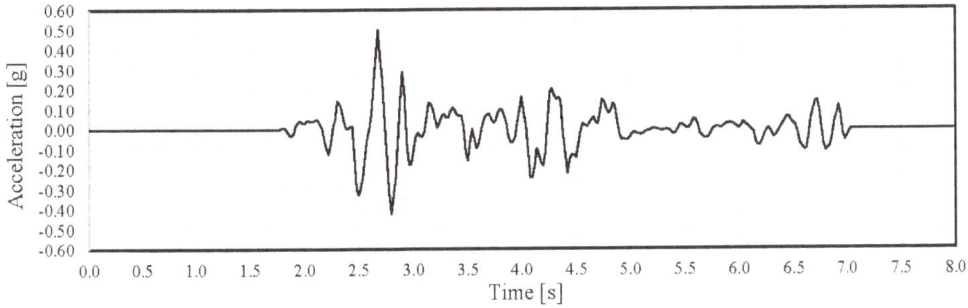

(**a**)

Figure 5. *Cont.*

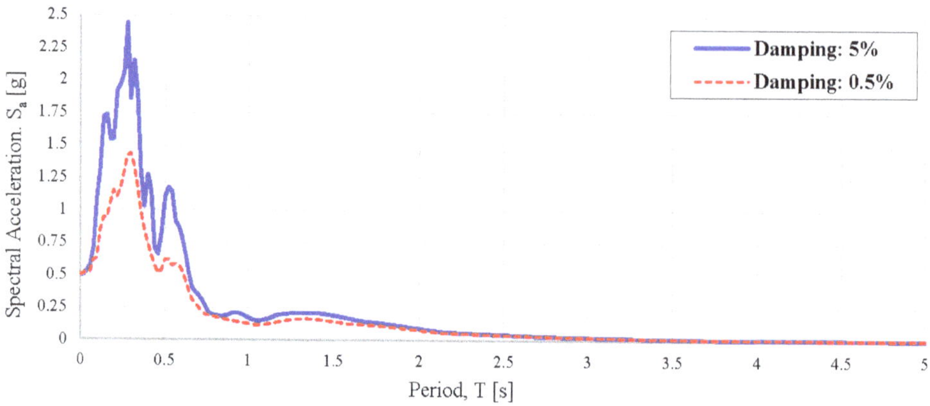

(**b**)

Figure 5. (**a**) Input motion considered for all case studies analysis; and (**b**) response spectral acceleration of input motion for 5% damping and 0.5% damping.

4. Discussions on FE Results and Parametric Sensibility

To study the effect of rigid and flexible wall tanks, the time history responses of the tank models were calculated, and a comparison was made between the obtained results. All the FEMs described earlier have been used to perform explicit nonlinear analyses with the seismic excitation applied in the form of the acceleration time history that was used for the experimental test by Manos and Clough [42]. The resulting hydrodynamic pressure profiles for the ten different tank configurations, characterized by various H/R slenderness ratios, are presented and were thoroughly analyzed. The envelopes of these pressure profiles provided a detailed understanding of how different geometric configurations influence the seismic response of the tanks particularly in terms of the distribution and magnitude of hydrodynamic pressures during the earthquake simulation. This analysis will help to assess the performance of both rigid and flexible wall tanks under seismic loading and highlight key differences between the two types of tank behavior.

We completed the analysis but decided to present the results for slenderness ratios of 0.5, 0.75, and 1.5, as 1.5 represents a transition between squat and slender tanks. We excluded 1.0 because it follows the trend of 0.75. Additionally, we focused on 3.0 and 6.0 while avoiding the intermediate ratios of 2.0, 2.5, 4.0, and 5.0. This approach provides a clear representation of the transition and extreme behaviors of squat and slender tanks, with both flexible and rigid walls, covering a wide range of thickness ratios.

4.1. Tank H/R = 0.5

To study the effect of both rigid and flexible wall tanks, the trend of pressures exerted by the fluid on the tank walls has been thoroughly investigated. The seismic response, expressed in terms of total wall pressure, of a tank with an H/R ratio of 0.5, obtained from numerical analyses, is compared with the design predictions of the AWWA [40], EC8-4 [34], and NZSEE [33] standards. These standards use the input motion response spectrum from Figure 5 and apply the combination rule for impulsive and convective pressure components specified in each respective code. Figure 6a shows the pressure profiles normalized by the tank height (z/H) for five different configurations of the R/t ratio, while Figure 6b compares the predictive expressions for impulsive and convective pressure distributions with the results of the FE simulations. As observed in Figure 6a, the pressure profiles exhibit a similar trend for all five different R/t ratios investigated. Pressure peaks occurred at the top of the wall due to the convective waves both for rigid and flexible wall tanks. These peaks were generally similar in magnitude, except for the rigid tank (R/t = 600), where the peak was smaller, occurring at approximately 0.9 normalized height (z/H). In Figure 6b, a comparison between the numerical results and the standard codes is presented. Generally, the results obtained from the FE analyses align well with the analytical predictions in the lower part of the tank wall. As expected from the preliminary analyses performed to calibrate the FE model, the impact of convective waves on the tank shell caused a pressure peak at the top of the wall, which was not captured by the design approaches proposed by the three regulations considered in this study. Under the assumption of flexible circular tanks, EC8-4 predicted the highest values for the pressure, as it recommended directly summing the impulsive and convective effects. On the other hand, when assuming rigid circular tanks, EC8-4 underestimated the pressure profiles compared with the FE simulations. The NZSEE standard provided more or less the same pressure values along the tank wall for both rigid and flexible tanks. However, the AWWA code tended to underestimate the pressure profiles compared to the FE results.

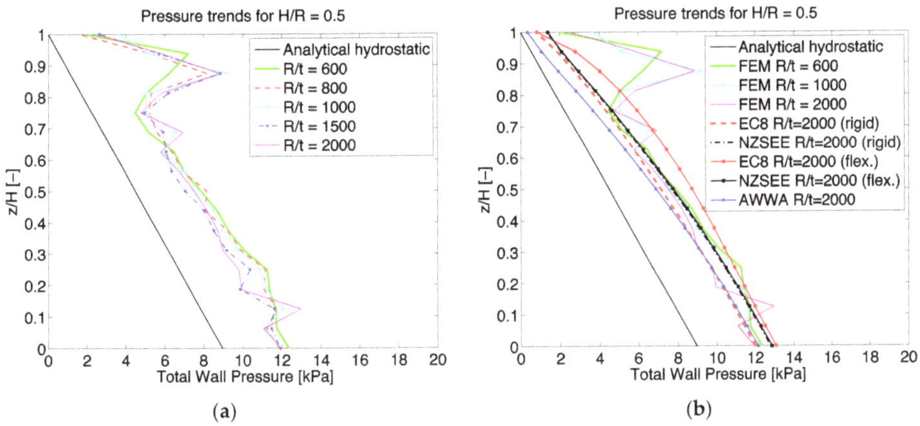

Figure 6. Horizontal pressure distribution of H/R = 0.5 tank ratio considering different R/t thickness ratios: (**a**) FE simulations and (**b**) predictive pressures computed according the international standards compared with FE simulations.

These observations highlight the limitations of current design codes in accurately predicting the pressure distribution for tanks subjected to seismic loading, especially at the top of the wall where convective waves play a significant role.

4.2. Tank H/R = 0.75

As observed in Figure 7a, the pressure profiles exhibit a similar trend across the five different R/t ratios investigated. Significant peaks of pressure occurred at the top of the wall due to convective waves both for rigid and flexible wall tanks. The largest peak occurred for the rigid tank (R/t = 600), at approximately 0.9 normalized height (z/H), and this peak decreased as the flexibility of the wall increased, reaching its lowest value for the flexible tank (R/t = 2000). In the lower part of the tank, the pressure profiles remained relatively uniform, showing no significant differences across the different R/t ratios. Figure 7b presents a comparison between the numerical results and the pressure distribution recommendations from the various standards along the wall. In general, the results obtained from the FE analyses aligned well with the analytical predictions in the lower part of the tank wall. However, as in previous cases, the impact of convective waves on the tank shell caused a pressure peak at the top of the wall that cannot be predicted by the design codes. Under the assumption of flexible circular tanks, EC8-4 predicted the highest pressure values, while for rigid circular tanks, EC8-4 underestimated the pressure profiles compared to the FE simulations. The NZSEE standard resulted in similar pressure values along the wall for both rigid and flexible tanks. In contrast, the AWWA code consistently underestimated the pressure profiles when compared to the FE results. These findings further emphasize the limitations of existing design codes, particularly in capturing the effect of convective waves at the top of the tank wall, which is critical for accurately predicting the seismic response of tanks.

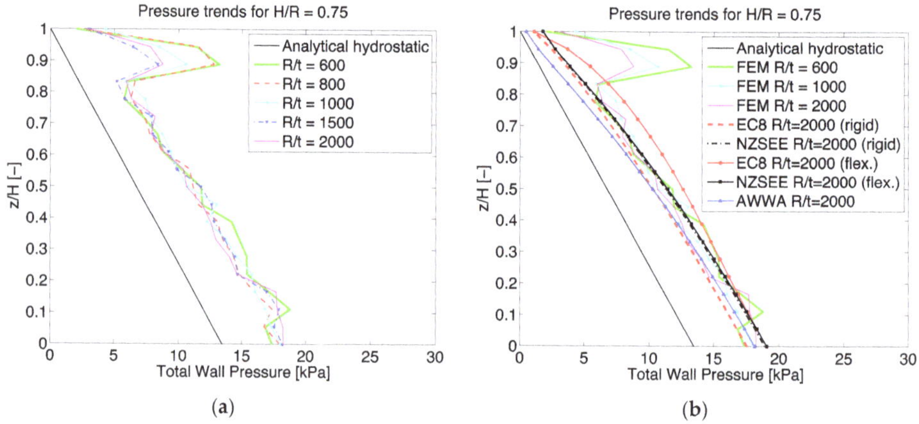

Figure 7. Horizontal pressure distribution of $H/R = 0.75$ tank ratio considering different R/t thickness ratios: (**a**) FE simulations and (**b**) predictive pressures computed according the international standards compared with FE simulations.

4.3. Tank H/R = 1.5

As observed in Figure 8a, the pressure profiles for the five different R/t ratios show a similar trend. The rigid tank displayed a very small peak of pressure at the top of the wall, around 0.9 normalized height (z/H), while the flexible tanks exhibited a much smoother profile with no noticeable pressure peaks. The results from the FE analyses are in good agreement with the analytical predictions in both the bottom and upper parts of the wall. In this specific case, compared to previous configurations where pressure peaks were absent in the upper part of the tank, the pressure profiles were very well predicted by the three design standard codes. Only the AWWA and EC8-4 standards, when applied to the rigid tank assumption, slightly underestimated the FE results, but the discrepancy was not significant. Looking at Figure 8b, it is immediately apparent that the pressure peaks observed in earlier analyses at the top of the shell were not present for this case study. This is because the convective contribution was considerably smaller in this case, which was due to the different slenderness ratio of the containment structure. The tank with an $H/R = 1.5$ ratio represents the transition point between squat tanks and slender tanks. It is important to note that despite the filling height of the tank being limited to 80% of the shell height, the FE analyses recorded pressures above the filling height as well. This is due to the free movement of the fluid inside the tank. When fluid moves vertically as part of the sloshing motion, it exerts horizontal pressures above the hydrostatic free surface.

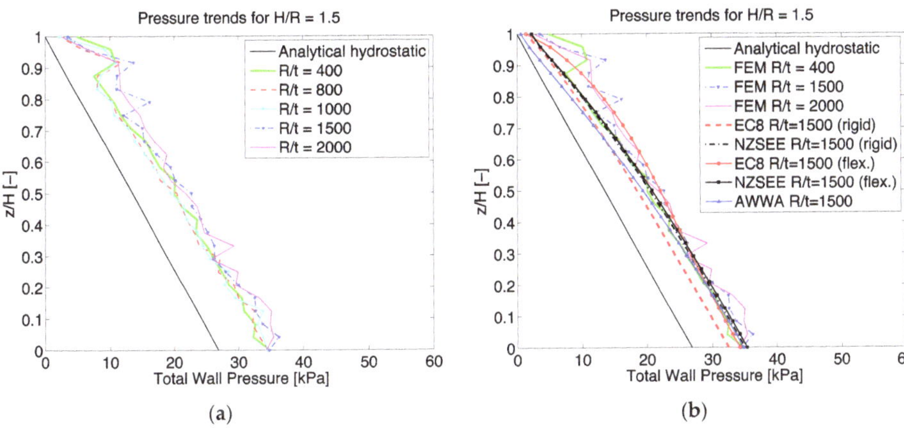

Figure 8. Horizontal pressure distribution of $H/R = 1.5$ tank ratio considering different R/t thickness ratios: (**a**) FE simulations and (**b**) predictive pressures computed according the international standards compared with FE simulations.

This phenomenon highlights the complexity of fluid behavior in seismic conditions particularly when sloshing effects contribute to pressure distributions above the typical water level in the tank. The results underscore the importance of accurately modeling fluid dynamics, especially for tanks with different slenderness ratios, to capture the full range of pressures exerted during seismic excitation.

4.4. Tank H/R = 3.0

As observed in Figure 9a, the pressure profiles for the five different R/t ratios show a similar trend. As in previous cases, significant pressure peaks occur at the top of the wall due to convective waves both for rigid and flexible wall tanks. The largest peak occurred for the flexible tank ($R/t = 1000$) at approximately 0.9 normalized height (z/H), and this peak decreased as the rigidity of the wall increased, reaching its lowest point for the rigid tank ($R/t = 250$—Figure 9b). In the lower part of the tank, the pressure profiles remained relatively consistent, except for the most flexible wall ($R/t = 1000$), which exhibited very large peaks due to the low resistance of the shell to horizontal excitations. It is evident that the $R/t = 1000$ tank configuration failed through the classical elastic–plastic elephant foot buckling mechanism just above the base of the wall, highlighting the vulnerability of highly flexible tanks under seismic loading. The predictions made using the international regulations (EC8-4, NZSEE, and AWWA) consistently underestimate the envelope of pressures observed from the FE analyses. While the small discrepancies in the lower part of the tank can be accounted for in design, the significant underestimation in the upper portion of the tank, caused by the convective pressure component, may lead to an unsafe design, particularly if a substantial reduction in shell thickness is implemented along the tank height. Finally, it is important to note that the contribution of sloshing motion increased clearly as the radius-to-thickness ratio (R/t) rose. However, current analytical proposals are not fully capable of accounting for this behavior, which is observed to become more pronounced as the flexibility of the tank increases. This underscores the need for a more accurate prediction of fluid–structure interaction in tanks particularly in cases involving highly flexible structures that are more susceptible to large convective forces.

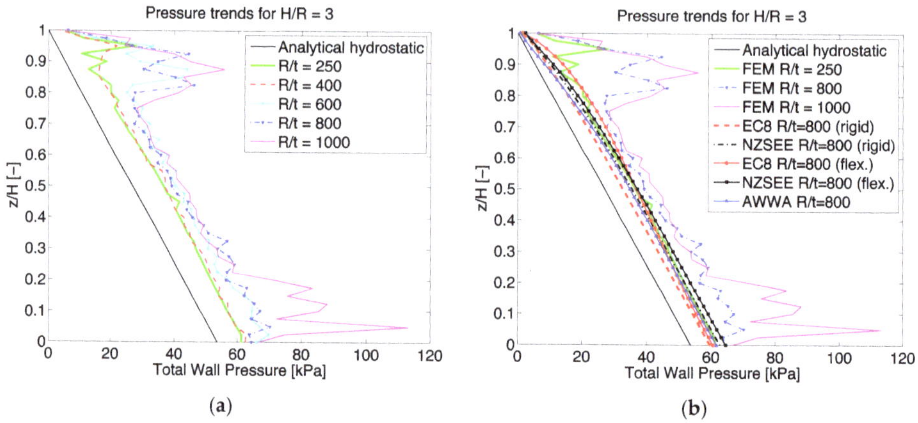

Figure 9. Horizontal pressure distribution of H/R = 3.0 tank ratio considering different R/t thickness ratios: (**a**) FE simulations and (**b**) predictive pressures computed according the international standards compared with FE simulations.

4.5. Tank H/R = 6.0

As observed in Figure 10a, the pressure profiles for the five different R/t ratios show a similar trend. For the more slender tanks, large pressure peaks occurred at the top of the wall due to convective motion waves both for rigid and flexible wall tanks. The largest peak occurred for the flexible tank (R/t = 250) at approximately 0.9 normalized height (z/H), and this peak decreased as the rigidity of the wall increased, reaching its lowest point at (R/t = 125). In the lower part of the tank, the pressure profiles remained relatively regular, except for the most flexible wall (R/t = 400), which exhibited very large peaks due to the low resistance of the shell to horizontal excitations. It is clear that the R/t = 400 configuration tank failed through the classical elastic–plastic elephant foot buckling mechanism just above the base of the wall.

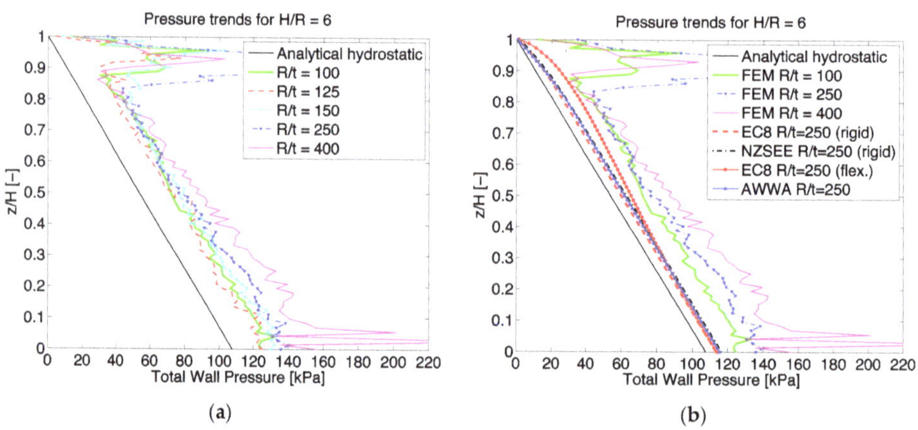

Figure 10. Horizontal pressure distribution of H/R = 6.0 tank ratio considering different R/t thickness ratios: (**a**) FE simulations and (**b**) predictive pressures computed according the international standards compared with FE simulations.

This case behaves similarly to the H/R = 5 case study, and the comparison between the numerical results and standard codes is shown in Figure 10b. As in previous analyses, the predictions made using international regulations underestimated the envelope of pressures observed in the FE analyses. These results further demonstrate the limitations of current design codes in accurately predicting the pressure distribution especially in cases where large convective waves and buckling mechanisms significantly affect the tank's behavior under seismic loading.

5. Numerical Results Considering a Set of Ten Scaled Recorded Accelerograms

To better understand how pressure trends may vary, a series of explicit dynamic nonlinear analyses were performed using a set of different accelerograms as input ground shaking motions. The dynamic response of the case study tanks was assessed by considering a set of ten scaled recorded accelerograms. These records were selected from the PEER NGA database, with magnitudes ranging from 6.2 to 7.6, ensuring compatibility with the EC8 design response spectrum. The reference response spectrum was defined with a peak ground acceleration (PGA) of 0.4 g and a soil category C, according to the European classification. Figure 11 shows the response spectra for the ten accelerograms used in the dynamic response analysis of the case study tanks, and it is complemented by Table 4, which provides details of the seismic input for nonlinear dynamic simulations. More specifically, PEER ID, event, component, moment magnitude (M_W), closest distance from recording site to fault area (D_r), duration (t_{tot}), and shear wave velocity over the first 30 m (V_s) are shown in Table 4. The matching between the mean response spectrum of the ten unscaled records and the EC8 design spectrum was verified for periods between 0.15 s and 4 s. This range is quite large because sloshing modes and tank impulsive modes have significantly different fundamental periods [45,46].

Table 4. Characteristics of ground motion records for nonlinear dynamic analyses.

Eq. ID	PEER ID	Event	Component	M_W [-]	D_r [km]	t_{tot} [s]	V_s [m/s]
01	1233	Chi-Chi, Taiwan	E	7.62	36	90	194
02	1153	Kocaeli	090	7.51	127	102	275
03	851	Landers	000	7.28	157	70	272
04	1810	Hector	090	7.13	92	60	345
05	1629	St Elias, Alaska	279	7.54	80	83	275
06	777	Loma Prieta	090	6.93	28	39	199
07	1043	Northridge-01	090	6.69	52	48	309
08	728	Superstition Hills-02	180	6.54	13	40	194
09	172	Imperial Valley-06	140	6.53	22	39	237
10	2615	Chi-Chi, Taiwan-03	N	6.20	40	107	273

For the analysis, three case studies were selected to represent squat and slender tanks, specifically with H/R = 0.75, 1.5, and 4 slenderness ratios. Figure 12a shows the pressure distributions for the squat tank (H/R = 0.75 and R/t = 2000) obtained from the FE simulations. A comparison between the results using shaking table excitations and those using the average of the accelerograms indicates that the pressure distributions for the average earthquakes overestimated the trends of pressures observed from the shaking table excitations. The discrepancies were small and are primarily observed at the bottom and upper parts of the tank, which were influenced by the severity of the selected accelerograms, especially for the "LC9" earthquake.

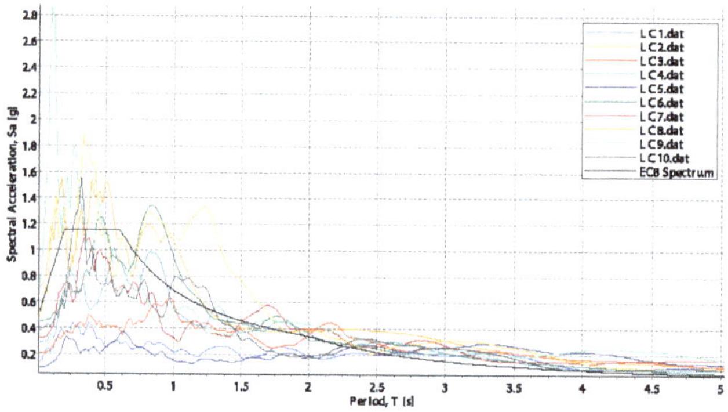

Figure 11. Comparison of the design acceleration response spectra with the spectra of ten real records selected and uniformly scaled to be spectrum compatible.

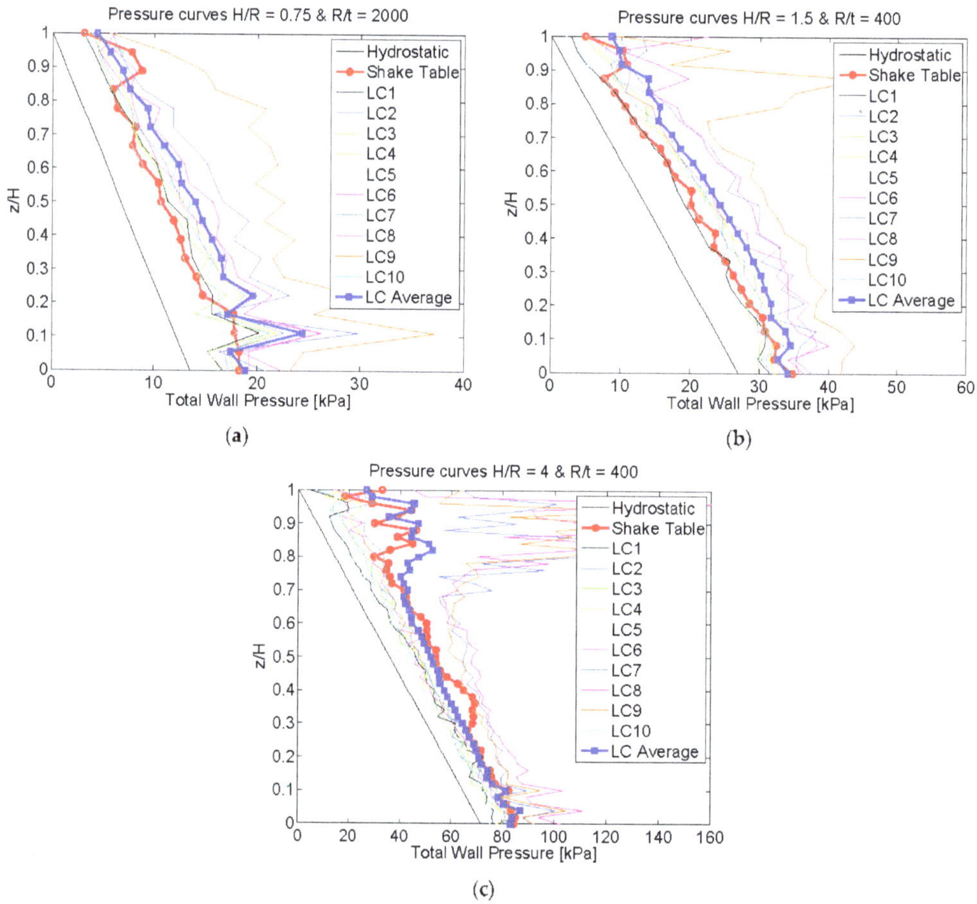

Figure 12. Horizontal pressure curves for $H/R = 0.75$ and $R/t = 2000$ (**a**), $H/R = 1.5$ and $R/t = 400$ (**b**), and $H/R = 4$ and $R/t = 400$ (**c**) tank considering a set of ten scaled recorded accelerograms and seismic excitations of shaking table.

The same accelerograms were applied to the $H/R = 1.5$ and $R/t = 400$ case study tank. Figure 12b shows the comparison of the total pressure profiles obtained from the shaking table excitations and the average of the ten accelerograms. Again, the pressures from the average earthquakes overestimated the trend of pressures obtained from the shaking table with the impact of the sloshing wave being mostly negligible. The "LC9" earthquake, however, had the most significant effect due to its pronounced sloshing wave impact.

For the slender tank case study ($H/R = 4$ and $R/t = 400$), the comparison of pressure profiles is shown in Figure 12c. In this case, the total pressure profile from the average earthquakes closely matched the profile obtained using the accelerogram from the shaking table. For slender tanks, it is important to note that the effect of sloshing wave impact is well captured by both input motions, whether using the accelerogram from the shaking table or the average of the set of accelerograms.

These findings demonstrate that for squat tanks, the average accelerograms tend to overestimate the pressures observed from shaking table excitations, while for slender tanks, both input motions accurately capture the sloshing wave effects, resulting in almost identical pressure profiles [47].

6. Conclusions

To determine the transition zone between rigid and flexible tank idealizations, the trend of hydrodynamic pressures exerted by the fluid on the tank wall during an earthquake has been extensively studied. Specifically, the seismic response of open-top steel tanks, both with rigid and deformable walls, fully anchored at the base, is analyzed. Understanding hydrodynamic pressures is critical because they directly influence the circumferential stresses in the steel shell, which can lead to shell buckling. The importance of selecting a rigid or flexible tank idealization lies in the potential for under- or over-estimating hydrodynamic pressures compared to actual values, which impacts the accuracy of the seismic analysis.

After calibrating the model with two experimental tests, 50 transient explicit analyses were performed using LS-DYNA software ver. 9 to generate the envelopes of hydrodynamic pressure profiles for ten different configurations of H/R slenderness ratios and five configurations of R/t ratios. The seismic responses obtained from the numerical analyses were then compared with the design predictions from the AWWA, EC8-4, and NZSEE standards. Given the differences in assumptions regarding rigid or flexible tanks and the complexity of the design methodologies in the international standards, the goal of this research was to investigate the prediction of hydrodynamic pressures to provide more accurate seismic analysis and design recommendations for above-ground cylindrical tanks. Key conclusions from the study are outlined below:

1. NZSEE Method: This method is relatively complex as it requires the use of advanced mathematical tools such as Bessel functions and the solution of "heavy" equations. While initially developed for rigid tanks, the NZSEE method was extended to flexible tanks with minimal modification, resulting in discrepancies between predicted and observed responses.
2. EC8-4 Method: This method is the only one that accounts for the influence of tank wall deformability. However, it is highly complex, as it requires an iterative procedure, and FE analyses are always necessary for accurate results.
3. AWWA Method: This is the simplest of the three methods but does not account for the flexibility of the tank wall. Additionally, it is based on an admissible stress design approach, and the choice of the force reduction factor used in the calculations remains debatable, as no unanimous consensus can be gathered on it.
4. The comparison of the methods proposed in the three standards reveals that differences between the results can be quite large when considering the various contributions. However, these differences become less significant when evaluating the

total pressures. This is due to the fact that the amplitude of hydrodynamic pressures is generally lower than that of the hydrostatic pressures. Specifically, for slender tanks, the influence of hydrodynamic pressures is sufficiently diminished. While variations of around 10% were observed between the total pressure profiles computed using the different methods, variations in the order of 25–30% were found in the impulsive contributions.

5. The previous comments, along with the subsequent considerations regarding the parametric sensitivities and the various ratios involved in the analyses, such as radius, height, thickness, as well as rigid vs. flexible and slender vs. squat configurations, lead to the most important result summarized in Figure 13. This figure provides guidance on the selection of flexible versus rigid tanks, depending on the H/R (height-to-radius) and R/t (radius-to-thickness) ratios. The figure offers valuable insights into how the tank's geometric properties influence the decision to adopt a flexible or rigid model for seismic design, helping to optimize safety and performance based on the tank's specific characteristics.
6. Explicit dynamic nonlinear analyses were conducted to accurately reproduce the FSI problem. These analyses were performed using the LS-DYNA software, treating the FSI problem according to the ALE formulation. The modeling approach was verified by comparing the results from a shake table test performed by other researchers with the numerical results obtained from the model developed in this study. The comparison demonstrated that the chosen modeling approach provides a sufficiently accurate prediction of the tank's seismic response, making it a reliable reference for future research in this area.
7. Nonlinear transient analyses were performed to assess the seismic response of ten case study tanks, which were designed to represent both squat and slender configurations. Each tank was analyzed using the seismic excitation corresponding to the acceleration time history from the experimental test by Manos and Clough [42]. The pressure profiles obtained from these analyses exhibited similar trends across the five different R/t ratios investigated. A noticeable peak in pressures occurred at the top of the wall for both rigid and flexible wall tanks: around 0.9 normalized height (z/H). However, the $H/R = 1.5$ case study showed a smaller peak due to a significantly reduced convective contribution, which is attributed to the different slenderness of the containment structure. This peak has been shown to be related to the sloshing motion, specifically the impact of the convective wave on the tank wall, which generates the pressure peak. The $H/R = 1.5$ case represents the transition point between squat and slender tanks. Based on the analysis of all the cases studied, the following conclusions can be drawn:

    a. For squat tanks ($0.5 < H/R < 1.5$), the largest pressure peak occurred in rigid tanks and decreased as the wall flexibility increased.
    b. For slender tanks ($1.5 < H/R < 6$), the largest pressure peak occurred in flexible tanks and decreased as the rigidity of the wall increased.

    These findings highlight the significant influence of tank flexibility and slenderness ratio on the seismic pressure distribution, emphasizing the need for careful consideration of these factors in the design and analysis of cylindrical tanks subjected to seismic excitation.
8. The pressure profiles obtained from the FE analyses were compared with the predictions from the three standard approaches (EC8-4, NZSEE, and AWWA) using the input motion response spectrum and the combination rule for impulsive and convective pressure components specified in each corresponding code. Generally, a good agreement was observed in the lower part of the wall. However, the impact

of the convective waves on the tank shell causes a peak in pressure at the top of the wall, which cannot be predicted by the design approaches proposed by the three regulations, except for the $H/R = 1.5$ case study. It is important to note that these peaks could lead to significant problems during seismic events and should be taken into account during the design of steel tanks. Although small discrepancies at the base can be easily accounted for in design, the considerable underestimation of pressures in the upper portion of the tank could lead to an unsafe and non-conservative design, particularly if there is a large reduction in thickness along the tank height. Comparing the pressure response values obtained from the FE simulations with those predicted by the standard codes, the following conclusions can be drawn:

a. AWWA underestimated the FE results for all studied cases, including both squat and slender tanks.
b. EC8-4 for rigid tanks underestimated the FE results across all cases: both squat and slender tanks. EC8-4 for flexible tanks overestimated the FE results only for squat tanks ($0.5 < H/R < 1.5$) but underestimated the results for slender tanks ($1.5 < H/R < 6$).
c. NZSEE recommendations, both for rigid and flexible tanks, led to similar pressure values for squat tanks ($0.5 < H/R < 1.5$) but underestimated the FE results for slender tanks ($1.5 < H/R < 6$).

These results highlight the limitations of the current design codes especially in predicting the impact of convective wave-induced pressure peaks at the top of the tank wall. The findings suggest a need for more accurate methods to capture the full seismic behavior of tanks—particularly those with slender designs.

9. The trend of pressures was investigated using a set of ten scaled recorded accelerograms, which were all compatible with the EC8 design response spectrum. It can be concluded that the pressure profiles obtained from these accelerograms were almost identical to those obtained using the seismic excitation from the shaking table test used in the experimental study by Manos and Clough [42]. This similarity indicates that the selected accelerograms provide a reliable representation of the seismic response and produce pressure profiles comparable to those observed in the experimental setup.

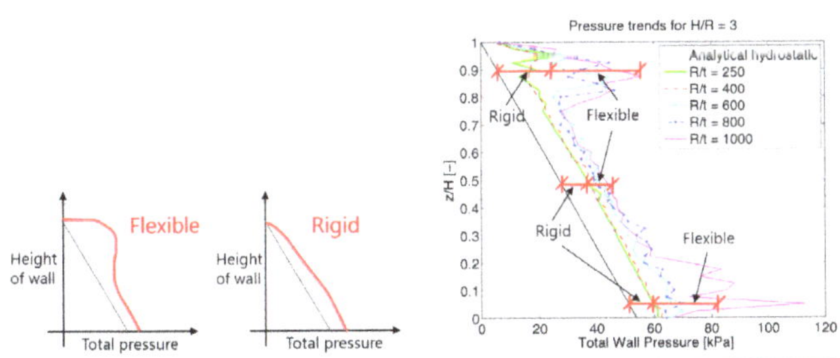

H/R		0.50	0.75	1.00	1.50	2.00	2.50	3.00	4.00	5.00	6.00
R/t	100	Rigid	Rigid	Rigid	Rigid	Rigid	Rigid	Rigid	Rigid	Rigid	Rigid
	125	Rigid	Rigid	Rigid	Rigid	Rigid	Rigid	Rigid	Rigid	Rigid	Rigid
	150	Rigid	Rigid	Rigid	Rigid	Rigid	Rigid	Rigid	Rigid	Rigid	Flexible
	250	Rigid	Rigid	Rigid	Rigid	Rigid	Rigid	Rigid	Rigid	Flexible	Flexible
	400	Rigid	Rigid	Rigid	Rigid	Rigid	Rigid	Rigid	Flexible	Flexible	Flexible
	600	Rigid	Rigid	Rigid	Rigid	Rigid	Rigid	Flexible	Flexible	Flexible	Flexible
	800	Rigid	Rigid	Rigid	Rigid	Rigid	Flexible	Flexible	Flexible	Flexible	Flexible
	1000	Rigid	Rigid	Rigid	Flexible	Flexible	Flexible	Flexible	Flexible	Flexible	Flexible
	1500	Rigid	Flexible	Flexible	Flexible	Flexible	Flexible	Flexible	Flexible	Flexible	Flexible
	2000	Flexible	Flexible	Flexible	Flexible	Flexible	Flexible	Flexible	Flexible	Flexible	Flexible

Figure 13. A guidance on the selection of flexible versus rigid tanks, depending on the *H/R* (height-to-radius) and *R/t* (radius-to-thickness) ratios.

Author Contributions: Conceptualization, E.B. and R.N.; methodology, E.B.; software, D.B.; validation, E.B. and D.B.; formal analysis, D.B.; writing—original draft preparation, R.N.; writing—review and editing, E.B. All authors have read and agreed to the published version of the manuscript.

Funding: This research received no external funding.

Institutional Review Board Statement: Not applicable.

Informed Consent Statement: Not applicable.

Data Availability Statement: The original contributions presented in this study are included in the article. Further inquiries can be directed to the corresponding author.

Conflicts of Interest: The authors declare no conflicts of interest.

References

1. Calvi, G.M.; Nascimbene, R. *Seismic Design and Analysis of Tanks*; John Wiley & Sons: Hoboken, NJ, USA, 2023.
2. Manos, G. Evaluation of the earthquake performance of anchored wine tanks during the San Juan, Argentina, 1977 Earthquake. *Earthq. Eng. Struct. Dyn.* **1991**, *20*, 1099–1114. [CrossRef]
3. Swan, S.W.; Miller, D.D.; Yanev, P.I. The Morgan Hill Earthquake of April 24, 1984—Effects on industrial facilities, buildings, and other facilities. *Earthq. Spectra* **1985**, *1*, 457–568. [CrossRef]
4. Stepp, J.C.; Swan, S.; Wesselink, L.; Haupt, R.W.; Larder, R.R.; Bachman, R.E.; Malik, L.; Eli, M.; Porush, A. Loma Prieta Earthquake reconnaissance report. Industrial facilities. *Earthq. Spectra* **1990**, *6* (Suppl. S1), 189–238.
5. Suzuki, K. Report on damage to industrial facilities in the 1999 Kocaeli Earthquake, Turkey. *J. Earthq. Eng.* **2002**, *6*, 275–296. [CrossRef]
6. Sezen, H.; Whittaker, A.S. Seismic performance of industrial facilities affected by the 1999 Turkey earthquake. *J. Perform. Constr. Facil.* **2006**, *20*, 28–36. [CrossRef]
7. Hatayama, K. Lessons from the 2003 Tokachi-oki, Japan, earthquake for prediction of long-period strong ground motions and sloshing damage to oil storage tanks. *J. Seism.* **2008**, *12*, 255–263. [CrossRef]
8. Krausmann, E.; Cruz, A.M.; Affeltranger, B. The impact of the 12 May 2008 Wenchuan earthquake on industrial facilities. *J. Loss Prev. Process Ind.* **2010**, *23*, 242–248. [CrossRef]
9. Zareian, F.; Sampere, C.; Sandoval, V.; McCormick, D.L.; Moehle, J.; Leon, R. Reconnaissance of the Chilean wine industry affected by the 2010 Chile offshore Maule Earthquake. *Earthq. Spectra* **2012**, *28* (Suppl. S1), S503–S512. [CrossRef]
10. Colombo, J.I.; Almazán, J.L. Experimental investigation on the seismic isolation for a legged wine storage tank. *J. Constr. Steel Res.* **2017**, *133*, 167–180. [CrossRef]
11. De Angelis, M.; Giannini, R.; Paolacci, F. Experimental investigation on the seismic response of a steel liquid storage tank equipped with floating roof by shaking table tests. *Earthq. Eng. Struct. Dyn.* **2010**, *39*, 377–396. [CrossRef]
12. Vakili, M.; Showkati, H. Experimental and numerical investigation of elephant foot buckling and retrofitting of cylindrical shells by FRP. *J. Compos. Constr.* **2016**, *20*, 4015087. [CrossRef]
13. Sezen, H.; Livaoglu, R.; Dogangun, A. Dynamic analysis and seismic performance evaluation of above-ground liquid-containing tanks. *Eng. Struct.* **2008**, *30*, 794–803. [CrossRef]
14. Paolacci, F.; Giannini, R.; De Angelis, M. Seismic response mitigation of chemical plant components by passive control techniques. *J. Loss Prev. Process Ind.* **2013**, *26*, 924–935. [CrossRef]
15. Reza, M.S.; Bursi, O.S.; Paolacci, F.; Kumar, A. Enhanced seismic performance of non-standard bolted flange joints for petrochemical piping systems. *J. Loss Prev. Process Ind.* **2014**, *30*, 124–136. [CrossRef]
16. Bursi, O.S.; Reza, M.S.; Abbiati, G.; Paolacci, F. Performance-based earthquake evaluation of a full-scale petrochemical piping system. *J. Loss Prev. Process Ind.* **2015**, *33*, 10–22. [CrossRef]
17. Bakalis, K.; Vamvatsikos, D.; Fragiadakis, M. Seismic risk assessment of liquid storage tanks via a nonlinear surrogate model. *Earthq. Eng. Struct. Dyn.* **2017**, *46*, 2851–2868. [CrossRef]
18. Veletsos, A.S.; Tang, Y.; Tang, H.T. Dynamic response of flexibly supported liquid-storage tanks. *J. Struct. Eng.* **1992**, *118*, 264–283. [CrossRef]
19. Houliara, S.; Karamanos, S.A. Buckling and post-buckling of long pressurized elastic thin-walled tubes under in-plane bending. *Int. J. NonLinear Mech.* **2006**, *41*, 491–511. [CrossRef]

20. Phan, H.N.; Paolacci, F.; Corritore, D.; Akbas, B.; Uckan, E.; Shen, J.J. Seismic vulnerability mitigation of liquefied gas tanks using concave sliding bearings. *Bull. Earthq. Eng.* **2016**, *14*, 3283–3299. [CrossRef]
21. Rotter, J.M. Elephant's foot buckling in pressurized cylindrical shells. *Stahlbau* **2006**, *75*, 742–747. [CrossRef]
22. Campedel, M.; Cozzani, V.; Garcia-Agreda, A.; Salzano, E. Extending the quantitative assessment of industrial risks to earthquake effects. *Risk Anal.* **2008**, *28*, 1231–1246. [CrossRef] [PubMed]
23. Abali, E.; Uçkan, E. Parametric analysis of liquid storage tanks base isolated by curved surface sliding bearings. *Soil Dyn. Earthq. Eng.* **2010**, *30*, 21–31. [CrossRef]
24. Spritzer, J.M.; Guzey, S. Nonlinear numerical evaluation of large open-top aboveground steel welded liquid storage tanks excited by seismic loads. *Thin-Walled Struct.* **2017**, *119*, 662–676. [CrossRef]
25. Bakalis, K.; Karamanos, S.A. Uplift mechanics of unanchored liquid storage tanks subjected to lateral earthquake loading. *Thin-Walled Struct.* **2021**, *158*, 107145. [CrossRef]
26. O'Reilly, G.J.; Shahnazaryan, D.; Dubini, P.; Brunesi, E.; Rosti, A.; Dacarro, F.; Gotti, A.; Silvestri, D.; Mascetti, S.; Ducci, M.; et al. Risk-aware navigation in industrial plants at risk of NaTech accidents. *Int. J. Disaster Risk Reduct.* **2023**, *88*, 103620. [CrossRef]
27. Ozsarac, V.; Brunesi, E.; Nascimbene, R. Earthquake-induced nonlinear sloshing response of above-ground steel tanks with damped or undamped floating roof. *Soil Dyn. Earthq. Eng.* **2021**, *144*, 106673. [CrossRef]
28. Ozdemir, Z.; Souli, M.; Fahjan, Y.M. Application of nonlinear fluid-structure interaction methods to seismic analysis of anchored and unanchored tanks. *Eng. Struct.* **2010**, *32*, 409–423. [CrossRef]
29. Rawat, A.; Mittal, V.; Chakraborty, T.; Matsagar, V. Earthquake induced sloshing and hydrodynamic pressures in rigid liquid storage tanks analyzed by coupled acoustic-structural and Euler-Lagrange methods. *Thin-Walled Struct.* **2019**, *134*, 333–346. [CrossRef]
30. Eshghi, S.; Razzaghi, M.S. Performance of cylindrical liquid storage tanks in Silakhor, Iran earthquake of 31 March 2006. *Bull. N. Z. Soc. Earthq.* **2007**, *40*, 173–182. [CrossRef]
31. Brunesi, E.; Nascimbene, R.; Pagani, M.; Beilic, D. Seismic performance of storage steel tanks during the May 2012 Emilia, Italy, Earthquakes. *J. Perform. Constr. Facil.* **2015**, *29*, 04014137. [CrossRef]
32. Yazdanian, M.; Ingham, J.M.; Kahanek, C.; Dizhur, D. Damage to flat-based wine storage tanks in the 2013 and 2016 New Zealand earthquakes. *J. Constr. Steel Res.* **2020**, *168*, 105983. [CrossRef]
33. Priestley, M.J.N. Seismic design of storage tanks. In *Recommendations of a Study Group of the New Zealand National Society for Earthquake Engineering*, 2nd ed.; NZSEE: Auckland, New Zealand, 2009.
34. *Eurocode 8—Design of Structures for Earthquake Resistance, Part 4: Silos, Tanks and Pipelines*; European Commitee for Standardization: Brussels, Belgium, 2006.
35. Jacobsen, L.S. Impulsive hydrodynamics of fluid inside a cylindrical tank and of fluid surrounding a cylindrical pier. *Bull. Seism. Soc. Am.* **1949**, *39*, 189–204. [CrossRef]
36. Housner, G.W. The dynamic behaviour of water tanks. *Bull. Seism. Soc. Am.* **1963**, *53*, 381–387. [CrossRef]
37. Housner, G.W. *Earthquake Pressures on Fluid Containers*; Technical Report California; Institute of Technology: Pasadina, CA, USA, 1954.
38. Veletsos, A.S.; Yang, J.Y. Earthquake response of liquid-storage tanks. In Proceedings of the 2nd Annual Engineering Mechanics Division Specialty Conference, ASCE Advanced in Civil Engineering Through Engineering Mechanics, North Carolina State University, Raleigh, NC, USA, 23–25 May 1977; pp. 1–24.
39. Malhotra, P.K.; Wenk, T.; Wieland, M. Simple procedure for seismic analysis of liquid storage tanks. *Struct. Eng. Int.* **2000**, *3*, 197–201. [CrossRef]
40. *AWWA D100-05*; Welded Carbon Steel Tanks for Water Storage. American Water Works Association: Denver, CO, USA, 2005.
41. Clough, D.P. *Experimental Evaluation of Seismic Design Methods for Broad Cylindrical Tanks*; Report UCB/EERC 77-10; Earthquake Engineering Research Center: Hyogo, Japan, 1977.
42. Manos, G.C.; Clough, R.W. *Further Study of the Earthquake Response of a Broad Cylindrical Liquid Storage Tank*; Report UCB/EERC 82-07; Earthquake Engineering Research Center: Hyogo, Japan, 1982.
43. Haroun, M.A. *Dynamic Analyses of Liquid Storage Tanks*; Research Report California Institute of Technology: Pasadena, CA, USA, 1980.
44. Belytschko, T.; Lin, J.; Tsay, C.S. Explicit algorithms for nonlinear dynamics of shells. *Comp. Methods Appl. Mech. Eng.* **1984**, *42*, 225–251. [CrossRef]
45. Merino, R.J.; Brunesi, E.; Nascimbene, R. Probabilistic evaluation of earthquake-induced sloshing wave height in above-ground liquid storage tanks. *Eng. Struct.* **2020**, *202*, 109870. [CrossRef]

46. Merino, R.J.; Brunesi, E.; Nascimbene, R. Derivation of floor acceleration spectra for an industrial liquid tank supporting structure with braced frame systems. *Eng. Struct.* **2018**, *171*, 105–122. [CrossRef]
47. Brunesi, E.; Nascimbene, R. Evaluating the Seismic Resilience of Above-Ground Liquid Storage Tanks. *Buildings* **2024**, *14*, 3212. [CrossRef]

applied sciences

MDPI

Article

Seismic Vulnerability Assessment of Existing Ground-Supported Liquid Storage Tanks with Deformed Shells

Raihan Rahmat Rabi [1,*] and Giorgio Monti [1,2]

[1] Department of Structural Engineering and Geotechnics, Sapienza University of Rome, Via A. Gramsci 53, 00197 Roma, Italy
[2] Department of Civil Engineering, Xi'an Jiaotong Liverpool University, Suzhou 215000, China
* Correspondence: raihan.rahmatrabi@uniroma1.it

Abstract: This study presents a novel methodology for the seismic vulnerability assessment of ground-supported liquid storage tanks, with a particular emphasis on tanks exhibiting deformed shells. The framework is built around the development of a simplified two-degree-of-freedom (2DoF) system, specifically designed to enable efficient probabilistic analysis through incremental dynamic analysis. The 2DoF model is calibrated using detailed nonlinear finite element simulations, which account for fluid–structure interaction and complex geometric effects. This approach allows for a significant reduction in computational effort while maintaining a high level of accuracy in predicting seismic fragility. A key feature of the proposed methodology is the explicit inclusion of deformed tank geometries, which are often encountered in real cases and can critically impact seismic performance. The study identifies and quantifies the effects of these deformations on various damage states, such as local buckling, overturning, and liquid overspill. Fragility curves are developed to characterize the probability of exceeding these damage states under different levels of seismic intensity. The proposed methodology is validated through a representative case study, demonstrating its ability to capture the key features of tank behavior and provide robust fragility estimates. The results highlight the potential of the 2DoF system to serve as a practical tool for seismic risk assessment, enabling engineers and researchers to evaluate tank vulnerabilities more efficiently.

Keywords: deformed tanks; seismic vulnerability; fragility curves; incremental dynamic analysis

1. Introduction

Liquid storage tanks are critical components in modern infrastructure, serving a wide range of functions, from water distribution and firefighting to storing oil and hazardous materials in industrial facilities. These tanks are indispensable for the extraction, processing, and distribution of vital resources. However, given the global prevalence of storage tanks, many are inevitably situated in seismically active regions. This has driven significant interest in developing earthquake-resistant designs and conducting seismic fragility analyses, with numerous experimental and analytical studies aimed at mitigating earthquake risks and evaluating tank vulnerability under seismic loads [1–9].

The failure of these tanks during earthquakes can result in catastrophic losses, environmental contamination, and disruption to essential services, as evidenced during recent seismic events in Italy (2012) [10] and Chile (2010), with buckling of the tank shell and failure of the anchor or support legs as primary causes. Similarly, Sezen and Whittaker [11] highlight the damage to larger on-grade storage tanks during earthquakes in the United States, Turkey, and Japan. In these cases, foundation instability and the sloshing of liquid inside the tanks also contributed to damage, with sloshing waves not only harming the upper portions of the tanks but also causing liquid overspill.

Over the past five decades, substantial research has been conducted to understand the seismic behavior of liquid storage tanks. Foundational studies by Jacobsen [12] and

Housner [13], followed by subsequent research [14–19], laid the groundwork by introducing the concepts of decoupling the tank–liquid system motion into two distinct components: the impulsive component, representing the dynamic behavior of the tank's lower part, and the convective component, which describes the movement of surface waves in the liquid. These studies have been widely adopted in seismic design codes, such as API 650 [20], ASCE/SEI 7-10 [21], CEN [22], and NZSEE [23], which adopt simplified models to estimate dynamic properties for idealized tank geometries.

More recently, fragility curves have emerged as a powerful tool for assessing seismic vulnerability, with significant research focused on developing fragility models for buildings and bridges [24–26] and, to a lesser extent, for liquid storage tanks. Notable early work by ALA [27,28] provided empirical fragility curves for on-ground supported liquid containers. HAZUS [29] later categorized tank damages into five distinct states and proposed corresponding fragility curves. Building on this, Iervolino et al. [30] developed fragility curves for unanchored liquid storage tanks using a surface response approach that considered a variety of tank geometries. Berahman and Behnamfar [31] used Bayesian methods to develop fragility curves. Phan et al. [32] incorporated uncertainty factors that influence their seismic performance and, using Gaussian process regression, developed fragility models reflecting these uncertainties [2]. Additionally, Merino Vela et al. [7,33] expanded on this by developing fragility functions for liquid tanks supported by special concentrically braced frames. In related work, Merino et al. [34] carried out dynamic analyses to optimize the freeboard height accounting for record-to-record variability in seismic excitation.

Despite these advances, most studies assumed an idealized, undeformed geometry for tanks, neglecting the potential influence of pre-existing deformations caused by prior seismic events. Such deformations, including shell denting and localized thinning, can substantially alter the dynamic behavior of tanks. The resulting changes in stiffness, mass distribution, and natural periods impact both impulsive and convective responses, rendering the assumptions in current design codes insufficient. This gap in the literature limits our ability to accurately assess the seismic performance of deformed tanks.

To address this critical gap, this study proposes a refined framework for the seismic vulnerability assessment of ground-supported cylindrical tanks with pre-existing deformations. The framework integrates finite element modeling (FEM) of deformed tank geometries with a simplified two-degree-of-freedom (2DoF) mechanical model to account for fluid–structure interactions. Incremental dynamic analysis (IDA) is then employed using a suite of earthquake ground motions to derive fragility curves for four critical failure modes: liquid spillage, elastic buckling, overturning moment failure, and "elephant's foot" buckling.

The proposed methodology joins simplified analytical models with the computation rigor of FEM analysis, offering a practical yet accurate approach to seismic vulnerability assessment. By focusing on deformed geometries, this study provides a more realistic representation of tank behavior under seismic loads.

It is worth emphasizing that this work represents an initial step toward establishing a consistent and practical framework for deriving fragility curves for deformed ground-supported liquid storage tanks. Building upon this foundation, ongoing and future studies aim to enhance the framework by incorporating additional critical factors that influence the seismic response of these strategic structures, such as more complex fluid–structure interaction, different deformation patterns, a broader range of failure mechanisms, and soil–structure interaction.

Figure 1 shows the proposed framework consisting of three phases for developing fragility curves of ground-supported cylindrical tanks with pre-existing deformations.

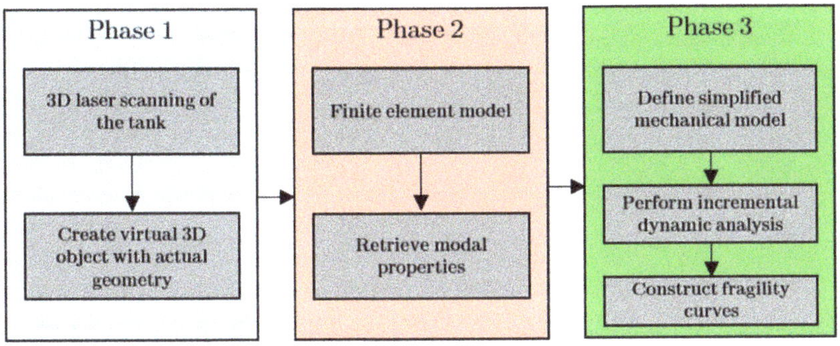

Figure 1. Proposed framework for developing fragility curves of fluid storage tanks.

2. Seismic Analysis of Liquid Storage Tanks

2.1. Mechanical Model

The seismic behavior of liquid storage tanks is complex due to the fluid–structure interaction inherent in their design. When subjected to ground motion, the dynamic response involves both the tank walls and the liquid within. This interaction creates significant computational challenges, particularly when the confined fluid and its intricate contact dynamics with the tank shell are explicitly modeled using *FEM*. These challenges are exacerbated in performance-based earthquake engineering (PBEE) frameworks, where numerous nonlinear analyses are required to capture the full range of seismic responses.

To address these computational demands, simplified surrogate models, such as the 2DoF spring-mass system, are commonly employed. These models provide an efficient alternative for representing the dynamic behavior of liquid storage tanks, capturing both the impulsive and convective components of motion. The impulsive component accounts for the liquid that moves rigidly with the tank wall, while the convective component represents the sloshing motion of the liquid free surface. Such models are widely used for both anchored and unanchored tanks and have been validated in previous studies [16,35].

In this study, a 2DoF spring-mass model is utilized to simplify the representation of the tank's dynamic behavior. As illustrated in Figure 2, the model consists of two masses (impulsive and convective) connected by springs and dampers, representing the stiffness and damping properties of the tank–liquid system. The equations of motion for this system are given as follows:

$$m_i\ddot{x}_i + (c_i + c_c)\dot{x}_i - c_c\dot{x}_c + (k_i + k_c)x_i - k_c x_c = 0 \tag{1}$$

$$m_c\ddot{x}_c - c_c\dot{x}_i + c_c\dot{x}_c - k_c x_i - k_c x_c = 0 \tag{2}$$

where m_i and m_c are the impulsive and convective masses, k_i and k_c are the impulsive and convective stiffness values, c_i and c_c are the impulsive and convective damping coefficients, x_i and x_c are the displacements of the impulsive and convective masses, and $\dot{x}$ and $\ddot{x}$ denote the first and second time derivatives of displacement.

The coupling between these differential equations highlights the interdependence of impulsive and convective motions, which significantly influences the tank's overall seismic response. In a matrix representation, Equations (1) and (2) are expressed as follows:

$$\mathrm{M}\,\ddot{x}(t) + \mathrm{C}\,\dot{x}(t) + \mathrm{K}\,x(t) = 0 \tag{3}$$

where M is the mass matrix, C the damping matrix, K the stiffness matrix, and x is the displacement vector, given as follows:

$$\mathrm{M} = \begin{bmatrix} m_i & 0 \\ 0 & m_c \end{bmatrix}$$

$$C = \begin{bmatrix} c_i + c_c & -c_c \\ -c_c & c_c \end{bmatrix}$$

$$K = \begin{bmatrix} k_i + k_c & -k_c \\ -k_c & k_c \end{bmatrix}$$

$$x = \begin{Bmatrix} x_i \\ x_c \end{Bmatrix}$$

The 2DoF model balances computational efficiency with accuracy, making it a practical choice for developing fragility curves. By capturing the primary dynamic characteristics of the tank–liquid system, the model serves as a reliable surrogate for more computationally demanding finite element analyses, particularly when requiring large-scale parametric studies or probabilistic seismic assessments.

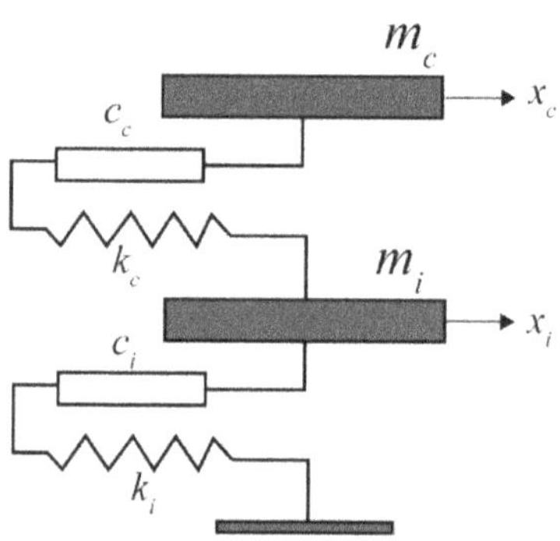

Figure 2. Schematic of the two-degree-of-freedom system (2DoF) representing impulsive and convective components of the tank.

2.2. Hydrodynamic Analysis

The seismic behavior of liquid storage tanks is governed by a complex interplay between the structural response of the tank shell and the dynamic response of the contained fluid. During an earthquake, seismic energy transfers from the ground to the fluid through the motion of the tank, resulting in two primary components of hydrodynamic response: the impulsive motion and the convective motion [36].

The impulsive component represents the portion of the liquid that moves rigidly with the tank wall, behaving as though it is fully attached to the structure. This motion is characterized by shorter periods and is driven by the rapid inertial forces transmitted from the ground motion. The convective component, on the other hand, corresponds to the sloshing motion of the liquid's free surface, which occurs at longer periods due to its oscillatory nature [37]. These distinct responses can be decomposed for analytical convenience, allowing separate evaluation of their effects on the tank wall and base.

2.2.1. Pressure Components

The total pressure p exerted on the tank walls during seismic events comprises three components: hydrostatic pressure p_h, due to the liquid's weight; hydrodynamic pressure made up of impulsive pressure p_i, associated with the rigid motion of the liquid;and convective pressure p_c, induced by sloshing. These are added up as follows:

$$p = p_h + p_i + p_c \tag{4}$$

The impulsive pressure is associated with the acceleration of the tank and the inertia of the attached liquid, while the convective pressure arises from surface waves and has a sinusoidal distribution over the height of the tank.

2.2.2. Convective Pressure Distribution

To estimate the convective pressure, a cylindrical coordinate system $(\xi,\ \zeta,\ \theta)$ is used as shown in Figure 3:

$$\xi = r/R \tag{5}$$

$$\zeta = z/H \tag{6}$$

where r is the radial distance, z is the height, R is the tank radius, and H is the fluid height.

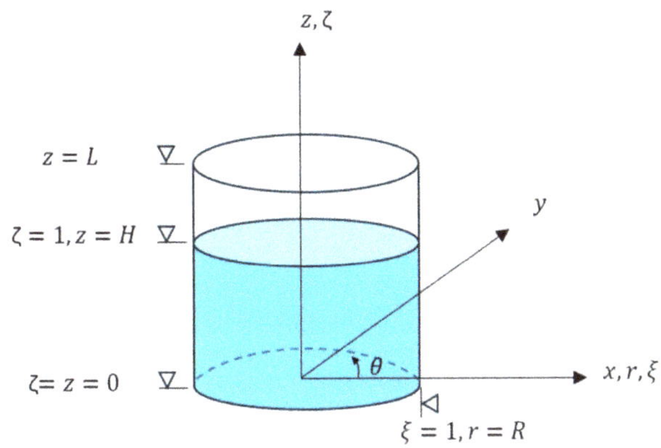

Figure 3. Coordinate system of cylindrical tanks.

The convective pressure p_c can be expressed as follows [38]:

$$p_c(\xi,\zeta,\theta) = C_{cn}(\xi,\zeta)\cdot\rho\, R\, \ddot{u}_{cn}\cos(\theta) \tag{7}$$

where $C_{cn}(\xi,\zeta)$ is the non-dimensional convective pressure distribution coefficient, ρ is the fluid density, $\ddot{u}_{cn}$ is the acceleration of the n-th convective mass mode, and θ is the angular position. The coefficient $C_{cn}(\xi,\zeta)$ is computed using the following expression:

$$C_{cn}(\xi,\zeta) = \sum_{n=1}^{\infty} \frac{2\cosh(\lambda_n\gamma\zeta)\ J_1(\lambda_n\xi)}{(\lambda_n^2-1)\ J_1(\lambda_n)\cosh(\lambda_n\gamma)} \tag{8}$$

where $J_1(\cdot)$ is the first-order Bessel function, $\gamma = H/R$ is the height-to-radius ratio, and λ_n is the n-th root of the Bessel function. The first three roots are approximately $\lambda_1 = 1.841$, $\lambda_2 = 5.331$, and $\lambda_3 = 8.536$, corresponding to the fundamental and higher-order modes of sloshing.

2.2.3. Impulsive Pressure Distribution

The impulsive pressure distribution is determined as follows [39]:

$$p_i(z,\ \theta,\ t) = (z)\cdot\rho\, H\, \ddot{u}_i\cos(\theta) \tag{9}$$

where $c_i'(z)$ is a height-dependent non-dimensional coefficient for impulsive pressure, and $\ddot{u}_i$ is the impulsive acceleration of the tank wall. The coefficient $c_i'(z)$ depends on the tank geometry and the ratio of tank mass to liquid mass, computed using the method used by Veletsos and Yang [39].

2.2.4. Simplified Hydrodynamic Representation

The segmentation of the hydrodynamic pressure into impulsive and convective components facilitates the lateral seismic analysis of tanks. Impulsive pressures are primarily influenced by the tank's structural response and ground acceleration, while convective pressures are governed by the sloshing behavior of the liquid's free surface. By decoupling these effects, the 2DoF model provides a computationally efficient means of estimating the internal forces and moments experienced by both the tank and its liquid contents. This analytical framework represents the basis for the derivation of failure modes and fragility curves in subsequent sections.

2.3. Failure Modes

2.3.1. Elastic Buckling

Elastic buckling is a critical failure mode, typically occurring in regions subjected to significant axial compression. This failure mode manifests as circumferential wrinkles or inward bending of the tank wall, either near the tank base or at intermediate heights, depending on the loading conditions and tank geometry. The primary drivers of elastic buckling include the axial compression forces induced by overturning moments during seismic excitation and the self-weight of the tank wall and roof.

The critical axial stress σ_e at which elastic buckling occurs in cylindrical shells is given by the following equation [22,40]:

$$\sigma_e = \frac{E_s\, t}{R \cdot \sqrt{3(1-\nu^2)}} \tag{10}$$

where E_s is the modulus of elasticity of the shell material, t is the thickness of the tank wall, R is the radius of the tank, and ν is the Poisson's ratio of the shell material.

For tanks with pre-existing deformations, the critical buckling stress (σ_e) is likely reduced due to local thinning, dents, or residual stresses in the shell. This highlights the importance of accounting for such imperfections when evaluating the seismic vulnerability of existing tanks, as the equations derived for idealized geometries may not fully capture the behavior of deformed structures.

The elastic buckling of the tank wall can be associated with critical values of the overturning moment (OM_{cr}) and the structural base shear (FS_{cr}), expressed as follows [40]:

$$OM_{cr} = \sigma_e\, \pi\, R^2\, t \tag{11}$$

$$FS_{cr} = \tau_e\, \pi\, R\, t \tag{12}$$

where τ_e is the critical shear buckling stress, which can be obtained using the following equation [40]:

$$\tau_e = 0.07708 \frac{\pi^2 E_s}{(1-\nu^2)^{\frac{5}{8}} \cdot \sqrt{\gamma}} \left(\frac{R}{t}\right)^{-5/4} \tag{13}$$

This relation incorporates the height-to-radius ratio $\gamma = H/R$ and the radius-to-thickness ratio R/t, which are critical geometric parameters influencing the tank's susceptibility to buckling.

2.3.2. Elephant's Foot Buckling

Elephant's foot buckling (EFB), which is an elastic–plastic buckling, is one of the most prevalent failure modes observed in cylindrical steel storage tanks during seismic events. This localized failure occurs near the tank base and is characterized by a high concentration of compressive meridional stress that results in outward bulging or buckling (Figure 4). This phenomenon is primarily driven by the interaction of vertical compressive stresses, hoop tensile stresses, and significant shear stresses, often exacerbated by base uplift during seismic loading. When these stresses exceed the critical buckling capacity of the tank material, EFB initiates. This localized failure is particularly dangerous as it can rapidly propagate around the tank's circumference, compromising its structural integrity.

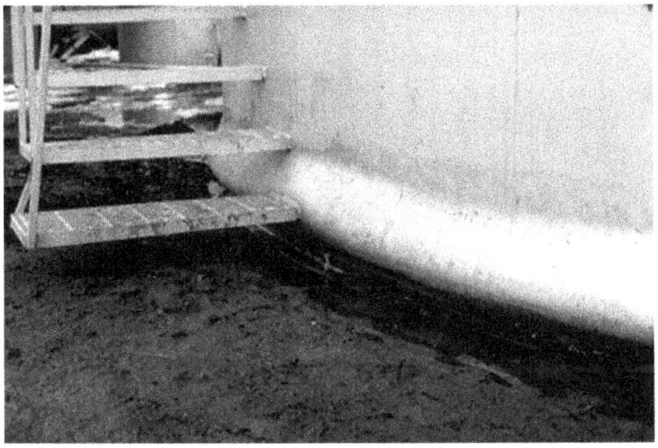

Figure 4. "Elephant's foot" buckling of storage tank during Northridge earthquake in California [16].

The critical buckling stress σ_{ep} for elastic–plastic behavior is expressed as follows [22,41]:

$$\sigma_{ep} = \frac{\sigma_e}{r+1}\left[1 - \left(\frac{pR}{tf_y}\right)^2\right]\left(1 - \frac{1}{1.12 + r^{1.5}}\right)\left(r + \frac{f_y}{250}\right) \tag{14}$$

where σ_e is the elastic critical buckling stress in Equation (10), $r = R/(400\,t)$ is a non-dimensional parameter based on the radius-to-thickness ratio, p is the total internal pressure (hydrostatic + hydrodynamic) acting on the tank wall in Equation (4), and f_y is the yield strength of the tank material.

Elephant's foot buckling occurs when the meridional stress (σ_m) in the tank wall exceeds the critical elastic–plastic buckling stress σ_{ep} in Equation (14). The meridional stress is calculated as follows:

$$\sigma_m = \frac{OM}{\pi R^2 t} + \frac{W}{2\pi R t} \tag{15}$$

where OM is the overturning moment and W is the total weight of the tank and its content.

For tanks with pre-existing shell deformations, the likelihood of EFB increases significantly. Localized dents, uneven wall thicknesses, and residual stresses amplify stress concentrations, reducing the critical buckling capacity.

2.3.3. Sloshing

Sloshing refers to the oscillatory motion of the liquid's free surface in response to seismic excitation. This phenomenon can significantly impact the structural performance of liquid storage tanks, particularly in tanks with insufficient freeboard (the distance between the liquid surface and the roof). Sloshing-induced forces are a critical design consideration, as they can cause roof damage, rupture of roof connections, and liquid spillage, compromising the tank's functionality after an earthquake.

The freeboard height (H_f) is defined as follows:

$$H_f = L - H \tag{16}$$

where L is the total height of the tank and H is the height of the liquid within the tank. According to API 650 [20], during seismic events, the maximum sloshing wave height (δ_s) can be calculated as follows:

$$\delta_s = R\,\frac{S_a(T_c)}{g} \tag{17}$$

where R is the radius of the tank, $g = 9.82\ \mathrm{m/s^2}$ is the gravitational acceleration, and $S_a(T_c)$ is the spectral acceleration at the convective period, T_c. When $\delta_s > H_f$, the liquid may spill over the tank's edge, posing environmental and operational risks. In tanks with pre-existing deformations, the sloshing response can deviate significantly from the predictions of idealized models. Deformations such as dents or uneven wall thicknesses alter the tank's natural frequencies, particularly the convective period T_c, which governs sloshing behavior. This results in changes to the spectral acceleration $S_a(T_c)$ and, consequently, the sloshing wave height δ_s.

3. Fragility Curves and Damage States

A fragility curve (FC) represents the probability that an engineering demand parameter EDP will exceed a specific capacity C_{DS_i} associated with a damage state DS_i, as function of a seismic intensity measure IM, such as Peak Ground Acceleration (PGA) or spectral acceleration or displacement, as follows:

$$P(EDP \geq C_{DS_i}|IM) = \Phi\left(\frac{\ln(EDP) - \ln(C_{DS_i})}{\beta_{DS_i}}\right) \tag{18}$$

where Φ is the standard normal cumulative distribution function, and β_{DS_i} is the logarithmic standard deviation of the capacity C_{DS_i}, representing the variability in structural capacity due to material, geometric, and modeling uncertainties.

The seismic performance of liquid storage tanks is evaluated through four critical damage states, each corresponding to the specific failure modes identified as critical. Table 1 summarizes these damage states, along with their associated capacities and fragility functions.

Table 1. Description of the damage states.

Damage State	Description	Capacity	Demand	Fragility Function
DS_{LO}	Liquid overspill	H_f in Equation (16)	δ_s in Equation (17)	$P(\delta_s \geq H_f|IM)$
DS_{BS}	Exceedance of critical base shear	FS_{cr} in Equation (12)	Base shear (FS)	$P(FS \geq FS_{cr}|IM)$
DS_{OM}	Exceedance of overturning moment	OM_{cr} in Equation (11)	Overturning moment (OM)	$P(OM \geq OM_{cr}|IM)$
DS_{EFB}	Exceedance of critical buckling limit	σ_{ep} in Equation (14)	σ_m in Equation (15)	$P(\sigma_m \geq \sigma_{ep}|IM)$

Incremental dynamic analysis (IDA) is used to derive the fragility curves for each damage state. The IDA process involves (1) selecting a suite of representative earthquake ground motions (Section 4.2); (2) scaling the ground motions incrementally to different intensity levels IM; and (3) simulating the tank response to each scaled ground motion using the 2DoF model developed in Section 2.

For each ground motion and IM, the $EDPs$ $(\delta_s, FS, OM, \sigma_m)$ are computed and compared to their respective capacities $\left(H_f, FS_{cr}, OM_{cr}, \sigma_{ep}\right)$ to determine whether a specific damage state is exceeded. The resulting fragility curves are plotted for each damage state, showing the probability of exceeding the capacity (C_{DS_i}) for varying IM levels. The flowchart of the procedure to obtain FCs using IDA is shown in Figure 5.

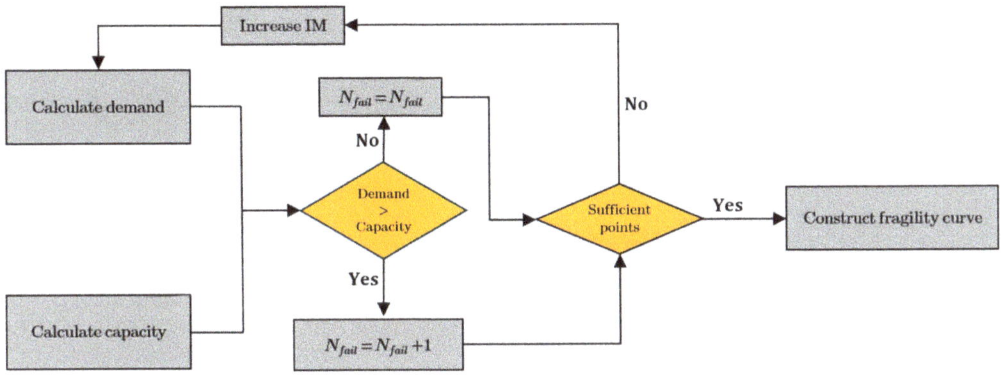

Figure 5. Procedure to obtain fragility curves using incremental dynamic analysis (IDA).

4. Case Study

4.1. Finite Element Modelling

The tank used in this study represents a typical design found in petroleum depots. Its key geometrical and mechanical properties are detailed in Table 2. The tank wall thickness varies linearly from 21 mm at the base to 9 mm at the top to account for axial and hoop stress distributions.

Table 2. Mechanical and geometrical properties of the cylindrical tank.

Diameter $D(\mathrm{m})$	Height $L(\mathrm{m})$	Fluid Height $H(\mathrm{m})$	Wall Thickness $t(\mathrm{mm})$	Fluid Density $\rho(\mathrm{kg/m^3})$	Steel Strength $f_y(\mathrm{MPa})$	Elastic Modulus $E_s(\mathrm{MPa})$
54.9	14.6	13.5	21 (base) to 9 (top)	980	235	210,000

The FE model was developed using ABAQUS v2021 software [42], as illustrated in Figure 6. The tank wall was modeled using S4R shell elements, four-node quadrilateral elements with reduced integration. These elements capture both bending and membrane behavior, essential for simulating buckling and stress distributions. The liquid inside the tank was represented using potential-based fluid elements, which allow for accurate modeling of hydrodynamic pressures without manual configuration of the fluid–structure interaction surface. The model includes all essential structural components, such as the roof, bottom plate, and cylindrical wall. Non-structural accessories, such as the nozzles, piping, fire extinguisher system, and stairs, have not been included in the model.

To accurately simulate the seismic behavior of an existing tank with pre-deformed geometry, a detailed inspection was performed using 3D laser scanning technology. This process captured the actual shape of the tank, identifying local deformations such as dents, uneven wall thicknesses, and residual out-of-plane displacements. The scanned data were used to create a virtual model of the tank's deformed shell. Figure 7 illustrates the deformation pattern identified, which is critical for accurately capturing the tank's modified stiffness, mass distribution, and dynamic properties.

The modal analysis of the FE model provided key dynamic characteristics of the tank–liquid system (Table 3); characteristics such as impulsive mass m_i, convective mass m_c, impulsive stiffness k_i, convective stiffness k_c, impulsive frequency f_i, and convective frequency f_c were retrieved, which were used to define the two degree of freedom system representing the simplified mechanical model of the tank.

Table 3. Dynamic characteristics of the tank–liquid system.

f_i(Hz)	f_c(Hz)	m_i(t)	m_c(t)	k_i(kN/m)	k_c(kN/m)
6.39	0.11	9,391	22,181	2,159,800	10,500

The following assumptions were made in developing the FE model: (1) the tank was assumed to be supported on a rock foundation, neglecting soil–structure interaction effects; (2) the fluid was considered incompressible and inviscid, consistent with standard hydrodynamic modeling approaches; and (3) non-linearities in the material's behavior, such as yielding, were limited to critical regions to maintain computational efficiency. While these assumptions provide a reasonable approximation for the tank's seismic response, future studies could incorporate soil–structure interactions and advanced material models for enhanced accuracy.

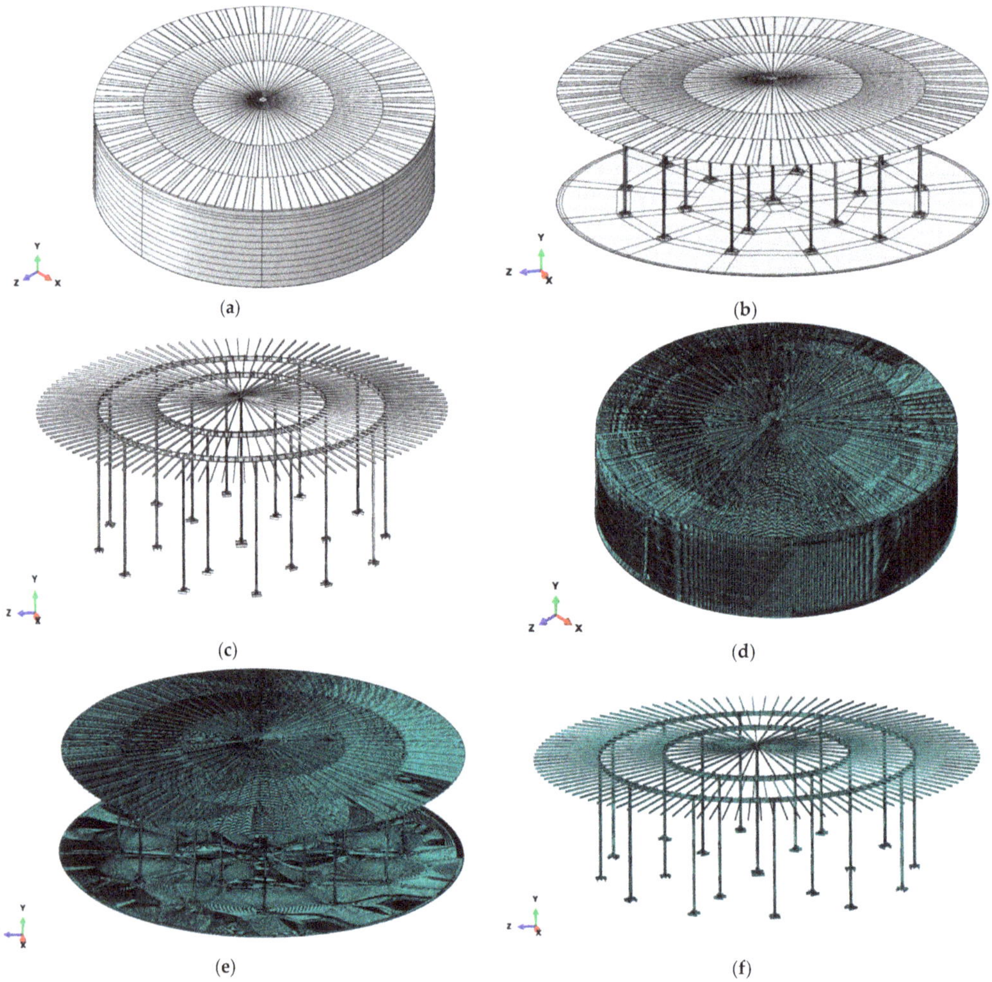

Figure 6. A finite element model of the tank: (**a**) a 3D view of the model; (**b**) the roof, base plate, and columns; (**c**) the columns and beams within the tank; (**d**) a 3D view of the mesh details of the model; (**e**) the mesh details of the base plate and roof; and (**f**) the mesh details of the beams and columns.

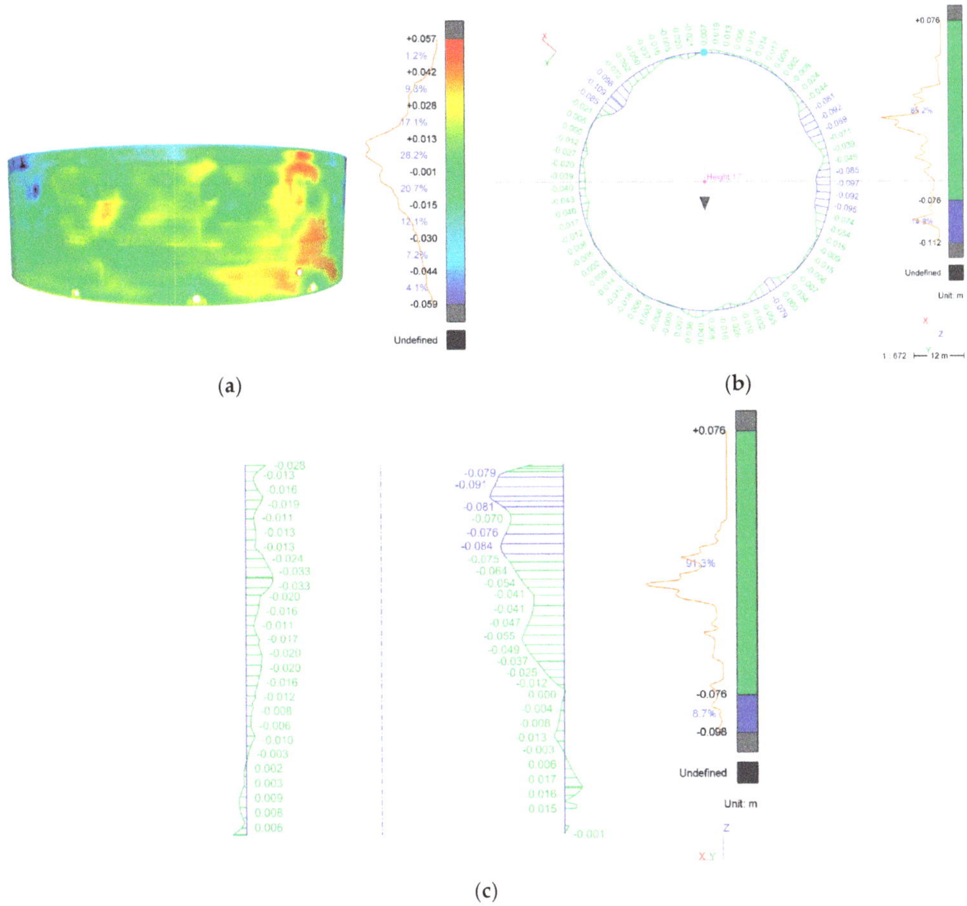

Figure 7. (**a**) Color map of deformed shell (unit: m); (**b**) top view (deviation magnification ×35); (**c**) side profile showing dents and out-of-plane displacements (deviation magnification ×38).

4.2. Selection of Ground Motion Records

A key step in evaluating the seismic vulnerability of structures is the selection of input earthquake ground motions. For this study, a suite of earthquake ground motions was carefully chosen to represent a wide range of seismic intensity levels and spectral characteristics. These records were used in incremental dynamic analysis (IDA) to evaluate the seismic response of the tank under varying intensity measures (IM). The ground motion records were selected based on the following criteria: (1) To capture a wide range of seismic effects, the records included both pulse-like and non-pulse-like motions. Pulse-like motions, often associated with near-fault effects, typically induce higher demands on structures due to long-period velocity pulses and permanent ground displacements. (2) Records were chosen to represent earthquakes of varying magnitudes and distances from the fault rupture to account for different seismic sources. (3) The records correspond to rock-like site conditions, consistent with the assumption of a rock foundation for the tank.

A total of 28 ground motion records were selected from the Next Generation Attenuation-West2 (NGA-West2) database [43], maintained by the Pacific Earthquake Engineering

Research (PEER) Center. Figure 8 illustrates the 2%-damping pseudo-acceleration response spectra of the selected records, along with the average spectrum.

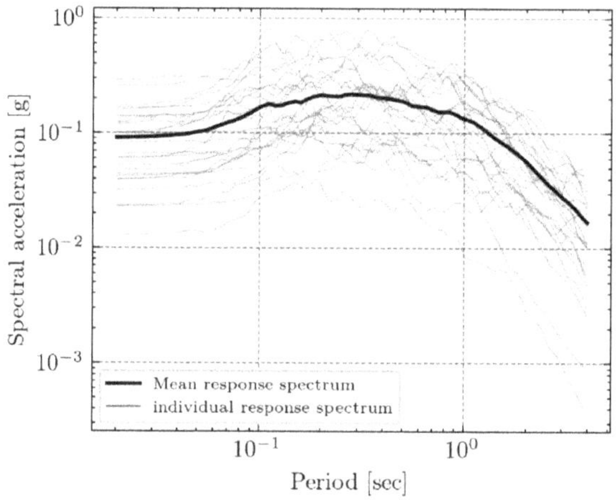

Figure 8. Pseudo-acceleration response spectra of selected ground motions.

4.3. Results and Discussion

The seismic response of the tank was evaluated through IDA, employing the selected suite of 28 ground motions (Section 4.2) applied to the 2DoF system (Section 2) modeled in OpenSeesPy [44] with the impulsive and convective characteristics determined from the FEM modal analysis (Section 4.1). Each ground motion was scaled incrementally from a PGA of 0.005 g up to 2.0 g, with a step size of 0.05 g. A total of 11,200 simulations were conducted to capture the tank response across various seismic intensities. For each simulation, the key $EDPs$ were extracted: (a) the sloshing wave height δ_s, (b) the base shear FS, (c) the overturning moment OM, and (d) the meridional stress σ_m. Figure 9 shows the variation of these parameters as a function of a_g. For the sloshing wave height, the increase is nearly linear at lower a_g values but becomes nonlinear at higher intensities ($a_g > 0.5$ g) due to complex fluid–tank interactions. Similarly, the base shear and overturning moment increase consistently with a_g, with the base shear showing a noticeable jump at moderate intensities.

The meridional stress increases steeply at moderate a_g levels, indicating that this parameter is more sensitive to moderate seismic intensities, which is particularly relevant for assessing the risk of elephant's foot buckling (DS_{EFB}).

These results were processed to derive fragility curves for the four identified damage states: liquid overspill DS_{LO}, base shear exceedance DS_{BS}, overturning moment exceedance DS_{OM}, and elephant's foot buckling DS_{EFB}, as presented in Figure 10. These curves represent the probability of exceeding each damage state as a function of seismic intensity.

The fragility curve for liquid overspill (DS_{LO}) (blue line) remains flat for lower a_g values, rising only after $a_g \approx 0.2$ g. This indicates that liquid overspill is highly resistant to low seismic intensities due to sufficient freeboard, with significant probability of occurrence under moderate seismic demands.

The curve for base shear exceedance (DS_{BS}) (green line) shows a steep rise at relatively low a_g values. This suggests that base shear is the most sensitive damage state, with high failure probabilities under low seismic demands.

For overturning moment exceedance (DS_{OM}) (red line), the fragility curve remains flat until $a_g \approx 0.8$ g and only begins to rise noticeably after that value. This behavior highlights the tank's resistance to overturning, likely due to its squat geometry and low center of gravity.

Finally, the fragility curve for elephant's foot buckling (DS_{EFB}) (yellow line) exhibits a steep slope, with failure probabilities rapidly increasing in the range of $a_g \approx 0.1$ g to 0.4 g.

This steep rise suggests that elephant's foot buckling is the second dominant failure mode at moderate seismic intensities, driven by stress concentrations near the base of the tank.

The comprehensive seismic fragility assessment of the tank is summarized in Table 4, where key statistical parameters, such as the median (μ) and standard deviation (β), are presented. These parameters are essential for constructing the fragility curves, as they quantify the central tendency and variability of the tank's response to seismic loads.

Table 4. Mean (μ) and standard deviation (β) of the fragility curves.

DS_{LO}		DS_{BS}		DS_{OM}		DS_{EFB}	
μ	β	μ	β	μ	β	μ	β
0.461	0.149	0.118	0.056	1.141	0.126	0.249	0.058

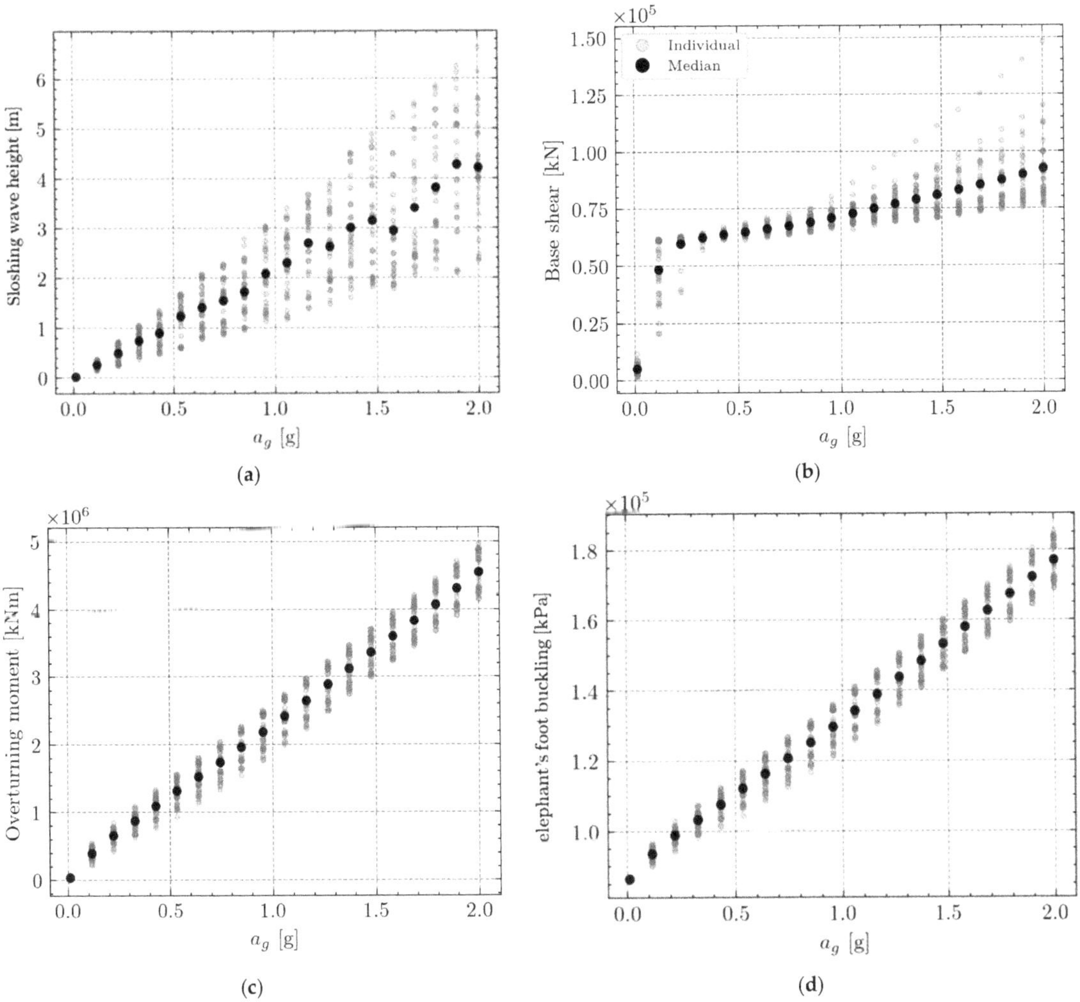

Figure 9. Results of incremental dynamic analysis from 28 accelerograms: (**a**) sloshing wave height, (**b**) base shear, (**c**) overturning moment, and (**d**) elephants foot buckling.

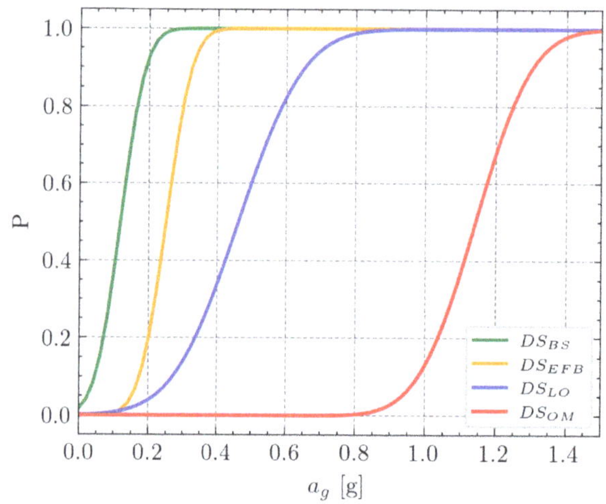

Figure 10. Fragility curves for the critical damage states of the liquid storage tank.

The median (μ) represents the intensity level at which there is a 50% probability of failure, while the standard deviation (β) provides insight into the spread or uncertainty in the tank's performance under varying levels of seismic demand. In terms of median values, DS_{OM} exhibits the highest μ value (1.141), indicating that the tank is highly resistant to overturning moments, requiring significant a_g to reach this damage state. DS_{LO} has a median capacity of 0.461, suggesting that liquid overspill occurs at moderate seismic intensities. DS_{EFB} demonstrates a much lower median capacity (0.249), indicating its susceptibility to elephant's foot buckling under relatively low PGA levels. DS_{BS} shows the lowest μ value (0.118), confirming that base shear exceedance is the most vulnerable damage state, occurring even under mild seismic excitations. In terms of logarithmic standard deviation, DS_{BS} (0.056) and DS_{EFB} (0.058) exhibit the smallest β values, indicating consistent behavior across simulations with minimal variability. DS_{LO} has a higher β value (0.149), suggesting greater variability in the occurrence of liquid overspill, likely influenced by geometric and operational factors. DS_{OM} shows a moderate β value (0.126), reflecting some variability in the overturning moment capacity.

5. Conclusions

This study introduces a novel methodology for the seismic vulnerability assessment of ground-supported liquid storage tanks, with a particular focus on tanks exhibiting deformed shells. A key contribution of this work is the development of a simplified two-degree-of-freedom (2DoF) system, which enables efficient IDA simulations for probabilistic analysis. The 2DoF model, calibrated using detailed nonlinear finite element (NL FEM) simulations, offers a balance between computational efficiency and accuracy, allowing for rapid evaluation of seismic fragility. The NL FEM model plays a critical role in capturing the complex fluid–structure interaction and calibrating the stiffness and damping properties of the 2DoF system, thereby bridging the gap between computational rigor and practical application.

Another distinguishing feature of this work is the explicit consideration of tank deformations, which are often overlooked in traditional seismic assessments. This inclusion highlights the significant influence of geometric imperfections on tank behavior, particularly in amplifying vulnerability to local buckling and other damage mechanisms. By integrating deformations into the analysis, the proposed framework provides a more realistic and comprehensive evaluation of tank performance under seismic loads.

The application of this methodology to a representative tank validated its ability to capture critical failure mechanisms and generate fragility curves associated with different damage states. The results confirmed the efficacy of the 2DoF system in reproducing seismic responses with sufficient accuracy while significantly reducing computational demands compared to full NL FEM simulations. The findings also underscored the importance of considering deformed geometries, which can exacerbate vulnerabilities in certain damage states.

This study represents a contribution to the seismic risk assessment of tanks by providing a practical yet rigorous framework that can be adapted to various tank configurations and seismic conditions. The proposed methodology offers engineers and researchers an effective tool for evaluating seismic performance, developing design improvements, and prioritizing retrofitting efforts through its simplified representation using a 2-DOF system. However, it is important to note that certain nonlinear effects, such as plastic deformation, buckling, or complex fluid–structure interactions (e.g., sloshing and impulsive forces), may be underrepresented in this approach. These limitations could affect the accuracy of the assessment, particularly under high-intensity seismic events or for tanks with complex geometries or boundary conditions. Future research should aim to incorporate these nonlinearities for a more comprehensive analysis.

Author Contributions: Conceptualization, R.R.R. and G.M.; methodology, R.R.R. and G.M.; software, R.R.R.; validation, R.R.R. and G.M.; writing—original draft preparation, R.R.R.; writing—review and editing, R.R.R. and G.M.; supervision, G.M. All authors have read and agreed to the published version of the manuscript.

Funding: This research was funded within the project DPC-ReLUIS 2024–26.

Institutional Review Board Statement: Not applicable.

Informed Consent Statement: Not applicable.

Data Availability Statement: Data are contained within the article.

Conflicts of Interest: The authors declare no conflicts of interest.

References

1. De Angelis, M.; Giannini, R.; Paolacci, F. Experimental Investigation on the Seismic Response of a Steel Liquid Storage Tank Equipped with Floating Roof by Shaking Table Tests. *Earthq. Eng. Struct. Dyn.* **2010**, *39*, 377–396. [CrossRef]
2. Phan, H.N.; Paolacci, F.; Di Filippo, R.; Bursi, O.S. Seismic Vulnerability of Above Ground Storage Tanks with Unanchored Support Conditions for Na-Tech Risks Based on Gaussian Process Regression. *Bull. Earthq. Eng.* **2020**, *18*, 6883–6906. [CrossRef]
3. Alessandri, S.; Caputo, A.C.; Corritore, D.; Giannini, R.; Paolacci, F.; Phan, H.N. Probabilistic Risk Analysis of Process Plants under Seismic Loading Based on Monte Carlo Simulations. *J. Loss Prev. Process Ind.* **2018**, *53*, 136–148. [CrossRef]
4. Bakalis, K.; Vamvatsikos, D.; Fragiadakis, M. Seismic Risk Assessment of Liquid Storage Tanks via a Nonlinear Surrogate Model. *Earthq. Eng. Struct. Dyn.* **2017**, *46*, 2851–2868. [CrossRef]
5. Bakalis, K.; Kohrangi, M.; Vamvatsikos, D. Seismic Intensity Measures for Above-Ground Liquid Storage Tanks. *Earthq. Eng. Struct. Dyn.* **2018**, *47*, 1844–1863. [CrossRef]
6. Colombo, J.I.; Almazán, J.L. Experimental Investigation on the Seismic Isolation for a Legged Wine Storage Tank. *J. Constr. Steel Res.* **2017**, *133*, 167–180. [CrossRef]
7. Merino Vela, R.J.; Brunesi, E.; Nascimbene, R. Derivation of Floor Acceleration Spectra for an Industrial Liquid Tank Supporting Structure with Braced Frame Systems. *Eng. Struct.* **2018**, *171*, 105–122. [CrossRef]
8. Sezen, H.; Livaoglu, R.; Dogangun, A. Dynamic Analysis and Seismic Performance Evaluation of Above-Ground Liquid-Containing Tanks. *Eng. Struct.* **2008**, *30*, 794–803. [CrossRef]
9. Campedel, M.; Cozzani, V.; Garcia-Agreda, A.; Salzano, E. Extending the Quantitative Assessment of Industrial Risks to Earthquake Effects. *Risk Anal.* **2008**, *28*, 1231–1246. [CrossRef] [PubMed]
10. Brunesi, E.; Nascimbene, R.; Pagani, M.; Beilic, D. Seismic Performance of Storage Steel Tanks during the May 2012 Emilia, Italy, Earthquakes. *J. Perform. Constr. Facil.* **2014**, *29*, 04014137. [CrossRef]
11. Sezen, H.; Whittaker, A.S. Seismic Performance of Industrial Facilities Affected by the 1999 Turkey Earthquake. *J. Perform. Constr. Facil.* **2006**, *20*, 28–36. [CrossRef]
12. Jacobsen, L.S. Impulsive Hydrodynamics of Fluid inside a Cylindrical Tank and of Fluid Surrounding a Cylindrical Pier. *Bull. Seismol. Soc. Am.* **1949**, *39*, 189–204. [CrossRef]
13. Housner, G.W. Dynamic Pressures on Accelerated Fluid Containers. *Bull. Seismol. Soc. Am.* **1957**, *47*, 15–35. [CrossRef]
14. Veletsos, A.S.; Tang, Y.; Tang, H.T. Dynamic Response of Flexibly Supported Liquid-Storage Tanks. *J. Struct. Eng.* **1992**, *118*, 264–283. [CrossRef]

15. Chen, W.; Haroun, M.A.; Liu, F. Large Amplitude Liquid Sloshing in Seismically Excited Tanks. *Earthq. Eng. Struct. Dyn.* **1996**, *25*, 653–669. [CrossRef]
16. Malhotra, P. Practical Nonlinear Seismic Analysis of Tanks. *Earthq. Spectra* **2000**, *16*, 473–492. [CrossRef]
17. Virella, J.C.; Godoy, L.A.; Suárez, L.E. Fundamental Modes of Tank-Liquid Systems under Horizontal Motions. *Eng. Struct.* **2006**, *28*, 1450–1461. [CrossRef]
18. Bakalis, K.; Fragiadakis, M.; Vamvatsikos, D. Surrogate Modeling for the Seismic Performance Assessment of Liquid Storage Tanks. *J. Struct. Eng.* **2016**, *143*, 04016199. [CrossRef]
19. Kalogerakou, M.E.; Maniatakis, C.A.; Spyrakos, C.C.; Psarropoulos, P.N. Seismic Response of Liquid-Containing Tanks with Emphasis on the Hydrodynamic Response and near-Fault Phenomena. *Eng. Struct.* **2017**, *153*, 383–403. [CrossRef]
20. *API 650*; Welded Tanks for Oil Storage. American Petroleum Institute: Washington, DC, USA, 2010.
21. *ASCE ASCE/SEI 7-16*; Minimum Design Loads and Associated Criteria for Buildings and Other Structures. American Society of Civil Engineers: Reston, VA, USA, 2016.
22. *ENV-1998-4*; CEN Eurocode 8: Design of Structures for Earthquake Resistance—Part 4: Silos, Tanks and Pipelines. Comité Européen de Normalisation: Brussels, Belgium, 2006.
23. *NZSEE Recommendations for Seismic Design of Storage Tanks*; New Zealand Society for Earthquake Engineering: Christchurch, New Zealand, 2009.
24. Vamvatsikos, D.; Cornell, C.A. Applied Incremental Dynamic Analysis. *Earthq. Spectra* **2004**, *20*, 523–553. [CrossRef]
25. Alavi, B.; Krawinkler, H. Behavior of Moment-Resisting Frame Structures Subjected to near-Fault Ground Motions. *Earthq. Eng. Struct. Dyn.* **2004**, *33*, 687–706. [CrossRef]
26. Choi, E.; DesRoches, R.; Nielson, B. Seismic Fragility of Typical Bridges in Moderate Seismic Zones. *Eng. Struct.* **2004**, *26*, 187–199. [CrossRef]
27. American Lifelines Alliance. *Seismic Fragility Formulation for Water Systems: Part 2-Appendices*; ASCE: Reston, VA, USA, 2001.
28. American Lifelines Alliance. *Seismic Fragility Formulation for Water Systems: Part 1-Guideline*; ASCE: Reston, VA, USA, 2001.
29. Agency, F.E.M. *Multi-Hazard Loss Estimation Methodology, Earthquake Model*; HAZUS-MH MR4 Technical Manual: Washington, DC, USA, 2003.
30. Iervolino, I.; Fabbrocino, G.; Manfredi, G. Fragility of Standard Industrial Structures by a Response Surface Based Method. *J. Earthq. Eng.* **2004**, *8*, 927–945. [CrossRef]
31. Berahman, F.; Behnamfar, F. Probabilistic Seismic Demand Model and Fragility Estimates for Critical Failure Modes of Un-Anchored Steel Storage Tanks in Petroleum Complexes. *Probabilistic Eng. Mech.* **2009**, *24*, 527–536. [CrossRef]
32. Phan, H.N.; Paolacci, F.; Alessandri, S. Enhanced Seismic Fragility Analysis of Unanchored Steel Storage Tanks Accounting for Uncertain Modeling Parameters. *J. Press. Vessel Technol. Trans. ASME* **2019**, *141*, 010903. [CrossRef]
33. Merino Vela, R.J.; Brunesi, E.; Nascimbene, R. Seismic Assessment of an Industrial Frame-Tank System: Development of Fragility Functions. *Bull. Earthq. Eng.* **2019**, *17*, 2569–2602. [CrossRef]
34. Merino, R.J.; Brunesi, E.; Nascimbene, R. Probabilistic Evaluation of Earthquake-Induced Sloshing Wave Height in above-Ground Liquid Storage Tanks. *Eng. Struct.* **2020**, *202*, 109870. [CrossRef]
35. Cortes, G.; Prinz, G.S. Seismic Fragility Analysis of Large Unanchored Steel Tanks Considering Local Instability and Fatigue Damage. *Bull. Earthq. Eng.* **2017**, *15*, 1279–1295. [CrossRef]
36. Seleemah, A.A.; El-Sharkawy, M. Seismic Response of Base Isolated Liquid Storage Ground Tanks. *Ain Shams Eng. J.* **2011**, *2*, 33–42. [CrossRef]
37. Moslemi, M.; Kianoush, M.R.; Pogorzelski, W. Seismic Response of Liquid-Filled Elevated Tanks. *Eng. Struct.* **2011**, *33*, 2074–2084. [CrossRef]
38. European Standardization. Eurocode 8-Design of Structures for Earthquake Resistance-Part 4: Silos, Tanks and Pipelines. *Specific Principles and Application Rules for Tanks*. 2006. Available online: http://bsb.co.in/EC_HTML/HTML/..%5Cpdf%5C30047498.pdf (accessed on 13 December 2024).
39. Veletsos, A.S.; Auyang, J. Earthquake Response of Liquid Storage Tanks. In *Advances in Civil Engineering Through Engineering Mechanics*; American Society of Civil Engineers: Raleigh, NC, USA, 1977.
40. Okada, J.; Iwata, K.; Tsukimori, K.; Nagata, T. An Evaluation Method for Elastic-Plastic Buckling of Cylindrical Shells under Shear Forces. *Nucl. Eng. Des.* **1995**, *157*, 65–79. [CrossRef]
41. Rotter, J.M. Local Collapse of Axially Compressed Pressurized Thin Steel Cylinders. *J. Struct. Eng.* **1990**, *116*, 1955–1970. [CrossRef]
42. Dassault Systemes Simulia ABAQUS 2023. Available online: https://www.3ds.com/products/simulia/abaqus (accessed on 13 December 2024).
43. Pacific Earthquake Engineering Research Center PEER NGA Database. Available online: https://www.ngawest2.berkeley.edu/ (accessed on 20 October 2024).
44. Zhu, M.; McKenna, F.; Scott, M.H. OpenSeesPy: Python Library for the OpenSees Finite Element Framework. *SoftwareX* **2018**, *7*, 6–11. [CrossRef]

applied sciences

MDPI

Article

Displacement-Based Seismic Design of Multi-Story Reinforced Concrete-Coupled Shear Wall Buildings with Energy-Dissipation Dampers

Zafira Nur Ezzati Mustafa and Taiki Saito *

Department of Architecture and Civil Engineering, Toyohashi University of Technology, Toyohashi 441-8580, Aichi, Japan; zafira.nur.ezzati.bt.mustafa.jp@tut.jp
* Correspondence: saito.taiki.bv@tut.jp

Abstract: This research aims to apply the displacement-based design method (DBDM) for the seismic design of reinforced concrete-coupled shear wall buildings equipped with energy dissipation dampers. The DBDM offers design simplicity by focusing on structural design based on a target design displacement, where the building converts into a single degree of freedom (SDOF) system. The implementation of dampers aims to reduce repair costs and downtime for buildings following significant seismic events. Two types of dampers are utilized in this study: metallic damper and viscoelastic damper. The DBDM procedure begins with determining the target displacement, which corresponds to the specific story drift ratio of the structural system, using a nonlinear static pushover analysis. For the structural wall system considered in this study, a target drift ratio of 1/250 is selected due to the inherent rigidity of the structure. The effective damping factor is then determined from the average energy absorption, which is based on the ductility factor of each structural member. Additionally, the effective period of the building is obtained from the displacement spectrum of the design-level earthquakes. Finally, the required damper shear capacity for the SDOF system is calculated based on the target deformation and effective stiffness. The design earthquakes are generated from the acceleration response spectrum for Level 2 earthquakes, as specified in the Japanese seismic code, utilizing three different sets of phase information: Kobe, El Centro, and random phase records. The effectiveness of the DBDM is scrutinized through a comparison with results obtained from time history analysis. The results obtained for 6-, 12-, and 18-story RC-coupled shear walls with energy dissipation dampers indicate that the proposed design methodology effectively meets the specified design objectives.

Keywords: coupled shear wall; seismic design; displacement-based design method; metallic damper; viscoelastic damper

1. Introduction

Over the last decade, Priestley devised the displacement-based design method (DBDM) with the goal of alleviating the shortcomings in the existing force-based design [1]. The primary issue with force-based design lies in the initial requirement to estimate the size of structural members based on dynamic force assumptions. When there is a change in the design force, the designer advocates for recalibration. The underlying principle of the design methodology is to create a structure that aims to achieve, rather than be constrained by predetermined performance limits under specific seismic intensities. Priestley outlines that the design procedure establishes the necessary strength at specified plastic hinge points to meet the design objectives in relation to defined displacement goals. Pennucci and Fox have employed the DBDM for designing RC-coupled shear walls (CSWs) with diagonal reinforced coupling beams and have implemented it for both designing and retrofitting various structures [2,3]. In the DBDM, the seismic force is approximated by substituting the multi-degree-of-freedom (MDOF) system with an analogous single-degree-of-freedom

(SDOF) system. The anticipated dissipation of energy from the inelastic deformation of the structure is accounted for by diminishing displacement demand spectrum through the application of a reduction factor derived from effective damping. In the DBDM, the crucial final step is a capacity design, which aims to ensure that inelastic action is confined to intended locations, preventing unintended plastic hinges from occurring elsewhere.

Generally, the seismic assessment and design of structures equipped with damping devices require nonlinear time history analysis (THA), which is highly time-consuming. The DBDM is a more direct method that can be applied to simplified design processes that being proposed by researchers to evaluate the seismic performance of passively controlled buildings. The improvised DBDM system proposed by Pennucci by adopting a rocking wall system with additional damper design methodology is effective and can be confidently applied to structures up to 20 stories high [4]. Yi Xiao proposed an efficient DBDM approach that is in accordance with the Chinese seismic design principle by establishing a relationship between the effective damping and structural period of the viscoelastic damper structural system [5]. Ying Zhou proposed a simplified DBDM by using the SDOF system that represents a damped structure by integrating the viscoelastic damper strain and stiffness ratio of the damper to the main structure to achieve predefined target performance [6]. Segovia and Ruiz evaluate cost efficiency using the DBDM approach for damage-controlled structure using the best combination of stiffness and strength ratios between a main structure and an added hysteresis energy dissipation system to achieve optimum performance [7].

Metallic dampers are widely recognized as effective mechanisms for dissipating seismic energy in structures during earthquakes, primarily through the inelastic deformation of metallic materials [8]. Integrating metallic dampers into structural systems has demonstrated a significant capacity for dissipating seismic energy, thereby mitigating earthquake-induced damage. Researchers and practitioners have increasingly adopted full-scale implementation of metallic dampers based on experimental evidence validating their performance. For instance, Whittaker conducted an extensive experimental program at the University of California, Berkeley, to evaluate the performance of an X-shaped plate damper, known as Added Damping And Stiffness (ADAS) elements [9].

Extensive analytical and experimental investigations have been conducted on the dynamic response of viscoelastic dampers and the seismic response of structures incorporating these dampers. The experimental studies utilized a 2/5-scale prototype structure subjected to various controlled ambient temperatures and recorded ground motions indicating that viscoelastic dampers significantly enhance seismic response reduction in structures [10,11]. Christopoulos replaced coupling beams with viscoelastic dampers to avoid consuming precious architectural space while addressing issues related to building occupants' perception of vibrations and diminishing both wind and earthquake effects on the structure [12]. Consequently, the building's structural efficacy is notably enhanced with viscoelastic damper implementation compared to conventional design methods.

Several researchers have proposed the DBDM for buildings with energy dissipation dampers; however, the amount of research is relatively limited. Therefore, this research examines the effectiveness of the DBDM for CSW structures with energy dissipation dampers. To refine the DBDM, a simplified approach was developed, involving the construction of a DBDM graph in which a capacity curve is derived from pushover analysis and a target performance line is calculated based on the structure's equivalent SDOF model. Structural design compatibility is confirmed when the target line touches the capacity curve. The analysis focused on CSWs at different heights with steel coupling beams and metallic and viscoelastic dampers. Using the proposed method, the necessary damping can be estimated to confirm adequate structural capacity. The efficacy of the DBDM is evaluated by comparing the story drifts obtained by the THA with the design target drift.

2. Design Procedure of DBDM

(Step 1) The flow chart in Figure 1 explains the procedure of the DBDM. The methodology starts with preparing structural prototypes. In this study, the structural system is classified as a structural wall system, where two shear walls are interconnected by coupling beams. Energy dissipation dampers are installed on these coupling beams.

(Step 2) As schematically shown in Figures 2 and 3, the step-by-step results of nonlinear push-over analysis are used to obtain the capacity curve of the equivalent SDOF system using Equation (1). The design target story drift is assumed to be 1/250.

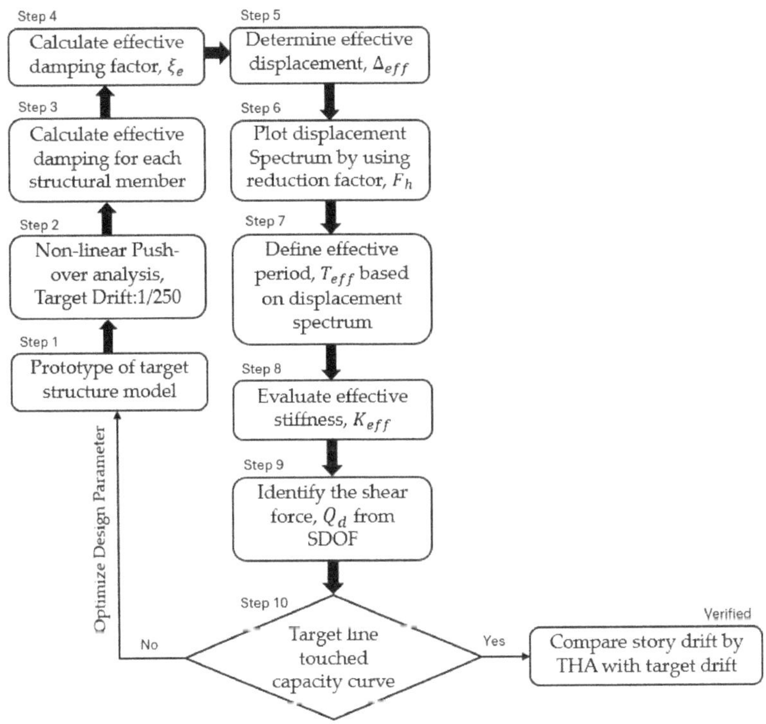

Figure 1. Displacement-based design method flowchart.

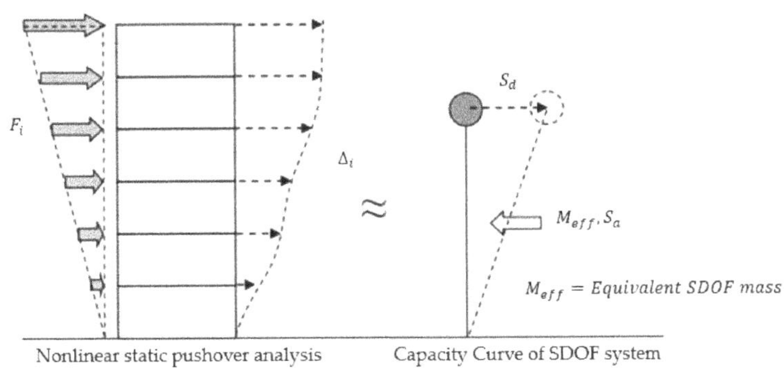

Figure 2. Pushover analysis to obtain equivalent SDOF.

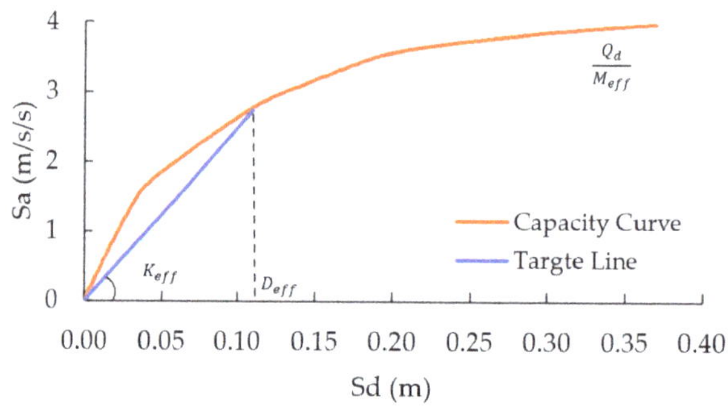

Figure 3. Displacement-based design method performance graph.

$$S_d = \frac{\sum m_i \Delta_i{}^2}{\sum m_i \Delta_i},\ S_a = \frac{\sum m_i \Delta_i{}^2}{(\sum m_i \Delta_i)^2} Q_d = \frac{\Delta_{eff}}{\sum m_i \Delta_i} Q_d \tag{1}$$

$$\text{Design target story drift}: \frac{1}{250} \times \text{inter-story height} \tag{2}$$

where

S_d: displacement spectrum;
S_a: acceleration spectrum;
Q_d: based shear force.
(Step 3) The effective damping factor for each member at the target drift is determined using the following formula based on the ductility factors [13].

$$\text{Damping of RC wall}: \xi_{e,wi} = 0.05 + 0.444((\mu - 1)/\mu\pi) \tag{3}$$

$$\text{Damping of metallic damper}: \xi_M = (0.8)2/\pi[(1 - 1/\mu)]\ [(1 - a)/(1 + a(\mu - 1))] \tag{4}$$

$$\text{Damping of viscoelastic damper}: \xi_{VE} = (G_c\omega)/(2G_E) \tag{5}$$

where

$\xi_{e,Wi}$: effective damping for RC shear wall;
ξ_M: effective damping for metallic damper;
ξ_{VE}: effective damping for viscoelastic damper;
μ: ductility factor;
a: stiffness ratio of metallic damper;
G_c: shear storage modulus of viscoelastic material;
ω: the circular frequency of the structure;
G_E: shear loss modules of viscoelastic material.
(Step 4) The damping factor of the entire structure is derived from the energy average

$$\xi_e = \frac{\sum E_{wi}\xi_{e,wi} + \sum E_{bi}\xi_{e,bi}}{\sum E_{wi} + \sum E_{bi}} \tag{6}$$

where

ξ_e: effective damping factor of structural system;
$\xi_{e,bi}$: effective damping factor of coupling beam
(=ξ_M for metallic damper, $= \xi_{VE}$ for viscoelastic damper);
E_{Wi}: potential energy of wall;
E_{bi}: potential energy of coupling beam
($= E_M$ for metallic damper, $= E_{VE}$ for viscoelastic damper).
(Step 5) The effective displacement of the SDOF system is determined based on the displacement distribution corresponding to the target displacement, as follows:

$$\Delta_{eff} = \frac{\sum m_i \Delta_i^2}{\sum m_i \Delta_i} \tag{7}$$

where

Δ_{eff}: effective displacement;
m_i: story mass;
Δ_i: story displacement at the target story drift.
(Step 6) Based on the target deformation and the effective damping factor, the effective period T_{eff} is derived from the displacement response spectrum, and the spectrum reduction factor F_h is calculated using the following formula:

$$F_h = \frac{1.5}{(1 + 10(\xi_e))} \tag{8}$$

(Step 7) Evaluate the effective stiffness, K_{eff}.

$$K_{eff} = \frac{4\pi^2 m_{eff}}{T_{eff}^2} \tag{9}$$

where

m_{eff}: effective mass;
T_{eff}: effective period.
(Step 8) Identify the shear force Q_d of the equivalent SDOF system based on effective stiffness and effective displacement.

$$Q_d = K_{eff} \Delta_{eff} \tag{10}$$

Figure 3 shows the target line of the equivalent linear system.
(Step 9) Design adequacy is evaluated based on whether the target line touches the capacity curve line, as shown in Figure 3. Additionally, if the target line surpasses the capacity curve, the design is considered under-designed. Meanwhile, if the target line is drawn below the capacity curve, it is considered as over-designed.
(Step 10) Assess the accuracy of the design by comparing the story drifts obtained by the THA with the target story drift.

3. Design Earthquakes and Coupled Shear Wall Structures

3.1. Earthquake Design Spectrum

The design earthquake for this study was characterized by the acceleration response spectrum for the Level 2 (safety limit state) earthquake on the stiff soil ground (soil profile Type 1) as specified in the Japanese seismic code [14].

The design acceleration response spectrum for the Level 2 earthquake is defined as

$$S_a(\mathrm{m/s/s}) = \begin{cases} 3.2 + 30T & for\ T < 0.16 \\ 8 & for\ 0.16 \le T < 0.64 \\ 5.12/T & for\ 0.64 \le T \end{cases} \tag{11}$$

where

T(s): the fundamental natural period of the building.

The surface soil amplification factor G_s for the soil profile Type 1 is defined as

$$G_s = \begin{cases} 1.5 & for\ T < 0.576 \\ 0.864/T & for\ 0.576 \le T < 0.64 \\ 1.35 & for\ 0.64 \le T \end{cases} \tag{12}$$

Then, the target acceleration response spectrum is obtained by $G_s \times S_a$. To generate artificial ground motions consistent with this spectrum, STERA Wave software V1.0 [15] was utilized. Three different artificial earthquake phases were implemented: a random phase with 120 s duration, the 1940 El Centro phase, and the 1995 Kobe phase. The acceleration response spectra of these artificial ground motions are depicted in Figure 4.

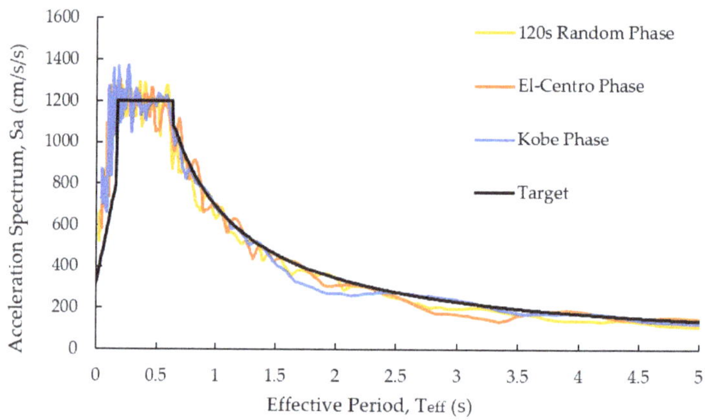

Figure 4. Acceleration of response spectra of artificial ground motions.

Utilizing Equation (8), displacement spectra for varying damping factors were computed and depicted in Figure 5. From the effective displacement values derived using Equation (7), the effective period of the structural system was determined from Figure 5.

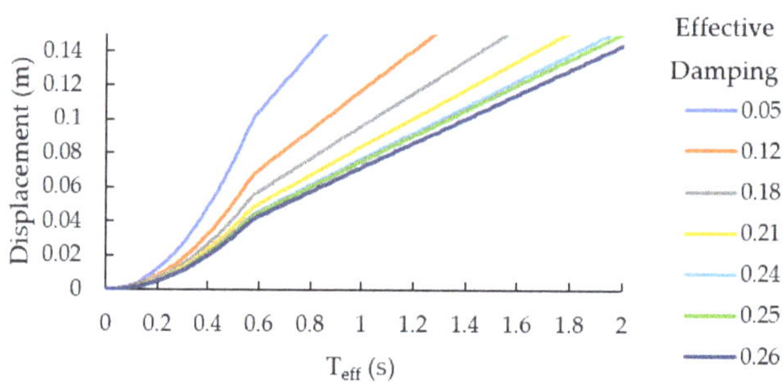

Figure 5. Displacement spectrum with various damping factors.

3.2. Coupling Shear Wall Structures with Energy Dissipation Dampers

This research examines the implementation of energy dissipation dampers, achieved by installing dampers in the middle of the coupling steel beams. Figure 6 depicts the placement of dampers for 6-story, 12-story, and 18-story CSWs, while Figure 7 shows cross-sectional views for each. The clear length of the RC walls is set to 5 m, with column sizes varying according to the building height. The total length of the coupling beams, including the dampers, is 2.1 m, with the first inter-story height set at 4.5 m and subsequent stories at 3.6 m. The structure properties of the CSWs are summarized in Table 1.

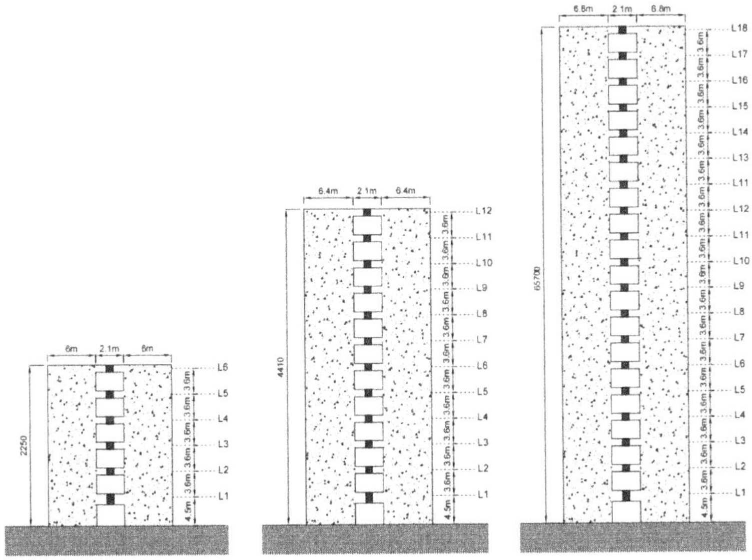

(a) 6 Stories (b) 12 Stories (c) 18 Stories

Figure 6. Elevation view of RC shear wall with different types of dampers.

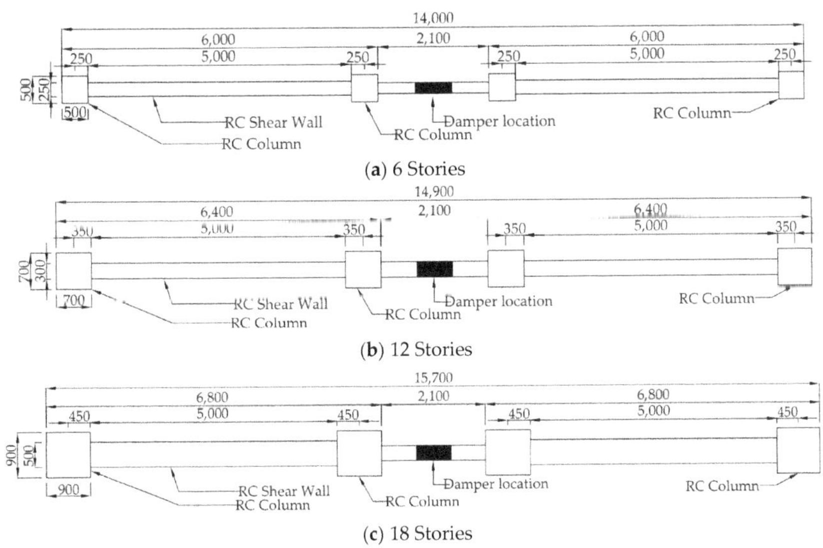

Figure 7. RC shear wall cross-section with different building story height.

Table 1. Structural member properties.

Structural Member	Size (mm)	Concrete Strength (MPa)	Steel Strength (MPa)	Rebar Size
		6 Stories		
Column	500 × 500	40	-	D25
Shear wall	5000 × 250	40	-	D10
Steel beam	I-600 × 300 × 10 × 10	-	325	-
		12 Stories		
Column	700 × 700	40	-	D29
Shear wall	5000 × 300	40	-	D10
Steel beam	I-600 × 300 × 10 × 10	-	325	-
		18 Stories		
Column	900 × 900	40	-	D35
Shear wall	5000 × 500	40	-	D10
Steel beam	I-600 × 300 × 10 × 10	-	325	-

4. DBDM of Coupled Shear Wall Structures with Metallic Dampers

The working mechanism of metallic dampers involves the dissipation of input energy through the plastic yielding of the metal. As illustrated in Figure 8, a metallic damper is positioned at the center of the coupling steel beam connecting two RC shear walls. The metallic damper was designed with an hourglass-shaped geometry, allowing the central section of the plate to undergo yielding and effectively dissipate energy.

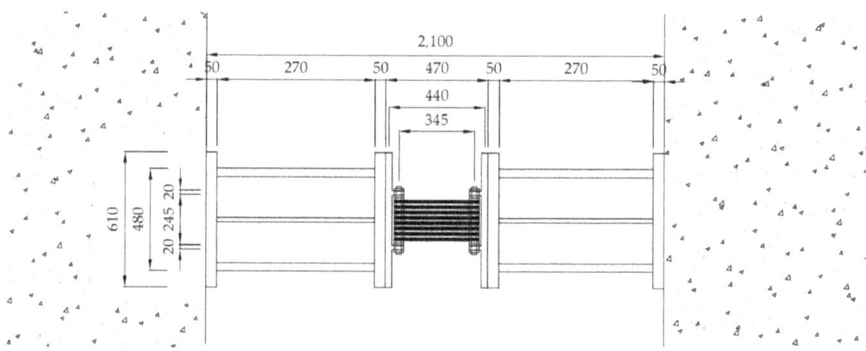

Figure 8. Metallic damper details.

Figure 9 illustrates the hysteresis loop of the metallic damper. The shape of the hysteresis loop is represented by the bilinear model. The symbols of elastic stiffness, yield force, and post-yield stiffness are defined as K_d, F_y, and K_{dd}. The properties of the metallic dampers are presented in Table 2. The maximum shear capacity (100%) of the damper was assumed to be 500 kN for 6-story CSW, 600 kN for 12-story CSW, and 700 kN for 18-story CSW. To determine the optimum amount of damper to achieve the target displacement, the analysis was performed by varying the damper capacity to 75%, 50%, and 25%. This ratio is referred to as damping force reduction (DFR) in this paper.

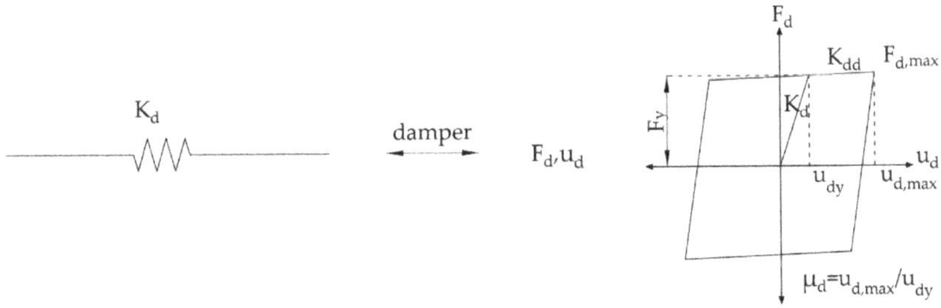

Figure 9. Metallic damper mathematical model.

Table 2. Metallic damper properties for 6, 12, and 18 stories.

		6 Stories	
Damper Type	Force, F_y (kN)	Stiffness, K_d (kN/mm)	Stiffness Ratio (K_{dd}/K_d)
100% Metallic	500	525	0.02
75% Metallic	375	394	0.02
50% Metallic	250	262	0.02
25% Metallic	125	131	0.02
		12 Stories	
Damper Type	Force, F_y (kN)	Stiffness, K_d (kN/mm)	Stiffness Ratio (K_{dd}/K_d)
100% Metallic	600	630	0.02
75% Metallic	450	472	0.02
50% Metallic	300	315	0.02
25% Metallic	150	157	0.02
		18 Stories	
Damper Type	Force, F_y (kN)	Stiffness, K_d (kN/mm)	Stiffness Ratio (K_{dd}/K_d)
100% Metallic	700	735	0.02
75% Metallic	525	551	0.02
50% Metallic	350	367	0.02
25% Metallic	175	184	0.02

Following the procedure of the DBDM outlined in Section 2, the capacity curves and target lines are compared, where the equivalent damping factor of the metallic damper is obtained using Equation (4).

Figure 10 illustrates the DBDM performance of the 6-story CSW system with metallic dampers with different DFRs. Figure 10a,b indicates that the 25% DFR and 50% DFR designs fail to meet the required design capacity, as evidenced by the target line surpassing the capacity curve line. Figure 10d illustrates the target line positioned below the capacity curve, indicating that the 100% DFR design is over-designed. On the other hand, as shown in Figure 10c, 75% DFR design meets the design requirements. The accuracy of the DBDM is verified by comparing the story drift obtained from the THA with the target story drift of 1/250 under three artificial ground motions as shown in Figure 11. In all cases, the maximum story drift for 75% DFR design is in good agreement with the target story drift, while the maximum story drifts for 50% DFR and 25% DFR exceed the threshold. This means that the accuracy of 75% DFR design, which was considered optimal in the DBDM, was confirmed.

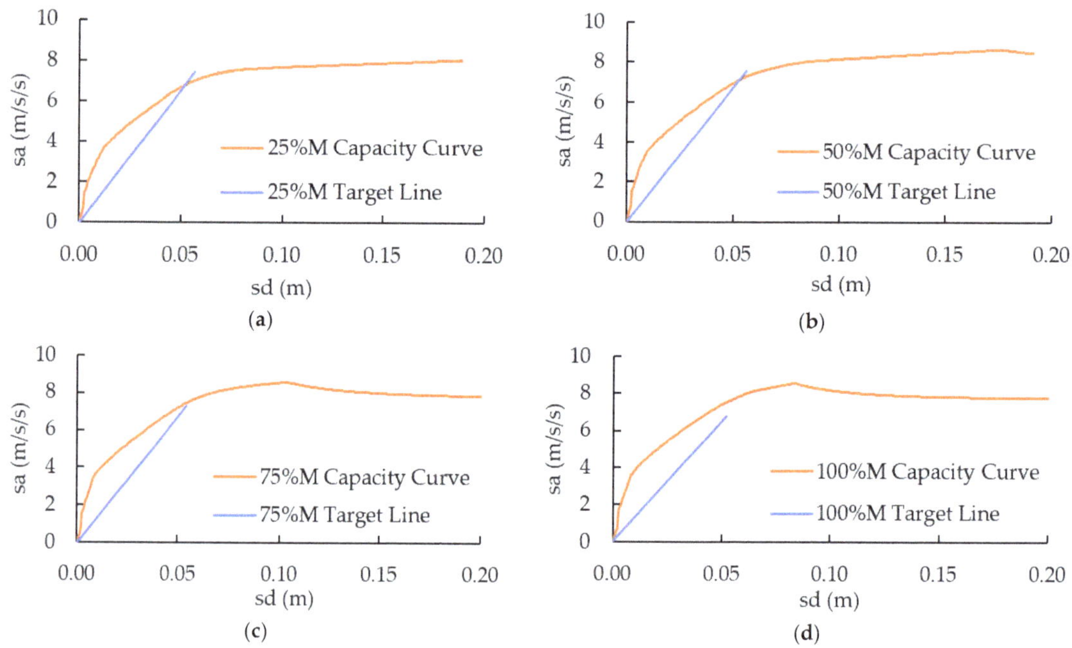

Figure 10. DBDM performance for 6 stories with metallic damper: (**a**) 25% DFR; (**b**) 50% DFR; (**c**) 75% DFR; (**d**) 100% DFR.

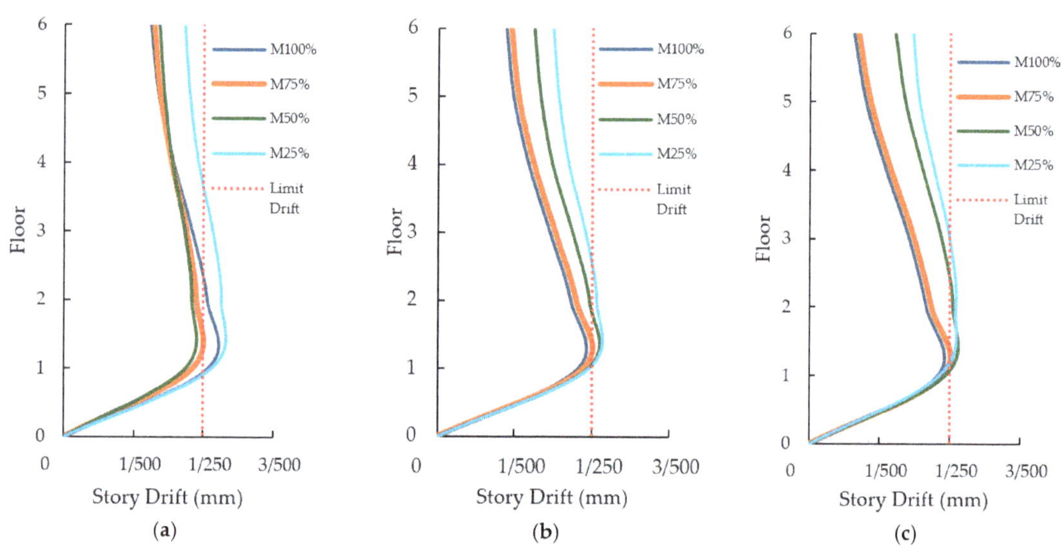

Figure 11. Story drift of 6 stories CSW with viscoelastic damper from THA: (**a**) random phase 120 s; (**b**) El Centro; (**c**) Kobe.

The results of the same study for 12-story CSW are shown in Figures 12 and 13, and the results for 18-story CSW are shown in Figures 14 and 15. As with the 6-story CSW, the 75% DFR design is considered optimal in the DBDM and it is found to best meet the target story drift in the THA.

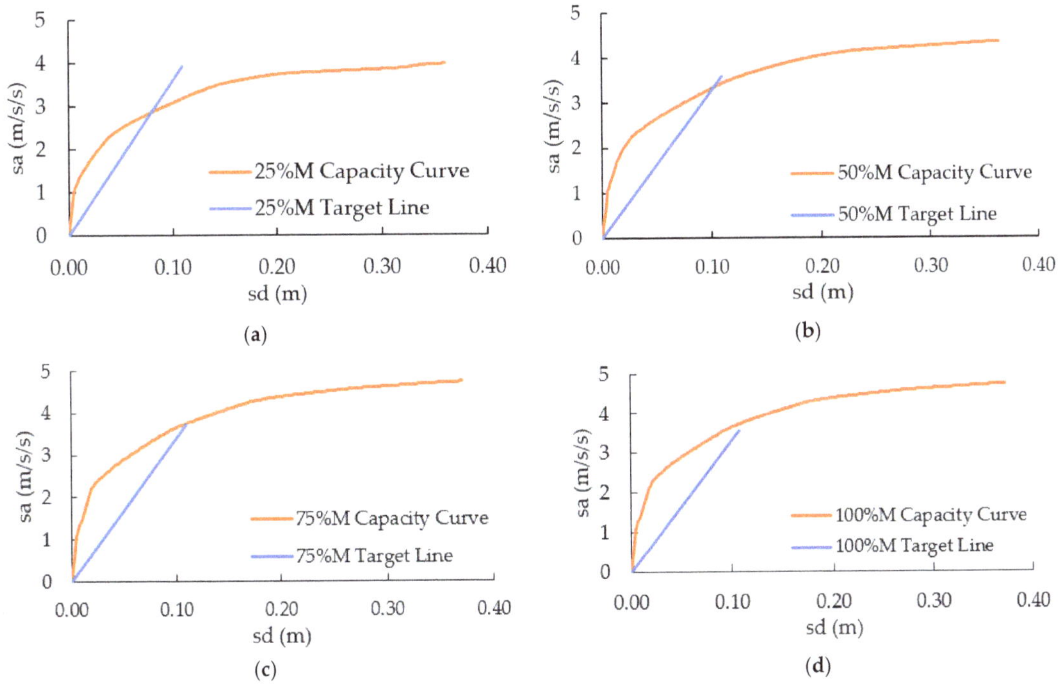

Figure 12. DBDM performance for 12 stories with metallic damper: (**a**) 25% DFR; (**b**) 50% DFR; (**c**) 75% DFR; (**d**) 100% DFR.

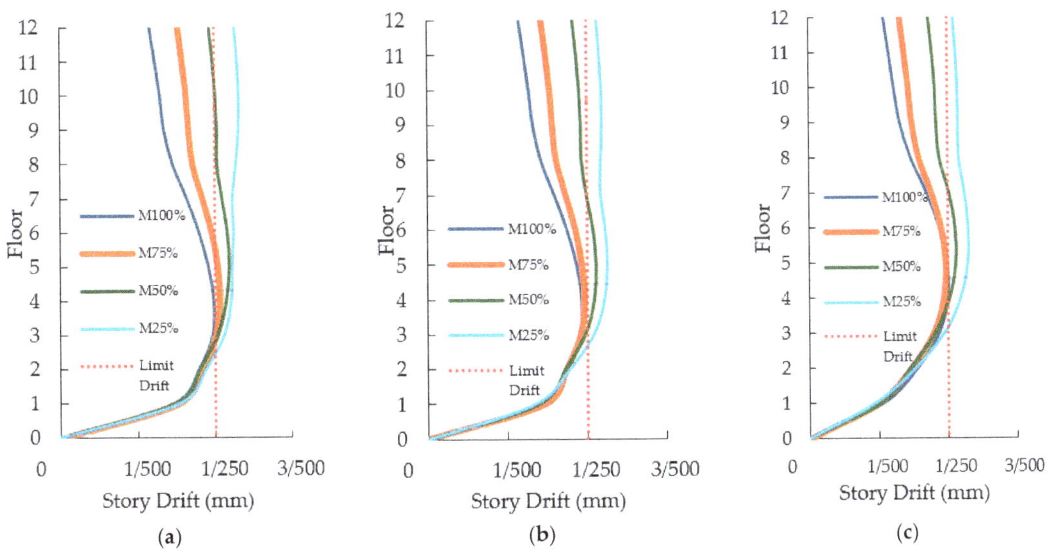

Figure 13. Story drift of 12 stories CSW with viscoelastic damper from THA: (**a**) random phase 120 s; (**b**) El Centro; (**c**) Kobe.

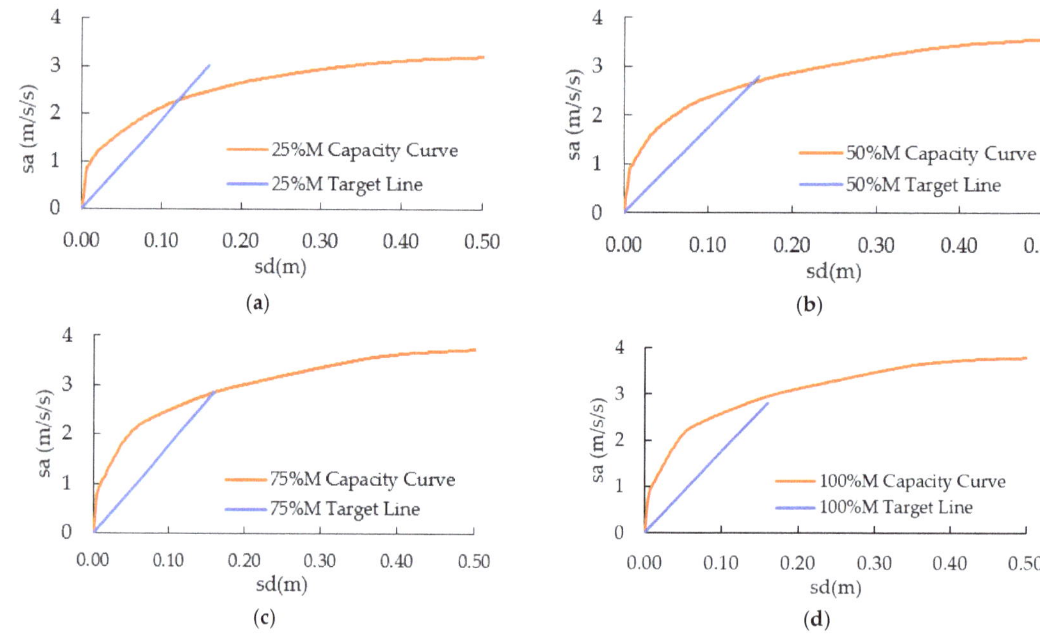

Figure 14. DBDM performance for 18 stories with metallic damper: (**a**) 25% DFR; (**b**) 50% DFR; (**c**) 75% DFR; (**d**) 100% DFR.

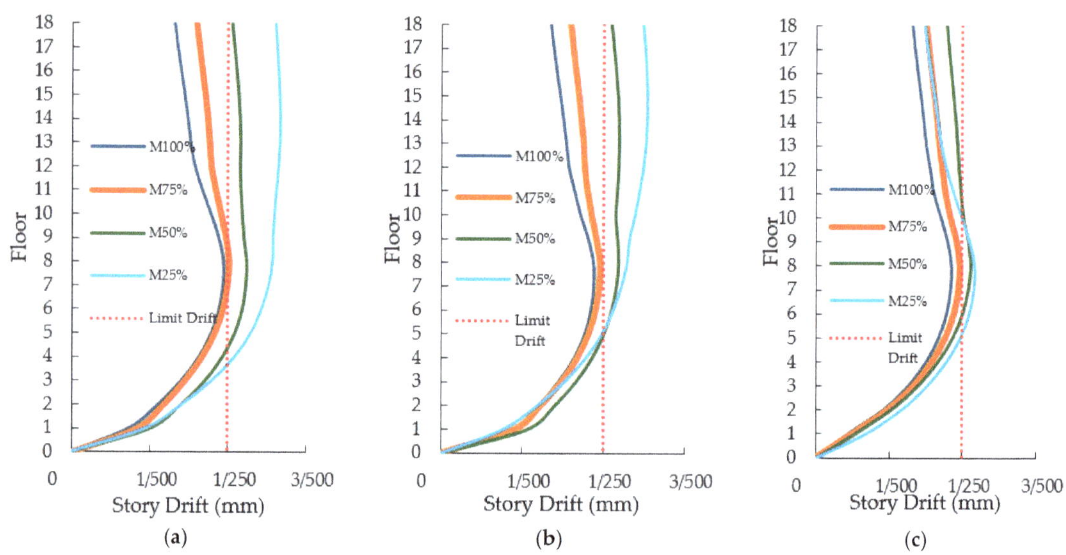

Figure 15. Story drift of 18 stories CSW with metallic damper from THA: (**a**) random phase 120 s; (**b**) El Centro; (**c**) Kobe.

5. DBDM of Coupled Shear Wall Structures with Viscoelastic Dampers

Viscoelastic dampers are utilized to mitigate vibrations by converting kinetic energy into heat, thereby attenuating system motion. Their design enables functionality across a wide frequency range. Figure 16 provides a schematic of the viscoelastic damper, illustrating the 5 mm thickness of the viscoelastic material, which accommodates a maximum shear displacement of up to 300% (15 mm). The area and the number of viscoelastic material layers are determined in accordance with the design capacity of the damper.

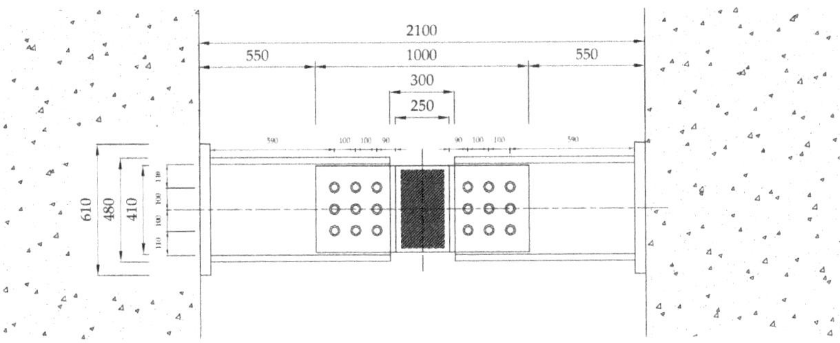

Figure 16. Viscoelastic damper details.

The hysteresis behavior of components and systems at a specific circular frequency for a vibration control structure with a viscoelastic damper is illustrated in Figure 17. The shape of the hysteresis loop of the viscoelastic damper is an inclined elliptical hysteresis curve. The symbols of the storage stiffness, damping coefficient, and an angular frequency are defined as K'_d, C_d, and ω. The properties of the viscoelastic damper are presented in Table 3. The maximum shear capacity (100%) of the damper was assumed to be 509 kN for the 6-story CSW, 605 kN for the 12-story CSW, and 770 kN for the 18-story CSW. To determine the optimum amount of damper to achieve the target displacement, the analysis was performed by varying the damper capacity to 75%, 50%, and 25% DFR.

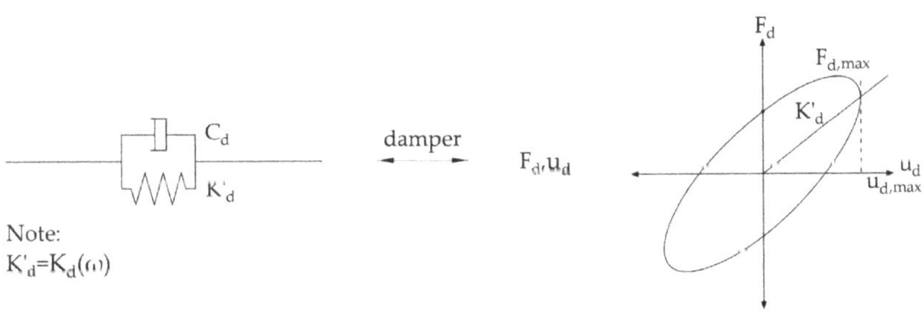

Figure 17. Viscoelastic damper mathematical model.

Table 3. Viscoelastic damper properties for 6, 12, and 18 stories.

		6 Stories		
Damper Type	Force, $F_{d,max}$ (kN)	Stiffness, K'_d (kN/mm)	Damping Coefficient, C_d (kNs/mm)	Angular Frequency, ω (mm/s)
100% Viscoelastic	509	37.318	2.265	15.325
75% Viscoelastic	328	27.989	1.583	16.448
50% Viscoelastic	255	18.659	1.077	16.111
25% Viscoelastic	127	9.330	0.552	15.708

Table 3. *Cont.*

12 Stories				
Damper Type	Force, $F_{d,max}$ (kN)	Stiffness, K'_d (kN/mm)	Damping Coefficient, C_d (kNs/mm)	Angular Frequency, ω (mm/s)
100% Viscoelastic	605	44.688	5.966	6.966
75% Viscoelastic	453	33.516	4.633	6.727
50% Viscoelastic	300	22.344	3.231	6.431
25% Viscoelastic	149	11.172	1.716	6.053
18 Stories				
Damper Type	Force, $F_{d,max}$ (kN)	Stiffness, K'_d (kN/mm)	Damping Coefficient, C_d (kNs/mm)	Angular Frequency, ω (mm/s)
100% Viscoelastic	770	56.448	12.132	4.327
75% Viscoelastic	579	42.336	9.462	4.161
50% Viscoelastic	385	28.224	6.646	3.949
25% Viscoelastic	192	14.112	3.580	3.666

Following the procedure of the DBDM outlined in Section 2, the capacity curves and target lines are compared, where the equivalent damping factor of the viscoelastic damper is obtained using Equation (5).

Figure 18 illustrates the DBDM performance of the 6-story CSW system with viscoelastic dampers with different DFRs. Figure 18a,b indicates that the 25% DFR and 50% DFR designs fail to meet the required design capacity, as evidenced by the target line surpassing the capacity curve line. Figure 18d illustrates the target line positioned below the capacity curve, indicating that the 100% DFR design is over-designed. On the other hand, as shown in Figure 18c, 75% DFR design meets the design requirements. The accuracy of the DBDM is verified by comparing the story drift obtained from the THA with the target story drift of 1/250 under three artificial ground motions, as shown in Figure 19. In all cases, the maximum story drift for 75% DFR design is in good agreement with the target story drift, while the maximum story drifts for 50% DFR and 25% DFR exceed the threshold. This means that the accuracy of 75% DFR design, which was considered optimal in the DBDM, was confirmed.

In the case of the 12-story CSW, the results exhibit similar trends to those observed in the 6-story CSW. As illustrated in Figure 20c, 75% DFR appears to be sufficient to meet the design target, with the DBDM indicating that the target performance line intersects the capacity curve. The accuracy of the DBDM is then verified as shown in Figure 21. In the case of the 18-story CSW, Figure 22b demonstrates that 50% DFR is sufficient, as the target curve intersects the capacity curve. As shown in Figure 23, the maximum story drift of the 18-story CSW with 50% DFR design obtained by the THA is found to best meet the target story drift. This fact implies the accuracy of the DBDM.

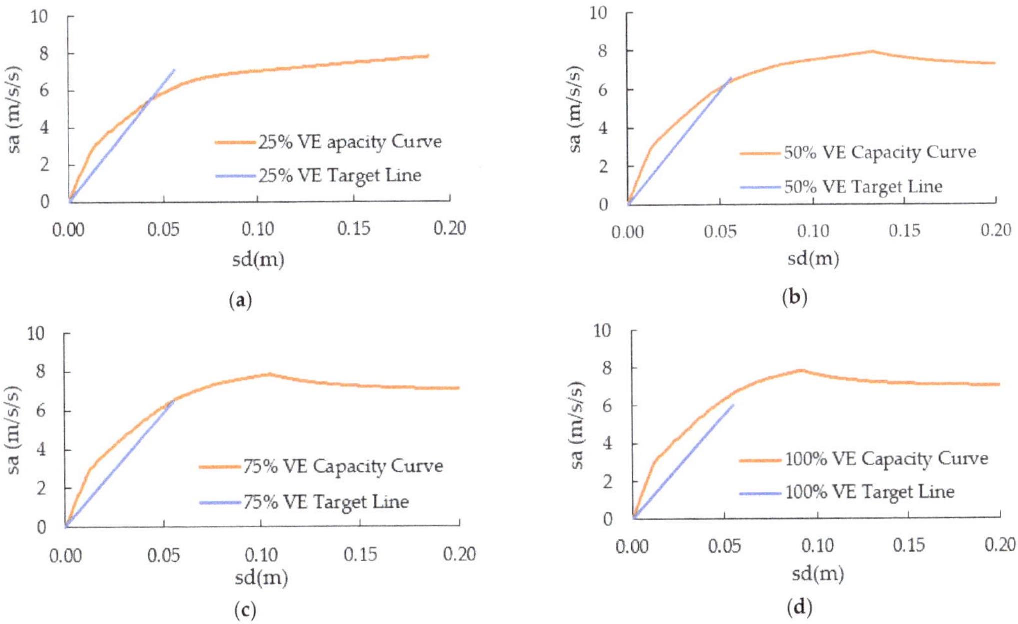

Figure 18. DBDM performance for 6 stories with viscoelastic damper: (**a**) 25% DFR; (**b**) 50% DFR; (**c**) 75% DFR; (**d**) 100% DFR.

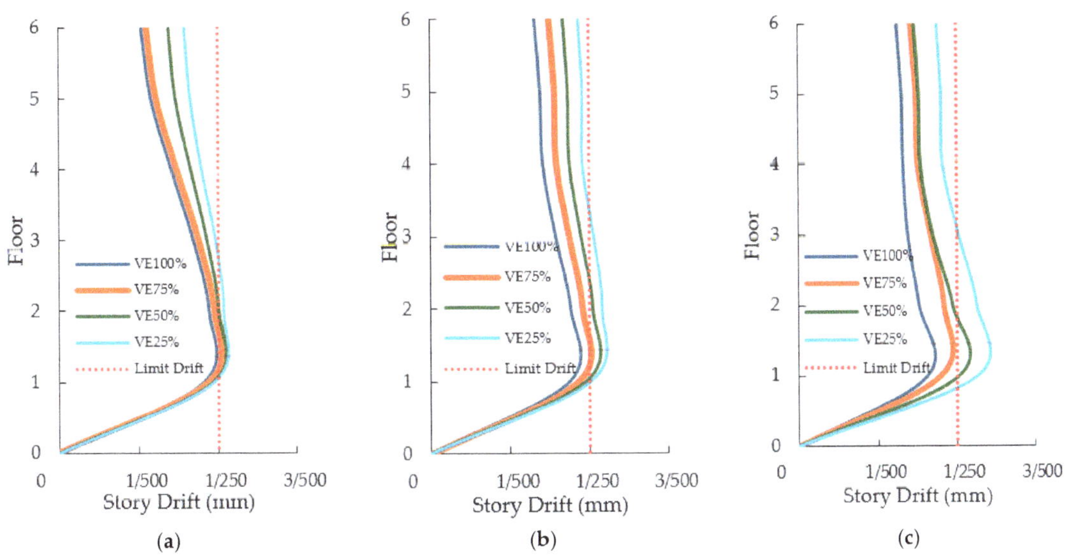

Figure 19. Story drift of 6 stories CSW with viscoelastic damper from THA: (**a**) random phase 120 s; (**b**) El Centro; (**c**) Kobe.

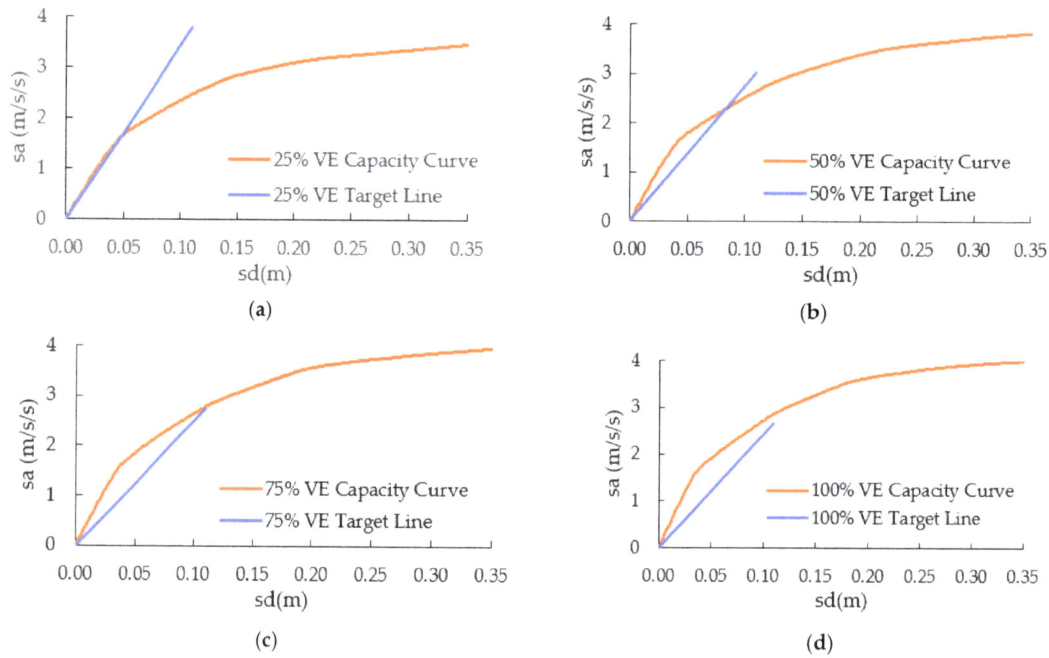

Figure 20. DBDM performance for 12 stories with viscoelastic damper: (**a**) 25% DFR; (**b**) 50% DFR; (**c**) 75% DFR; (**d**) 100% DFR.

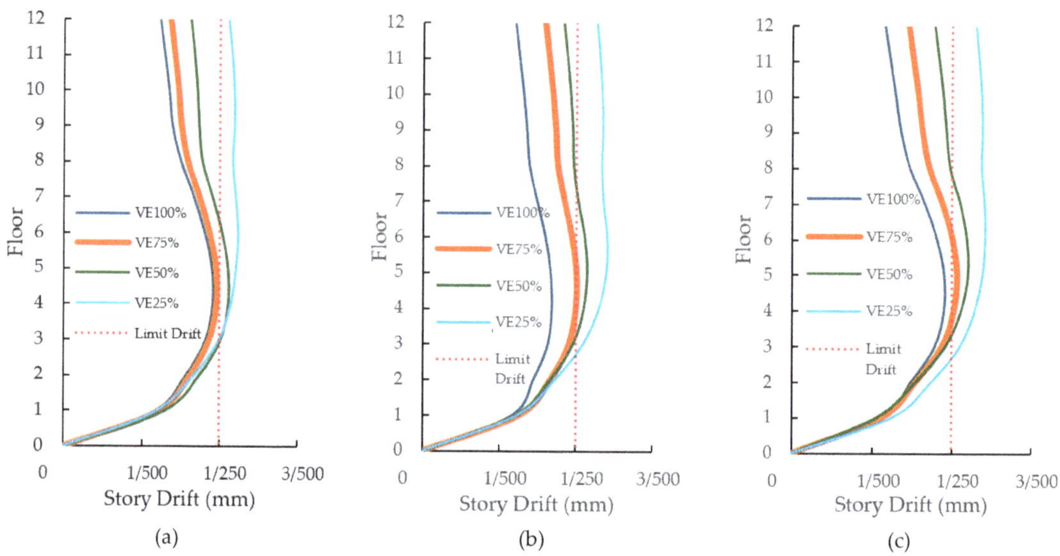

Figure 21. Story drift of 12 stories CSW with viscoelastic damper from THA: (**a**) random phase 120 s; (**b**) El Centro; (**c**) Kobe.

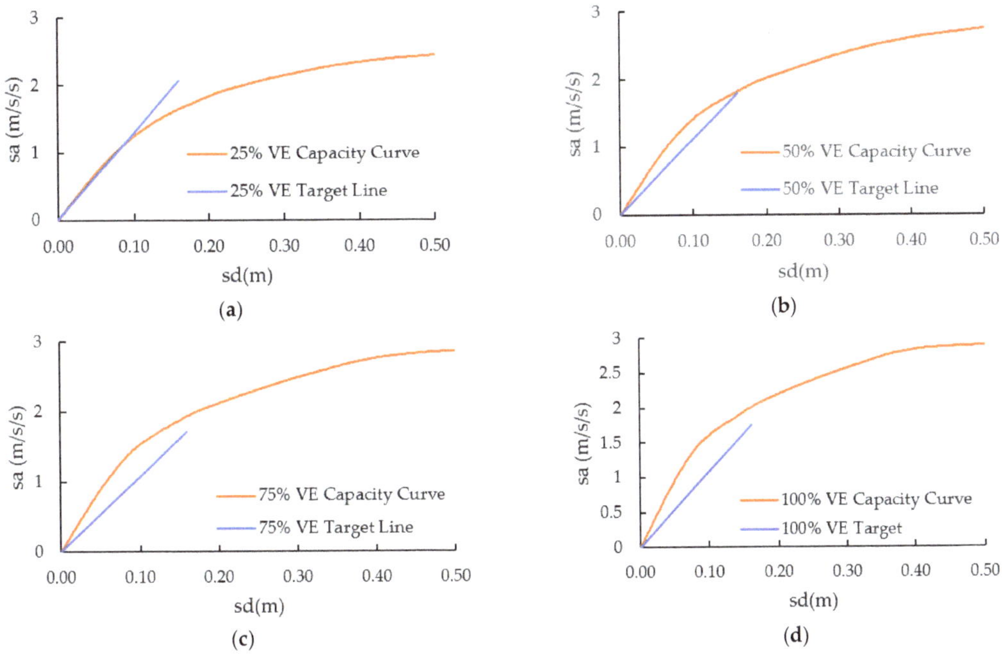

Figure 22. DBDM performance for 18 stories with viscoelastic damper: (**a**) 25% DFR; (**b**) 50% DFR; (**c**) 75% DFR; (**d**) 100% DFR.

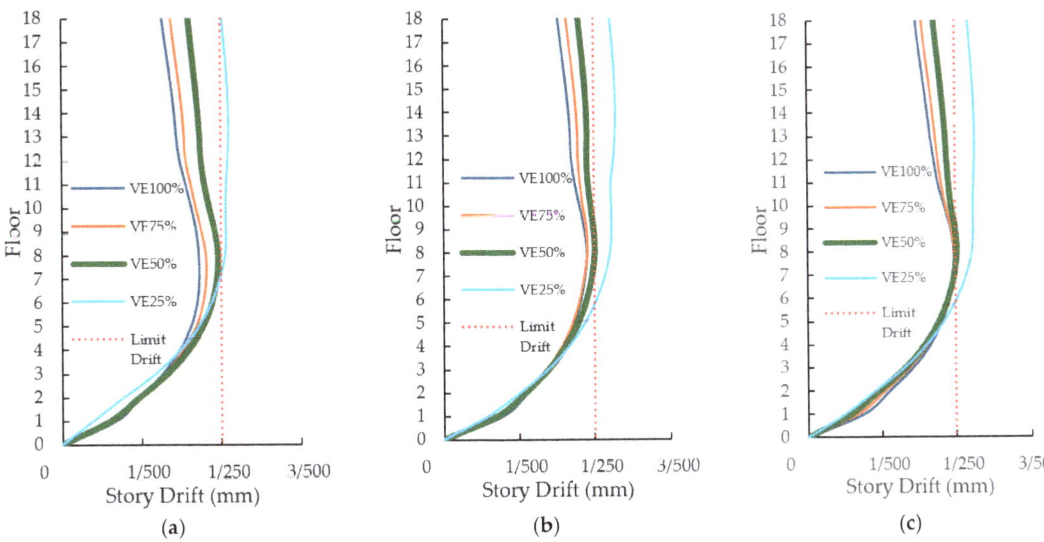

Figure 23. Story drift of 12 stories CSW with viscoelastic damper from THA: (**a**) random phase 120 s; (**b**) El Centro; (**c**) Kobe.

6. Conclusions

The displacement-based design method (DBDM) was introduced as an alternative to the traditional force-based design approach; however, its application to buildings with energy dissipation dampers has been limited. This paper presented the DBDM procedure for the coupled shear wall (CSW) systems with two types of energy dissipation dampers in the coupling beams; one is a metallic damper and the other is a viscoelastic damper.

The DBDM procedure initiates a pushover analysis to obtain the capacity curve of the equivalent SDOF system. Subsequently, the effective damping for each structural member at the target displacement is calculated. From the effective displacement and effective stiffness, the force-deformation relationship of the equivalent SDOF system is plotted as the target line together with the capacity curve. If the shear force of the target line intersects the capacity curve, the amount of the damper is considered adequate.

Firstly, the DBDM is applied for 6-story, 12-story, and 18-story CSWs with metallic dampers. DBDM results show that 75% damper force reduction (DFR) is adequate for all cases. The time history analysis (THA) verified the correctness of the DBDM. Next, the DBDM is applied to the CSWs with viscoelastic dampers. DBDM results show that 75% DFR is adequate for 6-story and 12-story CSWs, and 50% DFR is adequate for 18-story CSW. The accuracy of DBDM was also confirmed by the THA.

Author Contributions: Conceptualization, T.S. and Z.N.E.M.; methodology, Z.N.E.M.; software, T.S.; writing—original draft preparation, Z.N.E.M.; writing—review and editing, T.S.; supervision, T.S. All authors have read and agreed to the published version of the manuscript.

Funding: This research received no external funding.

Institutional Review Board Statement: Not applicable.

Informed Consent Statement: Not applicable.

Data Availability Statement: The data presented in this study are available on request from the corresponding author.

Acknowledgments: The first author appreciates MEXT for support through scholarship of doctoral degree.

Conflicts of Interest: The authors declare no conflicts of interest.

References

1. Priestley, M.J.N.; Calvi, G.M.; Kowalsky, M.J. Direct displacement-based seismic design of structures. *NZSEE Conf.* **2007**, *30*, 1–23.
2. Pennucci, D.; Sullivan, T.J.; Calvi, G.M. Displacement reduction factors for the design of medium and long period structures. *J. Earthq. Eng.* **2011**, *15*, 1–29. [CrossRef]
3. Fox, M.J.; Sullivan, T.J.; Beyer, K. Comparison of force-based and displacement-based design approaches for RC coupled walls in New Zealand. *Bull. N. Z. Soc. Earthq. Eng.* **2014**, *47*, 190–205. [CrossRef]
4. Pennucci, D.; Calvi, G.M.; Sullivan, T.J. Displacement-based design of precast walls with additional dampers. *J. Earthq. Eng.* **2009**, *13*, 40–65. [CrossRef]
5. Xiao, Y.; Zhou, Y.; Huang, Z. Efficient direct displacement-based seismic design approach for structures with viscoelastic dampers. In *Structures*; Elsevier: Amsterdam, The Netherlands, 2021; Volume 29.
6. Zhou, Y.; Aguaguiña, M.; Beskos, D.E.; Gong, S. A displacement-based seismic design method for building structures with nonlinear viscoelastic dampers. *Bull. Earthq. Eng.* **2021**, *19*, 4535–4585. [CrossRef]
7. Segovia, V.A.; Ruiz, S.E. Direct displacement-based design for buildings with hysteretic dampers, using best combinations of stiffness and strength ratios. *J. Earthq. Eng.* **2017**, *21*, 752–775. [CrossRef]
8. Soong, T.T. Passive energy dissipation systems in structural engineering. *John Wiley Sons Google Sch.* **1997**, *2*, 345–355.
9. Whittaker, A.S.; Bertero, V.V.; Thompson, C.L.; Alonso, L.J. Seismic Testing of Steel Plate Energy Dissipation Devices. *Earthq. Spectra* **1991**, *7*, 563–604. [CrossRef]
10. Chang, K.C.; Soong, T.T.; Oh, S.T.; Lai, M.L. Effect of ambient temperature on viscoelastically damped structure. *J. Struct. Eng.* **1992**, *118*, 1955–1973. [CrossRef]
11. Oh, S.T.; Chang, K.C.; Lai, M.L.; Nielsen, E.J. Seismic response of viscoelastically damped structure under strong earthquake ground motions. In Proceedings of the 10th World Conference on Earthquake Engineering, Madrid, Spain, 19–24 July 1992.
12. Christopoulos, C.; Montgomery, M. Viscoelastic coupling dampers (VCDs) for enhanced wind and seismic performance of high-rise buildings. *Earthq. Eng. Struct. Dyn.* **2013**, *42*, 2217–2233. [CrossRef]
13. Blandon, C.A.; Priestley, M.J.N. Equivalent viscous damping equations for direct displacement based design. *J. Earthq. Eng.* **2005**, *9*, 257–278. [CrossRef]

14. Building Center of Japan. *The Building Standard Law of Japan*; Building Center of Japan: Tokyo, Japan, 2004.
15. Saito, T. Structural Earthquake Response Analysis WAVE, Version 1.0 (STERA_WAVE V1.0). 2020. Available online: http://www.rc.ace.tut.ac.jp/saito/software-e.html (accessed on 20 October 2022).

applied sciences

MDPI

Article

A Consensus-Based Likert–LMBP Model for Evaluating the Earthquake Resistance of Existing Buildings

Burak Oz * and Memduh Karalar

Department of Civil Engineering, Zonguldak Bulent Ecevit University, 67100 Zonguldak, Turkey; memduhkaralar@beun.edu.tr
* Correspondence: burak.oz@beun.edu.tr

Abstract: Almost every year, earthquakes threaten many lives, so not only do developing countries suffer negative effects from earthquakes on their economies but also developed ones that lose significant economic resources, suffer massive fatalities, and have to suspend businesses and occupancy. Existing buildings in earthquake-prone areas need structural safety assessments or seismic vulnerability assessments. It is crucial to assess earthquake damage before an earthquake to prevent further losses, and to assess building damage after an earthquake to aid emergency responders. Many models do not take into account the surveyor's subjectivity, which causes observational vagueness and uncertainty. Additionally, a lack of experience or knowledge, engineering errors, and inconspicuous parameters could affect the assessment. Thus, a consensus-based Likert–LMBP (the Levenberg–Marquardt backpropagation algorithm) model was developed to rapidly assess the seismic performance of buildings based on post-earthquake visual images in the devastating Kahramanmaraş earthquake, which occurred on 6 February 2023 and had magnitudes of 7.7 and 7.6 and severely affected 11 districts in Türkiye. Vulnerability variables for buildings are assessed using linguistic variables on a five-point Likert scale based on expert consensus values derived from post-earthquake visual images. The building vulnerability parameters required for the proposed model are determined as the top hill–slope effect, weak story effect, soft story effect, short column effect, plan irregularity, pounding effect, heavy overhang effect, number of stories, construction year, structural system state, and apparent building quality. Structural analyses categorized buildings as no damage, slight damage, moderate damage, or severe damage/collapse. Training the model resulted in quite good performance ($mse = 7.26306 \times 10^{-5}$). Based on the statistical analysis of the entire data set, the mean and the standard deviation of the errors were 0.00068 and 0.00852, respectively.

Keywords: building damageability; consensus degree; Likert scale; Levenberg–Marquardt algorithm; ANN; backpropagation

1. Introduction

Many lives are threatened every year by earthquakes. It is not only developing countries that suffer from the negative effects of earthquakes on the economy [1] but also developed ones that experience significant economic losses, massive fatalities, and the suspension of occupancy as well as businesses every year [2]. A structural safety assessment or conducting a seismic vulnerability evaluation of existing buildings is crucial, since some structures may be susceptible to potential risks [3] in earthquake-prone areas, including some of them which might have been constructed before seismic codes were developed, which may mean they were not designed to resist earthquakes and therefore may not be strong enough to withstand them, and the condition of a building may also have declined as a result of a change in use or land with high liquefaction potential, or as a result of wear and tear over time [1]. Moreover, city planning and construction quality are problematic in developing countries due to insufficient expertise, imprecise legislation, insufficient funds, and unplanned urbanization, which results in low-quality materials and work causing earthquake deaths [4]. Estimating seismic damage before an earthquake is crucial to reducing future losses [5], and assessing building damage after an earthquake is an essential part of emergency response and recovery efforts as well. The level of building damage can range from superficial cracks to complete collapse depending on the building's properties, soil conditions, and earthquake and ground motion characteristics [6]. Based on the results of a seismic evaluation, a decision will be made as to whether the building needs to be repaired or renovated to increase its resistance to seismic forces, or if it should be demolished [7]. In recent years, alternative methods have been developed to rapidly identify deficient structures from large building stocks [8] and various rapid visual screening (RVS) techniques have been developed by different countries to deal with the devastating effects of earthquakes on buildings and humans [9]. RVS techniques all have their specific characteristics, but basic structural characteristics like irregularities are taken into consideration; however, the final scores of building vulnerability classifications differ significantly due to differences in conventional RVS techniques [10]. Bektaş and Kegyes-Brassai argue that RVS assessments can be used to evaluate the vulnerability of buildings in different regions, with appropriate adaptations [3]. It is therefore possible to calibrate existing RVS methods or to develop new RVS methods to overcome the deficiencies [3]. There are some limitations to traditional RVS methods, including surveyor bias, site-specific characteristics, and the uncertainty and vagueness of the data. Several computer algorithms can be used to overcome these kinds of disadvantages in evaluation, such as fuzzy logic, neural networks, and machine learning. It is generally agreed that further research is needed to implement or modify the method, according to Aldemir et al. [11] and Harith et al. [12], and that the methodology needs to be developed based on pre- or post-earthquake data, detailed seismic risk assessment methodologies, and/or soft RVS methods. With continuous innovation and refinement of the methodologies, the quality of the results is greatly improved [13]. Türkiye is located in an active earthquake zone that can cause very severe earthquakes. Türkiye experiences a severe earthquake every two years, while a very severe earthquake happens every three years, according to a statistical study [14]. Moreover, Mertol et al. came to the conclusion that almost all of Kahramanmaraş, Turkey's seriously damaged or fallen reinforced concrete structures were built between 1975 and 2000, when site inspections were rare or nonexistent, and the buildings were not designed and constructed according to the seismic codes that were in effect at the time [15]. There were several reasons for building failures and severe damage, including poor material quality, inadequate reinforcement, and framing problems related to their lateral load-carrying systems, deficiencies in construction materials and reinforcement configuration [14], inadequate development length, violations of bending stirrup ends at 135 degrees, violations of confinement zones, violations of the strong beam–stronger column analogy, and issues with building inspections [15]. Many models have no consideration for the surveyor's subjectivity, which causes observational vagueness and uncertainty [16]. It is also possible that the assessment would be affected by a lack of sufficient experience or knowledge, as well as many other factors, engineering errors, and inconspicuous parameters [17]. Therefore, a novel consensus-based Likert–LMBP model has been developed using post-earthquake visual data of 194 buildings in the Kahramanmaraş earthquake region for rapidly assessing the seismic performance of buildings before an earthquake. Building vulnerability variables are measured utilizing a five-point Likert scale with consensus values derived from linguistic data based on visual images from expert construction engineers. After structural analyses, each building is classified as no damage, slight damage, moderate damage, or severe damage/collapse. Similarly to studies published in the literature, this model was designed for reinforced concrete buildings of 1–7 stories; thus, it cannot be used for RC buildings higher than 7 stories.

Research gaps will be filled in the following ways:

- Seismic vulnerability can be assessed quickly before an earthquake occurs.
- Pre-earthquake visual images can be used to assess existing buildings.
- Experts use linguistic variables in their assessments.
- Consensus values combine differing opinions from experts.
- LMBP provides a high-performance solution to the proposed model.

2. Related Studies

Various seismic risk assessment approaches may be used to evaluate the structural integrity of existing structures [18], but they require complex analysis tools, detailed geometric and material data, as well as considerable time and expertise [4,8]. It is not feasible to use these detailed procedures to assess a large number of buildings at once due to a lack of time, resources, and manpower [4]. RVS is used to identify buildings most vulnerable to earthquake-induced damage by classifying them according to the seismic risk level and prioritizing them accordingly [19]. RVS procedures are divided into three main categories: national methods, statistics and machine learning methods, and rule-based methods [20]. These methods have been used in many studies. The following is a summary of a few studies. Various performance modifiers in the national methods including FEMA P-154, the EMS-98 scale, the RISK-UE Project, NRCC, RBTE-2019, etc., are used to quantify the vulnerability of buildings, such as building type, vertical irregularities, plan irregularities, construction quality, and year of construction [20]. The performance modifiers can be collected easily through observations and engineering drawings [21]. A statistical regression method was employed by Chanu and Nanda to analyze the relationship between the outcome variable (the damage grade of existing buildings) and the explanatory variables regarding building attributes that increase the vulnerability of buildings to earthquakes. For calculating the seismic performance scores of buildings, Ansal et al. proposed a general linear equation, and the weights of each vulnerability parameter were determined statistically [22]. In addition, predictor variables and damage grades were modeled as linear relationships [22]. The use of soft computing techniques in seismic assessments of existing buildings reduces human effort dramatically and improves accuracy [20]. In recent years, several soft rapid visual screening (S-RVS) approaches have been created. These methods can be quickly modified based on breakthroughs and prior earthquake data, and they provide more accuracy compared to traditional RVS methods [18]. S-RVS methods utilize artificial intelligence (AI) such as machine learning (ML) [9,17,23], Fuzzy Logic (FL) [23–27], and Artificial Neural Networks (ANNs) [28–31] in their development process. In recent years, AI has received significant attention in post-earthquake inspections because of its rapid advancement and potential to overcome the disadvantages of manual inspections [31] and has been extensively used in various scientific and technical fields. Studies on the use of these methods in civil engineering have shown promising outcomes and decreased reliance on human involvement, hence minimizing uncertainty and bias [9]. A study published by Mangalathu et al. [6] examined the effectiveness of four ML algorithms in predicting earthquake damage to buildings, including discriminant analysis, K-nearest neighbors, decision trees, and random forests. In the study of Harirchian et al., a comprehensive overview of different AI techniques for analyzing existing buildings and evaluating damage is provided [17]. Ogunjinmi et al. developed a web-based application to classify earthquake damage that can automatically and effectively categorize structural damage [32]. By using gradient-weighted class activation mappings and the convolutional neural network (CNN) model, the degree of damage was assessed [31]. FL evaluates linguistic and information uncertainties and is more realistic when it comes to qualitative assessment than crisp logic [26]. Şen proposed a supervised FL classification approach for assigning hazard categories to buildings similar to the classical fuzzy C-means approach [25]. A fuzzy inference system was proposed by Irwansyah et al. to determine the building damage hazard category following earthquakes based on a fuzzy inference system. A two-stage model architecture was developed: the first involved collecting expert advice for knowledge acquisition and the second involved building fuzzy rules to evaluate buildings. A total of nine input parameters were used, including ring balk, floor block, column, foundation, wall cracks, wall cover and floor cover, Tombak Layar, digital elevation model, peak ground acceleration, and distance from the fault. There were three categories of building damage hazards: slight, moderate, or severe [32]. Harirchian and Lahmer developed an RVS methodology based on an interval type-2 FL system to prioritize vulnerable buildings for detailed assessment while covering uncertainties and minimizing their effects [33]. Building vulnerability could be estimated using the proposed Damage Index, which combines inputs such as the number of stories, the age of a building, irregularities in the plan, irregularities in the vertical plane, quality of the building, and peak ground velocity with a single output variable, and the applicability of the proposed method was examined using a database of reinforced concrete building damage after an earthquake. Building damage states were divided into "None", "Low", "Moderate", "Severe", and "Collapse" [34]. The study by Ketsap et al. [33] examined how fuzzy earthquake risk assessments could be used to prioritize retrofitting buildings. A building's risk was assessed by evaluating several risk factors, including (1) building vulnerability determined by the RVS method based on FEMA 154 [35] and FEMA 155 [36], (2) seismic intensity of the site determined by peak ground acceleration, and (3) building occupancy (nine types) representing the significance of the building weighted by the Analytical Hierarchy Process. Utilizing a fuzzy rule-based model, the total risk scores were calculated by integrating the building importance and damageability scores [33]. Many studies have also used multi-criteria decision-making (MCDM) in seismic assessments [5,37–41]. The objective is to assess the severity of damage and analyze the correlation between observed damage levels and damage indices derived from various MCDM techniques. Harirchian, Jadhav, et al. applied several MCDM methods including the Weighed Sum Method (WSM), the Weighed Product Method (WPM), the Analytical Hierarchy Process (AHP), and the Technique for Order of Preference by Similarity to Ideal Solution (TOPSIS) [39]. To perform the MCDM method, several seismic parameters are used, including seismic zone, type of building, vertical irregularity, plan irregularity, pre-code, post-benchmark, and code details, soil type, number of stories, and appearance of building quality. The WSM and WPM methods achieved the highest accuracy among all MCDM methods followed by TOPSIS, while the AHP method performed worst [37]. It has been suggested by Harirchian, Jadhav, et al. that damage classifications may vary with a larger amount of data, so the results can be improved as a result [39]. More precise identification may be possible with the application of sophisticated MCDM techniques like ELECTRE, PROMOTHEE, and several hybrid techniques [37]. An innovative GIS (geographic information system)-based model has been developed by Sadrykia et al. using a modified combination of AHP, TOPSIS, and fuzzy sets theory to handle one of the important uncertainties in an incomplete data set, which is the vagueness of the existing knowledge about the factors affecting seismic vulnerability [5].

3. Methodology

Recently, several classic RVS approaches (FEMA 154 [35], EMS-98 [42], the RISK-UE Project [43], OASP [44], and NRCC [45]) have been established. Buildings constructed before the current standards may be susceptible to strong seismic effects. Thus, these fast evaluation methods are critical for current structures. Data from the 1989 Loma Prieta and 1994 Northridge earthquakes in California were used to calculate basic structural hazard (BSH) ratings in the first iteration of the FEMA 154 RVS method [35], which evaluated seismic hazard using damage probability matrices based on the modified Mercalli intensity (MMI). The Medvedev–Sponheuer–Karnik (MSK) scale-based developed EMS-98 scale is employed to classify buildings both qualitatively (building type and vulnerability) and quantitatively (indicating degrees of damage). Building type, susceptibility, and damage are classified subjectively and quantitatively using the MSK scale-based EMS-98 scale [46,47]. The 2003 damage estimate approach for the RISK-UE [43] had three steps: LM1, LM2, and LM3 methodologies. The LM1 method identifies vulnerability indices and classes. Based on the capacity and fragility model, the microseismic model (LM2) of the RISK-UE has been utilized to analyze the seismic risk in urban areas. The LM3 method calculates the structure seismic response utilizing thorough acceleration time histories and nonlinear dynamic analysis. The Earthquake Planning and Protection Organization (OASP) created the Greek RVS approach called (OASP-0). In the OASP, the structural system, including structural components and material qualities, is considered when using an RVS technique. The current NRCC [14] guideline evaluates structural and nonstructural aspects to assess threats to beams, columns, foundations, slabs, partition walls, ceilings, and building significance.

A consensus-based Likert–ANN approach has been adopted in this study to predict earthquake damage in residential buildings with RC frames. For this purpose, this research focused on selecting structures made of reinforced concrete with a maximum height of 7 stories. Furthermore, this assessment specifically focused on structures that were affected by the catastrophic Kahramanmaraş earthquake that occurred on 6 February 2023. In the analysis, 194 buildings up to seven stories with different structural characteristics were evaluated based on linguistic variables, followed by a feed-forward backpropagation (FFBP) neural network. The proposed model was developed based on structured analyses of the facade photographs of the selected buildings, and its seismic performance was evaluated. A Likert scale was used to assess the structural integrity of the buildings, and four damage levels were differentiated. Following that, ANN inputs from the building properties and output data from the structural analysis results were combined to create the proposed model. Lastly, the data were divided into two subsets to train and test the proposed model.

3.1. Obtaining Data for the Model

The ANN was designed using visual images of buildings damaged in the devastating Kahramanmaraş earthquake of 6 February 2023. It was of magnitudes 7.7 and 7.6 and severely affected 11 districts in Türkiye. The damage levels for 194 buildings were classified based on their visual appearance after the earthquake as no damage, slight damage, moderate damage, or severe damage or collapse. No earthquake damage means a structure was unharmed. There is no risk in using the building. Slight damage implies the earthquake caused paint, plaster, and tiny fractures on the building's walls. These are buildings with plaster. Moderate damage is defined as structures that may develop fine fractures in the carrying components and walls during an earthquake. Severe damage or collapse means that the earthquake damaged the building's load-bearing materials. Large and extensive shear fractures/separations occur in structures.

In the model, depending on their appearance following the earthquake, structures were classified as no damage, slight damage, moderate damage, severe damage, or collapse. After that, the consensus value for each parameter was used to generate the risk index (*RI*). Risk indices range from 0 to 1, whereas damage levels are 0.25, 0.50, 0.75, and 1. Except for building apparent quality, which is rated between 0 (good) and 1 (bad), building vulnerability parameters like soft stories, heavy overhangs, short columns, pounding effect, topography effect, vertical irregularity, and plan irregularity are quantified as 0 or 1. The details of the modeling are given in detail in the following sections.

3.2. The Evaluation Scale

Several research papers in the field of construction management use Likert scales to quantify both visible and non-observable characteristics. For example, the relative importance index was used by Datta et al. [48] to rank the critical project management success factors for the construction industry of Bangladesh, Memon et al. [49] identified the key challenges affecting construction project completion on time, several high-risk factors were identified by Yousri et al. [50] for Egyptian building construction projects, a study conducted by Oke et al. [51] examined the impact of the Internet of Things (IoT) on construction project performance, Ahmed Marey Alhammadi and Memon [52] identified factors that inhibit cost performance in United Arab Emirates (UAE) construction projects, and so on. A 5-point Likert scale is used in this study to evaluate input parameters to the proposed model.

3.3. *Building Vulnerability Parameters (Inputs to the Proposed Model)*

A total of eleven input vulnerability parameters are defined, each with five membership values. A list of linguistic vulnerability parameters and their values is provided in Appendix A. The following sections provide detailed explanations of these parameters.

3.3.1. Top Hill–Slope Effect

On top of hills, topographic amplification may increase ground motion intensity. Buildings on steep slopes steeper than 30 degrees also often have stopped foundations that are incapable of dispersing ground distortion equally to structural members [53]. It is advisable to avoid hillside slopes that are prone to sliding during an earthquake and instead choose only stable slopes for construction placement. Moreover, it would be more desirable to have many blocks on terraces rather than a single huge block with footings at significantly varying heights [24]. Seismic risk assessments should consider these two factors. A street survey can easily reveal both factors [53].

3.3.2. Vertical Irregularities

Buildings with columns or curtains that do not follow the height of the building create vertical irregularities, reflecting the effect of vertically discontinuous frames and changing floor areas [44]. Vertical irregularities are a major cause of building damage during strong earthquake excitation. There are a variety of factors that contribute to vertical irregularities, including steep slopes, significant wall apertures, structurally vulnerable or flexible floors, discrepancies in floor elevations, setbacks either within the same plane or outside of it [19], a column sitting on a console, vertically discontinuous columns, a curtain sitting on a column, and vertically discontinuous curtains.

Weak Stories

A weak story is usually characterized by vertical discontinuities or by reduced member sizes or reinforcements and tends to concentrate on inelastic activity, which may cause the story to collapse partially or completely [24].

Soft Stories

Buildings with soft stories have collapsed in past earthquakes around the world [53]. Buildings with soft stories have a larger share of severely damaged buildings than those with lower damage, regardless of their number of stories [54]. Soft stories typically occur when the ground floor is less stiff and stronger than the upper floors. Typically, this is the case in buildings located along the side of a main street in which the ground floor is used as a commercial or retail space while the upper floors are occupied by residences. The upper stories are strengthened and stiffened by many partition walls, but at the bottom, the commercial spaces are mostly left open between the frame members for customer circulation. Additionally, the ground stories may have taller clearances and a different axis system, causing irregularities in the construction. Euro code 8 stipulates that the displacement in the direction of seismic forces must not exceed the average story displacement by more than 20% at any given story [24].

Short Columns

In RC buildings, short columns are formed by semi-infilled frames, band windows in semi-buried basements, or mid-story beams around stairway shafts. A captive column usually sustains heavy damage during strong earthquakes due to its short length not being designed to handle high shear forces. A short column is easily identified from the outside because it usually forms along an exterior axis [53].

3.3.3. Plan Irregularities

A non-geometrically symmetrical plan and an irregular placement of vertical structural elements may cause torsion in a building. Due to its detrimental effects on the transmission of seismic loads, a large plan or diaphragm opening should be regarded as a serious kind of plan irregularity [55]. Plan irregularity describes the structural plan with re-entrant corners. E, L, T, and U are defined as complex building forms or shapes. E, H, L, T, U, Y, cross, or a mix of these shapes are common architectural configurations with projections or wings in the plan as re-entrant corners [56]. The irregularity of a structural plan is determined by plan shapes like E, L, T, U, and Z [19,56].

3.3.4. Pounding Effects

Buildings that are not sufficiently separated pound each other during an earthquake because vibration periods differ and vibration amplitudes are not synchronous. Pounding is most detrimental to adjacent buildings at corners, and uneven floor levels aggravate the effect of pounding. High-rise buildings are more likely to be damaged by pounding [53]. A misaligned floor level in an adjacent building is considered the worst-case scenario [27]. To avoid pounding, a building should not be located closer than 4% of its height to an adjacent building to avoid extensive damage [24].

3.3.5. Heavy Overhang Effect

Identifying the difference between ground and above-ground floors is necessary. Multistory reinforced concrete buildings with heavy balconies and overhanging floors shift their mass center upward, causing higher seismic lateral forces and overturning moments during a seismic event. Buildings with large, overhanging cantilever spans enclosed by concrete parapets are more likely to suffer heavy damage [53]. Furthermore, this effect is more noticeable in buildings with four, five, or six stories with heavy overhangs [54].

3.3.6. Number of Stories

A study by Sucuoğlu et al. concluded that building height is one of the most significant or perhaps even the most dominant determinants of seismic vulnerability in Turkish multistory concrete structures, and there is a strong correlation between damage and the number of stories [54], since higher buildings can suffer great deformation and damage during an earthquake [18]. In a report prepared for the Metropolitan Municipality of Istanbul following the Kocaeli and Düzce earthquakes of 1999, it was found that building damage correlated strongly with the number of stories [53]. Şen also found that the most vulnerable buildings in Istanbul were medium-rise (4–7 stories) reinforced concrete frame buildings designed based on outdated regulations [24]. It is clearly seen that there is almost a linear increase in damage with the number of stories. The ratio of moderately and severely damaged buildings increases steadily with the number of stories [53]. However, if buildings are designed to comply with modern seismic codes, damage will be distributed uniformly regardless of the number of stories [54].

3.3.7. Construction Year

The assessment of building vulnerability is based on the construction year, which serves as an indicator of construction quality, design methodologies used, and seismic details included in the project. Old structures cannot be expected to function well because ancient construction procedures ignore the seismic details of modern building requirements [18]. Construction years provide information about building design regulations as well as design standards that are classified as low, moderate, and high. A study by Bektaş and Kegyes-Brassai examined the effects of the construction year on seismic design codes in Albania, and the authors concluded that this existing classification technique could be used by modifying the thresholds based on the seismic standards for the region under consideration [18].

3.3.8. Structural System States

Different types of buildings are more or less vulnerable to earthquakes. Structural load-carrying systems determine a building's lateral strength and ductility during a seismic event. Several parameters affect this parameter, including structural geometry, member dimensions, and design quality. It was stated in the Ministry of Environment and Urban Planning's (2019) method for evaluating existing reinforced concrete buildings of 1 to 7 floors that the buildings could have reinforced concrete frames (RCF) or reinforced concrete frame and curtain (RCFC) systems for load bearing. As stated, the load-bearing system should be determined from the inside of a basement floor if there is one, or from inside a store if there is one, or if it cannot be determined, the RCF should be assumed to be ineffective or zero.

3.3.9. Apparent Building Quality

Building apparent quality and earthquake damage are closely related. Construction quality, workmanship, and maintenance reflect the apparent quality of the building [53]. Sucuoğlu et al. [54] found that severely damaged buildings in three- to five-story buildings were of lower quality than those in other damage groups after the Düzce earthquake. Additionally, apparent quality became more significant as building height increased [54]. The apparent quality of a building can generally be classified by a trained observer [53]. Workmanship, material quality, and existing damage and deterioration are used to determine whether a building's construction is poor, moderate, or good. If a building's flooring, ceilings, brick walls, uneven surfaces, scratches, or efflorescence have surface or appearance problems, the building's construction quality is considered average [18].

3.4. Building Seismic Performance Scores (Outputs of the Proposed Model)

A damage score indicates the level of damage likely to be sustained by a structure. Building damageability is computed based on the building vulnerability parameters. Some studies categorize building damage into five damage levels: very low, low, moderate, high, and very high [25,53]. There are, however, three damage levels according to some studies: low risk, medium risk, and high risk [57], or four damage grades: none, light, moderate, and severe/collapse [54]. Damage grades are generally defined as follows. No damage: the structure is not significantly damaged; slight damage: capillary bending or shear cracks in columns, beams, and slabs, as well as capillary shear cracks in walls; moderate damage: local or longitudinal cracks in most columns and beams; severe damage: poured concrete covers, buckled longitudinal reinforcement bars, and wide cracks; and collapse: excessive deformation causing a structure to collapse. Damage grades were classified into four categories in this study: none, slight, moderate, and severe or collapse, as in the study by Sucuoğlu et al. [54]. As explained in detail in the introduction, determining building seismic performance varies depending on the proposed model. The use of linear functions is suggested by some studies, while the use of fuzzy functions, neural networks, and multi-criteria decision-making methods is suggested by others. A consensus-based Likert–ANN model is recommended in this study.

3.5. The Consensus-Based Likert–ANN Model

The classical methods linguistically mark pounding effects as Yes or No [24]. With classical modeling, the building vulnerability parameters such as soft stories, heavy overhangs, short columns, pounding effects, topography effects, vertical irregularities, and plan irregularities are quantified as 1 or 0 depending on crisp two-valued logic, except for the building apparent quality, which is rated as 0 (good), 1 (moderate), and 2 (poor) [53]. As a result, no distinction is made between "slightly", "moderate", and "very" quantities in this case. Buildings on steep slopes steeper than 30 degrees, for example, are all quantified as 1; however, a building on top of a hill may be intensified by topographic amplification. FL involves fuzzifying inputs and outputs into multiple fuzzy sets. In some cases, FL with multiple values is more advantageous than crisp logic with two values. The proposed model has been developed to predict seismic risk levels of existing buildings based on linguistic information gathered from expert opinions based on a 5-point Likert scale. Figure 1 illustrates the overall structure of the proposed model.

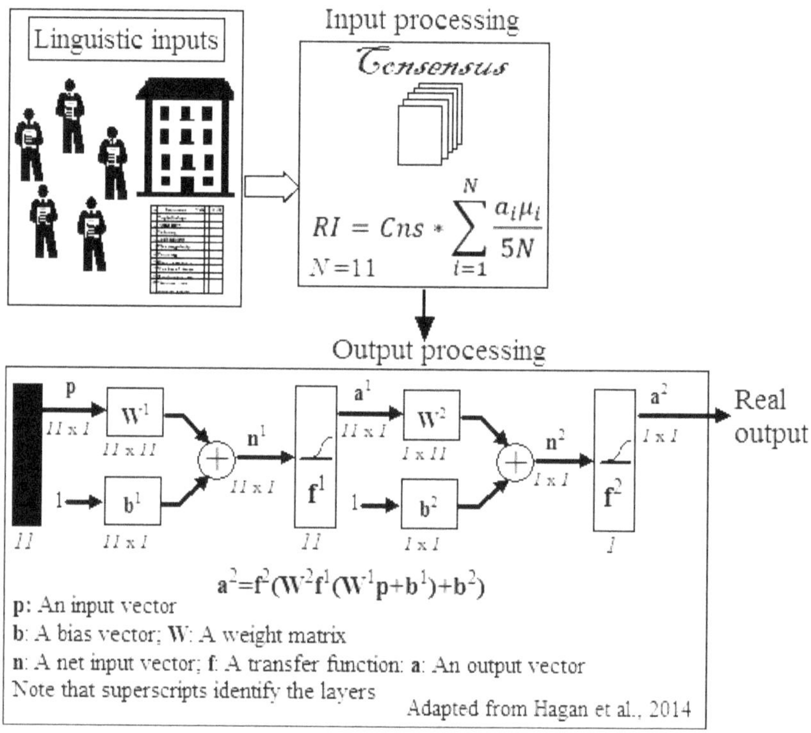

Figure 1. The structure of the consensus-based Likert–ANN model, only the output processing frame was adapted from Hagan et al., 2014 [58].

3.5.1. Consensus Degree

Experts will evaluate each vulnerability parameter individually using the form (Appendix A) based on a five-point Likert scale. Following that, the risk index (*RI*) based on the consensus value for each parameter will be calculated using the equation shown below.

$$RI = Cns * \sum_{i=1}^{N} \frac{a_i}{5N} \tag{1}$$

where *Cns* stands for the consensus degree of responses to each parameter, *N* indicates the total number of experts (suggested five), and a_i is the constant that represents the weight assigned to each response (in this case, very low = 1; low = 2; moderate = 3; high = 4; and very high = 5). Individuals in a sample group are said to be in consensus when they agree on a declarative statement [59]. Based on this definition, if an equal number of participants choose their responses in the two extreme categories on a Likert scale, i.e., strongly disagree and strongly agree, there is no consensus; however, this group shows full consensus if all participants choose the same category on the Likert scale. Therefore, for all combinations of response patterns, the consensus degree varies between zero and one. Tastle and Wierman [59] concluded that the consensus provides a value for comparing different Likert distributions and matches human intuition, and suggested calculating a consensus (*Cns*) degree as follows:

$$Cns(X) = 1 + \sum_{i=1}^{n} p_i log_2\left(1 - \frac{|x_i - \mu_x|}{d_x}\right) \quad (2)$$

where X is the response, μ_x is the mean of X, n represents the number of categories on an ordinal scale, x_i is the degree of agreement in category i, p_i is the probability of the occurrence of x_i, and d_x is the width of X, i.e., $d_x = X_{max} - X_{min}$ (in this case, $d_x = 5 - 1 = 4$).

3.5.2. Neural Network (NN) Modeling

This study can be viewed as pattern classification, and feedforward networks can be used to classify patterns. The non-linearly separable classification problem can only be solved by multilayer networks, and a multilayer perceptron trained using a backpropagation algorithm is the most widely used neural network [58]. There is no general knowledge regarding how many layers or how many neurons are necessary for adequate performance of multilayer networks. Most neural networks have two or three layers, whereas four or more layers are rarely used [58]. When choosing the fastest training algorithm for a given problem, several factors need to be taken into account, including how complex the problem is, the number of data points in the training set, the weights and biases within the network, and the error goal [58]. However, LMBP, despite a large number of computations, is the fastest neural network training algorithm for moderate numbers of parameters [58,60], and well-suited for neural network training [58]. A two-layer feedforward artificial neural network was considered. Therefore, this study uses an 11-11-1 ANN model with LMBP to predict the seismic performances of existing buildings, and the MATLAB NN toolbox (nntool) is used to train and test the model because the fastest way to train a feedforward network of small or medium size is to use "trainlm" in MATLAB [61]. The proposed approach to model training is shown in Figure 2.

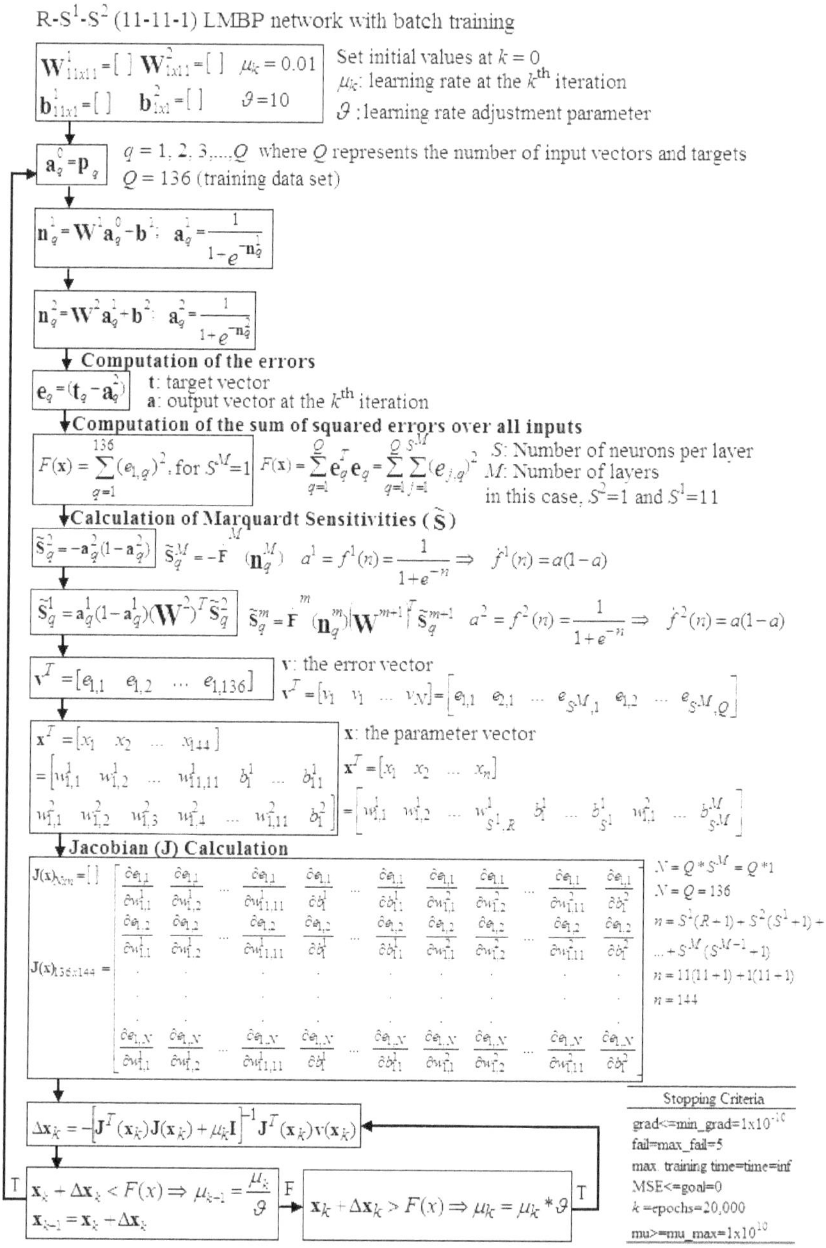

Figure 2. Model's LMBP training flow adapted from Hagan et al. [58] and Demuth et al. [61].

4. Results

An evaluation of 194 buildings was conducted with post-earthquake visual images and 11 linguistic variables to categorize damage levels following the Kahramanmaraş earthquake. Thus, it is thought that pre-earthquake visual images can be used to estimate the possible damage levels of existing buildings before a similar earthquake. The results of the designed model are described in detail below.

4.1. Inputs and Outputs of the Model

A visual analysis of 194 buildings affected by the Kahramanmaraş earthquake was conducted using linguistic variables based on a five-point Likert scale. Based on their outward appearance after the earthquake, 194 structures were categorized as having no damage, slight damage, moderate damage, severe damage, or collapse. According to their damage levels, 48 of the buildings had no damage, 49 had slight damage, 48 had moderate damage, and 49 had severe damage or collapse. A list of linguistic variables and their values is included in Appendix A. After that, Equations (1) and (2) were used to calculate the risk index (*RI*) based on the consensus value for each parameter. The model inputs consist of risk indices ranging from 0 to 1, while its outputs consist of damage levels varying between 0.25, 0.50, 0.75, and 1. An overview of the model's input–output values is shown in Figure 3.

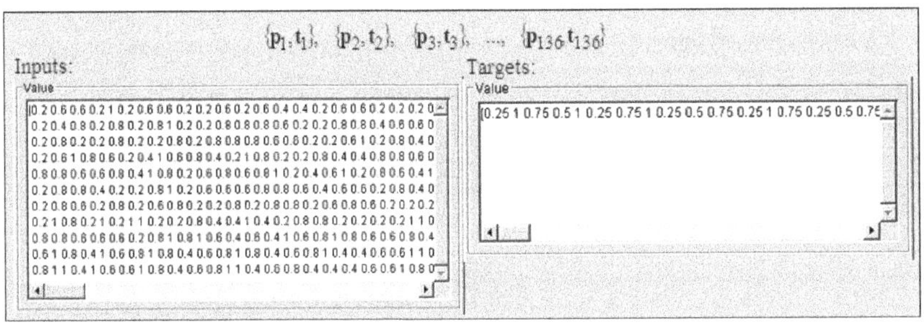

Figure 3. A view of the model's input and output values.

4.2. Architecture of the Model

In most cases, a network with fewer parameters than data points in the training set can sufficiently represent the training data for the network to generalize. Several approaches have been used to create simple networks: growing, pruning, global searches, regularization, and early stopping. A network is grown by starting with no neurons and adding neurons until its performance is adequate. It is also possible to stop training before the network overfits. A simple method for improving generalization is early stopping. This method aims to minimize overfitting by stopping training before the minimum is reached. Additionally, when the backpropagation algorithm converges, it is not certain that it is an optimum solution, so trying several initial conditions ensures an optimal solution. As a step-by-step approach, we compared the performance of various architectures by adding neurons to the hidden layer from 11-4-1 to 11-11-1 at various initial parameters to find the ideal solution, as described above by Hagan et al. [58]. The final architecture and training parameters are shown in Figure 4.

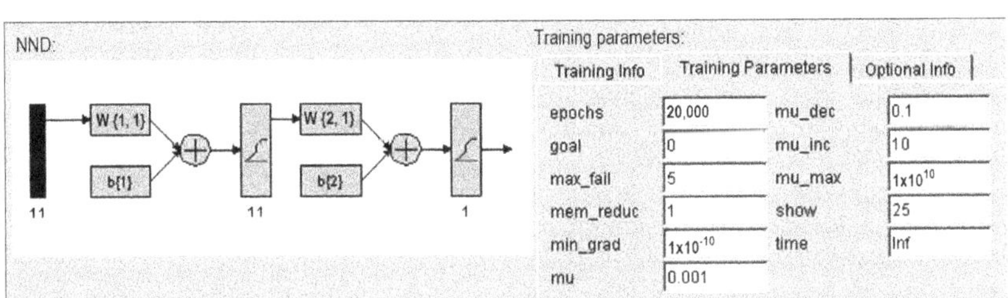

Figure 4. NND design, training parameters, and stopping criteria.

4.3. Training, Testing, and Validation

The model was trained using the MATLAB neural network toolbox. As a first step, the data set was divided into training, testing, and validation subsets. There are various ratios recommended for this. For example, the MATLAB neural network toolbox (nntool) normally divides a network's data automatically when the network is trained, and the default ratios for training, testing, and validation are 0.7, 0.15, and 0.15, respectively [61]. In this study, the default ratios were not changed. The model was run many times with various initial parameters to find the global minimum point, and finally, the global minimum point sufficient for the model's performance was reached with the initial values shown in Figure 5. Figure 5 shows the final parameters as well.

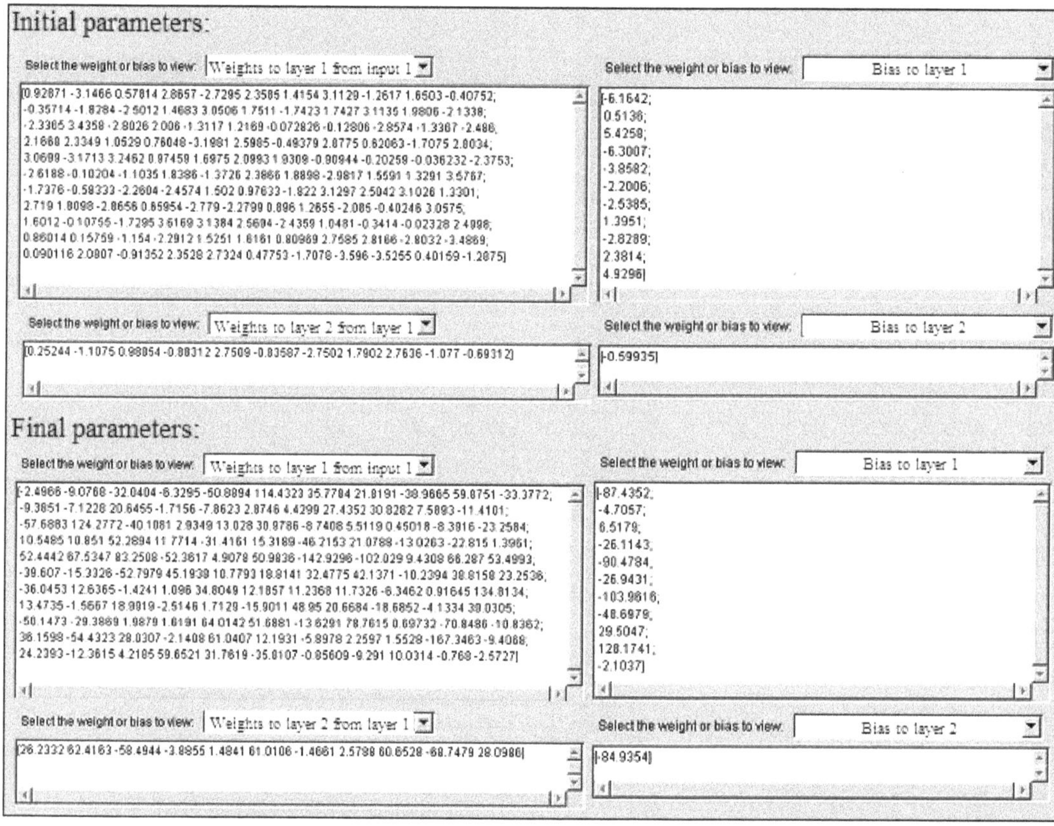

Figure 5. The model's initial and final parameters.

The error generally reduces as the number of epochs of training increases; however, the error may start to increase on the validation data as the network starts overfitting to the training data [62]. In the default Levenberg–Marquardt training parameters, five consecutive increases in validation error stop the training [61], and the epoch with the lowest validation error results in the best performance [62]. The proposed model reached max_fail (five in this case) in its 794th iteration and completed training with an adequate performance value (mean squared error), as shown in Figure 6. There are three different colored lines in Figure 6. The blue line represents the mean squared error (mse) of the training set, the green line represents the mse of the validation set, and the red line represents the mse of the test set. As the red and green lines overlap, the red line cannot be seen because it lies beneath the green line. Note that the test curve and validation curve overlap in the figure.

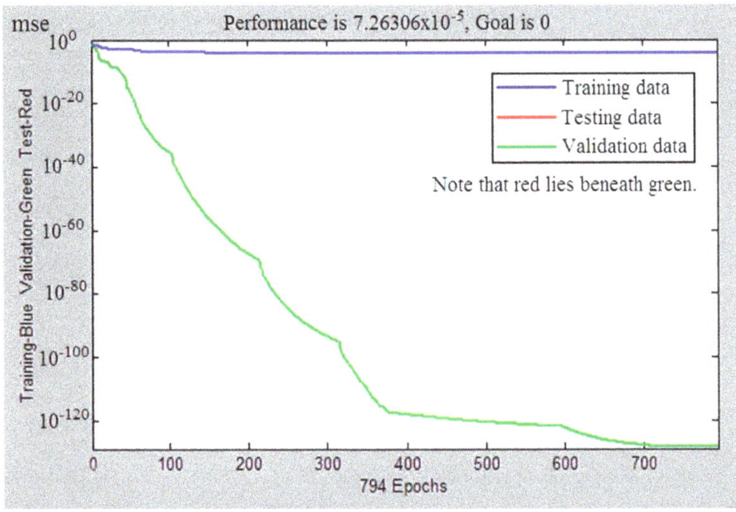

Figure 6. Model's neural network performance graph.

The statistical value for errors in the entire data set (e = t − a) is $M = 0.00068 \pm STD = 0.00852$. Additionally, we obtained 0.02 on one data point, 0.06 on one data point, and 0.07 on two data points. The model shows quite good performance. All of the data points are included in Figure 7's target–output graph. There were 194 structures evaluated for damage levels: 48 were completely unaffected (represented by a 0.25), 49 were slightly damaged (represented by a 0.50), 48 were moderately damaged (represented by a 0.75), and 49 were severely damaged (represented by a 1.00).

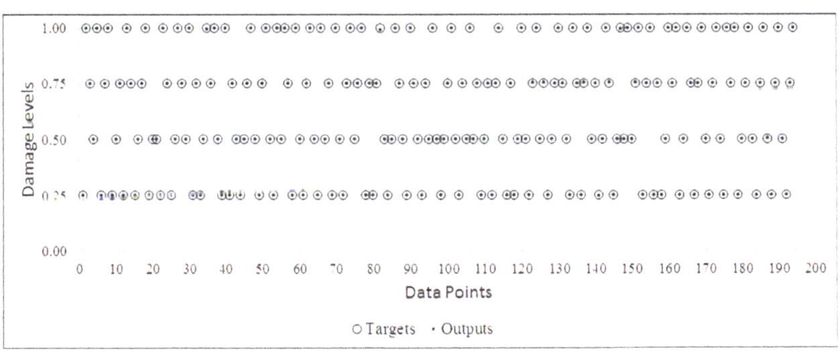

Figure 7. Target–output graph after training.

5. Discussion

Ruggieri et al. developed a framework using ML for the analysis of the seismic vulnerability of existing buildings, which identifies buildings in an area and uses their photos to generate a vulnerability index, thereby reducing the time and effort required for surveys and interviews [13]. Similarly, a proposed model was developed based on the structural analysis of post-earthquake facade photographs in the Kahramanmaraş earthquake zone, and its seismic performance was evaluated. A five-point Likert scale with consensus values derived from linguistic data on the visual images was used to measure building vulnerability levels. Several predictor variables were used as input parameters for the ML algorithms, including the closest distance to the surface projection of fault rupture and the spectral acceleration at the site of interest. Variables such as building age (in years), number of stories, plan irregularities, floor area, and replacement cost were also used to predict building performance. Green, Yellow, and Red were the damage tagging categories [6]. In another study, Allali et al. scaled component damage levels and global damage levels according to no damage, slight damage, moderate damage, severe damage, and collapse. Similarly, our study defined eleven input vulnerability parameters with five membership values each. Based on the structural analysis, buildings were classified as having no damage, minor damage, moderate damage, or severe damage/collapse.

Various AI techniques have been used in the literature to design models, and their performances have been analyzed. Kaplan et al. presented a method in which a building's performance level was categorized as low risk, medium risk, or high risk, claiming that the method showed 83% accuracy, compared to the long and hectic code-based method [47]. An approach based on FL with a genetic algorithm was used by Allali et al. to assess the damage to buildings with different lateral load-resisting components post-earthquake, and approximately 90% accuracy was achieved in their results [63]. Harirchian et al. examined the effectiveness of ML applications in damage prediction using a support vector machine model as the damage classification technique [33]. According to the study, all 22 input parameters resulted in a 52% accuracy rate, which is acceptable given the sample size [64]. According to the results of the study by Mangalathu et al., predictions made using the random forest algorithm had a 66% accuracy in assigning tags and 79% accuracy in actual yellow tags [6]. In the study by Özkan et al., a generalized regression neural network and an FFBP neural network were used for the network architecture [44]. The nonlinear incremental mode combination method was used to evaluate the earthquake performance of these building models. They claimed that an ANN could successfully predict the earthquake behavior of reinforced concrete buildings [65]. Chen and Zhang developed a learning model that enables early disaster risk reduction decisions by automating building vulnerability assessments in seismic areas [54]. The proposed model here, however, uses an ANN with two layers (11-11-1) and LMBP to predict the seismic performances of existing buildings. The structural integrity of the buildings was assessed using a Likert scale, and four levels of damage were differentiated. In the following step, input data from the building properties and output data from the structural analysis results were combined to create the proposed model. Training the model resulted in quite good performance. Based on the statistical analysis of the entire data set, the mean and the standard deviation of the errors were 0.00068 and 0.00852, respectively. Four data points differed from the others but were quite acceptable: 0.02 on one data point, 0.06 on one data point, and 0.07 on two data points.

6. Conclusions

Earthquakes threaten many lives every year. They affect not only developing countries but also developed ones and they cause significant economic losses, massive deaths, as well as the suspension of occupancy and businesses. For future losses to be minimized, it is crucial to estimate seismic damage before an earthquake occurs. Existing buildings in earthquake-prone areas need structural safety or seismic vulnerability assessments. Various RVS techniques have been developed by different countries in recent years to deal with the devastating effects of earthquakes on buildings and humans. However, RVS methods have some limitations, such as surveyor bias and site-specific characteristics, as well as data uncertainty and vagueness. Fuzzy logic, neural networks, and ML are all computer algorithms that can be used to overcome these kinds of disadvantages in evaluation.

Türkiye is located in an active earthquake zone that can cause very severe earthquakes; every two years, it experiences severe earthquakes, and every three years, it experiences very severe earthquakes. Therefore, a new model was developed to rapidly assess building seismic performance. A consensus-based Likert–ANN approach has been used in this study to predict earthquake damage levels in residential buildings with RC frames. The ANN was designed by analyzing the visual images of buildings damaged by the devastating Kahramanmaraş earthquake of 6 February 2023, which measured 7.7 and 7.6 on the Richter scale and severely affected 11 districts in Türkiye. Vulnerability variables for buildings are measured using linguistic data on a five-point Likert scale based on expert consensus values derived from post-earthquake visual images. An extensive literature review was conducted to determine the building vulnerability parameters required for the proposed model, including the top hill–slope effect, weak story effect, soft story effect, short column effect, plan irregularity, pounding effect, heavy overhang effect, number of stories, construction year, structural system state, and apparent building quality. Structural analyses categorized buildings as no damage, slight damage, moderate damage, or severe damage/collapse. In terms of damage levels, 48 of the buildings had no damage, 49 had slight damage, 48 had moderate damage, and 49 had severe damage or collapse. Model inputs consist of risk indices ranging from 0 to 1, while outputs consist of damage levels varying between 0.25, 0.50, 0.75, and 1.

A two-layer neural network (11-11-1) with LMBP was used to train the model after 194 buildings up to seven stories were evaluated. Max_fail (five) was reached in the proposed model's 794th iteration, and training was completed with an adequate performance value ($mse = 7.26306 \times 10^{-5}$). Statically, errors in the entire data set (e = t − a) are $M = 0.00068 \pm STD = 0.00852$, with 0.02 on one data point, 0.06 on one data point, and 0.07 on two data points. Using this model, it is possible to assess the seismic vulnerabilities of existing buildings by analyzing pre-earthquake facade images with 11 parameters. Experts can use linguistic variables on a five-point Likert scale, and consensus values combine differing opinions from experts. LMBP provides a high-performance solution to the proposed model. Although the model shows good performance and it can be used for rapid visual analysis, a detailed analysis may still be needed. There is room for improvement. A further evaluation of the model's reliability should be performed with multiple expert groups and with data from a variety of earthquake zones or should be remodeled with more data. This requires much time, but if an AI application evaluates relevant parameters from photographs, this time will be much shorter, and the proposed model will become more useful.

Author Contributions: Conceptualization, B.O. and M.K.; Methodology, B.O.; Validation, B.O. and M.K.; Formal analysis, B.O.; Investigation, B.O.; Resources, M.K.; Data curation, M.K.; Writing—original draft, B.O.; Writing—review & editing, B.O. and M.K. All authors have read and agreed to the published version of the manuscript.

Funding: This research received no external funding.

Institutional Review Board Statement: Not applicable.

Informed Consent Statement: Not applicable.

Data Availability Statement: The original contributions presented in the study are included in the article, further inquiries can be directed to the corresponding author.

Conflicts of Interest: The authors declare no conflict of interest.

Appendix A. Form for Evaluating Building Vulnerability

n	Building Vulnerability Parameters	Very Low	Low	Medium	High	Very High
1	Top hill–slope effect: (Topographic amplification can intensify ground motion on hills. Buildings on steep slopes steeper than 30 degrees often have foundations that are unable to disperse ground distortion evenly).					
2	Weak story effect: (Weak stories usually have vertical discontinuities or reduced member sizes or reinforcements, which can lead to partial or total collapse).					

3	Soft story effect: (Ground floors typically have softer structures than upper floors. Commercial spaces at the bottom are mostly left open (large wall openings) for customer circulation, or the ground stories have taller clearances and different axis systems).					
4	Short column effect: (Short columns formed by semi-infilled frames, band windows in semi-buried basements, or mid-story beams around stairway shafts).					
5	Plan irregularity: (A non-geometrically symmetrical plan, an irregular placement of vertical structural elements, or a large plan or diaphragm opening; plan shapes like E, L, T, U, +, and Z).					
6	Pounding effect: (A building should not be located closer than 4% of its height to an adjacent building. High-rise buildings are more likely to be damaged by pounding. A misaligned floor level in an adjacent building is considered the worst-case scenario).					
7	Heavy overhang effect: (Buildings with large, overhanging cantilever spans enclosed by concrete parapets are more likely to suffer heavy damage).					
8	Number of stories:	1–2	3	4	5	6–7
9	Construction year: (Construction years provide information about building design regulations as well as design standards).	0–10	11–20	21–30	31–40	>40
10	Structural system state: (Parameters such as structural geometry, member dimensions, and design quality affect this parameter. If it cannot be determined, the RCF should be assumed to be medium).	Very good	Good	Medium	Poor	Very poor
11	Apparent building quality: (Very poor if the structure has severe damage; very good if the structure has no visible cracks or damage; medium if there are cracks in some structural members).					

References

1. Haryanto, Y.; Hu, H.-T.; Han, A.L.; Hidayat, B.A.; Widyaningrum, A.; Yulianita, P.E. Seismic vulnerability assessment using rapid visual screening: Case study of educational facility buildings of Jenderal Soedirman University, Indonesia. *Civ. Eng. Dimens.* **2020**, *22*, 13–21. [CrossRef]
2. Rashid, M.; Ahmad, N. Economic losses due to earthquake-induced structural damages in RC SMRF structures. *Cogent Eng.* **2017**, *4*, 1296529. [CrossRef]
3. Bektaş, N.; Kegyes-Brassai, O. Conventional RVS methods for seismic risk assessment for estimating the current situation of existing buildings: A state-of-the-art review. *Sustainability* **2022**, *14*, 2583. [CrossRef]
4. Coskun, O.; Aldemir, A.; Sahmaran, M. Rapid screening method for the determination of seismic vulnerability assessment of RC building stocks. *Bull. Earthq. Eng.* **2020**, *18*, 1401–1416. [CrossRef]
5. Sadrykia, M.; Delavar, M.R.; Zare, M. A GIS-based fuzzy decision making model for seismic vulnerability assessment in areas with incomplete data. *ISPRS Int. J. Geo-Inf.* **2017**, *6*, 119. [CrossRef]
6. Mangalathu, S.; Sun, H.; Nweke, C.C.; Yi, Z.; Burton, H.V. Classifying earthquake damage to buildings using machine learning. *Earthq. Spectra* **2020**, *36*, 183–208. [CrossRef]
7. El-Betar, S.A. Seismic vulnerability evaluation of existing RC buildings. *HBRC J.* **2018**, *14*, 189–197. [CrossRef]
8. Bhalkikar, A.; Kumar, R.P. A comparative study of different rapid visual survey methods used for seismic assessment of existing buildings. *Structures* **2021**, *29*, 1847–1860. [CrossRef]
9. Kumari, V.; Harirchian, E.; Lahmer, T.; Rasulzade, S. Evaluation of machine learning and web-based process for damage score estimation of existing buildings. *Buildings* **2022**, *12*, 578. [CrossRef]

10. Doğan, T.P.; Kızılkula, T.; Mohammadi, M.; Erkan, İ.; Kabaş, H.T.; Arslan, M. A comparative study on the rapid seismic evaluation methods of reinforced concrete buildings. *Int. J. Disaster Risk Reduct.* **2021**, *56*, 102143. [CrossRef]
11. Aldemir, A.; Guvenir, E.; Sahmaran, M. Rapid screening method for the determination of regional risk distribution of masonry structures. *Struct. Saf.* **2020**, *85*, 101959. [CrossRef]
12. Harith, N.S.H.; Jainih, V.; Ladin, M.A.; Adiyanto, M.I. Assessing the vulnerability of Kota Kinabalu buildings. *Civ. Eng. Archit.* **2021**, *9*, 68–77. [CrossRef]
13. Ruggieri, S.; Cardellicchio, A.; Leggieri, V.; Uva, G. Machine-learning based vulnerability analysis of existing buildings. *Autom. Constr.* **2021**, *132*, 103936. [CrossRef]
14. Cenan Mertol, H.; Tunc, G.; Akis, T. Damage observation of reinforced concrete buildings after 2020 Sivrice (Elazığ) earthquake, Turkey. *J. Perform. Constr. Facil.* **2021**, *35*, 04021053. [CrossRef]
15. Mertol, H.C.; Tunç, G.; Akış, T.; Kantekin, Y.; Aydın, İ.C. Investigation of RC Buildings after 6 February 2023, Kahramanmaraş, Türkiye Earthquakes. *Buildings* **2023**, *13*, 1789. [CrossRef]
16. Wu, D.; Mendel, J.M. Recommendations on designing practical interval type-2 fuzzy systems. *Eng. Appl. Artif. Intell.* **2019**, *85*, 182–193. [CrossRef]
17. Harirchian, E.; Kumari, V.; Jadhav, K.; Rasulzade, S.; Lahmer, T.; Raj Das, R. A synthesized study based on machine learning approaches for rapid classifying earthquake damage grades to RC buildings. *Appl. Sci.* **2021**, *11*, 7540. [CrossRef]
18. Bektaş, N.; Kegyes-Brassai, O. Development in fuzzy logic-based rapid visual screening method for seismic vulnerability assessment of buildings. *Geosciences* **2022**, *13*, 6. [CrossRef]
19. Bektaş, N.; Lilik, F.; Kegyes-Brassai, O. Development of a fuzzy inference system based rapid visual screening method for seismic assessment of buildings presented on a case study of URM buildings. *Sustainability* **2022**, *14*, 16318. [CrossRef]
20. Harirchian, E.; Hosseini, S.E.A.; Jadhav, K.; Kumari, V.; Rasulzade, S.; Işık, E.; Wasif, M.; Lahmer, T. A review on application of soft computing techniques for the rapid visual safety evaluation and damage classification of existing buildings. *J. Build. Eng.* **2021**, *43*, 102536. [CrossRef]
21. Barfield, W. *The Cambridge Handbook of the Law of Algorithms*; Cambridge University Press: Cambridge, UK, 2020.
22. Ningthoujam, M.; Nanda, R.P. Rapid visual screening procedure of existing building based on statistical analysis. *Int. J. Disaster Risk Reduct.* **2018**, *28*, 720–730. [CrossRef]
23. Mora, E.; Ordóñez Bueno, M.; Gómez, C. Structural Vulnerability Assessment Procedure for Large Areas Using Machine Learning and Fuzzy Logic. *IRECE* **2021**, *12*, 358. [CrossRef]
24. Şen, Z. Rapid visual earthquake hazard evaluation of existing buildings by fuzzy logic modeling. *Expert Syst. Appl.* **2010**, *37*, 5653–5660. [CrossRef]
25. Şen, Z. Supervised fuzzy logic modeling for building earthquake hazard assessment. *Expert Syst. Appl.* **2011**, *38*, 14564–14573. [CrossRef]
26. Amiri Shahmirani, M.R.; Akbarpour Nikghalb Rashti, A.; Adib Ramezani, M.R.; Golafshani, E.M. Application of fuzzy modelling to predict the earthquake damage degree of buildings based on field data. *J. Intell. Fuzzy Syst.* **2021**, *41*, 2717–2730. [CrossRef]
27. Mangir, A.; Okumus, V. Pre-earthquake fuzzy logic-based rapid hazard assessment of reinforced concrete buildings. *Case Stud. Constr. Mater.* **2023**, *19*, e02534. [CrossRef]
28. Gao, Y.; Mosalam, K.M. Deep transfer learning for image-based structural damage recognition. *Comput.-Aided Civ. Infrastruct. Eng.* **2018**, *33*, 748–768. [CrossRef]
29. Liang, X. Image-based post-disaster inspection of reinforced concrete bridge systems using deep learning with Bayesian optimization. *Comput.-Aided Civ. Infrastruct. Eng.* **2019**, *34*, 415–430. [CrossRef]
30. Pan, X.; Yang, T. Postdisaster image-based damage detection and repair cost estimation of reinforced concrete buildings using dual convolutional neural networks. *Comput.-Aided Civ. Infrastruct. Eng.* **2020**, *35*, 495–510. [CrossRef]
31. Ogunjinmi, P.D.; Park, S.-S.; Kim, B.; Lee, D.-E. Rapid post-earthquake structural damage assessment using convolutional neural networks and transfer learning. *Sensors* **2022**, *22*, 3471. [CrossRef] [PubMed]
32. Irwansyah, E.; Hartati, S. Three-stage fuzzy rule-based model for earthquake non-engineered building house damage hazard determination. *J. Adv. Comput. Intell. Intell. Inform.* **2017**, *21*, 1298–1311. [CrossRef]
33. Ketsap, A.; Hansapinyo, C.; Kronprasert, N.; Limkatanyu, S. Uncertainty and fuzzy decisions in earthquake risk evaluation of buildings. *Eng. J.* **2019**, *23*, 89–105. [CrossRef]
34. Harirchian, E.; Lahmer, T. Improved rapid visual earthquake hazard safety evaluation of existing buildings using a type-2 fuzzy logic model. *Appl. Sci.* **2020**, *10*, 2375. [CrossRef]
35. Agency, F.E.M. *Rapid Visual Screening of Buildings for Potential Seismic Hazards: A Handbook*; Government Printing Office: Washington, DC, USA, 2017.
36. Agency, F.E.M. *Rapid Visual Screening of Buildings for Potential Seismic Hazards: Supporting Documentation*; Government Printing Office: Washington, DC, USA, 2015.
37. Harirchian, E.; Jadhav, K.; Mohammad, K.; Aghakouchaki Hosseini, S.E.; Lahmer, T. A comparative study of MCDM methods integrated with rapid visual seismic vulnerability assessment of existing RC structures. *Appl. Sci.* **2020**, *10*, 6411. [CrossRef]
38. Fortunato, A.; Fabbrocino, F.; Angelillo, M.; Fraternali, F. Limit analysis of masonry structures with free discontinuities. *Meccanica* **2018**, *53*, 1793–1802. [CrossRef]

39. Fabbrocino, F.; Farina, I.; Modano, M. Loading noise effects on the system identification of composite structures by dynamic tests with vibrodyne. *Compos. Part B Eng.* **2017**, *115*, 376–383. [CrossRef]
40. Modano, M.; Fabbrocino, F.; Gesualdo, A.; Matrone, G.; Farina, I.; Fraternali, F. On the forced vibration test by vibrodyne. In Proceedings of the COMPDYN 2015 Conference Proceedings, Crete Island, Greece, 25–27 May 2015; pp. 25–27.
41. Peng, Y. Regional earthquake vulnerability assessment using a combination of MCDM methods. *Ann. Oper. Res.* **2015**, *234*, 95–110. [CrossRef]
42. Grünthal, G. *European Macroseismic Scale 1998 (EMS-98)*; European Seismological Commission, Sub Commission on Engineering Seismology, Working Group Macroseismic Scales. Conseil de l'Europe, Cahiers du Centre Européen de Géodynamique et de Séismologie: Luxembourg, 1998; Volume 15, p. 99.
43. Milutinovic, Z.V.; Trendafiloski, G.S. *RISK-UE Project: An Advanced Approach to Earthquake Risk Scenarios with Applications to Different European Towns*; Contract: EVK4-CT-2000-00014, WP4: Vulnerability of Current Buildings; European Commission: Brussels, Belgium, 2003; pp. 1–111. Available online: http://www.civil.ist.utl.pt/~mlopes/conteudos/DamageStates/Risk%20UE%20WP04_Vulnerability.pdf (accessed on 23 October 2022).
44. Earthquake Planning and Protection Organization (OASP). *Provisions for Pre-Earthquake Vulnerability Assessment of Public Buildings (Part A)*; Technical Report; Greek Society of Civil Engineers: Athens, Greece, 2000.
45. NRC. *Manual for Screening of Buildings for Seismic Investigation*; NRC Publications Archive: Ottawa, ON, Canada, 1993.
46. Pittore, M.; Graziani, L.; Maramai, A.; Haas, M.; Parolai, S.; Tertulliani, A. Bayesian estimation of macroseismic intensity from post-earthquake rapid damage mapping. *Earthq. Spectra* **2018**, *34*, 1809–1828. [CrossRef]
47. Musson, R.M.; Grünthal, G.; Stucchi, M. The comparison of macroseismic intensity scales. *J. Seismol.* **2010**, *14*, 413–428. [CrossRef]
48. Datta, S.D.; Sobuz, M.H.R.; Nafe Assafi, M.; Sutan, N.M.; Islam, M.N.; Mannan, M.B.; Akid, A.S.M.; Hasan, N.M.S. Critical project management success factors analysis for the construction industry of Bangladesh. *Int. J. Build. Pathol. Adapt.* 2023, *ahead-of-print*. [CrossRef]
49. Memon, A.H.; Memon, A.Q.; Khahro, S.H.; Javed, Y. Investigation of Project Delays: Towards a Sustainable Construction Industry. *Sustainability* **2023**, *15*, 1457. [CrossRef]
50. Yousri, E.; Sayed, A.E.B.; Farag, M.A.; Abdelalim, A.M. Risk identification of building construction projects in egypt. *Buildings* **2023**, *13*, 1084. [CrossRef]
51. Oke, A.E.; Arowoiya, V.A.; Akomolafe, O.T. Influence of the Internet of Things' application on construction project performance. *Int. J. Constr. Manag.* **2022**, *22*, 2517–2527. [CrossRef]
52. Alhammadi, A.S.A.M.; Memon, A.H. Inhibiting factors of cost performance in uae construction projects. *Int. J. Sustain. Constr. Eng. Technol.* **2020**, *11*, 126–132.
53. Ansal, A.; Özaydın, K.; Edinçliler, A.; Saglamer, A.; Sucuoglu, H.; Özdemir, P. *Earthquake Master Plan for Istanbul, Metropolital Municipality of Istanbul, Planning and Construction Directorate*; Geotechnical and Earthquake Investigation Department: Istanbul, Turkey, 2003.
54. Sucuoğlu, H.; Yazgan, U.; Yakut, A. A screening procedure for seismic risk assessment in urban building stocks. *Earthq. Spectra* **2007**, *23*, 441–458. [CrossRef]
55. İlerisoy, Z. Discussion of the structural irregularities in the plan for architectural design within the scope of earthquake codes. *Period. Polytech. Archit.* **2019**, *50*, 50–62. [CrossRef]
56. Mendi, H.E. Evaluation of Architectural and Consciousness Exploration Architecture-Based Issues in Seismic Design. Master's Thesis, Middle East Technical University, Ankara, Turkey, 2005.
57. Kaplan, O.; Guney, Y.; Topcu, A.; Ozcelikors, Y. A rapid seismic safety assessment method for mid-rise reinforced concrete buildings. *Bull. Earthq. Eng.* **2018**, *16*, 889–915. [CrossRef]
58. Hagan, M.T.; Demuth, H.B.; Beale, M.H.; De Jesús, O. *Neural Network Design*, 2nd ed.; Martin Hagan: Stillwater, OK, USA, 2014.
59. Tastle, W.J.; Wierman, M.J. Consensus and dissention: A measure of ordinal dispersion. *Int. J. Approx. Reason.* **2007**, *45*, 531–545. [CrossRef]
60. Hagan, M.T.; Menhaj, M.B. Training feedforward networks with the Marquardt algorithm. *IEEE Trans. Neural Netw.* **1994**, *5*, 989–993. [CrossRef]
61. Demuth, H.B.; Beale, M.H. *Neural Network Toolbox; For Use with Matlab; Computation, Visualization, Programming; User's Guide, Version 4*; Math Works: Natick, MA, USA, 2000.
62. Lopez, C.P. *Deep Learning with Matlab. Neural Networks by Examples*; Cesar Perez; Lulu.com: Morrisville, NC, USA, 2020.
63. Allali, S.A.; Abed, M.; Mebarki, A. Post-earthquake assessment of buildings damage using fuzzy logic. *Eng. Struct.* **2018**, *166*, 117–127. [CrossRef]
64. Harirchian, E.; Lahmer, T.; Kumari, V.; Jadhav, K. Application of support vector machine modeling for the rapid seismic hazard safety evaluation of existing buildings. *Energies* **2020**, *13*, 3340. [CrossRef]
65. Özkan, E.; Demir, A.; Turan, M.E. A new ANN based rapid assessment method for RC residential buildings. *Struct. Eng. Int.* **2023**, *33*, 32–40. [CrossRef]

applied sciences

MDPI

Article

Comparison between the Dynamic Responses of Steel Buildings with Medium and Deep Columns under Several Seismic Intensities

Federico Valenzuela-Beltrán, Mario D. Llanes-Tizoc, Eden Bojorquez, Juan Bojorquez, J. M. Leal-Graciano, Victor Baca, Robespierre Chavez and Alfredo Reyes-Salazar *

Faculty of Engineering, Autonomous University of Sinaloa, Ciudad Universitaria, Culiacán 80000, Mexico; fvalenzuelab@uas.edu.mx (F.V.-B.); mariollanes@uas.edu.mx (M.D.L.-T.); eden@uas.edu.mx (E.B.); juanbm@uas.edu.mx (J.B.); jesusleal@uas.edu.mx (J.M.L.-G.); victorbaca@uas.edu.mx (V.B.); robespierre@uas.edu.mx (R.C.)

* Correspondence: reyes@uas.edu.mx

Abstract: Structural engineers often use deep columns in high seismic areas to reduce drifts, yet this somehow contradicts what is stated in some tests in the sense that even though deep columns may satisfy current seismic provisions, they can suffer premature twisting; this is an indication that a lot of research is needed in this area. Numerical and experimental studies have been conducted to estimate the response of steel buildings with medium and deep columns under the action of static and cyclic loading; however, studies accounting for the dynamic characteristics of buildings and strong motions are not common. In addition, responses in terms of local parameters have not been considered either. In this study, the nonlinear seismic responses of steel buildings with perimeter moment-resisting frames and medium (W14) columns are numerically calculated and compared to those of similar steel buildings with equivalent deep columns in terms of cost (W27 and larger). Low-, mid-, and high-rise steel building models with different dynamic characteristics, as well as several strong motions with different frequency contents, are considered. Results indicate that the drifts of the models with medium columns may be up to 140% greater than those of the models with deep columns. Significant reductions are also observed for top displacements, normalized interstory shears, and combined normalized axial loads and bending moments. Hence, the seismic demands of the buildings with deep columns may be much smaller than those of the buildings with medium columns and, therefore, the buildings with deep columns exhibit a superior behavior, which results in more economical designs. The reduction is greater for the case of low- and mid-rise buildings than for high-rise buildings. One of the reasons for this is that as medium columns are replaced by deep columns, the stiffness and the strength increase, which are lower in the tallest model.

Keywords: moment-resisting steel frames; deep columns; low-, mid-, and high-rise buildings; seismic loading; local and global response

1. Introduction and Literature Review

Different arrangements of structural members are continuously studied with the aim of improving the structural behavior of buildings subjected to the action of various types of loads. In the case of steel buildings under the action of severe seismic loads, moment-resisting frames (MRF) are widely used due to their great ductility capacity. In some developed countries, like the United States of America, the common practice is to use in these buildings two MRF in each direction, usually located at the perimeter (PMRF) and gravity frames (GF) at the interior. The first are designed to resist the total seismic load and the second to resist the gravity loads. Hence, for seismic design purposes, the buildings are usually modeled as two-dimensional (2D) structures.

There are many numerical and experimental studies related to the seismic behavior evaluation of steel buildings with MRF. Before the Northridge Earthquake of 1994, a limited number of laboratory tests at reduced scale were conducted to evaluate the seismic behavior of the columns. After the mentioned earthquake, the cyclic behavior of W beams was extensively investigated; however, for columns only a few cases were considered. Historically speaking, the reported studies regarding the seismic behavior evaluation of columns in steel buildings with MRF have mostly been for W14 columns or smaller. Gupta and Krawinkler [1] studied the behavior of several steel building models designed according to design provisions of the United States. Lee and Foutch [2] studied the seismic nonlinear behavior of a set of post-Northridge buildings which represent typical buildings with MRF. Foutch and Yun [3] investigated the accuracy of modeling methods used in the design of MRF. Mele et al. [4] compared the seismic behavior of steel buildings with perimeter MRF with that of steel buildings with spatial MRF. In another study, Fathi et al. [5] proposed modifications for the behavior factor of moment-resisting steel frames modeled as multi-degree-of-freedom systems. Lee and Foutch [6] studied the seismic behavior of three-, nine-, and twenty-story steel moment-resisting frames with W14 columns designed using different force reduction factors. Krishnan et al. [7] determined the damage to two 18-story steel moment-frame buildings, one existing and one new, located in southern California. Black [8] proposed two stability coefficients that can be used to quantify the P–Δ effect during the elastic and inelastic lateral displacement of regular steel moment-resisting frames. Sejal et al. [9] compared the responses of a steel moment-resisting frame designed by the Performance-Based Plastic design method and conventional elastic design method based on the seismic evaluation investigated with both nonlinear static and nonlinear dynamic analysis under different ground motions. Teran-Gilmore et al. [10] studied the pertinence of using an importance factor to promote an adequate structural and nonstructural performance of essential facilities. Athanasios et al. [11] formulated a procedure to estimate lateral displacements at the first yielding of plane steel frames to be used in a performance-based seismic design. Formisano et al. [12] evaluated the robustness assessment of two methods for steel-framed buildings under catastrophic events by using two steel-framed buildings designed according to old and new seismic Italian codes. Liu et al. [13] assessed the structural reliability of low-to-mid-rise steel frames taking into account the system-based design approach. Samanta and Huang [14] investigated the impact of several ground-motion scaling procedures on the response of high-rise moment-resisting frame buildings. Loulelis et al. [15] developed explicit expressions for different values of strength reduction factors for the first significant modes of steel moment-resisting frames where strength deterioration and panel zone effects were considered. Ruiz-García et al. [16] studied the response of three-dimensional steel moment-resisting buildings having three-, nine-, and twenty-story heights under the bidirectional attack of real seismic sequences with different earthquake ground-motion features. Many other important studies aimed at meeting specific objectives, but following the general goal of evaluating the seismic behavior of steel frames with medium columns, can be found in the literature [17–21].

Despite the large amount of research developed in the area of the seismic behavior of steel buildings with MRF and the important contributions of the above-mentioned papers, only a few studies have been developed concerning the performance evaluation of steel buildings with MRF and deep columns. However, because of the compactness of the webs and flanges of a vast number of deep sections of grade 50, and because of the necessity of higher bending and shear stiffness to reduce drifts, structural engineers often use deep columns in high seismic regions resulting in more economical designs. Furthermore, it is stated that using W14 columns or smaller may result in unnecessarily heavy and more expensive columns. This seemingly contradicts what has been stated in some investigations [22] where it is argued that even though deep columns satisfy current seismic provisions, they can suffer premature twisting, which occurs at late stages of loading. Similarly, Chi and Uang [23] conducted an experimental study by using beam–

column subassemblies with W27 columns, finding that lateral–torsional buckling of the beams produces twisting of the columns. Another study [24], however, encouraged the use of deep columns, stating that column twisting might have occurred primarily because of the way the specimen was tested where no top flange bracing was used at the beams, which in actual buildings is normally provided by the floors. Hence, if lateral–torsional buckling is avoided, the columns may remain stable even for significant levels of seismic demands. This stable behavior is expected to be better for the case of steel buildings that are symmetrical in plan. Zhang et al. [25] concluded that the behavior of moment connections with reduced beam sections and deep wide-flange columns satisfied the criteria of the AISC seismic provisions. Shao and Hale [26] tested three full-scale beam–column assemblies using W36 × 256 beams and W30 × 261 columns, concluding that the proposed connections satisfy the two interstory displacement cycles of 0.04 and the inelastic rotation of 0.03 required for the Office of Statewide Health Planning and Development (OSHPD). Ricles and Zhang [27] and Zhang and Ricles [28,29] showed that, in steel frames with deep columns, a composite floor slab provides restraint to the top flange of the beams reducing (1) the magnitude of lateral movement of the beam top and bottom flanges in the RBS, (2) column twist, and (3) strength degradation due to beam instability. Elkady and Lignos [30], by using Finite Element simulations, demonstrated that column sections with large web slenderness, close to the web compactness limit requirements, have a high rate of cyclic deterioration in strength and axial shortening due to severe web local buckling. In many other studies [31–33], particular aspects of the response of steel buildings with deep columns were also addressed.

Despite the important contributions of the above-mentioned studies concerning the behavior of steel buildings with medium and deep columns, as further stated below, there are many issues that deserve our attention. It is worth mentioning that in the experimental cyclic tests developed to evaluate and compare the behavior of steel buildings with medium and deep columns, the effect of the dynamic properties of buildings as well as that of the frequency content of the strong motions cannot be considered. It seems to be an assumption that the structural demands, namely maximum interstory drifts or top displacements, will be the same for the buildings with medium and deep columns; if so, obviously the buildings with deep columns are more likely to fail. However, as it is shown in this paper, the seismic demands on the buildings with deep columns may be much smaller than those of the buildings with medium columns and, therefore, buildings with deep columns present a superior behavior.

2. Justification and Objectives

It is worth mentioning that most of the studies concerning deep columns have been focused on individual beam–column elements, or on subassemblies of beams and columns representing a part of the building, subjected to the action of static and cyclic loads. In addition, as stated earlier, it seems there is an implicit assumption in many of these studies in the sense that the structural demands, namely maximum interstory drifts or top displacements, are the same for the buildings with both column sizes. The response of steel buildings with deep columns under the action of different strong motions with different frequency contents has not been studied; it is known that the dynamic response of a structure depends not only on the structural properties, but also on the loading properties, such as those of earthquake loading.

It is emphasized that it is not an objective of this paper to compare the seismic responses of steel buildings with medium and deep columns for the same drift demand, but rather to compare their dynamic responses numerically obtained considering models and seismic records with different dynamic characteristics. The comparison is made in terms of displacements and resultant stresses (member forces), which includes overall (global) and local as well as single and combined response parameters. The equivalence of the two column sizes is expressed in terms of weight (equal cost). It is assumed in this study that drifts of up to 3.5% can be reached without the occurrence of lateral buckling of beams or columns twisting.

It is important to mention that a small part of a big issue is addressed in this research; the specific objectives are as follows:

(1) Calculate the nonlinear seismic responses of the building models with medium columns in terms of interstory drifts and roof displacements, which represent overall response parameters. These parameters are compared with those of the same buildings, but with equivalent deep columns in terms of cost (W27 and larger).

(2) Compare the seismic responses of the buildings with medium and deep columns for the case of single global resultant stresses (interstory shears) and local resultant stresses (axial load and bending moments) at the exterior and interior columns located on the ground level.

(3) Compare the responses, as in Objective (2), but now for normalized single global and normalized combined local resultant stresses.

3. Procedure, Structural Models, and Strong Motions

3.1. Parameters of the Study

Three plane frame models, representing steel building structures of low-, mid- and high-rise steel building structures with different dynamic characteristics, and fifteen strong seismic motions with different frequency contents, are considered in the study. The buildings are modeled as complex 2D MDOF systems, which are assumed to have, firstly, columns of medium size (W14), and then columns of deep size. A medium column, in the context of this paper, is a column with a depth of 14 inches; in the same manner, a deep column is a column with a depth of 27 inches or greater. The building responses with the two column sizes are compared in terms of displacements and member forces, which include overall (global), local, and single and combined response parameters. The seismic records are scaled up to obtain several levels of structural demands, which range from elastic to significant inelastic behavior. The vertical seismic component as well as the gravity loads are considered in the seismic analyses.

The Ruaumoko computer program [34] is used to perform the required nonlinear time-history seismic analyses. The Newmark Constant Average Acceleration Method is used to numerically evaluate the seismic response. The lumped mass matrix, Rayleigh Damping, the P-δ effect, and large displacement are also considered. A bilinear hysteresis rule with 5% of post-yielding stiffness as well as the lumped plasticity approach are assumed in the analysis. The axial load–bending moment interaction is given by the yield interaction surface proposed by Chen and Atsuta [35].

3.2. Structural Models with W14 Columns

The three-, nine-, and twenty-story buildings with PMRF considered in the SAC steel project [36] located in the Los Angeles area are used in this study; they are assumed to satisfy the code requirements existing at the time of the project (Uniform Building Code) [37]. These models are used for the following reasons: (a) they represent a standard design practice in the United States; (b) they were designed for experienced structural engineers in the United States for a very important project; and (c) they are widely used in numerical investigations. They will be referred to, in general, as the SAC models and will be particularly denoted hereafter as Models M1, M2, and M3, respectively. The first three lateral vibration periods of Model M1 are 1.05 s, 0.29 s, and 0.15 s, while those of Model M2 are 2.36 s, 0.79 s, 0.54 s. The corresponding values for Model M3 are 4.15 s, 1.34 s, 0.75 s. Their design base shear coefficients are 0.08, 0.04, and 0.03, for Models M1, M2, and M3, respectively, according to ASD-UBC 94. The damping is considered to be 3% of the critical (ζ = 3%). The plan and the elevations of Models M1, M2, and M3 are given in Figures 1–3, respectively. In such Figures (plan), the I-shaped symbol indicates the orientation of the columns; the term "4 bays @ 30′" means "four bays spaced every 30 feet". The axial loads and bending moments are studied for some exterior (EXT) and interior (INT) columns of the PMRF located at the ground-floor level, as indicated in the figures. The sections, including girders and columns, of Models M1 and M2 can be seen in Table 1, while those of Model M3 are shown in Table 2. It is to be noted that that all column sections used in these models are compact.

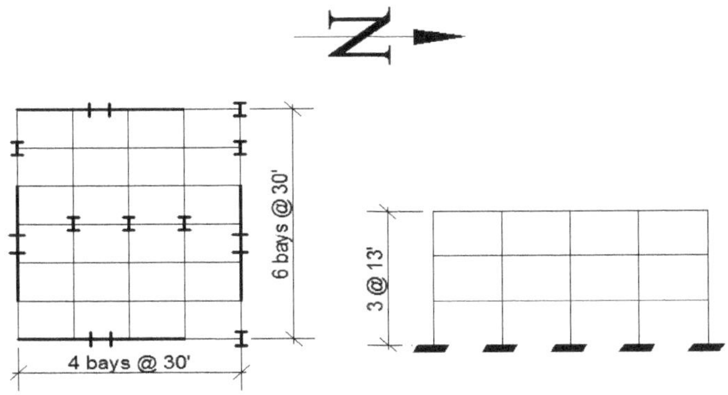

Figure 1. Plan and elevation, 3-story model.

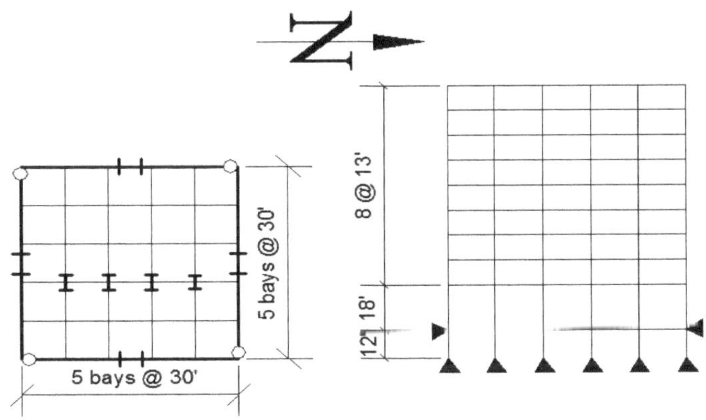

Figure 2. Plan and elevation, 9-story model.

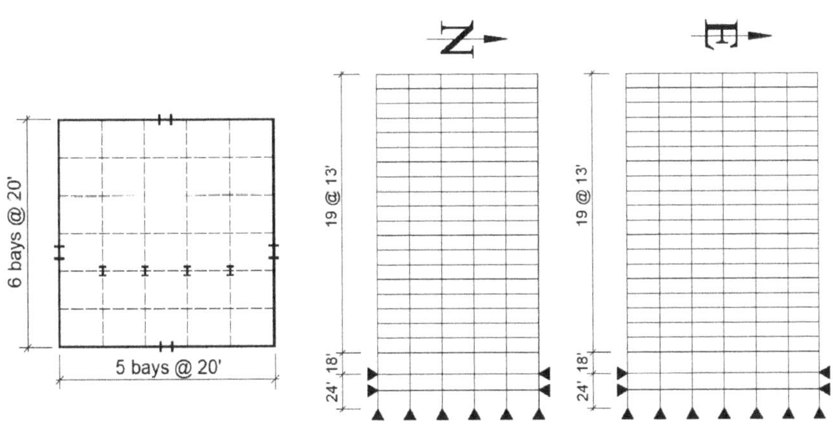

Figure 3. Plan and elevation, 20-story model.

Table 1. Beam and column sections, PMRF of Models M1 and M2.

Model	Story	Columns		Girders
		Outside	Inside	
3-story	1	W14 × 257	W14 × 311	W33 × 118
	2	W14 × 257	W14 × 311	W30 × 116
	3/Roof	W14 × 257	W14 × 311	W24 × 68
9-story	Basement-1	W14 × 370	W14 × 500	W36 × 160
	1	W14 × 370	W14 × 500	W36 × 160
	2	W14 × 370	W14 × 500	W36 × 160
	3	W14 × 370	W14 × 455	W36 × 135
	4	W14 × 370	W14 × 455	W36 × 135
	5	W14 × 283	W14 × 370	W36 × 135
	6	W14 × 283	W14 × 370	W36 × 135
	7	W14 × 257	W14 × 283	W30 × 99
	8	W14 × 257	W14 × 283	W27 × 84
	9/roof	W14 × 233	W14 × 257	W24 × 68

Table 2. Beam and column sections, PMRF of Model M3.

Story	Columns		Girders
	Outside	Inside	
Basement-1	15 × 15 × 2.00	W24 × 335	W14 × 22
Basement-2	15 × 15 × 2.00	W24 × 335	W30 × 99
1	15 × 15 × 2.00	W24 × 335	W30 × 99
2	15 × 15 × 2.00	W24 × 335	W30 × 99
3	15 × 15 × 1.25	W24 × 335	W30 × 99
4	15 × 15 × 1.25	W24 × 335	W30 × 99
5	15 × 15 × 1.25	W24 × 335	W30 × 108
6	15 × 15 × 1.00	W24 × 229	W30 × 108
7	15 × 15 × 1.00	W24 × 229	W30 × 108
8	15 × 15 × 1.00	W24 × 229	W30 × 108
9	15 × 15 × 1.00	W24 × 229	W30 × 108
10	15 × 15 × 1.00	W24 × 229	W30 × 108
11	15 × 15 × 1.00	W24 × 229	W30 × 99
12	15 × 15 × 1.00	W24 × 192	W30 × 99
13	15 × 15 × 1.00	W24 × 192	W30 × 99
14	15 × 15 × 1.00	W24 × 192	W30 × 99
15	15 × 15 × 0.75	W24 × 131	W30 × 99
16	15 × 15 × 0.75	W24 × 131	W30 × 99
17	15 × 15 × 0.75	W24 × 131	W27 × 84
18	15 × 15 × 0.75	W24 × 117	W27 × 84
19	15 × 15 × 0.75	W24 × 117	W24 × 62
20/Roof	15 × 15 × 0.50	W24 × 84	W21 × 50

3.3. Structural Models with Deep Columns

The geometry and the beams are the same for the models with medium and deep columns. Obviously, the columns of the PMRF are different. The models with deep columns will be referred to hereafter, in general, as the deep models and, in particular, as Models L1, L2, and L3, respectively. The first three lateral vibration periods of Model L1 are 0.88 s, 0.17 s, and 0.12 s, while those of Model L2 are 1.95 s, 0.40 s, 0.21 s. The corresponding values for Model M3 are 3.87 s, 0.75 s, 0.40 s. The column sections are shown in Table 3 for Models L1 and L2, while those of Model L3 can be seen in Table 4.

Table 3. Deep columns of Models L1 and L2.

Model	Story	Columns	
		Outside	Inside
1	1/2	W27 × 258	W27 × 307
	2/3	W27 × 258	W27 × 307
	3/roof	W27 × 258	W27 × 307
2	−1/1	W27 × 368	W27 × 494
	1/2	W27 × 368	W27 × 494
	2/3	W27 × 368	W27 × 494, W27 × 448
	3/4	W27 × 368	W27 × 448
	4/5	W27 × 368, W27 × 281	W27 × 448, W27 × 368
	5/6	W27 × 281	W27 × 368
	6/7	W27 × 281, W27 × 258	W27 × 368, W27 × 281
	7/8	W27 × 258	W27 × 281
	8/9	W27 × 258, W27 × 235	W27 × 281, W27 × 258
	9/roof	W27 × 235	W27 × 258

Table 4. Deep columns of Model L3.

Model	Story	Columns Inside
3	Basement-1	W40 × 328
	Basement-2	W40 × 328
	1/2	W40 × 328
	2/3	W40 × 328
	3/4	W40 × 328
	4/5	W40 × 328
	5/6	W40 × 328
	6/7	W40 × 221
	7/8	W40 × 221
	8/9	W40 × 221
	9/10	W40 × 221
	10/11	W40 × 221
	11/12	W40 × 221
	12/13	W40 × 192
	13/14	W40 × 192
	14/15	W40 × 192
	15/16	W40 × 131
	16/17	W40 × 131
	17/18	W40 × 131
	18/19	W40 × 117
	19/20	W40 × 117
	20/roof	W40 × 84

The models used in the study are subject to different levels of deformation, including one associated with the formation of plastic hinges. In this regard, lateral–torsional buckling (LTB), flange local buckling (FLB), and web local buckling (WLB) should be avoided. In this research, it is assumed that the slab provides continuous lateral bracing in such a way that LTB may not occur. FLB and WLB may not occur either since the flanges and webs of the section are compact (see Section 3.5). It is to be pointed out that, as for the case of the models with medium columns (Models M1, M2, and M3), only compact sections were used for Models L1 and L2. In order to have a similar increment in stiffness for the case of the twenty-story building (Model L3) as for that of Models L1 and L2, while replacing medium columns with deep columns, whenever it was possible the largest W compact sections (W40) given in the AISC Construction Manual [38] were selected. In fact, equivalent deep compact columns were found for all sections except for the last three upper ones (W24 × 131, W24 × 117 and W24 × 84). However, hypothetical equivalent sections, which are assumed to be compact, were considered (W40 × 131, W40 × 117 and W40 × 84) in such a way that the increment in stiffness and in plastic moment capacity were in the same proportion as that of the other columns. To clarify, the increment in stiffness, as well as in plastic moment, while replacing medium columns with deep columns for the twenty-story building, is smaller (by about 15%) than those of the three- and nine-story models. Another factor contributing to the smaller increment in lateral stiffness in the case of the twenty-story model is due to the fact that deep columns are used only inside since the others (outside) are HSS box columns. It is worth mentioning that there is experimental evidence that if compact sections are used in a moment-resisting steel frame, drifts of up to 5% can be reached and still vibrate in a stable manner [39–43]. Despite that, these examples do not mean that all the connections with compact sections can have the expected seismic behavior. In general, the post-earthquake connection details and highly ductile members are required by the AISC seismic provisions and have been adopted in many new steel constructions in regions of high seismicity.

3.4. Strong Motion Records

The models are excited by fifteen strong motion records, which are representative of the area where the structural models are located; their main characteristics are given in Table 5 and are denoted as Strong Motions 1 to 15. It is worth mentioning that, for example, ASCE 7–16 establishes a minimum of eleven motion pairs to achieve reliable estimates of mean structural response in nonlinear dynamic analysis. The data of the seismic records were obtained from the Data Set of the National Strong Motion Program (NSMP) of the United States Geological Survey (USGS). The horizontal seismic component with the greater peak ground acceleration (PGA), as well as the vertical component and the tributary gravity load, are simultaneously applied to the PMRF oriented in the *EW* structural directions. Then, the horizontal component with the greater PGA is replaced by the other horizontal component (with the lower PGA), which together with the vertical component and the gravity loads are applied to the PMRF oriented in the *NS* direction. It must be noted that the PMRF oriented in the *EW* direction of the three-story building, or of the nine-story building, is identical to that of the *NS* direction, but they are not identical for the twenty-story building. The pseudo-acceleration response spectra for both horizontal components and 3% of critical damping are shown in Figure 4a,b for the *NS* and *EW* directions, respectively. The spectra are scaled to have a $S_a/g = 1.0$ at a period of 1.03 s.

Table 5. Seismic records.

Designation	Record Information	Magnitude (Mw)	PGA (m/s^2) NS	PGA (m/s^2) EW	Period (s) NS	Period (s) EW
LA1	Imperial Valley, 1940	6.9	4.52	6.63	0.53	0.46
LA2	Imperial Valley, 1979	6.5	3.86	4.78	0.16	0.34
LA3	Landers, 1992	7.3	4.14	4.17	0.73	0.33
LA4	Kern, 1952	7.3	5.11	3.53	0.25	0.23
LA5	Loma Prieta, 1989	7	6.53	9.50	0.21	0.2
LA6	Northridge, 1994, Newhall	6.7	6.65	6.45	0.31	0.31
LA7	Northridge, 1994, Rinaldi	6.7	5.23	5.69	0.39	0.29
LA8	Northridge, 1994, Sylmar	6.7	5.59	8.03	0.31	0.36
LA9	North Palm Springs, 1986	6	10.01	9.68	0.17	0.21
LA10	Coyote Lake, 1979	5.7	5.79	3.28	0.15	0.21
LA11	Morgan Hill, 1984	6.2	3.12	5.36	0.18	0.16
LA12	Parkfield, 1966, Cholame 5W	6.1	7.65	6.20	0.37	0.3
LA13	Parkfield, 1966, Cholame 8W	6.1	6.81	7.75	0.17	0.21
LA14	North Palm Springs, 1986	6	5.08	3.71	0.13	0.21
LA15	Whittier, 1987	6	7.54	4.70	0.7	0.28

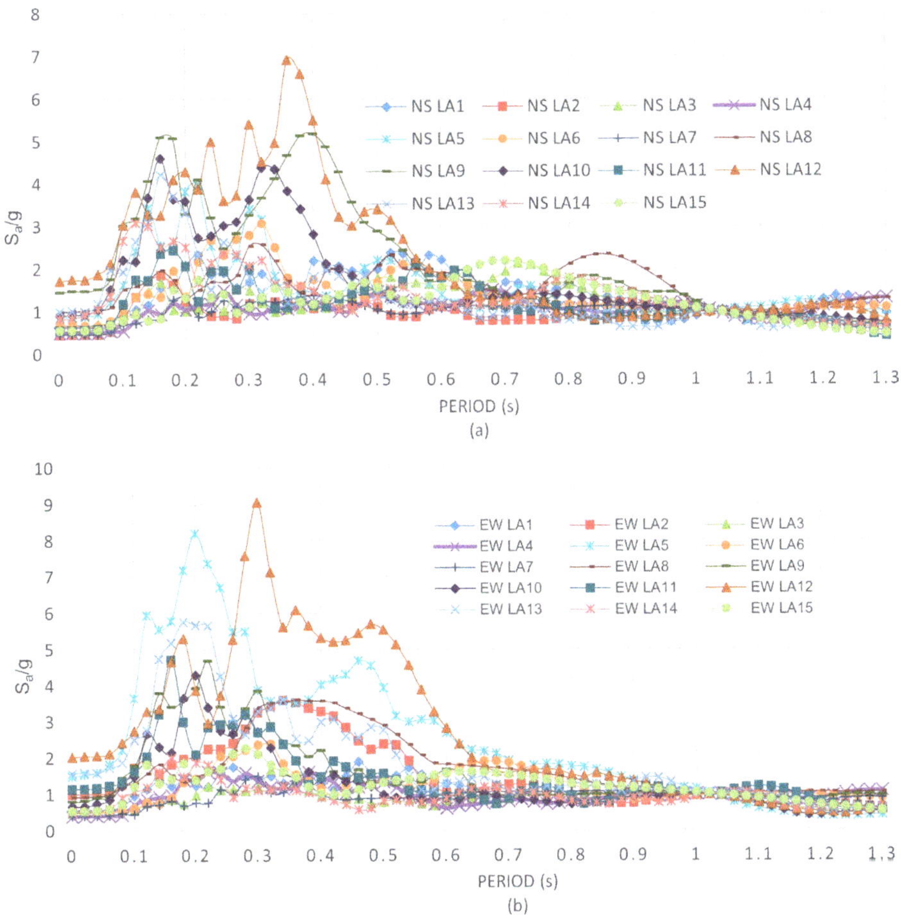

Figure 4. Response spectra: (**a**) NS direction, (**b**) EW direction.

In order to have different levels of deformation, which range from elastic to significant inelastic behavior, the strong motions are scaled in terms of the spectral pseudo-acceleration evaluated at the fundamental vibration period of the structure (S_a (T_1)). The values of S_a vary from 0.2 g to 1.2 g with increments of 0.2 g for the three-story building, from 0.1 g to 0.5 g for the nine-story building with increments of 0.1 g, and from 0.1 g to 0.25 g for the twenty-story building with increments of 0.05 g. For the lowest strong motion intensities mentioned above, the buildings' behavior was essentially elastic (maximum drift of about 0.5%), particularly for the three- and nine-story buildings; for the intermediate seismic intensities, moderate yielding occurred (maximum drift of about 1.6%), while for the highest intensities, a deformation state very close to collapse was developed for some strong motions; the maximum drifts were about 3.5% for a few particular strong motions. Therefore, the largest seismic intensities of 1.2 g, 0.5 g, and 0.25 g produced, for some seismic records, a maximum drift of about 3.5% in the three-, nine-, and twenty-story models, respectively.

The gravity loads used are as follows [36]: (1) the dead load at the floors to calculate the weight is 4.5 kN/m^2; (2) the dead load at the floors to calculate the mass is 4.04 kN/m^2; (3) the dead load at the roof is 3.9 kN/m^2; and (4) the reduced live load at the floors and roof is 0.94 kN/m^2. In addition, the seismic mass for the whole structure is as follows:

(1) 1023.09 kN-s^2/m for the roof of the three-story model; (2) 945.6 kN-s^2/m for floor 2 of the three-story building; (3) 1054.83 kN-s^2/m for the roof of the nine-story building; (4) 996.25 kN-s^2/m for floor 2 of the nine-story building; (5) 979.22 kN-s^2/m for floors 3 to 9 of the nine-story building; and (6) for the twenty-story building, the values are 584.88 kN-s^2/m, 564 kN-s^2/m, and 551.29 kN-s^2/m, for the roof, floor 2, and floors 3 to 20, respectively.

3.5. Local, Shear, and Lateral–Torsional Buckling of Deep Columns

The main incentive of structural engineers in using deep columns is to reduce the maximum drifts. Similarly, using deep columns easily helps to achieve the weak-beam strong-column requirements. However, width/thick ratios of webs (h_w/t_w) and flanges ($b_f/2t_f$) are relatively large so revisions of the LTB, FLB, and WLB, as well as of the web shear buckling (WSB), are required. Columns in the MRF of a steel building subjected to significant deformations produced by earthquakes are usually under the action of moderate axial loads but very high bending moments, so plasticization is mainly produced by bending, and, consequently, compactness in bending needs to be revised.

As stated earlier, in this research, it is assumed that the slab provides continuous lateral bracing in such a way that LTB may not occur. Regarding FLB and WLB, from a revision of the properties and dimensions of the deep column sections used in Models L1, L2, and L3 [41], it is observed that the $b_f/2t_f$ and h/t_w ratios do not exceed the limit values given in Chapter B (Table B4.1b of the AISC Specification [44]), so the flanges and webs are both compacts. Similarly, regarding the web shear, it is revised so that the sections satisfy the requirements given in Section G2.1 of the AISC Specifications [42] in such a way that WSB does not occur and the web shear yielding (WSY) is the controlling limit state.

4. Results and Discussion

4.1. Objective 1: Results in Terms of Displacements

In this part of the paper, the seismic responses of the building models with medium columns, in terms of drifts and roof displacements, are compared to those of the models with deep columns. To make the comparison, the following ratios (R_D and R_{DT}) are used:

$$R_D = \frac{D_M}{D_L} \tag{1}$$

$$R_{DT} = \frac{DT_M}{DT_L} \tag{2}$$

where the symbols R, D, DT, represent the words ratio, drift, top displacement, respectively. Similarly, M and L, represent the words medium and large, which in turn stand for medium and deep columns, respectively. Hence, for a direction and a story of a given model, D_M and D_L represent the maximum drifts when medium and deep columns are used, respectively. DT_M and DT_L have a similar meaning except that top displacements are used instead. A value of R_D, or R_{DT}, greater than unity will indicate a reduction in the drift, or in the top displacement, when deep columns are used in the buildings. On the other hand, values of R_D or R_{DT} smaller than unity will indicate an increment in the above-mentioned response parameters.

Results for drifts are discussed first. Some values of the R_D parameter for the *NS* direction of the three-story model are presented in Figure 5a,c,e for seismic intensities of 0.4 g, 0.8 g, and 1.2 g, respectively; the corresponding results for the *EW* direction are shown in Figure 5b,d,f. In these graphs, the term *ST* stands for the number of stories.

A significant variation in the R_D parameter can be observed from one strong motion to another without showing any trend, indicating a significant effect of the frequency content of the strong motions and the higher modes of vibration on the drifts of the buildings with medium and deep columns. The values are greater than unity in most cases. For a few cases, however, the values are smaller than unity. These results illustrate the significant differences that exist between the responses of buildings subjected to seismic loading and to static or cyclic loading. If static lateral loading were applied, the structural drifts would decrease in all stories for all cases.

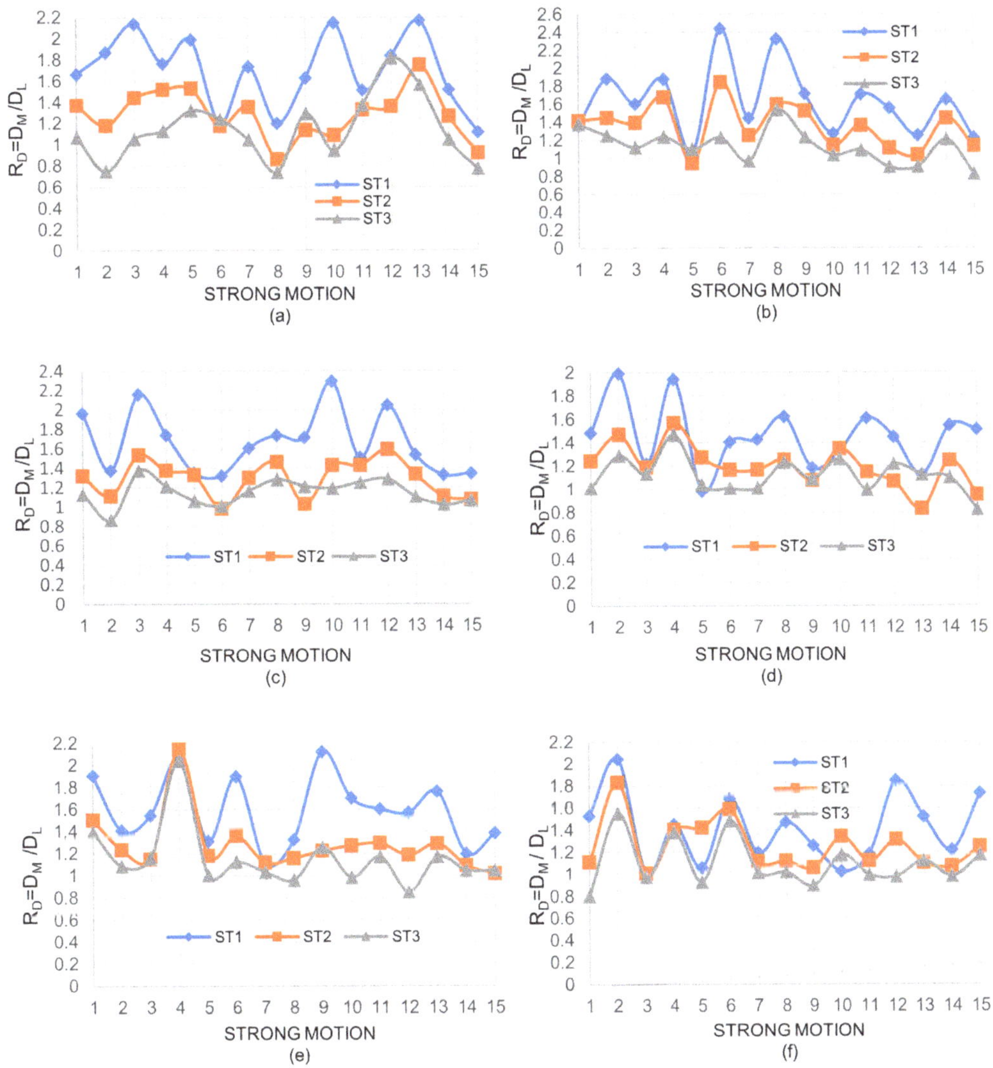

Figure 5. Values of the R_D parameter for the 3-story building: (**a**) S_a = 0.4 g *NS*, (**b**) S_a = 0.4 g *EW*, (**c**) S_a = 0.8 g *NS*, (**d**) S_a = 0.8 g *EW*, (**e**) S_a = 1.2 g *NS*, and (**f**) S_a = 1.2 g *EW*.

It is also observed that the closer the story is to the base, the greater the R_D values will be; however, no clear trend is observed for R_D as the seismic intensity increases. The most important observation that can be made so far is that the R_D parameter takes values of up to 2.4, which represents a considerable reduction in the structural drifts when deep columns are used.

In addition to the six graphs given in Figure 5, another six were generated for R_D considering the other seismic intensities (S_a = 0.2 g, 0.6 g, and 1.0 g) and both directions of the three-story model. Twelve graphs were also developed for roof displacements (R_{DT}). Similar sets of plots were also developed for the nine-story and twenty-story models.

However, the total results are presented and discussed in all cases only in terms of the mean values (MVs) of the R_D and R_{DT} ratios.

The MVs of R_D for the three-story model are shown in Figure 6a,b for the *NS* and *EW* directions, respectively. The corresponding values for the nine-story and the twenty-story models can be seen in Figure 7a,b and Figure 8a,b, respectively. It is worth mentioning that the formation pattern of plastic hinges (PHs) can vary significantly from one seismic record to another and from one model to another. For example, for the highest intensity of seismic record number 9 and the three-story model, PHs developed first on the beams of the first floor and then on the exterior columns of the ground level. However, for the case of the highest intensity of seismic record number 3 and the twenty-story model, PHs developed first on some beams located on the middle third of the building height and then on the exterior columns of the second floor.

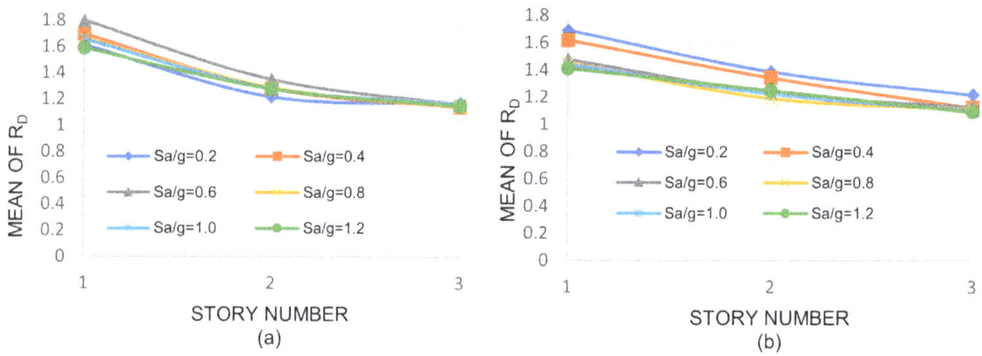

Figure 6. Mean values of R_D, 3-story building: (**a**) *NS* direction, (**b**) *EW* direction.

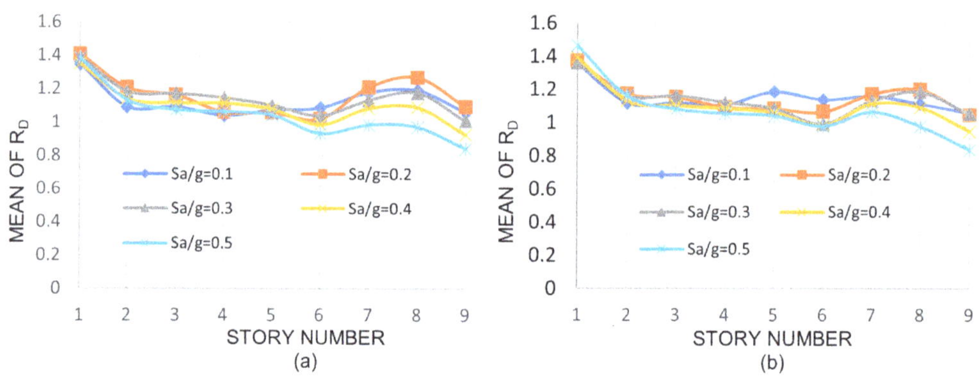

Figure 7. Mean values of R_D, 9-story building: (**a**) *NS* direction, (**b**) *EW* direction.

The results of the three-story model (Figure 6) indicate that, on an average basis and as observed from the plots of individual strong motions, the reduction in the drifts may be significant when deep columns are used; mean values of R_D greater than 1.6 can be seen in many cases for the first story. It is also noted that the values tend to monotonically decrease as the story number increases, but no trend is observed with the seismic intensity. The maximum R_D mean values are quite similar for the *NS* and *EW* directions.

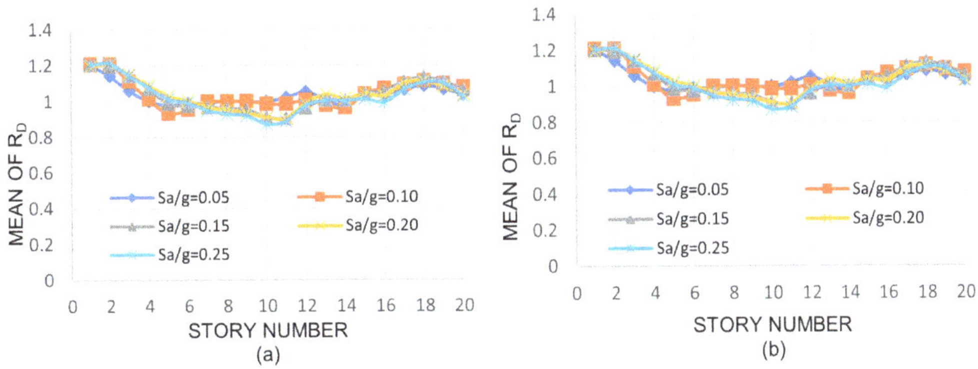

Figure 8. Mean values of R_D, 20-story building: (**a**) *NS* direction, (**b**) *EW* direction.

From the results of the nine- and twenty-story buildings (Figures 7 and 8), it is shown that the drift reductions, although smaller than those of the three-story building, may still be considerable; the maximum values are 1.4 and 1.2 for the nine- and twenty-story model, respectively. For the intermediate stories (8–12) of the twenty-story model, values very close to and even smaller than unity (implying an increment) can be observed; however, it does not correspond to the maximum structural demands in terms of drifts in most of these cases.

It is worth mentioning that the mean values of R_D for the twenty-story building were also calculated by using hypothetical W44 columns (assuming that they were compact) instead of W40 sections in such a way that the ratios of the moments of inertia, or plastic moment, of the deep columns to those of the medium columns, were approximately the same for the three building models. It is important to say that the maximum observed mean values of R_D were only larger by about 4% with respect to those of Figure 8; the maximum observed mean values of R_D were about 1.25. Thus, the trend regarding the reduction in R_D as the height of the building increases remains unchanged.

As stated earlier, for a given building, the mean values of R_D tend to decrease with the story number. In the same manner, the R_D mean values tend to also decrease as the building becomes taller. One of the reasons for this is that as medium columns are replaced by deep columns, the stiffness and the strength increase, which are lower in the tallest model. Another reason is that the effectiveness of deep columns in reducing the drifts becomes less significant for the stories away from the base since their stiffness relative to the base decreases. This is similar to the transversal stiffness associated to a point of a cantilever beam, which is inversely proportional to the distance from the fixed support (base) to the point under consideration.

The MVs for the R_{DT} parameter are shown in Column (2) of Table 6 for the three models. The results are similar to those of R_D for the case of the three- and nine-story models in the sense that the reduction in the roof displacements is significant when deep columns are used and that no trend is observed with the seismic intensity. However, the maximum values of R_{DT} are smaller than those of R_D. For the case of the twenty-story building, the reduction is negligible.

As in the case of R_D, the maximum mean values of R_{DT} tend to decrease with the building height: the maximum values are observed to be 1.38, 1.17, and 1.02 for the three-, nine-, and twenty-story models, respectively. These results indicate that the observed reduction in the response in terms of two parameters, which are considered to be similar and widely used to represent the maximum overall structural deformation, may be quite different when deep columns are used.

Table 6. MVs for the R_{DT}, R_A, R_M parameters.

Model	Direction	S_a/g (1)	R_{DT} (2)	R_A EXT (3)	R_A INT (4)	R_M EXT (5)	R_M INT (6)
1	NS	0.2	1.21	0.99	0.97	0.66	0.70
		0.4	1.24	0.98	0.95	0.66	0.68
		0.6	1.33	0.95	0.94	0.60	0.62
		0.8	1.29	0.93	0.94	0.57	0.59
		1.0	1.29	0.91	0.93	0.56	0.58
		1.2	1.27	0.91	0.93	0.55	0.57
	EW	0.2	1.38	1.01	0.98	0.68	0.73
		0.4	1.29	0.96	0.95	0.61	0.65
		0.6	1.20	0.94	0.94	0.58	0.60
		0.8	1.17	0.93	0.94	0.56	0.58
		1.0	1.19	0.93	0.93	0.55	0.57
		1.2	1.20	0.93	0.93	0.54	0.56
2	NS	0.1	1.22	0.99	0.98	0.78	0.84
		0.2	1.16	0.98	0.97	0.78	0.83
		0.3	1.17	0.98	0.97	0.74	0.79
		0.4	1.11	0.97	0.97	0.71	0.75
		0.5	1.06	0.98	0.97	0.69	0.72
	EW	0.1	1.15	0.99	0.98	0.78	0.85
		0.2	1.13	0.98	0.98	0.77	0.82
		0.3	1.13	0.97	0.97	0.71	0.76
		0.4	1.08	0.96	0.97	0.68	0.72
		0.5	1.06	0.96	0.97	0.68	0.70
3	NS	0.05	0.97	0.93	0.99	0.67	0.70
		0.10	1.02	0.95	0.98	0.68	0.70
		0.15	1.01	0.97	0.99	0.66	0.68
		0.20	1.01	0.96	0.98	0.64	0.65
		0.25	1.01	0.96	0.98	0.65	0.66
	EW	0.05	0.99	0.95	1.00	0.70	0.74
		0.10	1.00	0.96	0.99	0.69	0.71
		0.15	1.00	0.99	0.99	0.66	0.68
		0.20	1.01	0.98	0.99	0.66	0.67
		0.25	1.01	0.98	0.98	0.65	0.65

4.2. Objective 2: Results in Terms of Resultant Stresses: Interstory Shears, Axial Loads, and Bending Moments

The structural demands in terms of resultant stresses are now discussed. The ratios R_V, R_A, and R_M, calculated according to Equations (3)–(5) given below, are used to compare the responses of the buildings for the two column depths in terms of interstory shears, axial loads, and bending moments, respectively.

$$R_V = \frac{V_M}{V_L} \quad (3)$$

$$R_A = \frac{A_M}{A_L} \quad (4)$$

$$R_M = \frac{M_M}{M_L} \quad (5)$$

In the above equations, V_M and V_L, A_M and A_L, or M_M and M_L have a similar meaning as that of D_M and D_L in Equation (1), except that interstory shears, axial loads, and bending moments are now considered. Only the mean values of these ratios are given and discussed. The mean values of R_V are shown in Figures 9–11 for the three-, nine-, and twenty-story models, respectively, while those of R_A and R_M are given in Columns (3) through (6) of Table 6. As stated earlier, for a given model, only the exterior column at the base and the one following it (interior column) are considered (see plan of models in Figures 1–3).

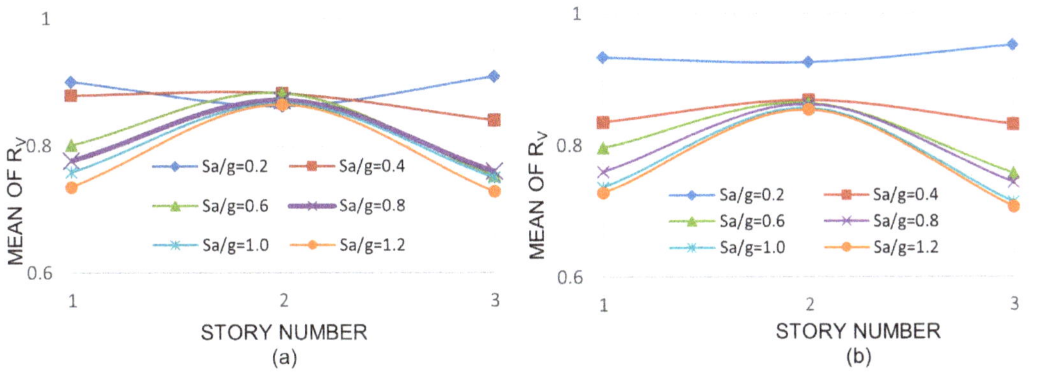

Figure 9. Mean values of R_V, 3-story building: (**a**) *NS* direction, (**b**) *EW* direction.

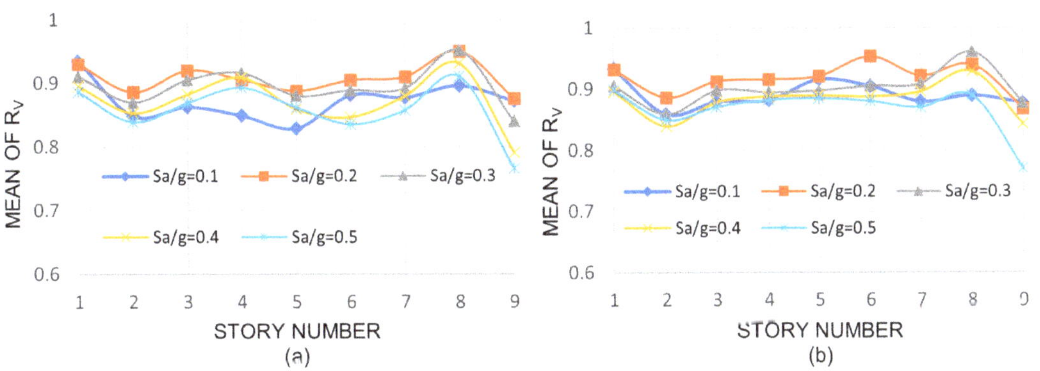

Figure 10. Mean values of R_V, 9-story building: (**a**) *NS* direction, (**b**) *EW* direction.

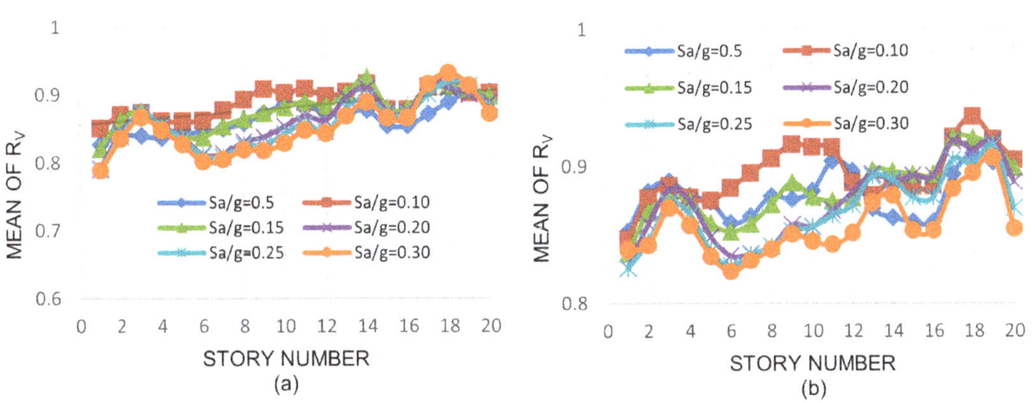

Figure 11. Mean values of R_V, 20-story building: (**a**) *NS* direction, (**b**) *EW* direction.

It can be observed that the mean values of R_V are smaller than unity in all cases, indicating that the interstory shears are larger for the models with deep columns. This was expected since the shear stiffness of deep columns is larger than that of medium columns; for example, the design shear strength, $\phi_V V_n$, of the exterior base column of the three-story building (W14 × 257) is 572 kips (2545 kN), while that of the corresponding equivalent deep column (W27 × 258) is 852 kips (3791 kN). The values ϕ_V and V_n are the shear reduction factor and the nominal shear strength, respectively. A little or moderate variation is also observed in the mean values of R_V with the seismic intensity, with the story number, or with the model height; the values range from 0.75 to 0.96, from 0.77 to 0.96, and from 0.79 to 0.94 for the three-, nine-, and twenty-story buildings, respectively.

Results of Table 6 indicate that the R_A mean values for the three-story building are very close to unity, tending to slightly decrease with the seismic intensity; the values range from 0.99 to 0.91. For the other two models, they are essentially equal to unity; this is explained as follows: the axial load in the columns of the buildings with medium columns is produced by three sources: the horizontal seismic component (A_{MH}), the vertical seismic component (A_{MV}), and the gravity load (A_{MG}). The corresponding parameters for the buildings with deep columns are A_{LH}, A_{LV}, and A_{LG}. The axial loads produced by the combined effect of the vertical seismic component and the gravity load for the models with medium columns (A_{MV} + A_{MG}) are observed to be quite similar to that of the models with deep columns (A_{LV} + A_{LG}). In addition, in both cases (medium or deep columns) the above-mentioned combined effect is much larger than the corresponding effect of the horizontal component (A_{MH} or A_{LH}) in such a way that values of R_A very close to unity are expected. In all cases, the values are quite similar for the *NS* and *EW* directions as well as for the interior and exterior columns. Hence, it can be stated that the axial load demands are essentially the same for the models with medium and deep columns.

For the case of R_M, unlike the R_A parameter, the mean values may be much smaller than unity, indicating that the bending moments are greater for the models with deep columns. The reason for this is that, as for the case of R_V, the bending stiffness is greater for deep columns; the design flexural strength ($\phi_b M_n$) is 1830 kip-ft and 3200 kip-ft for the W14 × 257 and W27 × 258 columns, respectively. It is also observed that the values, in general, tend to slightly decrease as the seismic intensity increases but no clear variation is observed with the model height. The values range from 0.54 to 0.73, from 0.68 to 0.85, and from 0.64 to 0.74 for the three-, nine-, and twenty-story models, respectively.

4.3. *Objective 3: Results in Terms of Normalized Single Overall and Normalized Combined Local Resultant Stresses*

In Sections 4.1 and 4.2 of the paper, the seismic demands in terms of displacements and resultant stresses, respectively, were calculated for the models with medium columns and compared to those of the models with deep columns. It is worth mentioning that drifts and top displacements are commonly used to represent the overall behavior of steel buildings; however, interstory shears, axial loads, and bending moments are not. In addition, the comparison between the responses of the buildings with the two columns sizes in terms of these three response parameters may not be representative of the structural response since the shear, the axial load, and the bending capacities are not considered. In this sense, the comparison in terms of the seismic responses normalized with respect to the design strengths is more appropriate.

4.3.1. Normalized Shear Strength

The normalized interstory shear ratio, R_{VN}, is calculated as follows:

$$R_{VN} = \frac{\frac{V_M}{\phi_V V_{M,n}}}{\frac{V_L}{\phi_V V_{L,n}}} \tag{6}$$

where $\phi VVM, n$ and $\phi VVL, n$ are the design shear strength of the medium and deep columns, respectively. All other terms were defined earlier. As for the other response parameters, the discussion is made only in terms of the mean values; they are shown in Figures 12–14 for the three-, nine-, and twenty-story models, respectively. For the three- and nine-story models, as for the case of the RV ratio, in general, little or moderate variation is observed in the mean values of RVN with the seismic intensity, with the story number, or with the model height. However, the mean values are much larger for the case of R_{VN} indicating that the normalized interstory shears are reduced when deep columns are used; the values are greater than unity in all cases, ranging from 1.05 to 1.40 and from 1.13 to 1.40 for the three- and nine-story buildings, respectively. For the case of the twenty-story building, unlike the three- and nine-story buildings, the mean values of R_{VN} vary from one story to another; they are essentially equal to unity from stories 1 through 11, then they increase from stories 12 to 17, and finally are equal to unity again from stories 18 to 20. The range in variation extends from 0.96 to 1.34. One of the reasons for this is that the ratio of the nominal shear strength of deep to medium columns for stories 1–11 and 18–20 is only a little larger than unity, while for the other stories (12–17), the ratio is considerably larger than unity.

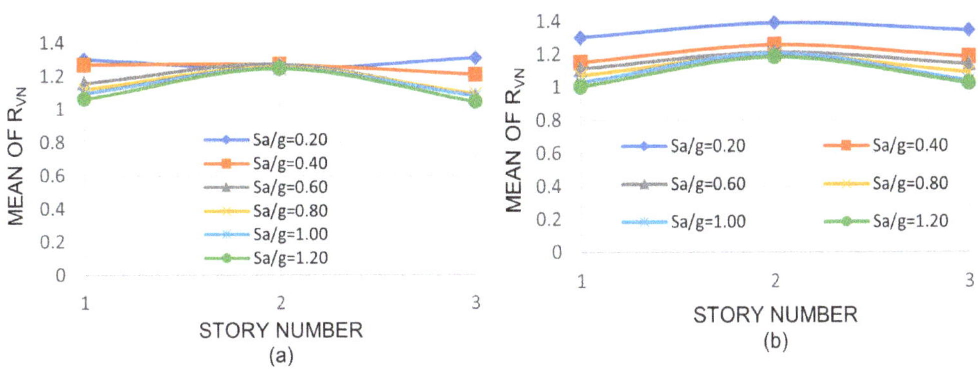

Figure 12. Mean values of R_{VN}, 3-story building: (**a**) *NS* direction, (**b**) *EW* direction.

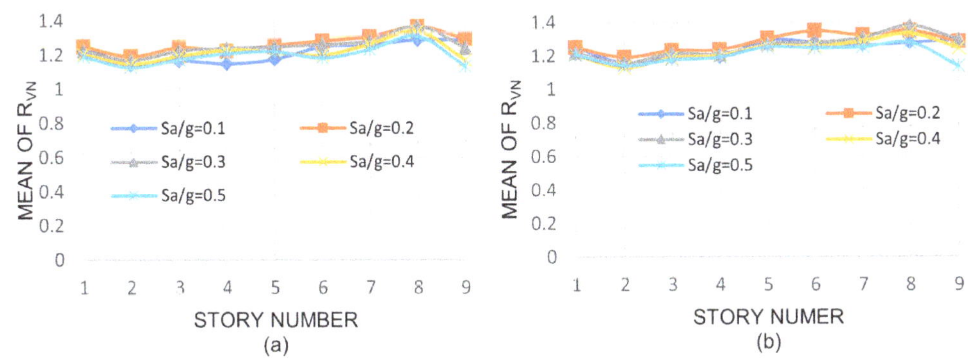

Figure 13. Mean values of R_{VN}, 9-story building: (**a**) *NS* direction, (**b**) *EW* direction.

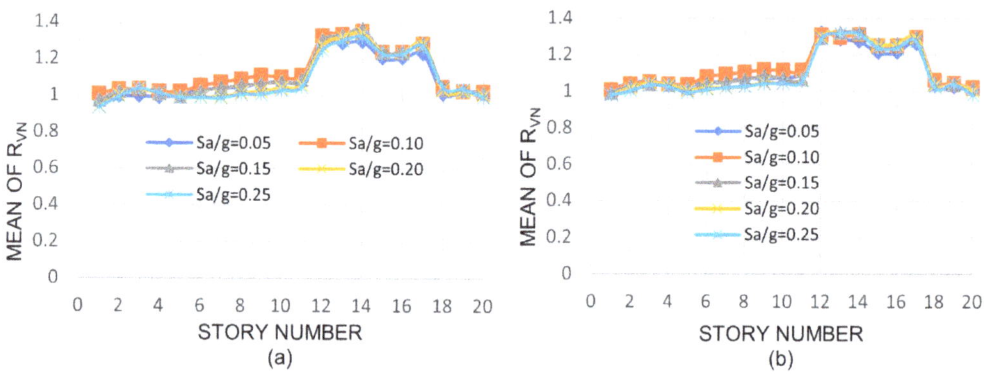

Figure 14. Mean values of R_{VN}, 20-story building: (**a**) NS direction, (**b**) EW direction.

4.3.2. Combined Normalized Axial Loads and Bending Moments

For the case of combined axial loads and bending moments, the AISC [42] interaction equations represent a useful tool to make the comparison between the responses of the buildings with the two different columns since they relate the required strength (demand) and the available strength (capacity). For doubly symmetric steel members under compression and flexure with respect to the major axis, which is the case addressed in this study, these equations are as follows [42]:

$$\text{for} \frac{P_u}{\phi_c P_n} \geq 0.2 \quad (7)$$

$$\frac{P_u}{\phi_c P_n} + \frac{8}{9} \frac{M_{ux}}{\phi_b M_{nx}} \leq 1 \quad (8)$$

$$\text{for} \frac{P_u}{\phi_c P_n} < 0.2 \quad (9)$$

$$\frac{P_u}{2\phi_c P_n} + \frac{M_{ux}}{\phi_b M_{nx}} \leq 1 \quad (10)$$

where the following apply:

P_u = required axial compressive strength

$\phi_c P_n$ = available axial compressive strength

M_{ux} = required flexural strength with respect to the major axis

$\phi_b M_{nx}$ = available flexural strength with respect to the major axis

To make the comparison, the R_{AM} parameter is used, which is defined as follows:

$$R_{AM} = \frac{R_{CM}}{R_{CL}} \quad (11)$$

where R_{CM} represents the combined axial load and bending moment demands on the medium columns according to Equation (8) or Equation (10); and R_{CL} represents the same, but deep columns are considered instead.

The mean values of RAM are presented in Table 7 for all the models. For the case of the three-story building, the values tend to decrease monotonically as the seismic intensity increases, being larger than unity only for the first three values of Sa/g. The greater observed value is 1.23 indicating that, on an average basis, the combined normalized axial loads and bending moments are significantly reduced when deep columns are used. For the nine-story building, as for the three-story building, the mean values of RAM tend to decrease monotonically with the seismic intensity; however, unlike for the three-story model, the values are larger than unity in all cases, with the maximum values being 1.31.

For the twenty-story building, unlike the three- and nine-story buildings, the values are slightly greater or slightly lower than unity, implying the same normalized combined response for the models with medium and deep columns. This behavior was observed before for other parameters (RDT) for the twenty-story building. The results are shown to be similar for the interior and exterior columns as well as for the NS and EW directions.

Table 7. MVs for the R_{AM} parameter.

Model	Direction	S_a/g	R_{AM} EXT	R_{AM} INT
1	NS	0.2	1.14	1.19
		0.4	1.13	1.15
		0.6	1.05	1.05
		0.8	0.99	1.00
		1.0	0.98	0.99
		1.2	0.96	0.97
	EW	0.2	1.17	1.23
		0.4	1.06	1.10
		0.6	1.01	1.02
		0.8	0.97	0.98
		1.0	0.96	0.97
		1.2	0.95	0.96
2	NS	0.1	1.24	1.31
		0.2	1.24	1.31
		0.3	1.19	1.24
		0.4	1.13	1.19
		0.5	1.10	1.14
	EW	0.1	1.24	1.32
		0.2	1.22	1.30
		0.3	1.14	1.19
		0.4	1.08	1.13
		0.5	1.08	1.11
3	NS	0.05	1.03	1.03
		0.10	0.98	1.05
		0.15	0.99	1.03
		0.20	0.97	0.99
		0.25	0.98	0.99
		0.30	0.98	0.99
	EW	0.05	1.07	1.10
		0.10	1.00	1.06
		0.15	1.01	1.02
		0.20	1.00	1.01
		0.25	0.99	0.99
		0.30	0.99	0.99

5. Conclusions

Because of the compactness of the webs and flanges of a vast number of deep steel sections of grade 50, and because of the necessity of higher bending and shear stiffness to reduce drifts, structural engineers often use deep columns in high seismic areas, which results in more economical designs. Yet this contradicts what is stated in some experimental investigations in the sense that even though deep columns satisfy current seismic provisions, they may suffer premature twisting. In other studies, however, which encourage the use of deep columns, it is stated that deep column twisting in the tests might have occurred primarily because there was no lateral bracing at the top flange of the beams, which is normally provided by the floors in actual buildings. This is an indication that much research needs to be developed in this area.

In this study, the nonlinear seismic responses, in terms of several parameters, of steel buildings with perimeter moment-resisting frames and medium (W14) columns are numerically calculated and compared to those of similar steel buildings with equivalent deep columns in terms of cost (W27 and larger). Three building models and fifteen strong motions with different dynamic characteristics are used in the study. The main findings are as follows:

(1) The interstory drifts may be significantly reduced when deep columns are used. For a given model, such reduction tends to decrease through the height of the model; similarly, such reduction becomes less relevant as the height of the building increases. The drifts of the models with medium columns may be up to 140%, 80%, and 60% larger than those of the models with deep columns, for the three-, nine-, and twenty-story models, respectively.

(2) The corresponding percentages for top displacements are 38%, 25%, and 2%, indicating that the maximum reductions in terms of top displacements are smaller than those of drifts. The implication of this is that the observed reduction in the response when deep columns are used in terms of two parameters, which are considered to be similar and widely used to represent the maximum overall structural deformation, may be quite different.

(3) Significant reductions are also observed for normalized interstory shears. However, for the case of normalized combined axial loads and bending moments, the reduction, as for the case of top displacements, is significant only for the three- and nine-story models.

(4) Hence, the seismic demands on the buildings with deep columns may be much smaller than those of the buildings with medium columns and, therefore, buildings with deep columns exhibit a superior behavior, which may result in more economical designs. The reduction is greater for the case of low- and mid-rise buildings than for high-rise buildings. One of the reasons for this is that as the medium columns are replaced by deep columns, the stiffness and the strength increase, which are lower in the tallest model.

Author Contributions: Conceptualization, J.M.L.-G. and A.R.-S.; Methodology, E.B., J.B., V.B., R.C. and A.R.-S.; Software, J.B. and V.B..; Formal analysis, F.V.-B., M.D.L.-T. and J.M.L.-G.; Investigation, V.B.., R.C. and A.R.-S.; Data curation, F.V.-B., M.D.L.-T., J.B., V.B.. and R.C.; Writing – original draft, F.V.-B. and A.R.-S.; Writing – review & editing, E.B. and A.R.-S.; Visualization, M.D.L.-T. and J.M.L.-G.; Supervision, E.B.; Funding acquisition, E.B. and A.R.-S. All authors have read and agreed to the published version of the manuscript.

Funding: This research was funded by the National Council of Science and Technology (CONACyT) under grant number CF-2023-G-1636, and by the Autonomous University of Sinaloa under grant PRO_A8_013_PROFAPI 2022.

Institutional Review Board Statement: Not applicable.

Informed Consent Statement: Not applicable.

Data Availability Statement: The data presented in this study are available in the article.

Conflicts of Interest: The authors declare no conflict of interest.

References

1. Gupta, A.; Krawinkler, H. Behavior of Ductile SMRFs at Various Seismic Hazard Levels. *J. Struct. Eng.* **2000**, *126*, 98–107. [CrossRef]
2. Lee, K.; Foutch, D. Performance Evaluation of New Steel Frame Buildings for Seismic Loads. *Earthq. Eng. Struct. Dyn.* **2001**, *31*, 653–670. [CrossRef]
3. Foutch, A.; Yun, S. Modeling of Steel Moment Frames for Seismic Loads. *J. Constr. Steel Res.* **2002**, *58*, 529–564. [CrossRef]
4. Mele, E.; Di Sarno, L.; De Luca, A. Seismic Behavior of Perimeter and Spatial Steel Frames. *J. Earthq. Eng.* **2004**, *8*, 457–496. [CrossRef]
5. Fathi, M.; Daneshjoo, F.; Melchers, R. A Method for Determining the Behaviour Factor of Moment-Resisting Steel Frames with Semi-Rigid Connections. *Eng. Struct.* **2006**, *28*, 514–531. [CrossRef]
6. Lee, K.; Foutch, D. Seismic Evaluation of Steel Moment Frames Buildings Designed Using Different R-Values. *J. Struct. Eng. Div. ASCE* **2006**, *132*, 1461–1472. [CrossRef]

7. Krishnan, S.; Ji, C.; Komatitsch, D.; Tromp, J. Performance of Two 18-Storey Steel Moment-Frame Building in Southern California during Two Large Simulated San Andres Earthquakes. *Earthq. Spectra* **2006**, *22*, 1035–1061. [CrossRef]
8. Black, E. Use of Stability Coefficients for Evaluating the P–Δ Effect in Regular Steel Moment Resisting Frames. *Eng. Struct.* **2011**, *33*, 1205–1216. [CrossRef]
9. Sejal, P.; Vasanwala, S.; Desai, A. Comparison of Steel Moment Resisting Frame Designed by Elastic Design and Performance Based Plastic Design Method Based on the Inelastic Response Analysis. *Int. J. Civ. Struct. Eng.* **2012**, *2*, 1081–1097.
10. Teran-Gilmore, A.; Diaz, G.; Reyes, C. Displacement-Based Conception of Moment-Resisting Frames That House Essential Facilities. *Soil Dyn. Earthq. Eng.* **2013**, *46*, 96–113. [CrossRef]
11. Athanasios, I.; Dimopoulos, A.; Bazeos, N.; Beskos, D. Seismic Yield Displacements of Plane Moment Resisting and X-Braced Steel Frames. *Soil Dyn. Earthq. Eng.* **2012**, *41*, 128–140.
12. Formisano, A.; Landolfo, R.; Mazzolani, F. Robustness Assessment Approaches for Steel Framed Structures under Catastrophic Events. *Comput. Struct.* **2015**, *147*, 216–228. [CrossRef]
13. Liu, W.; Rasmussen, K.J.R.; Zhang, H. Systems Reliability for 3D Steel Frames Subject to Gravity Loads. *Structures* **2016**, *8*, 170–182. [CrossRef]
14. Samanta, A.; Huang, Y. Ground-Motion Scaling for Seismic Performance Assessment of High-Rise Moment-Resisting Frame Building. *Soil Dyn. Earthq. Eng.* **2017**, *94*, 125–135. [CrossRef]
15. Loulelis, D.; Papagiannopoulos, G.; Beskos, D. Modal Strength Reduction Factors for Seismic Design of Steel Moment Resisting Frames. *Eng. Struct.* **2018**, *154*, 23–37. [CrossRef]
16. Ruiz-García, J.; Yaghmaei-Sabegh, S.; Bojórquez, E. Three-Dimensional Response of Steel Moment-Resisting Buildings under Seismic Sequences. *Eng. Struct.* **2018**, *175*, 399–414. [CrossRef]
17. Fogarty, J.; Wu, T.Y.; El-Tawil, S. Collapse Response and Design of Deep Steel Columns Subjected to Lateral Displacement. *J. Struct. Eng.* **2017**, *143*, 04017130. [CrossRef]
18. Elkady, A.; Lignos, D.G. Full-Scale Testing of Deep Wide-Flange Steel Columns under Multiaxis Cyclic Loading: Loading Sequence, Boundary Effects, and Lateral Stability Bracing Force Demands. *J. Struct. Eng.* **2018**, *144*, 04017189. [CrossRef]
19. Sediek, O.A.; Wu, T.-Y.; McCormick, J.; El-Tawil, S. Prediction of Seismic Collapse Behavior of Deep Steel Columns Using Machine Learning. *Structures* **2022**, *40*, 163–175. [CrossRef]
20. Wu, T.-Y.; Pal, P.S.; Wang, H.-C. Collapse Risk of Steel Framed Buildings with Deep Columns under Tri-Directional Excitation. *J. Constr. Steel Res.* **2023**, *208*, 108030. [CrossRef]
21. Sepulveda, C. Seismic Behavior of Deep and Slender Steel Columns through Full-Scale Cyclic and Hybrid Testing. Ph.D. Thesis, University of California San Diego, La Jolla, CA, USA, 2024.
22. Gilton, C.; Chi, B.; Uang, C. *Cyclic Response of RBS Moment Connections: Weak-Axis Configuration and Deep Column Effects*; University of California San Diego: La Jolla, CA, USA, 2000.
23. Chi, B.; Uang, C. Cyclic Response and Design Recommendations of Reduced Beam Section Moment Connections with Deep Columns. *J. Struct. Eng.* **2002**, *128*, 464–473. [CrossRef]
24. Shen, J.; Astaneh-Asl, A.; McCallen, D. Use of Deep Columns in Special Steel Moment Frames. *Steel Tip Rep.* **2002**.
25. Zhang, X.; Ricles, J.; LuL, W.; Fisher, J. Analytical and Experimental Studies on Seismic Behavior of Deep Column-to-Beam Welded Reduced Beam Section Moment Connections. In Proceedings of the 13th World Conference on Earthquake Engineering, Vancouver, BC, Canada, 1–6 August 2004. Paper No 1599.
26. Shao, D.; Hale, T. Full Scale Testing and Project Application of Sideplate Moment Connection for SMRF Using Deep Columns. In Proceedings of the 2004 SEAOC Convention, Monterey, CA, USA, 25–28 August 2004.
27. Ricles, J.; Zhang, X. Seismic Performance of Reduced Beam Section Moment Connections to Deep Columns. In Proceedings of the Structures Congress, St Louis, MI, USA, 18–21 May 2006.
28. Zhang, X.; Ricles, J. Experimental Evaluation of Reduced Beam Section Connections to Deep Columns. *J. Struct. Eng.* **2006**, *132*, 346–357. [CrossRef]
29. Zhang, X.; Ricles, J. Seismic Behavior of Reduced Beam Section Moment Connections to Deep Columns. *J. Struct. Eng.* **2006**, *132*, 358–367. [CrossRef]
30. Elkady, A.; Lignos, D. Collapse Assessment of Steel Moment Resisting Frames Designed with Deep Members. In Proceedings of the Vienna Congress on Recent Advances in Earthquake Engineering and Structural Dynamics, Vienna, Austria, 28–30 August 2013; p. 249.
31. Islam, A.; Imanpour, A. Stability of Wide-Flange Columns in Steel Moment-Resisting Frames: Evaluation of the Canadian Seismic Design Requirements. *Bull. Earthq. Eng.* **2022**, *20*, 1591–1617. [CrossRef]
32. Monjardin-Quevedo, J.G.; Reyes-Salazar, A.; Tolentino, D.; Gaxiola-Camacho, O.D.; Vazquez-Becerra, G.E.; Gaxiola-Camacho, J.R. Seismic Reliability of Steel SMFs with Deep Columns Based on PBSD Philosophy. *Structures* **2022**, *42*, 1–15. [CrossRef]
33. Boushehri, K.; Tsavdaridis, K.D.; Cai, G. Seismic Behaviour of RWS Moment Connections to Deep Columns with European Sections. *J. Constr. Steel Res.* **2019**, *161*, 416–435. [CrossRef]
34. Carr, A. RUAUMOKO, Inelastic Dynamic Analysis Program 2016.
35. Preserve, L.; Chen, W.F.; Atsuta, T. Interaction Equations for Biaxially Loaded Sections. *J. Struct. Div.* **1972**, *98*, 1035–1052.

36. Federal Emergency Management Agency. State of the Art Report on Systems Performance of Steel Moment Frames Subjected to Earthquake Ground Shaking. 2000. Available online: https://www.atcouncil.org/pdfs/FEMA355Ctoc.pdf (accessed on 11 December 2023).
37. UBC. Structural Engineering Design Provisions, Uniform Building Code. In Proceedings of the International Conference of Building Officials, Whittier, CA, USA, 16–19 April 1997.
38. *AISC Steel Construction Manual*; American Institute of Steel Construction: Chicago, IL, USA, 2010.
39. Roeder, C.W.; Schneider, S.P.; Carpenter, J.E. Seismic Behavior of Moment-Resisting Steel Frames: Analytical Study. *J. Struct. Eng.* **1993**, *119*, 1866–1884. [CrossRef]
40. Leon, R.T.; Shin, K.J. Performance of Semi-Rigid Frames. In Proceedings of the Structures XIII Congress, Boston, MA, USA, 2–5 April 1995; ASCE: New York, NY, USA, 1995; pp. 1020–1035.
41. Nader, M.N.; Astaneh, A. Dynamic Behavior of Flexible, Semirigid and Rigid Steel Frames. *J. Constr. Steel Res.* **1991**, *18*, 179–192. [CrossRef]
42. Osman, A.; Ghobarah, A.; Korol, R.M. Implications of Design Philosophies for Seismic Response of Steel Moment Frames. *Earthq. Eng. Struct. Dyn.* **1995**, *24*, 127–143. [CrossRef]
43. Schneider, S.P.; Roeder, C.W.; Carpenter, J.E. Seismic Behavior of Moment-Resisting Steel Frames: Experimental Study. *J. Struct. Eng.* **1993**, *119*, 1885–1902. [CrossRef]
44. *AISC Specification for Structural Steel Buildings*; American Institute of Steel Construction: Chicago, IL, USA, 2010.

applied sciences

MDPI

Article

A Damage Index for Assessing Seismic-Resistant Designs of Masonry Wall Buildings Reinforced with X-Bracing Concrete Frames

Yossakorn Chonratana and Vuttichai Chatpattananan *

School of Engineering, King Mongkut's Institute of Technology Ladkrabang, Bangkok 10520, Thailand
* Correspondence: vuttichai.ch@kmitl.ac.th

Abstract: In this paper, a suitable damage index is demonstrated to assess the seismic-resistant design of masonry wall buildings reinforced with double x-bracing concrete frames. As a criterion indicative of the damage level that might occur after an earthquake, the damage index can be calculated by using analytic results, by using the Park–Ang formula on masonry wall buildings reinforced with concrete structures, and by adjusting the index values in accordance with the results of the analysed models. The data used in this study are collected from the results of four-storey concrete structures with masonry walls under cyclical lateral forces. To simulate the masonry walls' structural behaviours for damage assessment, x-bracings placed as crosses on each bracing are used to support the compressive strength. Then, the analysis results are used to assess the damage that occurs to the masonry wall building structure by considering deformation and energy decay; additionally, a suitable damage index is calculated for each damage level. The damage index can be considered in the seismic-resistant design of masonry walls reinforced with x-bracing concrete frames.

Keywords: damage index; masonry wall; cyclic load; equivalent strut

1. Introduction

An earthquake is a force that occurs in the horizontal direction and spreads to every floor of a structure. Moreover, earthquakes damage the masonry walls of reinforced concrete buildings. This damage affects the safety of the people in the building. For that reason, there are many studies on the evaluation of the damage levels of structures under earthquake forces to reduce disaster losses. In Thailand, there is a risk of earthquakes. The most damaged buildings are small- and medium-sized buildings, such as residential houses, commercial buildings, and school buildings. In addition, masonry is used to build partition walls in these buildings [1].

Masonry walls are parts of buildings that are essential for their behaviours under active forces from earthquakes. Most structural masonry buildings in Thailand feature walls built using antique masonry and block masonry. These techniques are used to support the weight of the beams, floors, and roofs, which are constructed with other materials, such as wood and steel. Most of the destruction to structural masonry buildings in Thailand results from the buildings leaning away from the masonry walls' planes. The damage in the beginning appears as diagonal cracking of the wall caused by the initial vibration force. If the force is severe, the cracking of the wall occurs. Finally, the wall cannot stand as usual, and it is destroyed. In brief, the damage that occurs to the wall interacts with the building frame when the building moves under an earthquake force. The level of damage can be evaluated using a structural analysis if appropriate models and criteria are used.

This research shows an appropriate damage index definition for the damage assessment of masonry wall buildings reinforced with x-bracing concrete frames; these frames are used as indicative criteria of the damage level that might occur under an earthquake force. The damage index can be calculated by using analytical results, by using the Park–Ang formula on models of masonry wall buildings reinforced with concrete structures, and by adjusting the index values in accordance with the results of the analysed models [2].

2. Literature Review

2.1. Park–Ang Damage Index

The damage index is a criterion for quantitatively measuring damage. Additionally, the index suggests that a structure has an external force acting on it that leads to damage, such as earthquake and lateral forces. The index can be used to assess the damage by determining the level of damage. Generally, the damage index is defined as a value. If the value is equal to 0, no damage occurs. If the value is equal 1, high-to-extreme damage occurs. In this research, the Park–Ang damage index is used to evaluate the damage levels of masonry wall buildings. The total damage caused by inelastic deflection and the damage due to energy collapse in the structure from back-and-forth responses are shown. The damage index is written as the following equation [2]:

$$DI = \frac{\delta_M}{\delta_u} + \frac{\beta}{Q_y \delta_u} E \qquad (1)$$

where

δ_M is the deformation value after alternating the building force.
δ_u is the maximum deformation capacity value under unidirectional force before a disaster.
β is a coefficient in the relationship between the damage level and accumulated energy, with its value depending on the type of building being analysed.
Q_y is the strength at the yield point of the building.
E is the accumulated energy under the reverse action force.

The displacement value δ_M and accumulated energy value under the reverse action force E can be analysed by using this model under an earthquake force; the model considers the maximum deformation capacity value δ_u, and the coefficient β depends on the type of building being analysed. This coefficient is calibrated with values from laboratory experiments.

2.2. Structural Analysis of Model of Masonry Wall Building Reinforced with X-Bracing Concrete Frame

The deformation value and accumulated energy value under the reverse action force in Equation (1) can be determined by analysing the building structure under an earthquake force, which is examined by using an appropriate model. A masonry wall is modelled by using a multiple-strut model. For masonry walls within reinforced concrete building frames with an equivalent x-bracing form; the diagonal compressive strength is obtained as shown in Figure 1A. This paper investigates the response of masonry wall buildings subjected to horizontal forces. The structural behaviour of the walls is represented through x-braces, modelled according to the approach proposed by Saneinejad and Hobbs (1995). The damage index proposed by Park and Ang is adopted to examine the damage level of the masonry walls, and an infilled RC frame is designed and used as a case study. The cross symbol on the first bracing is between the corner of the beam and the column. Furthermore, the bracing model is defined by concrete section elements, with relational values between the stress and strain, reducing the lateral force, as shown in Figure 1B.

The stress–strain relationship of the model is defined by four parameters: peak strength (f_{mi}), peak strain (ε_{mi}), residual strength (f_{mri}), and residual strain (ε_{mri}). These parameters can be obtained from equations calibrated with values from laboratory experiments. The cross-sectional size of the bracing that determines whether the thickness is equal to the actual thickness of the original wall (t_w) and the effective width is equal to $0.7a$ in Equation (2) or W1 in Equation (4). According to α, the regulation from the DPT standard 1301/1302-61 [3] is as follows:

$$a = 0.175(\gamma_1 h_{col})^{-0.4} + \gamma_{inf} \qquad (2)$$

where h_{col} is the height of the column measured from the base of the pole to the centre point of the beam, r_{inf} is the diagonal length of the masonry wall, and γ_1 is a coefficient used to find the equivalent bracing width of the masonry wall. Then, the equation can be written as follows:

$$\gamma_1 = [\frac{E_{me}t_{inf}\sin(2\theta)}{4E_{fe}I_{col}h_{inf}}]^{0.25} \tag{3}$$

where
E_{fe} is the modulus of elasticity of the rigid frame material.
E_{me} is the modulus of elasticity of the masonry wall material.
I_{col} is the moment of inertia of the column's cross-sectional area.
h_{inf} is the height of the masonry wall.
t_{inf} is the thickness of the masonry wall.
θ is the angle between the height and length of the masonry wall.

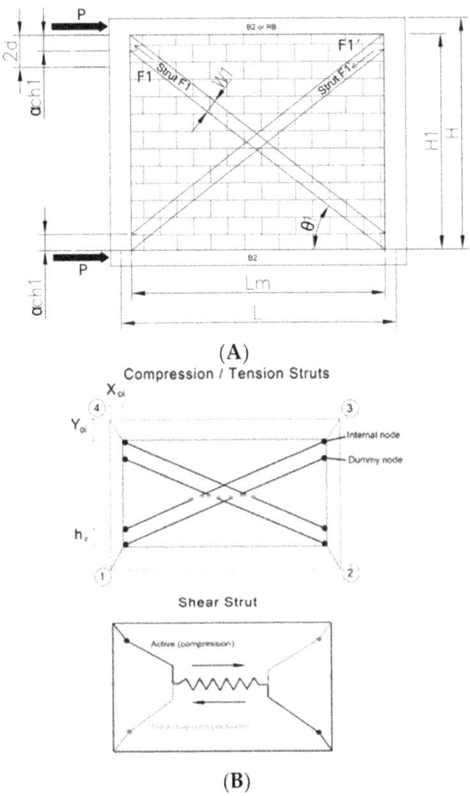

Figure 1. Model of masonry wall building reinforced with x-bracing frame. (**A**) Model of masonry wall building reinforced with x-bracing concrete frames for compressive strength. (**B**) Model of masonry wall building reinforced with x-bracing concrete frame for compression/tension and shear strut.

In addition, the compressive strength of the reinforced masonry wall f_m is proportional to the specific surface area of the masonry walls S_r and can be calculated from the relationship between f_m and S_r [4]. The width of the compressive strength W_1 is calculated from the touch surface length of the force unit, which can spread between the column and masonry wall [5], as follows:

$$W_1 = \alpha_c \frac{l_m}{\sqrt{h_1.^2 + l_m.^2}} = \alpha_c h_1 \cos\theta_1 \tag{4}$$

$$\alpha_c = \frac{1}{h}\sqrt{\frac{2M_{pj} + 2\beta_c M_{pc}}{\sigma_c.^t}} \tag{5}$$

$$\sigma_c = \alpha_c \frac{f_m}{\sqrt{1 + 3\mu^2 r^4}} \tag{6}$$

where
μ is the friction coefficient between the building frame and brick wall;
r is the ratio between the height and width of the building frame $(r = h/l)$;
β_c is the reduction factor of the pole $(\beta_c = 0.2)$.

2.3. Coefficient Calibration of the Relationship between the Damage Level and Energy Decay

As mentioned above, the coefficient is featured in the relationship between the damage level and accumulated energy. β depends on the type of building analysed, and it is calibrated with values from experiments with the analysis models. Moreover, an appropriate coefficient for masonry walls is found that samples the testing results from several previous studies and creates a model sample for testing by simulating x-bracing characteristics, as mentioned before.

Then, the analysis results are taken to assess the damage to the wall structure by considering the deformation value and energy decay and by adjusting the coefficient that relates the damage level to the accumulated energy to obtain a value that corresponds with the testing results.

2.4. Damage Index of Study Sample to Assess Seismic-Resistant Design of Masonry Wall Buildings

In this research, previous testing results are selected from various sources for study. Furthermore, these previous experimental results have a variety of designs. Some designs consider earthquake resistance and others do not. However, the various structural systems from the eight testing samples are as follows: Mehrabi's sample is a rigid frame and hollow block concrete brick wall [6], Wararuksajja's sample is an anti-bending frame and hollow block concrete brick wall [7], Jiang's sample is a lightweight brick anti-bending rigid frame [8], Morandi's sample is a hollow brick wall rigid frame [9], and Huang's sample is a brick wall and hollow block concrete rigid frame [10]. The size and detail characteristics of the model samples are shown in Figure 1 and Tables 1 and 2.

Table 1. Details of the model samples from the literature review.

Sample Name	H (mm)	L (mm)	D (mm)	W1 (mm)	t_w (m)
		Mehrabi et al. [6]			
No. 4	1537	2312	147	187	0.092
		Wararuksajja et al. [7]			
WS01	2800	4000	250	315	0.100
WS02	2800	4000	250	331	0.100
WS03	2800	4000	250	334	0.100
		Jiang et al. [8]			
AFKJ1	2950	5940	350	200	0.427

Table 1. *Cont.*

Sample Name	H (mm)	L (mm)	D (mm)	W1 (mm)	t_w (m)
		Morandi et al. [9]			
TA2	3125	4587	300	350	0.370
		Huang et al. [10]			
IF-1	1375	2250	218	120	0.200
IF-2	1375	2250	218	180	0.204

Table 2. Mechanical properties of the reinforcement x-bracing concrete frames used in the model samples from the literature review.

Sample Name	Maximum Compressive Strength (Mpa)	Pending Strength (Mpa)	Compressive Stress	Pending Stress (mm)
No. 4	5.5	0.19	0.0010	0.0060
WS01	9.4	0.20	0.0017	0.0050
WS02	7.8	0.20	0.0017	0.0050
WS03	6.8	0.20	0.0014	0.0018
AFKJ1	3.8	0.47	0.0016	0.0060
TA2	2.9	0.38	0.0011	0.0060
IF-1	5.4	0.22	0.0020	0.0090
IF-2	2.8	0.46	0.0014	0.0100

3. Structural Analysis of Reinforcement X-Bracing Concrete Frames Used in the Model Samples from the Literature Review

A masonry wall building is simulated according to an analysis of the models in each study by using the SeismoStruct 2023 program 3D model [11]. By simulating a masonry wall building reinforced with x-bracing concrete frames at the head of the columns with shear spring failure, each bracing property is determined regarding the masonry wall building failure behaviours from the experimental results of the analysed models. Then, the monotonic loading and cyclic loading characteristics of the deformed samples are analysed by defining displacement control according to the force pattern used to test each sample. Figure 2 shows a sample that defines the displacement value in the model [12–29]. Then, the analysis results are used to calculate and determine a suitable coefficient and damage index.

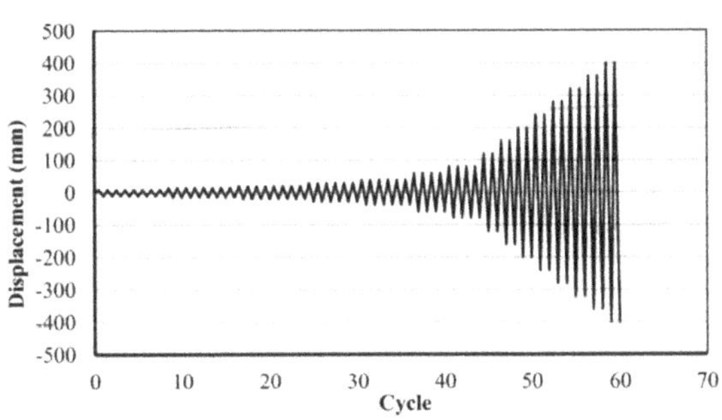

Figure 2. Monotonic loading and cyclic loading characteristics of the model samples [12–29].

The results of the analysis of the deformation behaviour are modelled. The results are shown in Figures 3 and 4, which show the relationship between the relative lateral displacement and lateral force of each model sample. In the figure, the lateral force continuously increases until structural failure occurs, and the maximum lateral force from the analysis is consistent with the testing results of the analysed models, as shown in Table 3.

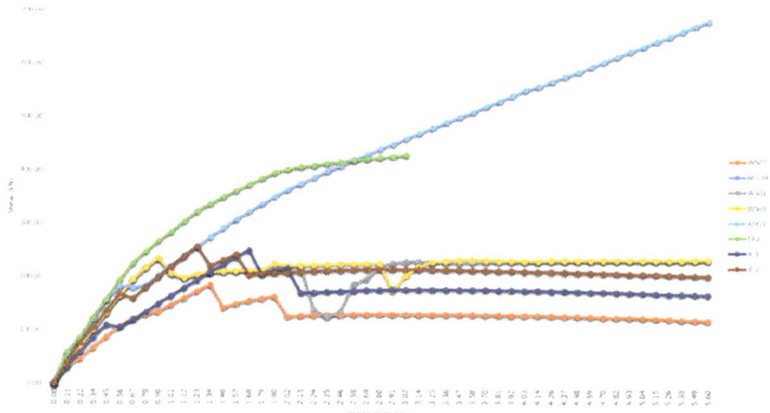

Figure 3. Relationship between the lateral force and lateral displacement of the model samples.

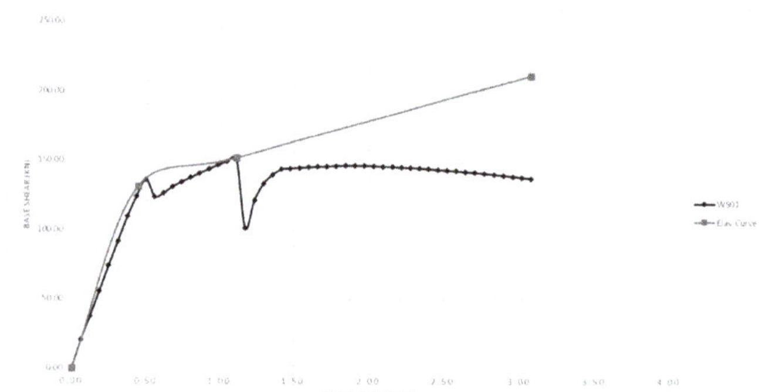

(**A**) Analysis results of No. 4.

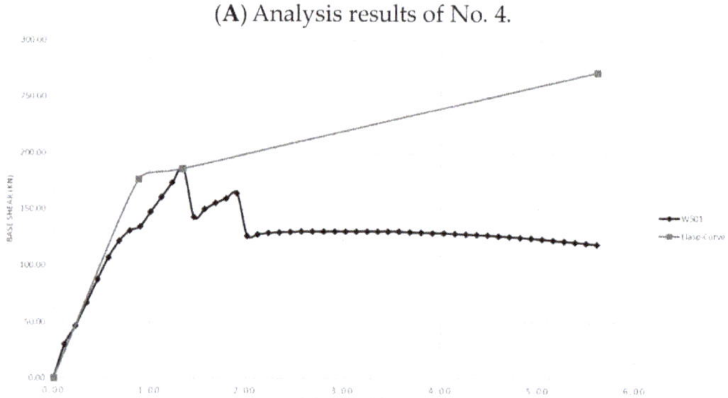

(**B**) Analysis results of WS01.

Figure 4. *Cont.*

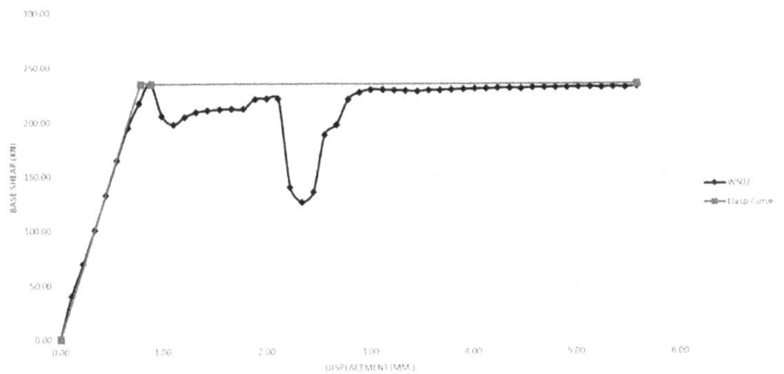

(C) Analysis results of WS02.

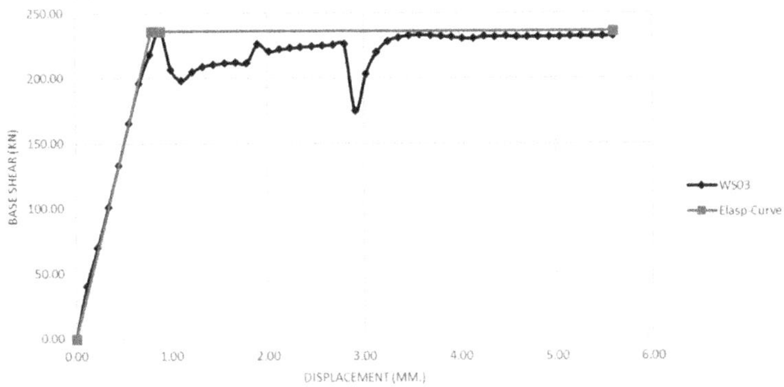

(D) Analysis results of WS03.

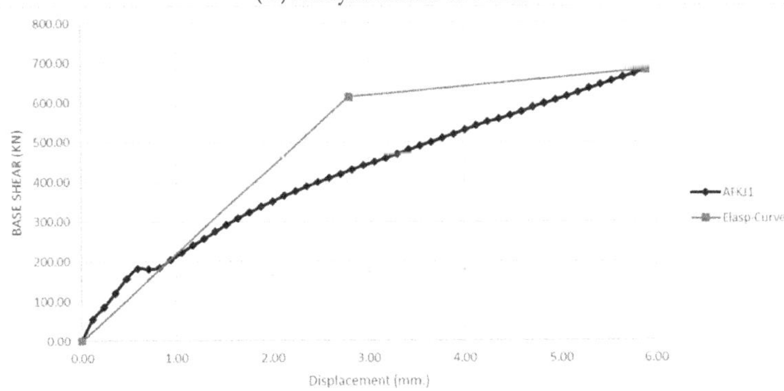

(E) Analysis results of AFKJ1.

Figure 4. *Cont.*

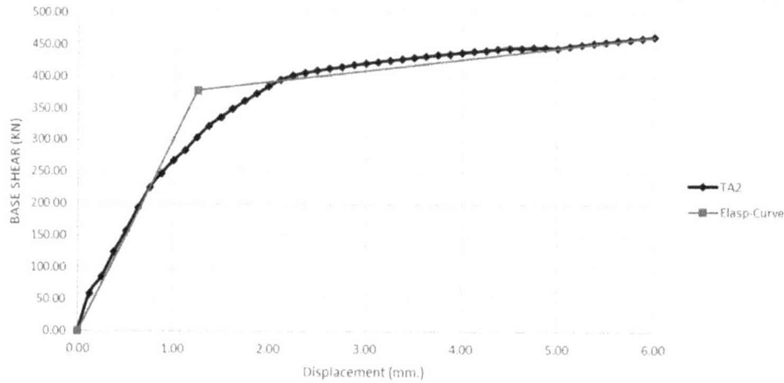

(**F**) Analysis results of TA2.

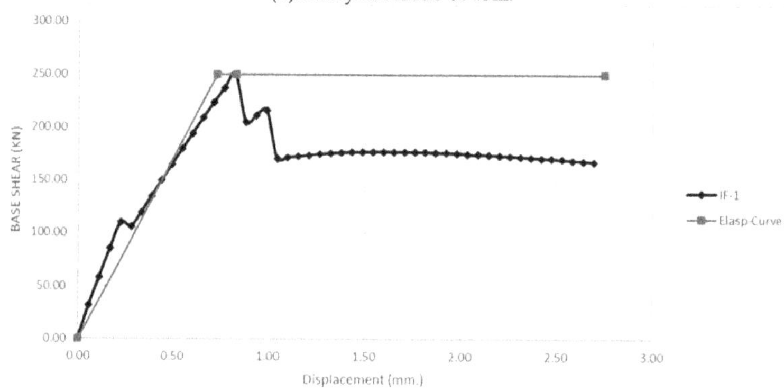

(**G**) Analysis results of IF-1.

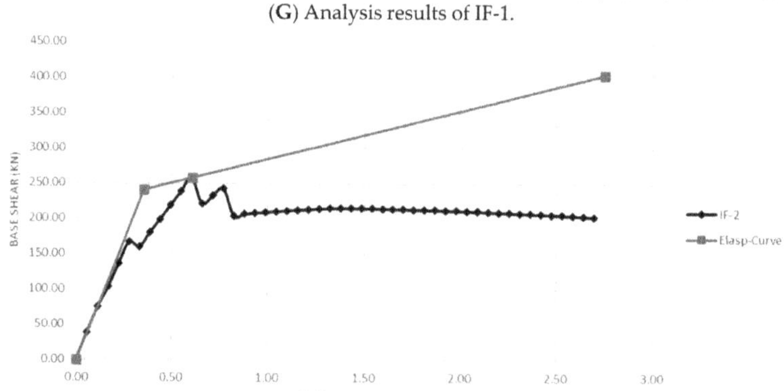

(**H**) Analysis results of IF-2.

Figure 4. Analysis results from alternating the lateral forces of the model samples.

Table 3. Maximum lateral force and maximum lateral displacement of the model samples (cyclic loading).

Sample	Load Target	Result Model Structure Analysis	
	Lateral Force (kN)	Maximum Lateral Force to Displacement (kN)	Maximum Lateral Displacement (mm)
No. 4	502.39	210.74	3.07
WS01	502.39	186.11	5.60
WS02	502.39	234.690	5.60
WS03	502.39	235.920	5.60
AFKJ1	502.39	613.140	5.90
TA2	502.39	378.470	6.25
IF-1	502.39	250.110	2.75
IF-2	502.39	239.860	2.75

3.1. Damage Index Calculation Used in the Structural Analysis of Model Samples from the Literature Review

The calculation of the damage index must involve variables obtained from the model's analysis. The variable values can be found as detailed below.

The accumulated energy under cyclic loading (E) calculated from the relationship between the lateral force and bracing axis deformation can be obtained from the area under compressive strength loading conditions, with two instances of bracing deformation under cyclic loading conditions. The energy is used to calculate the total accumulated energy of the masonry wall. The stress strength at yield point (Q_y) is found from the lateral resistance of the x-bracing, and it is determined to be 70% of the maximum total resistance. The maximum deformation capacities under cyclic loading (δ_M) conditions from the analysis of structural deformation are found; the elements have two bracings under cyclic loading conditions at the maximum lateral displacement of each cycle. The value is defined as the damage index in the positive direction.

The maximum deformation capacity under monotonic loading (δ_u) has a value set to 1.25 times the deformation value at the destruction point, according to the model structure analysis results. The deformation value at the structural destruction point is defined as equal to the deformation value at the total strength, which is 25% of the reduction from the maximum strength. Results of model samples of masonry wall buildings reinforced with x-bracing concrete frames as shown in Table 4.

Table 4. Results of model samples of masonry wall buildings reinforced with x-bracing concrete frames.

Sample Name	Q_y (kN)	δ_M (m)	δ_u (m)
No. 4	151.48	0.0111	0.014
WS01	186.11	0.0134	0.017
WS02	234.69	0.009	0.011
WS03	235.92	0.009	0.011
AFKJ1	680.86	0.059	0.074
TA2	467.93	0.0625	0.078
IF-1	250.11	0.0083	0.010
IF-2	257.22	0.0061	0.008

The correlation coefficient between the damage level and accumulated energy (β) is optimized by adjusting the coefficient of determination from 0.10 to 0.35 and by calculating the damage index from the failure point. Then, a damage index value is obtained that is different for each model sample. For the optimization of a suitable coefficient, statistical calculations are used to determine an average damage index from the samples. Table 5 shows the damage index values calculated from each masonry wall building reinforced with x-bracing concrete frames, and the average damage index of the sample of masonry wall buildings reinforced with x-bracing concrete frames indicates the failure point. By using other coefficient values in the table, the appropriate coefficient is found to be 0.10–0.35. The coefficient can be used to calculate the average damage index; this is equal to 0.935 at the destruction point, which is close to 1.00.

Table 5. Damage index (DI) using various coefficients (β) of the model samples.

Sample Name	β					
	0.10	**0.15**	**0.20**	**0.25**	**0.30**	**0.35**
No. 4	0.818	0.837	0.846	0.846	0.855	0.864
WS01	0.818	0.836	0.845	0.845	0.854	0.863
WS02	0.810	0.819	0.824	0.824	0.829	0.834
WS03	0.810	0.819	0.824	0.824	0.829	0.833
AFKJ1	0.822	0.843	0.854	0.854	0.865	0.876
TA2	0.833	0.867	0.883	0.883	0.900	0.917
IF-1	0.808	0.817	0.821	0.821	0.825	0.829
IF-2	0.806	0.812	0.815	0.815	0.818	0.821
Average	0.816	0.831	0.839	0.839	0.847	0.855
SD.	0.009	0.018	0.023	0.023	0.027	0.032
Avg. + SD.	0.825	0.849	0.862	0.862	0.874	0.887
Avg. − SD.	0.806	0.813	0.816	0.816	0.819	0.823

3.2. Determination of Damage Level for Assessing Seismic-Resistant Design of Masonry Wall Buildings

The appropriate coefficient DI is equal to 0.24. Therefore, this coefficient is used to calculate the damage indices at various levels of deformation of the model sample as a guideline to indicate the level of damage that may occur. The damage index calculation results of the three experimental samples used in Wararuksajja's research, namely WS01, WS02, and WS03, which provide the damage index at various damage points, are shown in Table 6. The characteristics of masonry wall damage from the test and damage index, which can define criteria for determining the damage level in four sessions, are shown in Table 7.

Table 6. Damage index at each damage level.

Damage Characteristics	Damage Index (DI)		
	WS01	**WS02**	**WS03**
Total collapse of infill	1.57	1.52	1.16
Extensive large cracks	1.07	1.03	0.84
Corner crushing	0.77	0.74	0.52
First diagonal crack	0.48	0.47	0.24

Table 7. Criteria for determining the level of Park–Ang damage index.

Damage Level	Physical Characteristics	Damage Index
Collapse	Total collapse of infill	>1.00
Severe	Extensive large cracks	0.80–1.00
Moderate	Corner crushing	0.50–0.80
Minor	First diagonal crack	0.25–0.50

4. Case Study of Masonry Wall Building Reinforced with X-Bracing Concrete Frames

A reinforced rigid frame model is shown in Figure 5A. The columns were reinforced with steel bars and the x-bracing concrete frame technique, because the shear strength acts on columns in buildings. The beams were reinforced with steel bars, as shown in Table 8. This is equivalent to using plastic hinges, which may be placed at the end of beams.

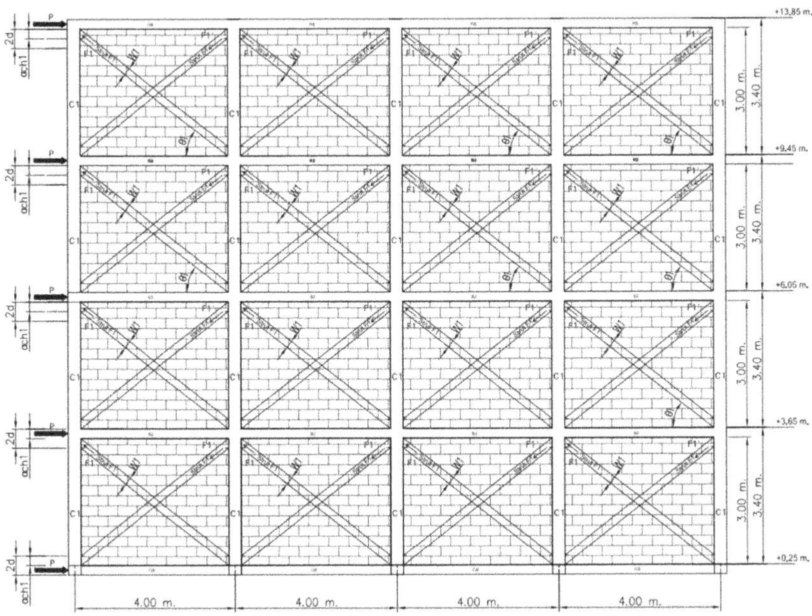

(A)

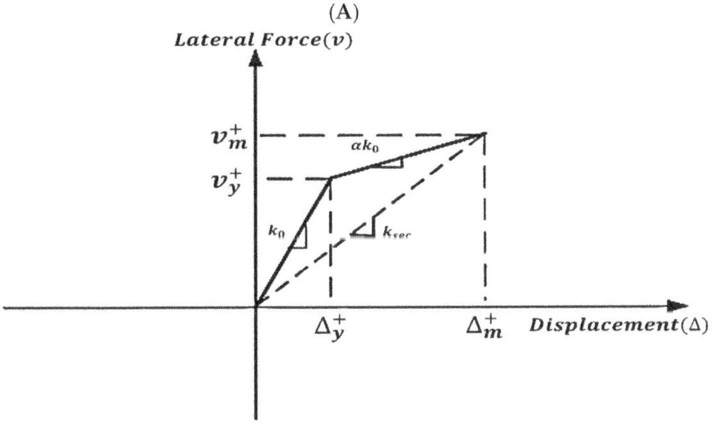

(B)

Figure 5. Part of the wall frame model. (**A**) An analytical model of the masonry walls according to the equivalent compressive x-brace concrete frame method. (**B**) Relation of the lateral force to the displacement value.

Table 8. The details of the reinforced steel bars.

Building	Size (mm.)	Main Reinforcement (mm.)	Stirrup Reinforcement (mm.)
C1	200 × 200	8DB16	RB6@150
B2	200 × 400	5DB16	RB6@150
RB	200 × 400	5DB16	RB6@150
W1	125 × 200	2RB9	RB6@150

The column's moment resistance (M_S) can be calculated from the sum of the original column's moment resistance (M_c) and the x-bracing concrete frames' moment resistance (M_B) as follows:

$$M_S = M_C + M_B \tag{7}$$

The x-bracing concrete frames' moment resistance (M_B) can be calculated from the sum of the concrete brace moment resistance and the steel bar moment resistance as follows:

$$M_B = \sum_{i=1}^{n} \left(f_{ys}(S_s)i_i\right)+\sum_{i=1}^{n}\left(f_{\prime c}(S_c)i_i\right) \tag{8}$$

where
fys and $f'c$ = yield the strength of the steel bar and compressive strength of the concrete brace;
Ss and Sc = cross-section of the steel bar and concrete brace;
V_{BF} = lateral load resistance of the reinforced rigid frame.

$$V_{BF} = 2(Mpj + Ms)/h = 2(Mpj + Mc + Mb)/h \tag{9}$$

where M_{pj} = the joint connecting plastic moment, which is considered to include the smallest moment of the column M_{pc} plastic moment, the beam M_{pb} plastic moment, and the joint connecting moment of the column–beam. The shear strength of the column V_C is half that of V_{BF}:

$$V_C = \left(M_{pj} + M_s\right)/h \tag{10}$$

Case Study of Masonry Wall Building Reinforced with X-Bracing Concrete Frames

The concrete building structure has a first floor level 0.25 m, and the building has a width of 4.00 m/bay, with four spaces; a height of 13.85 m; a general floor thickness of 0.10 m; and a live load of 3.00 kN/m^2. Moreover, the front of the building frame, as shown in Figure 5A, consists of masonry walls, with the width of the entrance being equal to the column spacing and cross-sectional dimensions. The details of the reinforced steel bars of the columns and beams are shown in Table 8. The compressive strength of the cylinder concrete is 24 MPa, and the tensile strength of the reinforced steel bars is 400 MPa. An analytical model of a masonry wall building reinforced with x-bracing concrete frames is used with the equivalent compressive bracing frame method. Here, the width of the diagonal compressive strength of the masonry wall is W_1. Moreover, the diagonal compressive strength of the masonry wall is the lateral force resistance of the wall, which can be written in a graph expressing the relation of the lateral force to the displacement value, as shown in Figure 5B. When v_m^+, v_y^+, Δ_m^+, and Δ_y^+ are the maximum resistance, resistance at the yield point, maximum movement, and movement at the yield point in the direction of the acting force, K_0 and K_{sec} are the initial stiffness value and cross-sectional stiffness value, respectively.

The maximum resistance of the compressive strength V_m of the masonry wall building reinforced with x-bracing concrete frames can be calculated as

$$V_m = f_1 \cos\theta_1 + f_1' \cos\theta 1 = 2W_1 t f_a \cos\theta_1 \tag{11}$$

where
f_a is the allowable compressive stress of the masonry wall prism, $f_a = 0.6\varnothing f_m$;
f_m is the maximum compression of the masonry wall prism;
$\varnothing$ = 0.65, and t = the thickness of the masonry wall;
θ_1 is the tilt angle of the diagonal compressive strength of the masonry wall.

The balance of the lateral force P is equal to the sum of the resistance frame and compressive strength f_1,f_1':

$$P = V_{BF} + f_1 \cos\theta_1 + f_1' \cos\theta_1 \tag{12}$$

$$P = V_{BF} + 2W_1 t f_a \cos\theta_1 \quad (13)$$

The reinforced concrete building's rigid frame was calculated and designed for seismic resistance. We calculated the concrete building structure using a computer design in the SeismoStruct 2023 program to identify the requirements for the seismic resistance of the concrete building frame using the nonlinear static force method (a pushover analysis) with response spectrum acceleration. The building was located in Chiang Rai Province (DPT Standard 1301/1302-61) [3]. The lateral force base shear of the concrete building's rigid frame design was 502.39 kN, and the code used for the lateral displacement safety factor was ASCE41-17 [30]. For the reinforced design of the rigid frame, with steel bars along the length of the columns, to reinforce the beams along their length, the size of the main reinforcement was DB16, and the size of the stirrup reinforcement was RB6; the size of the main reinforcement of the steel bars of the x-bracing concrete frames was RB9, and the size of the stirrup reinforcement was RB6 at the end and base of the columns, spanning a length 2.0 times that of the beams' depth. The parts of the masonry walls using x-bracing concrete frames are shown in Figure 6.

The moment resistance and shear resistance of the original column were compared with those of the reinforced column, and they were calculated using Equations (7), (8), and (10), as shown in Table 9. Here, V_c is the value of the resistance of a single column. The resistance of the reinforced x-bracing concrete frames in the masonry walls of the building can be calculated using Equations (7)–(13), as shown in Table 10, where the P value is the resistance of a single span of the case study concrete building structure, as shown in Figure 6A. The design lateral force, which we also calculated for the four-storey concrete building structure, was 502.39 kN for both the original wall structure and reinforced x-bracing structure.

Table 9. Resistance of the column of the masonry wall building reinforced with x-bracing concrete frames.

Parameter	Original Column	Reinforced X-Brace Concrete Frames
M_c (kNm)	30.05	30.05
M_b (kNm)	-	518.00
M_S (kNm)	-	548.05
M_{pj} (kNm)	23.65	23.65
V_c (kNm)	11.19	190.57

Table 10. Resistance of the masonry wall building reinforced with x-bracing concrete frames.

Parameter	Original Walls	Reinforced X-Brace Concrete Frames
α_C	0.085	0.1176
W_1 (mm)	-	124.26
$2W_1 tfa \cos\theta_1$ (kN)	-	75.52
V_{BF} (kN)	22.38	155.79
P (kN)	22.38	231.31
Design lateral force (kN)	502.39	502.39

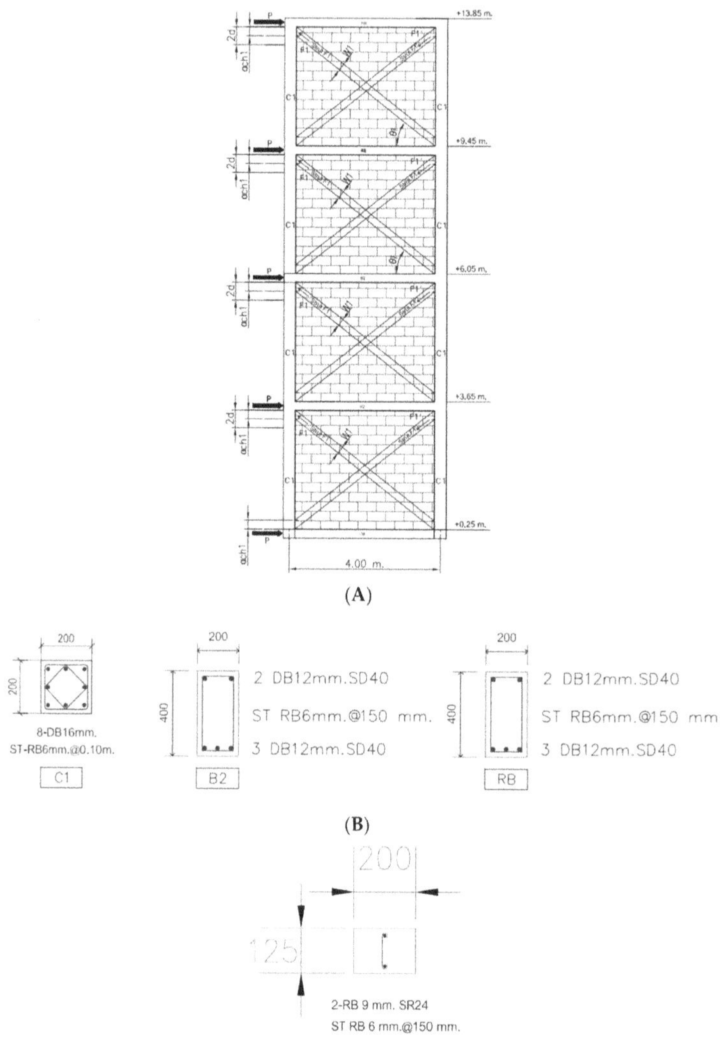

Figure 6. Masonry walls, columns, and beams reinforced with x-bracing concrete frame. (**A**) Single-span reinforced building frame; (**B**) cross-section of the reinforced column and beam; (**C**) cross-section of the reinforced x-bracing concrete frame.

5. Analysis of Masonry Wall Building Reinforced with X-Bracing Concrete Frames

For the four-storey concrete building structure, as shown in Figure 5A, we created a model for a structural analysis using the SeismoStruct 2023 program 3D model, as shown in Figure 6. The data used for the structural analysis of the masonry wall building reinforced with x-bracing concrete frames are shown in Tables 9 and 10. The moment resistance of the columns and lateral force resistance of the walls were calculated, as shown in Equations (1)–(13).

An analysis of structure of the four-storey masonry wall building reinforced with x-bracing concrete frames was performed using the nonlinear static force method (pushover analysis method), calculating the target storey drift value (δ_t) with the DPT Standard 1301/1302-61. We used response spectrum acceleration for the building, which was located in Chiang Rai Province. The target design storey's drift value δ_u was increased by 1.25 times ($1.25\delta_m$) to assess the damage index (DI). The parameter results of the target storey's drift value (β) equalled 0.1–0.35, as shown in Table 11.

Table 11. Results of masonry walls with and without x-bracing concrete frames used in the four-storey concrete building structure models.

Case Study	Q_y (kN)	δ_M (m)	δ_u (m)	β					
				0.10	0.15	0.20	0.25	0.30	0.35
Four-storey concrete building with x-bracing concrete frames	478.780	0.048	0.06	0.83	0.85	0.86	0.86	0.88	0.89
Four-storey concrete building without x-bracing concrete frames	67.160	0.240	0.30	1.69	2.59	3.03	3.03	3.48	3.93

Figure 7 shows the maximum lateral force and maximum lateral displacement of the four-storey concrete building structure with masonry walls reinforced with x-bracing concrete frames; when the lateral force is 478.78 kN, the initial starting point of the lateral displacement is equal to 4.80 mm, and the yield maximum lateral displacement is approximately 24.00 mm. The results of the model structure analysis are shown in Table 11. The damage index (DI) indicates extensive large cracks, but the original walls without x-bracing concrete frames show a damage index criterion of more than 1.0, as shown in Table 7.

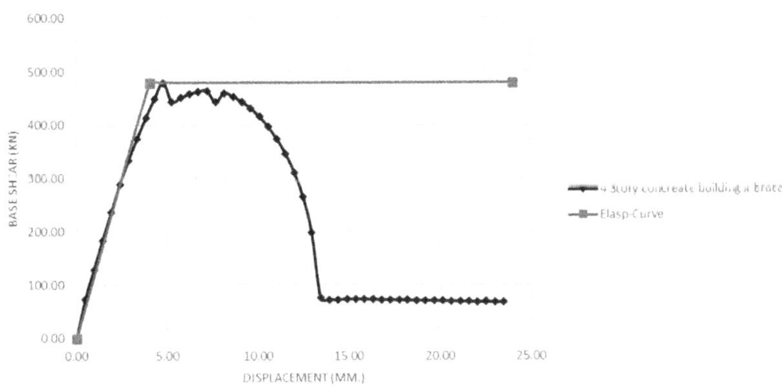

Figure 7. Model of masonry wall building reinforced with x-bracing concrete frame.

In Figure 8, we can see that the masonry wall building reinforced with x-bracing concrete frames resists higher maximum lateral force and yield lateral displacement values than the original walls without x-bracing concrete frames. For the original wall without x-bracing concrete frames, when the lateral force is 67.16 kN, the initial starting point of the lateral displacement is equal to 14.99 mm. By comparison, the structure reinforced with x-bracing concrete frames resists a higher lateral force value than the original wall without an x-bracing concrete frame structure, with an approximately 7.13 times difference in lateral forces.

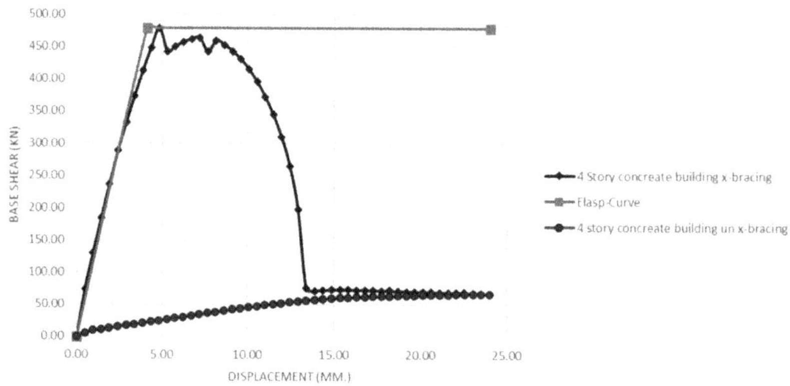

Figure 8. Comparison between masonry wall building with x-bracing and original wall building without x-bracing concrete frame.

6. Conclusions

In this study, a suitable damage index is determined. Masonry wall buildings reinforced with x-bracing concrete frames are assessed by analysing various models to study the damage behaviours under monotonic loading and cyclic loading conditions. Results, the case study demonstrates that the masonry wall building, reinforced with x-bracing concrete frames, exhibits enhanced seismic resistance, reduced potential for severe damage, and improved overall structural performance compared to the original structure. The detailed analysis and calculations support the effectiveness of x-bracing in mitigating the impact of seismic forces will affect to building collapse. The study results are summarized in the following main points:

1. The monotonic loading and cyclic loading conditions are analysed at the maximum lateral force. The analysis reveals that the results are consistent in the same direction and with the model structure analysis results.
2. The correlation coefficient between the damage level and accumulated energy (β) is appropriate for use in the calculation of the damage index using the Park–Ang equation. The Park–Ang equation for a masonry wall building (β) is equal to 0.10–0.35.
3. The criteria for the consideration of the damage level are defined as follows: when the damage index is between 0.25 and 0.50, the specimen has a minor damage level; when the damage index is between 0.50 and 0.80, the specimen has a moderate damage level; when the damage index is between 0.80 and 1.00, the specimen has a severe damage level; and when the damage index is more than 1.00, the specimen has a collapse damage level, resulting in the structure of the building experiencing a disaster.
4. The damage index for assessing the seismic-resistant design of masonry wall buildings reinforced with x-bracing concrete frames shows that they resist higher lateral force base shear values before lateral displacement than the original walls without x-bracing concrete frame structures, with values of 478.78 kN and at 67.16 kN, respectively, demonstrating an approximately 7.13 times difference in lateral forces. The masonry wall buildings reinforced with x-bracing concrete frames show extensive large cracks, and the damage index criterion (DI) is equal to 0.8 in the maximum lateral force base shear design.

Therefore, the use of reinforced x-bracing concrete frames in masonry wall buildings offers a holistic approach to seismic resistance. This strategy addresses the vulnerabilities of traditional masonry structures, providing improved stability, increased shear strength, controlled de-formation, and enhanced energy dissipation. The adaptability of x-bracing systems makes them valuable not only in new construction but also in retrofitting existing buildings for better seismic performance.

Author Contributions: Conceptualization, Y.C. and V.C.; methodology, Y.C. and V.C.; software, Y.C. and V.C.; validation, Y.C. and V.C.; formal analysis, Y.C. and V.C.; investigation, Y.C. and V.C.; resources, Y.C. and V.C.; data curation, Y.C. and V.C.; writing—original draft preparation, Y.C. and V.C.; writing—review and editing, Y.C. and V.C.; visualization, Y.C. and V.C.; supervision, Y.C. and V.C.; project administration, Y.C. and V.C. All authors have read and agreed to the published version of the manuscript.

Funding: This research received no external funding for the cost of publication.

Institutional Review Board Statement: Not applicable.

Informed Consent Statement: Not applicable.

Data Availability Statement: Data is contained within the article.

Conflicts of Interest: The authors declare no conflict of interest.

References

1. Lukkunaprasit, P.; Ruangrassamee, A.; Boonyatee, T.; Chintanapakdee, C.; Jankaew, K.; Thanasisathit, N.; Chandrangsu, T. Performance of structures in the MW 6.1 mae lao earthquake in Thailand on May 5, 2014, and implications for future construction. *J. Earthq. Eng.* **2015**, *20*, 219–242.
2. Park, Y.J.; Ang, A.H.S. Mechanistic seismic damage model for reinforced concrete. *J. Struct. Eng.* **1985**, *111*, 723–732.
3. *DPT's Standard 1301/1302. (2561)*; Earthquake Vibration Resistance Building Design Standards. Department of Public Works and Town & Country Planning: Bangkok, Thailand, 2018.
4. Amornpunyapat, R.; Panyakapo, P.; Panyakapo, M. Development of Lightweight Concrete Interlocking Block Panel with Water Treatment Sludge and Expanded Metal Ferrocement. *Eng. J.* **2021**, *25*, 81–97.
5. Saneinejad, A.; Hobbs, B. Inelastic design of infilled frames. *J. Struct. Eng. ASCE* **1995**, *6682*, 634–660.
6. Mehrabi, A.B.; Shing, P.B.; Schuller, M.P.; Noland, J.L. Experimental evaluation of masonry-infilled RC frames. *J. Struct. Eng.* **1996**, *122*, 228–237.
7. Wararuksajja, W.; Srechai, J.; Leelataviwat, S. Seismic design of RC moment-resisting frames with concrete block infill walls considering local infill-frame interactions. *Bull. Earthq. Eng.* **2020**, *18*, 6445–6474.
8. Jiang, H.; Liu, X.; Mao, J. Full-scale experimental study on masonry infilled RC momentresisting frames under cyclic loads. *Eng. Struct.* **2015**, *91*, 70–84.
9. Morandi, P.; Hak, S.; Magenes, G. Performancebased interpretation of in-plane cyclic tests on RC frames with strong masonry infills. *Eng. Struct.* **2018**, *156*, 503–521.
10. Huang, Q.; Guo, Z.; Kuang, J.S. Designing infilled reinforced concrete frames with the 'strong frame-weak infill' principle. *Eng. Struct.* **2016**, *123*, 341–353.
11. SeismoStruct user manual. *Seismostruct User Manual*; Seisomsoft Ltd.: Pavia, Italy, 2021.
12. Kazemi, M.T.; Morshed, R. Seismic shear strengthening of R/C columns with ferrocement jacket. *Cem. Concr. Compos.* **2005**, *27*, 834–842.
13. Leeanansaksiri, A.; Panyakapo, P.; Ruangrassme, A. Seismic capacity of masonry infilled RC frame strengthening with expanded metal ferrocement. *Eng. Struct.* **2018**, *159*, 110–127.
14. Longthong, S.; Panyakapo, P.; Ruangrassamee, A. Seismic Strengthening of RC Frame and Brick Infill Panel using Ferrocement and Expanded Metal. *Eng. J.* **2020**, *24*, 45–59.
15. Li, B.; Lam, E.S.S. Influence of interfacial characteristics on the shear bond behavior between concrete and ferrocement. *Constr. Build. Mater.* **2018**, *176*, 462–469.
16. Panyamul, S.; Panyakapo, P.; Ruangrassame, A. Seismic shear strengthening of reinforced concrete short columns using ferrocement with expanded metal. *Eng. J.* **2019**, *23*, 175–189.
17. Salinas, S.; Guzman, A.; Carrillo, J. Performance evaluation of structures with reinforced concrete columns retrofitted with steel jacketing. *J. Build. Eng.* **2021**, *33*, 101510.
18. Panyakapo, P. Seismic analysis of RC frame with brick infill panel strengthened by steel cage and expanded metal. *Eng. J.* **2021**, *25*, 29–44.
19. Saiidi, M.; Sozen, M.A. *Simple and Complex Models for Nonlinear Seismic Response of Reinforced Concrete Structures*; Report UILU–ENG-79-2031; Department of Civil Engineering, University of Illinois: Urbana, IL, USA, 1979.
20. Stewart, W.G. The Seismic Design of Plywood Sheathed Shear Walls. Ph.D. Thesis, Department of Civil Engineering, University of Canterbury, Christchurch, New Zealand, 1987.
21. Naeim, F.; Kelly, J.M. *Design of Seismic Isolated Structures: From Theory to Practice*; John Wiley: Hoboken, NJ, USA, 1999.
22. Symans, M.D.; Charney, F.A.; Whittaker, A.S.; Constantinou, M.C.; Kircher, C.A.; Johnson, M.W.; McNamara, R.J. Energy Dissipation Systems for Seismic Applications: Current Practice and Recent Developments. *J. Struct. Eng.* **2007**, *134*, 3–21. [CrossRef]
23. Kelly, J.M. The role of damping in seismic isolation. *Earthq. Eng. Struct. Dyn.* **1999**, *28*, 3–20.

24. Engle, T.; Mahmoud, H.; Chulahwat, A. Hybrid Tuned Mass Damper and Isolation Floor Slab System Optimized for Vibration Control. *J. Earthq. Eng.* **2015**, *19*, 1197–1221. [CrossRef]
25. Matta, E.; De Stefano, A. Robust design of mass-uncertain rolling-pendulum TMDs for the seismic protection of buildings. *Mech. Syst. Signal Process.* **2009**, *23*, 127–147. [CrossRef]
26. De Domenico, D.; Ricciardi, G. Optimal design and seismic performance of tuned mass damper inerter (TMDI) for structures with nonlinear base isolation systems. *Earthq. Eng. Struct. Dyn.* **2018**, *47*, 2539–2560. [CrossRef]
27. Mantakas, A.G.; Kapasakalis, K.A.; Alvertos, A.E.; Antoniadis, I.A.; Sapountzakis, E.J. A Negative Stiffness Dynamic Base Absorber for Seismic Retrofitting of Residential Buildings. *Struct. Control Health Monit.* **2022**, *29*, e3127. [CrossRef]
28. Kapasakalis, K.A.; Antoniadis, I.A.; Sapountzakis, E.J. Constrained Optimal Design of Seismic Base Absorbers Based on an Extended KDamper Concept. *Eng. Struct.* **2020**, *226*, 111312. [CrossRef]
29. Kapasakalis, K.A.; Antoniadis, I.A.; Sapountzakis, E.J. Performance Assessment of the KDamper as a Seismic Absorption Base. *Struct. Control Health Monit.* **2019**, *27*, e2482. [CrossRef]
30. *Standard ASCE 41-17;* Seismic Evaluation and Retrofit of Existing Buildings (41-17). American Society of Civil Engineers: Reston, VA, USA, 2017.

applied sciences

MDPI

Article

Seismogram Rearrangement as a Perspective Basis for Defining New Attributes

Ranko Babić [1,*] and Lidija Babić [2]

1 Department of Electrical Engineering, Faculty of Technical Sciences, University of Priština in Kosovska Mitrovica, Knjaza Miloša 7, 38220 Kosovska Mitrovica, Serbia
2 Department of Civil Engineering, Faculty of Technical Sciences, University of Priština in Kosovska Mitrovica, Knjaza Miloša 7, 38220 Kosovska Mitrovica, Serbia; lidija.babic@pr.ac.rs
* Correspondence: ranko.babic@pr.ac.rs; Tel.: +381-63416063

Abstract: Proper regulative in earthquake risk reduction, hazard assessment, earthquake resistant design and construction of structures, should be under perpetual improvement. They are mainly the result of earthquake study and analysis of post-seismic effects. In this context, a new frame for seismogram analysis is proposed, based on its decomposition into elementary and single (positive and negative) fluctuations (EF/SF). Decomposition enables rearrangement, both providing the frame for defining new attributes, where several main ones are proposed and elaborated in the paper: fan diagram of EFs structural axes; R- or R_Q-envelope, formed over positive and over negative peaks of ordered SFs; two forms of difference functions of positive and negative R/R_Q-envelopes, as a measure of seismogram's (a)symmetry, in respect to t-axis; distributions of SFs durations. Several others are in perspective. Analogy between R-envelope and seismogram's spectrum is considered from the aspects of arranged elements, and asymptotic behavior of these two functional attributes. More objective definition of seismogram duration is proposed, determined by linear regression across the scatter plot of SFs peaks. R-envelope showed itself as an invariant in the seismogram population, and the distribution of SFs durations as well. Results are presented through extensive graphics, backed with mathematical description. Thorough discussion of the concept and results is given.

Keywords: earthquake; seismogram; fluctuation; seismogram attribute; envelope; spectrum; invariant; asymptote; seismogram duration; linear regression; earthquake resistance regulative

1. Introduction

Specific for its unpredictability and covering vast areas, a strong earthquake inflicts heavy damage, which can be a tremendous shock for the engaged society. Various disciplines include methods of earthquake risk reduction and hazard assessment [1], which consider social and economic consequences of earthquakes in studying the earthquake vulnerability of the built environment [2,3].

Material damage can even be assessed by planning, designing, and constructing structures according to standards of earthquake resilience [1,4–6]. There are various levels of earthquake resistance, which determine the cost of structures and construction. The design and construction of earthquake-resistant structures are based on regulations and standards derived from earthquake studies and the effects they inflicted on existing structures.

Civil engineers strive to improve regulations of seismic resistance of structures. Experts from other branches derive similar rules and recommendations. These efforts are based on a better comprehension of seismic processes, which are usually described through seismograms and their various characteristics, often called attributes.

In this paper, it is propose a new frame for defining attributes, with the hope to define new features, which could make the picture of earthquakes more complete and more suitable, in addition to existing ones. A better understanding and description of earthquakes surely means better and more refined regulations and standards.

1.1. Earthquake "Speaks" through Seismogam(s)

Seismicity is the main source of our knowledge about the Earth's structure. For a more elaborate general view, see [7] "Vol. 4, paper 4.01, p. 1924". A seismogram is almost the only source of information about an earthquake, apart from tangible effects on society and the (built) environment. It is a record of seismic effects at the site, which are the result of many complex processes in the seismic source, during wave propagation through the Earth, the influence of seismic sensors, recorders, and processing techniques applied on the seismogram [8] "Ch. 1, Figure 1.2, p. 6; Ch. 11".

On the other hand, the above title is just half of the story: valuable data on earthquake stem also from the cross between the seismogram itself and seismic effects on the built environment.

Various experts are interested in earthquakes and their effects, including geophysicists, seismologists and observatory personnel, civil engineers, disaster managers and risk planners, urban planners, to whole governments, not to mention related researchers, students, training personnel [8] "Ch. 1, p. 29", [1,9].

For the sake of this analysis, all complex processes standing behind a seismogram will be temporarily put aside, and attention will be focused on its waveform only. The reasons for such a formal approach will be elaborated in Section 2.

The seismogram is by all means fluctuational in nature, representing recorded fluctuation of acceleration, velocity, or displacement. Its waveform looks like a wave train randomly composed of wavelets (from the source) mutually shifted in time [7] "Vol. 1, paper 1.03.1.3, p. 73, p. 131". The term 'wavelets' is used here just as a language item, not as a mathematical tool. If the plural of "fluctuation" is used, then, from the more formal point of view, one can consider such waveform as a sequence of impulses of random amplitude, duration, and shape, where the shape proves to be very important. This issue can be further formalized, and a seismogram be considered as the sequence of elementary fluctuations (EF) [10].

1.2. Purpose and Categories of Attributes

A seismogram is usually a very complex transient. It contains, or hides, all the available information about an earthquake but taken at the (recording) site. As such, it is cumbersome and intricate for direct use. In addition, not all experts are interested in the information the seismogram waveform brings with them, but they seek some aspects that are the most important for their needs. So, it asks for processing, to a certain extent, to find out needed characteristics, often called attributes or features, which carry particular significance for users and reveal interesting content from raw recordings. Here it is accepted the widest definition of seismic attribute, as any information of interest that can be derived from raw seismic data [11].

There are many references involving the notion of 'seismic attribute' [12–16], often giving the definition of the issue. What can be noticed is a slight evolution in the meaning of the seismic attribute [12,13]. In close relation to the definition stands the issue of attribute classification [12,16]. Either with respect to definition or classification, this research relies on the broader meaning of both notions.

Only a few characteristics can be extracted directly from a seismogram, but it does not mean simply picking up the item(s) because it is not the result of a passive perception of details, but rather depends on the researcher's recognition ability, backed by his/her knowledge and experience about waveform background, and many other conditions [8] "Ch. 1, Section 1.3, Ch. 11".

It may seem that the reading is burdened by subjective factors, but provided a skilled analyst instead of, e.g., a novice, this makes the description objective, quantified, and comparable to other cases because the examiner performs according to the prescribed procedure. In [8] "Ch. 1, Section 1.3", it is explicitly stated that to "provide properly processed and interpreted seismograms under these diverse conditions, it is necessary to establish protocols for all aspects of observatory operation that may affect the seismological data itself". There are other interesting proposals regarding procedures of seismogram interpretation [17].

Even for the most skillful researcher, much of the content in seismograms remains hidden. To extract such information, various (post)processing techniques are used, either performed by humans (graphically numerical methods) or by machines (usually computer programs), up to artificial intelligence [18]. In both cases, techniques stem from respective algorithms. In the first case, the human is functioning as a 'machine' (following the strict protocol).

There are (too?) many attributes [19], and each one is dedicated to something particular in seismogram or any other random signal. Such a heap of attributes needs categorization. The grade "too" and question mark definitely prove that the seismogram is a very complex matter.

They can be roughly classified into parametric and functional attributes. Parametric ones can be further divided into single-parametric and group-parametric. Single-parametric can be local and general, with the latter determined from the whole seismogram. Examples of local single-parametric are the peak-to-peak max span (*B*), and the period (*T*), given in, e.g., [8] "Ch. 3, p. 47, Figure 3.19, Figure 3.28" while general single-parametric are, e.g., 5–95% duration [20–22], seismogram energy [8] "Ch. 3" etc.

The root-mean-square acceleration

$$a_{rms} = \left[\frac{1}{T_2 - T_1} \int_{T_1}^{T_2} a^2(t) dt \right]^{1/2} \tag{1}$$

is an important, and also interesting single-parametric attribute, which, implicitly, contains other attributes (parameters): T_1 and T_2, where the size and location of the interval $T_2 - T_1$ strongly influence the value of a_{rms} for a transient signal like seismogram, so T_1 and T_2 commonly chosen to exclude insignificant shaking. There are several proposals, e.g., for T_1 as the time of S-wave arrival, and for T_2: $T_2 = T_1 + 10$ s, or $T_2 = T_1 + T_d$ (T_d is the duration of faulting, or $T_2 = 1/f_0$) [1] "p. 76".

This is not the final word about duration (see [8] "Ch. 3, p. 135, Figure 3.62"), showing the complexity of the issue. The attribute a_{rms} only shows how the matters in seismology are far from simple.

As for functional attributes, a common one is the envelope, in simple and more complex forms. For others, see [8] "Ch. 3". Simple or peak envelope is related to seismogram waveform in an apparent way.

The known Husid diagram is a functional attribute:

$$H(t) = \frac{\int_0^t a^2(t)\,dt}{\int_0^{T_D} a^2(t)\,dt} = \frac{1}{\int_0^{T_D} a^2(t)\,dt} \int_0^t a^2(t)\,dt = \frac{1}{A_D} \int_0^t a^2(t)\,dt \tag{2}$$

as the integral as a function of its upper limit, normalized by the total integral. The T_D is the total duration of the earthquake, and *a*(*t*) is the instantaneous earthquake ground acceleration [23,24], where the A_D represents the total integral of squared acceleration.

In direct relationship with *H*(*t*) and a_{rms} is the Arias intensity

$$I_{Ar} = \frac{\pi}{2g} \int_0^{T_D} a^2(t)\,dt \tag{3}$$

where T_D is, again, the total duration of the record, and *a*(*t*) is the acceleration time series [23,24], so that one can write:

$$a_{rms} = \left[\frac{1}{T_D} \frac{2g}{\pi} I_{Ar} \right]^{1/2} \qquad I_{Ar} = \frac{\pi\, T_D}{2g} a_{rms}^2 \tag{4}$$

No wonder that all these attributes, parametric or functional, stem from acceleration seismograms. The Hurst exponent is a useful statistical method for inferring the properties of a random time series without making assumptions about stationarity. Within voluminous literature about the Hurst exponent, a quick and insight can be found in [25]. It is also applied to seismograms [26].

Another important functional attribute is the spectrum of seismograms, as the result of the Fourier transform. There are also parametric attributes of the spectrum (see [27]). It is interesting to note that the spectrum, although functional, i.e., continuous, can also be defined as a group-parametric attribute, which will consider later in detail.

The envelope, as a functional attribute in various forms, is widely used, and it is at the focus in this paper.

1.3. Envelope of Seismogram—Issue of Complexity

If one draws the peak envelope over positive elementary fluctuations (EF) of a seismogram, one gets better insight into peak amplitude variation, which would not be that much visible through mere spikes (see, [14,28]). It has something to do with gestalt principles of perception—if spikes along a seismogram are not so dense as necessary, one's vision cannot gather them into the shape to perceive a variation. Just the fact that the envelope is drawn as a line over the peaks, making visual continuation, helps in the perceptive grouping of impulses [29].

The envelope is a widely used attribute of seismograms or other signals. Besides the plain (peak) envelope, there is also the envelope defined as a mixture of a signal itself and its Hilbert transform [11]

$$E(t) = \sqrt{s^2(t) + s_H^2(t)} \quad s_H(t) = H[s(t)] \; \{= Hs(t)\} \; \{= H(s)(t)\} \tag{5}$$

where various labels of the operator's action are quoted and direct and smoothed envelopes as well.

Apart from the information about P- and S-phase, seismologists often focus on the characteristics of seismogram envelopes (rms envelope in [30]). Extensive use of envelopes can be found in many papers [29,31]. Another interesting kind of envelope is the signal-to-noise ratio:

$$STA(t) = \frac{1}{T_{STA}} \int_{t}^{t+T_{STA}} SNR(\tau)\, d\tau \quad LTA(t) = \frac{1}{T_{LTA}} \int_{t-T_{LTA}}^{t} SNR(\tau)\, d\tau \tag{6}$$

which is calculated for short- and long-term averages (*STA, LTA*) to detect phases as [32]

$$[STA - LTA]_C(t) = \int_0^t [STA(\tau) - LTA(\tau)]\, d\tau \tag{7}$$

Seismograms are recorded for various physical quantities (acceleration, velocity displacement), orientations (e.g., Hor-1, Hor-2, and Vert, in PEER database nomenclature, or equivalent ones), and filtered (low-, band, or high-frequency) [33].

It is obvious that the envelope, as an attribute of seismogram, gains in importance with waveform complexity or fluctuational density. Different recordings (acceleration, velocity, or displacement) of the same seismic event at the same site show different waveform densities [33], and the same record is different with different filtering ([7] "article 1.01, p. 22, Figure 13").

The seismogram gains in complexity with propagation distance, which is documented in many references [29], but the distance is not the only cause. Just to mention processes during fault rupture, the size, and geometry of the fault, emitted (delivered) energy, etc.

Seismic processes, in the whole range from the fault to the site, are 'buried', so they are difficult for open study. In Section 2, a brief analysis of the specific open 'fault' mechanism of the Chelyabinsk event is given, and the following consequences, are as an illustration of said complexity.

Rise of seismogram complexity with the distance has its coin reverse. Comparison of accelerograms for a range of event sizes, chosen to have a hypocentral distance of approximately 25 km, shows an increase of complexity with magnitude ([7] "article 4.18, Figure 4, p. 2437").

In one word, the envelope seeks a critical "mass" of a waveform to justify its applicability. Otherwise, if a seismogram waveform is heavily and randomly indented (contains only a few fluctuations), its envelope is of little use (e.g., displacement record).

With all these various types of envelopes, it seems there is no room left for a new one. Nevertheless, in this article it is proposed a new envelope, better to say a new type of envelope, which is neither the result of some complicated mathematical transformation, such as, Fourier or Hilbert, nor it is that simple to stem from the raw waveform. The proposed new envelope is mainly based on the rearrangement (recomposition) of single fluctuations (SFs) from a raw seismogram into a new sequence, ordered by increasing or decreasing peak amplitudes. Because of that, the new envelope is named as the R-envelope (from "Rearranged", "Recomposed").

1.4. Article Aims and Findings

A seismogram is the result of many complex physical processes. It is very difficult to relate every detail in a seismogram with the processes standing behind. Even if it might look like an obvious fact, it is important to remember that the seismogram is the picture of seismic processes at the site. Therefore, the formalization of seismograms is inevitable—one considers what it has at hand. How, then to peer through the fluctuations of the seismogram and try to see something more behind the wiggles?

The lowest structural level of a seismogram is at the waveform elements, impulses, or wavelets (considered as small waves, not as mathematical tools). In this paper, such chunks of seismogram are defined as elementary or single fluctuations (EF and SF). The next structural level is the short train of SFs/EFs. Here, the authors perceive the next stage of research. The SFs are simpler but easier for manipulation, while the EFs are more complex but also more natural as elementary building blocks of the seismogram, as the simplest shapes still carry alternating change, which actually causes shaking (quaking) in the 'push-pull' pattern.

Getting to EFs/SFs means prior parsing of the seismogram. Segmentation also provides rearrangement of EFs/SFs to create new waveform structures, which can reveal new information previously invisible. As the simplest regrouping, the overlapping of EFs gave a 'wood', a structure which prevents 'not see the wood for the trees'. To simplify the mess, the EF is approximated by triangles or even by structural axis, making the 'wood' more transparent.

Sorting the durations of SFs by the length, for both positive and negative ones, led to very interesting results, which peered not only into the seismogram waveform but also into its spectrum.

More organized, more systematic rearrangement, the ordering of SFs or of EFs by peak amplitudes, yielded the so-called R-envelope, either in natural (R) or in regular (quantized, R_Q) forms. It turned out that the R-envelope represents a kind of invariant in the population of seismograms. Moreover, there is a direct correlation between the R-envelope and ordered durations of SFs.

The difference between R-envelopes, drawn over positive SFs and negative SFs, is expected to produce a rule about seismogram symmetry but enabled, for the present, only individual interpretations.

Where this approach of seismogram segmentation and rearrangement showed its potential is a new definition of seismogram duration in objective way, which excludes subjective factors but outcomes from the waveform.

2. Materials and Methods

2.1. Elementary Fluctuation—Formal and Real Meaning

The term "fluctuation" (to fluctuate) is irregular rising and falling (change) of a physical quantity in an alternating pattern, although random, which is consisted of "ups" and "downs", separated by crossings of a certain axis (mostly the t-axis), or some constant value, or even some monotonic trend curve (e.g., change of stock price or currency exchange rate). So, one can see some regularity in such irregular behavior: alternating positive-negative pattern, in brief, random cyclic behavior.

In this research, the term "elementary fluctuation" (EF) is defined, and the seismogram is considered a sequence of EFs. Further, the EF is composed of two subsequent alternate random impulses, which are called single fluctuations (SF), as in [10]. So, the EF, or the SF, is the basic structural element of a seismogram.

When the fault slips, the two sides of the fault move in opposite directions, which creates a different polarity of the first motion P in different directions (see [34] "p. 217, Figure 7.3"), so one should distinguish two conventions in defining the EF composition: (1) the leading impulse is positive. Hence the other is negative, and vice versa (Figure 1). To put it more clearly [10],

$$EF^{+} = SF^{+} + SF^{-} \qquad EF^{-} = SF^{-} + SF^{+} \tag{8}$$

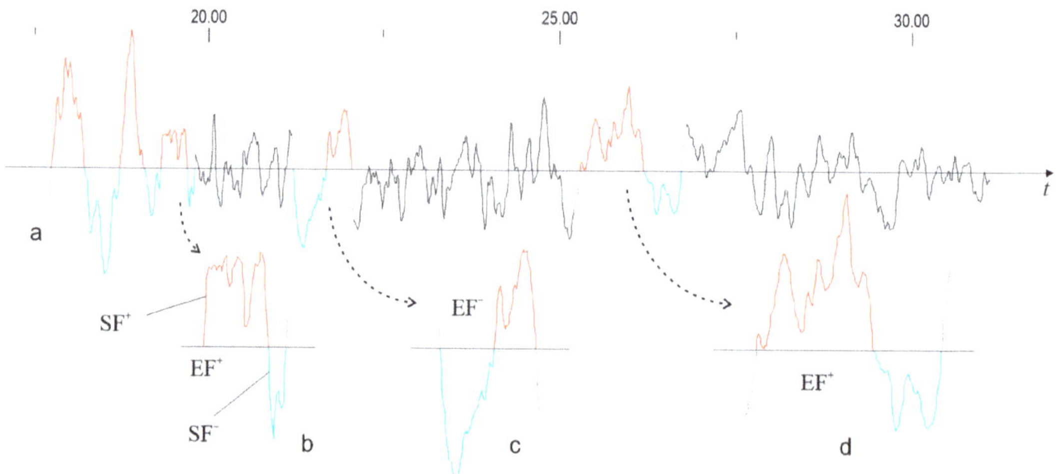

Figure 1. A few examples of EFs (depicted as pairs of positive and negative SFs) were extracted from a fragment of RSN 1616 seismogram (**a**). The cases (**b**,**d**) are of positive polarity (the leading SF is positive), while the case (**c**) is inverted, i.e., of negative polarity (the leading SF is negative). Positive SFs are in red solid line while negative SFs are in blue solid line.

The EFs are separated by zero-crossings (Figure 1), the points where the random waveform crosses the actual axis, "the t-axis, or some constant, or even some monotonic trend curve".

Within repetitive fluctuational patterns one can distinguish regular and irregular fluctuations. Regular EFs can mean quasi-periodicity, while irregular EFs mean just repeating a general pattern [35].

The generation of a seismogram is caused by processes during fault rupture in the hypocenter, which picture is conveyed by seismic waves to the seismograph location. In-between, seismic waves propagate through a (spherically) multilayered, inhomogeneous, anisotropic medium, which yields intricate combination of many reverberations of initial waves [8] "Ch. 2", [36]. Even with the seismic kinematics (ray formalism), the picture at the site remains pretty entangled. Standard Nomenclature of Seismic Phases speaks for itself, and the distribution of seismic sources as well [37].

Global Earth models are just statistical pictures of real mantle structure for study of teleseismic events. Top-down mantle convection brings much more complicated models and, hence, propagation pictures [7,38] "article 1.16, Figure 12, p. 555/576".

At closer distances, propagation conditions can be considered as even more complex, caused by the usually very irregular structure of lithosphere, since the main propagation in these cases takes place mainly in the crust [8] "Ch. 2; article 1.08.7, Figures 5–12, pp. 260–281". It is not only a matter of internal but also of external irregularities since there are effects of the Earth's surface topography [7] "article 4.18.6.13, Figure 15, p. 2453".

The events observed at closer distances also send seismic waves toward all seismographs around the world, where usually are recorded only the basic transients. For teleseismic observations, subsequent reverberations are also recorded. This research is restricted only to close-range earthquakes.

The lithosphere is a layered structure, often corrugated (due to tectonic movements), where basins (bedrocks) are infilled with stratified deposits (soil). Bedrock-soil boundaries and soil surfaces affect the propagation of seismic waves in several ways, causing reflections, refractions, and scatterings. The geometry of the infilled basin makes a real resonator, which can create resonances of incident seismic waves and, thus, change the fluctuation pattern in the seismogram. What is interesting to note and to add to such a model of resonator is the urban (built environment) layer, situated on the soil surface, consisting mainly of concrete structures, which changes the wave impedance of the soil surface and affects the seismic picture around the site, even with possibility to amplify seismic effects [1] "p. 29, Figure 3.1; p. 54, Figure 4.4; p. 145, Figure 5.7", [8] "Ch. 3, p. 224, Figure 3.113". To get an image of possible effects which can stand in the background of seismic phenomena, see [39,40] "Figure 4".

The term 'hypocenter' suggests a point-like source of seismic waves. It is actually a relative "point", which size depends on distance. Fault interfacing surfaces create a large system ranging from a (few) kilometers (s) to tens or hundreds of kilometers in length. If recorded closely, then each place on the rupturing fault acts as a point source of (its own) seismic waves, so all of them create a very complex picture at the seismic sensor. What's more, an earthquake often starts at one place of the fault, and then the perturbation also propagates along the fault, triggering other strained places to rupture, highly influencing recording conditions and the waveform of seismogram [7] "article 4.07, Figure 5, p. 2132".

Now, we may have an idea of what all (might) stand behind "wiggles" in seismogram, remembering the words of Professor Wiechert, "understand each wiggle" [8] "Ch. 1, p. 6".

It is very difficult to peer through the soil into the seismic processes. In many cases, it turns into a kind of guesswork.

A suitable case to study the seismic effects more transparently, not buried in the Earth depths, and to illustrate the effect of increasing seismogram's complexity with the distance is Chelyabinsk/Ural meteorite (small asteroid) impact of 15 February 2013, 17 m in diameter, mass approximately 11,000 tons, which struck Earth's atmosphere at 19.16 km/s speed and exploded in a meteor air burst at about 30 km above Earth's surface, resulting in explosion equivalent to over 470 kilotons of TNT. The air-friction heat evaporated debris, so just 0.05% of the original rock reached the ground as meteorites. This means that the main "seismic" mechanism was the air shockwave, of a pretty simple waveform covering vast soil area and triggering detections from monitoring stations as far away as Antarctica. Seismic records, 26 in total, compiled at the GFZ, taken within the span of 227 km to 2197 km, represent a remarkable example of seismogram waveform change with distance.

Instead of complex fault rupture processes superimposed to create an intricate waveform, here is a simple "fault" mechanism, providing something similar to delta short-impulse excitation. From recording distances, one can infer that the propagation of resulting seismic waves took place mostly through the mantle [41–50].

Details about the shockwave 'fault' mechanism can be found in [48]. Apart from the GFZ graphical account of the event [43,45], there are recordings, mostly taken as teleseismic, with arrays in China and the USA [49,50]. This 'fault' mechanism is, by all means, very specific. It is directional, even to say one-dimensional, without direct material impact, where the whole energy is transferred by the air. It is a kind of 'soft' rupture. Quite the skew angle of arrival, and almost graze angle, in combination with shockwave conical pressure front, indicate the area of the 'fault', and time profile of perturbation [43], which is easily seen comparing scales on the t-axis of this event and any other usual seismogram, e.g., RSN1616. A completely Earth-originating case can be seen in [51] "Figure 6".

But whatever the processes, in the source or during propagation, they keep the rudimentary and general shape of a random alternation. One should stress this apparent fact just to point out fundamental phenomena and principles that generate basically the same shape. After all processes in the said seismic 'information chain' took their part, what a structure or, generally, the built environment experiences is written in a seismogram, i.e., in the sequence of EFs. To say it otherwise, a structure is not 'interested' in the origin of wiggles that shake it, but it is 'interested' in how to survive and withstand the incoming EFs train.

2.2. Envelope of Rearranged Individual Fluctuations

A plain (peak) envelope, drawn over a seismogram's peak amplitudes of fluctuations, clearly shows general variation along the waveform (Figure 2).

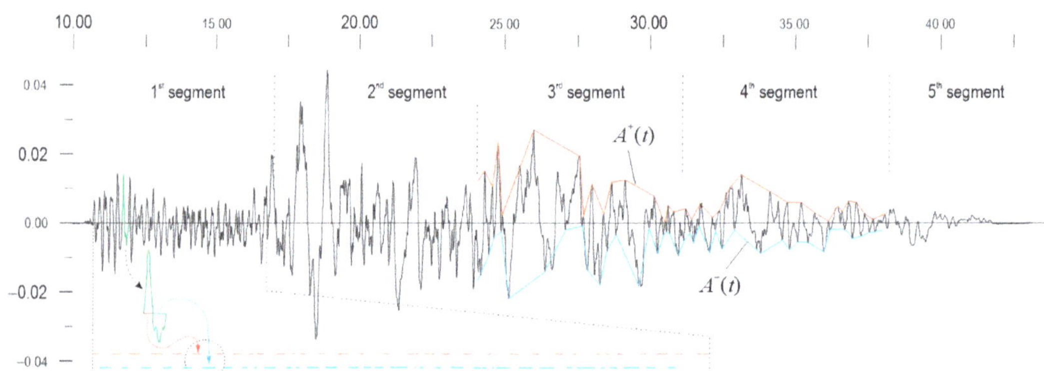

Figure 2. RSN 1616 seismogram is composed of miscellaneous shapes of EFs. It is divided into five segments for easier referencing. The plain envelopes, positive $A^+(t)$ (red solid line) and negative $A^-(t)$ (blue solid line), are drawn over the waveform only within the 3rd and 4th segments. The durations of SF^+ and SF^- within the 1st segment are measured and drawn by red (positive SFs) and blue (negative SFs) lines and shown downwards with a zoomed-in time scale. The insert showing an EF (green solid line) designates durations of corresponding SF^+ and SF^-, where arrows (dashed red and blue lines) graphically explain the procedure.

The segmentation of the seismogram into single fluctuations (SEs) results in sequences of random shape impulses, both positive and negative ones. Recomposition of these sets of SEs, separately for positive and negative SEs, by increasing or decreasing peak amplitudes, will produce another rearranged waveform. Thus, one gets two ordered "halves" of the original seismogram, one composed of positive SFs and the other of negative SFs (see Figure 3). The rearrangement processes are schematically represented in [10] "Figures 3, 5 and 11".

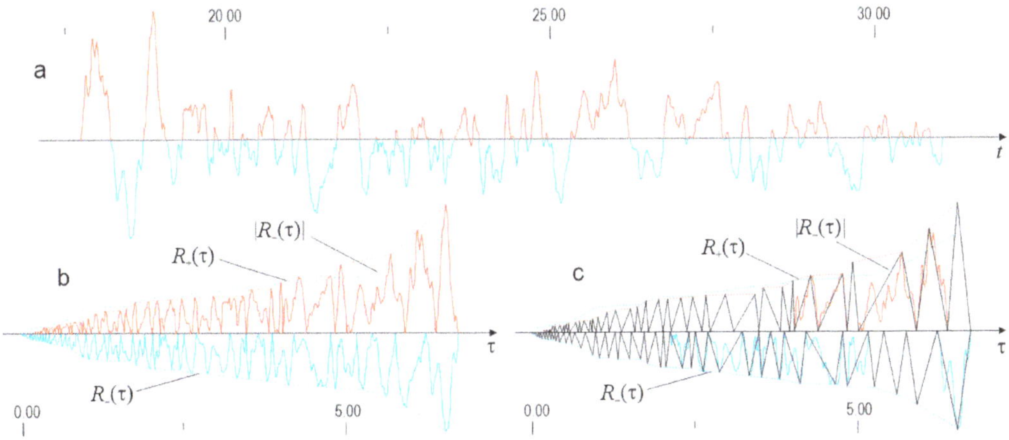

Figure 3. (**a**) To explain the idea of rearrangement (ordering) of SFs, a fragment of seismogram RSN 1616 is taken, and all SFs are separated (by colors, red line for positive SFs, blue line for negative SFs). (**b**) Positive SFs and negative SFs are ordered by increasing maximum amplitude, and positive $R_+(\tau)$ and negative $R_-(\tau)$ envelopes are drawn over their peaks as polygonal lines. Note that the duration of $R(\tau)$ is about twofold shorter than the fragment (**a**). (**c**) The same as in (**b**), where SFs are approximated by asymmetric triangles, where some SFs with complex shapes were left for comparison.

Now, it is routine to draw the R-envelope along arranged SFs, both over positive and over negative SF sequences [28,52] "Figure 3", which contains new information about seismogram. Also, the previous correlative content in the original seismogram is heavily disrupted by recomposition, but new information appeared, telling us something new about fluctuating content in the seismogram, which is much more easily perceived than from the raw one.

Therefore, the R-envelope can be designated as a new functional attribute of the actual seismogram, which also provides a basis for defining other new parametric attributes.

The (peak) envelope "hovers" over amplitude peaks along a waveform, making the shape of fluctuations irrelevant. This means it is sufficient to replace whatever complex form of the SF by only two numbers, one for peak amplitude and another for SF duration, which in geometrical terms means a triangle replacing (or approximating) the respective SF (see Figure 3c).

The replacement of a SF by the approximating triangle is explained in more detail in Figure 4. Similar idea is applied to the source time function $S(t)$ for many earthquakes having body-wave magnitudes less than 6.5, which often can be adequately represented by a simple triangle-shaped pulse, where the triangle width is related to the corner frequency and fault length [7] "article 1.05.6, Figure 20, pp. 190–182".

In the case of approximation by triangles, one can use symmetric and asymmetric forms (Figure 4) [10]. There are also other approximating shapes, but no further elaboration of this issue will be given here. The triangle approximation of any shape of SF, whatever complex, evidently facilitates numerics about the R-envelope. This is equivalent to a sort of non-uniform sampling of seismograms, where samples are peak amplitudes of SFs.

To put it more formally, an elementary fluctuation EF, within a seismogram $s(t)$, is decomposed into two single fluctuations, SF^+ and SF^-, and their waveforms $sf_m(t)$ are approximated by corresponding triangles $\varphi_m(t)$, which are derived from generative (isosceles) triangular impulse $i(t)$, scaled by respective parameters p/n (amplitude), b (duration factor), and t_m (time shift):

$$\mathrm{EF}_m = ef_m(t) \begin{matrix} \rightarrow \mathrm{SF}_m^+ \rightarrow sf_m^+(t) \rightarrow \phi_m^+(t) = \phi_m(t, p_m, \Delta t_m^+) = p_m \cdot i\,(a_m^+ \cdot t - t_m^+) \quad m = 1, 2, \cdots, N \\ \rightarrow \mathrm{SF}_m^- \rightarrow sf_m^-(t) \rightarrow \phi_m^-(t) = \phi_m(t, n_m, \Delta t_m^-) = n_m \cdot i\,(a_m^- \cdot t - t_m^-) \end{matrix} \tag{9}$$

This relation can be considered as a parametric transform of seismograms. The time-scaling factor of the triangle base is:

$$a_m = \frac{\Delta t_m}{t_0} \tag{10}$$

where t_0 is the duration of the referent (generative) triangle (see Figure 4).

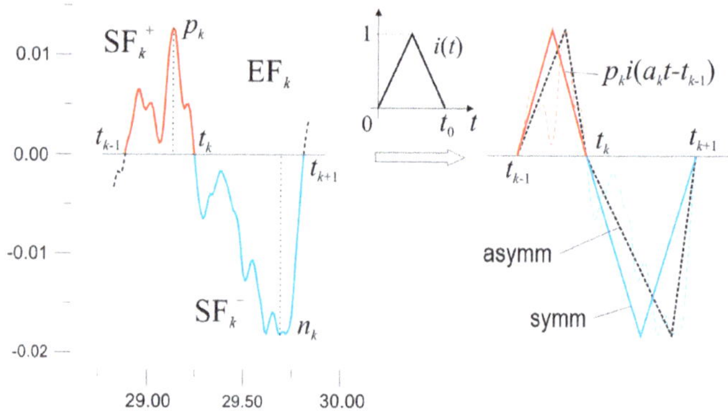

Figure 4. An elementary fluctuation (EF) is decomposed into two alternate single fluctuations (SF), positive SF$^+$ (in red solid line) and negative SF$^-$ (in blue solid line). Key parameters of EF are indicated: peak amplitudes p_k (positive) and n_k (negative) and t-axis crossings (t_{k-1}, t_k, t_{k+1}). The shape function $i(t)$, in this case the triangle impulse, approximates both SFs, renormalized by SFs parameters: peak amplitudes (p_k, n_k) and durations ($\Delta t_k{}^+ = t_k - t_{k-1}$, $\Delta t_k{}^- = t_{k+1} - t_k$). Symmetrical and asymmetrical cases of approximating triangles are presented, depending on the position of peak amplitudes of SFs within $\Delta t_k{}^+$ and $\Delta t_k{}^-$. Numerical marks on both axes show the position of this wavelet (EF) in the seismogram on Figure 2.

There is no essential difference in approximating the simple SFs or complex-shape SFs by symmetric (isosceles) or asymmetric triangles, particularly in so-called dense seismograms. Also, the plain envelope, anyway, cancels all the SF shape details beneath it, reducing the whole complex shape to one number, peak amplitude, as a peak of an approximating triangle. Would it be symmetric or asymmetric is a nuance. All said, it simplifies mathematical expressions and numerical calculations.

This does not mean the issue of approximating impulse form is not worth further research, from the aspect of e.g., pattern recognition and minimum-error description of original SF by regular shape.

2.3. Conversion of Seismogram into R-Envelope

Having a sequence of elements (e.g., numbers), which differ from each other only in size, it is possible to apply a mapping that arranges the elements from a random sequence into an ordered one, which seems to be an interesting mathematical aspect, elaborated in [10].

The first step of the procedure is the decomposition of a raw seismogram $s(t)$ into EFs and SFs, according to (9)

$$s(t) = \sum_{m=1}^{N} \mathrm{EF}_m = \sum_{m=1}^{N} (\mathrm{SF}_m^+ + \mathrm{SF}_m^-) = \sum_m^N sf_m^+(t) + \sum_m^N sf_m^-(t) \tag{11}$$

Approximating SFs by triangular impulses, as described earlier, we get (Figure 3):

$$s(t) = \sum_m^N p_m \cdot i(a_m^+ \cdot t - t_m^+) + \sum_m^N n_m \cdot i(a_m^- \cdot t - t_m^-) \quad (12)$$

Graphical and numerical Approximations are equivalent. The graphical is more apparent to humans, while the numerical is more "apparent" to machines (computers). Graphical procedure is actually a combination of graphical and numerical one, very suitable in developing computer programs.

Scale factor a_m (12) is closely related to the "pitch" [53] of seismic signals. Variations of a_m indicate a repeating rate of fluctuations.

The numerical variant of Equation (12) is a sequence of numbers:

$$S_t = \{(t_{S1}, a_1), (t_{S2}, a_2), (t_{S3}, a_3), \cdots, (t_{SK}, a_K)\} = \{(t_{Sk}, a_k)\}_{k=1}^{K=2N} \quad (13)$$

representing time shifts (t_{Sk}) and peak amplitudes (a_k) of respective triangles:

$$S_{\Delta t} = \{(\Delta t_1, a_1), (\Delta t_2, a_2), (\Delta t_3, a_3), \cdots, (\Delta t_K, a_K)\} = \{(\Delta t_k, a_k)\}_{k=1}^{K} \quad (14)$$

with durations and peak amplitudes of respective triangles.

The upper limit of index k is $2N$, since there are N EFs and $2N$ SFs.

Combining the Equations (13) and (14) we get:

$$S = \{(a_1, 0), (a_2, \Delta t_1), (a_3, \Delta t_1 + \Delta t_2), \cdots\}_{k=1}^{K} = \left\{\left(a_k, t_{Sk} = \sum_{j=1}^{k-1} \Delta t_j\right)\right\}_{k=1}^{K} \quad (15)$$

We then separate impulses with positive and negative amplitudes, making two subsets:

$$S_{\Delta t} = \{(\Delta t_k, a_k)\}_{k=1}^{K} = S_+ + S_- = \left\{\left(\Delta t_k^+, a_k^+\right)\right\}_{k=1}^{N} + \left\{\left(\Delta t_k^-, a_k^-\right)\right\}_{k=1}^{K} = P + N \quad (16)$$

The central step of transformation is rearrangement of positive and negative triangles in respective ordered sets by increasing or decreasing peak amplitudes (Figure 3):

$$S_+^{(a)} = \{a_1, a_2, a_3, \cdots, a_K\} \overset{sort}{\rightarrow} \left\{a_{(1)}, a_{(2)}, a_{(3)}, \cdots, a_{(K)}\right\} = \{p_1, p_2, \ldots, p_K\} = P \quad (17)$$

$$S_-^{(a)} = \{a_1, a_2, a_3, \cdots, a_L\} \overset{sort}{\rightarrow} \left\{a_{(1)}, a_{(2)}, a_{(3)}, \cdots, a_{(L)}\right\} = \{n_1, n_2, \ldots, n_L\} = N \quad (18)$$

where only amplitudes (superscript (a) in S) are considered.

Mathematical details describing only the rearrangement algorithm are omitted here for brevity.

To draw the peak R-envelopes, $R_+(\tau)$ and $R_-(\tau)$, over arranged triangles, positive and negative, as polygonal lines connecting triangles' peaks, we need time locations of triangles' peaks, i.e., middle points on t-axis of each triangle:

$$t_k^{(p)} = (\Delta t_k^+/2) + \sum_{j=1}^{k-1} \Delta t_j^+ \quad (19)$$

where the symmetric triangles are used. The coordinates of R-envelope are (Figure 3):

$$R_+(\tau) \leftrightarrow \left\{\left(p_k, t_k^{(p)}\right)\right\}_{k=1}^{N} \quad R_-(\tau) \leftrightarrow \left\{\left(n_k, t_k^{(n)}\right)\right\}_{k=1}^{N} \quad (20)$$

2.4. Order—Key Feature of the R-Envelope and of the Spectrum

The seismogram, labeled with $s(t)$, is a raw information about an earthquake, while its spectrum is a chief seismogram's functional attribute, usually given in earthquake databases, e.g., [33]:

$$S(\omega) = \int_{-\infty}^{\infty} s(t)\, e^{-i\omega t} dt = F[s(t)] \quad s(t) = \frac{1}{2\pi}\int_{-\infty}^{\infty} S(\omega)\, e^{i\omega t} d\omega = F^{-1}[S(\omega)] \tag{21}$$

Quite a decent survey of FT, elaborate but brief, can be found in [54]. Being the Fourier transform of seismic signal $s(t)$, the spectrum $S(\omega)$ is fully equivalent but not equal to $s(t)$. Another notation, s and $\hat{s}$, emphasizes their equivalence:

$$s(t) \to \hat{s}(\omega) = s(t) \overset{F}{\to} \hat{s}(\omega) = (Fs)(\omega) = \int_{-\infty}^{\infty} \left[e^{-i\omega t} dt \cdot s(t)\right] \tag{22}$$

Since $s(t)$ and $\hat{s}(\omega)$ are equivalent but not equal, they do not contain the same quantity of information (Q)

$$Q[s(t)] > Q[\hat{s}(\omega)] \tag{23}$$

To get $s(t)$ from $S(\omega)$, one must add sinusoids (through the F^{-1}) to the spectrum values:

$$s(t) = \frac{1}{2\pi}\int_{-\infty}^{\infty} S(\omega)\, e^{i\omega t} d\omega = \int_{-\infty}^{\infty} \left[\frac{1}{2\pi}|S(\omega)|d\omega\right] e^{i[\omega t + \theta(\omega)]} = \int_{-\infty}^{\infty} \left[\frac{1}{2\pi}|S(\omega)|d\omega\right] \cdot \cos[\omega t + \theta(\omega)] \tag{24}$$

which are previously removed time components in operation (21).

Whatever the "raw" domain, where one originally observes certain physical phenomena (time, length, angle, etc.), for all of them, the transformed domain will always be the frequency, or the ω-domain, because the Fourier operator itself contains sinusoids, either in complex ($e^{i\omega t}$) or in plain ($\cos\omega t$) forms (see Equation (24)), which the most characteristic parameter is the frequency [55].

It might seem pretty obvious, but the ω-domain (ω-axis) is arranged by frequency. The same stands for the t-domain (the t-axis). In both cases, there is a continuous increase of numeric values along both axes. What makes the important difference between these two arrangements is the fact that a point on the ω-axis is not a simple numerical quantity, as on the t-axis—it is the basic parameter of a sinusoid component—its frequency. Figuratively, each point at the ω-axis hides a sinusoid waveform of assigned frequency since the term frequency corresponds only to sinusoid shape so that one can ascribe arrangement of sinusoid shapes along the ω-axis, with respective frequencies, and make a parallel to the arrangement in the R-envelope, which is, however, simpler and more obvious.

Moreover, there is a general tendency in the spectrum (amplitude or power, not in phase), nevertheless, its complexity, to steadily decrease with frequency, see, e.g., [54], so that one can treat the ω-axis as the spectrum's horizontal asymptote. In similar way, the R-envelope not only steadily, but monotonically approaches the t-axis. This issue of R-envelope depends on several, mostly practical factors, and will be considered in Section 4.

Although the R-envelope proposed in this paper, seems to be a continuous function (functional attribute), it is actually a group-parametric attribute (set of parameters) since it abstracts real shapes of SFs, and relies on peak amplitudes and durations of SFs, i.e., "equivalent triangles". So, it is not possible to reconstruct the original waveform of a seismogram just from this envelope (see Figure 3). A similar fact holds for the spectrum, which parameters need adding sinusoid waveforms to create the original signal.

Also the spectrum can be represented as a group-parametric attribute, as the line spectrum, given by the Fourier series. If a seismogram is formally periodized [56], the result is a line (discrete) spectrum. Having on mind the general feature of any spectrum to decrease with frequency, it can be truncated to extract its significant part, and to yield a (discrete) finite set of parameters describing the seismogram. It is essentially numerical compression of seismograms, i.e., from huge recording files (thousands of numbers), having thousands of numerical values to a set of several ones, much more suitable to manipulate with. All that being said, it is obvious how organized information is more compact than in its raw form. Implications for the R-envelope are apparent.

Similarly, the R-envelope can also be considered as numerical compression of the seismogram, because it represents the impulses, i.e., SFs, by real numbers or even integers. Yet, it still contains remnants of raw structural information.

2.5. Attributes Based Directly on EFs/SFs

The original seismogram looks identical to its parsed version. Nevertheless, the latter one provides additional degree of freedom for manipulations with the waveform, just because it is segmented. There is no need at once to go up to rearrangement to get the R-envelope, but to apply more simple operations on EFs/SFs to attain some attributes.

One idea is simply to shift the EFs along the *t*-axis and overlap them to coincide their inflection points (Figure 5a,b).

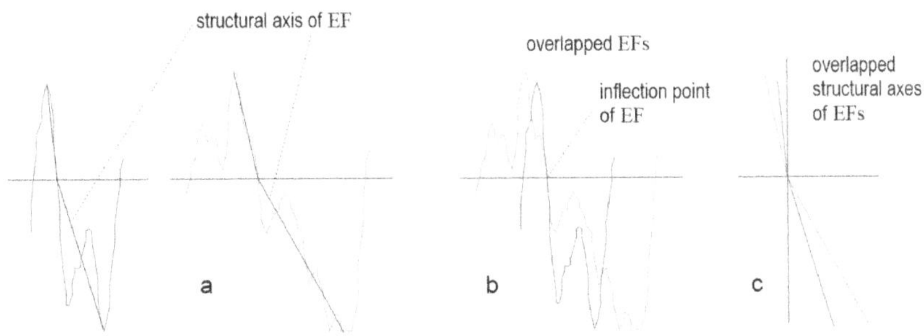

Figure 5. (**a**) Graphical explanation of the idea of EF structural axis: the two-segment line connecting the peaks and *t*-axis crossing skeletonizes entangled shape of EF. (**b**) Already two overlapped EFs, whose inflection points coincide, indicate what would look like a picture of tens of EFs. (**c**) Overlapped structural axes make the representation on (**b**) much more transparent. Different EFs are represented by dark red and green solid lines in (**a**) to distinguish them in (**b**), and their structural axes in (**c**).

The same picture could be attained with the SFs, but with the SFs, one should use all of them to create a proper picture, while with the EFs, it is possible to get the contours of the final picture with a sample of limited size.

At last, the EFs is a more natural form of earthquake acting, as is said above, reflecting the very basic form of change in the seismogram—the 'push-pull' action, more closely to the pattern in which an earthquake acts on structures. On the other hand, the SFs are slightly artificially separated, to give additional room for manipulations.

It is not difficult to comprehend the picture resulting from overlapping of all or from a considerable number of EFs.

The pretty entangled mesh of overlapped EFs can show a certain average picture but is very blurred. More transparency is attained by introducing, let us call it, the structural axis of the EF, as explained in Figure 5a,c. This axis connects three characteristic points of an EF: both peaks and zero crossing (the inflection point). Note that the term 'inflection point' refers to the function, which would formally describe the actual EF. All these are point-like, pretty local features, so the structural axis form highly depends on the shape of EF, which is very variable.

A more adequate feature would be the slope (tangent) through the inflection point (where the EF crosses the *t*-axis) (see Figures 6a and 7). Notice the difference between ascending (blue lines) and descending (red lines) tangents, which shows a kind of statistical asymmetry of SFs.

It determines the maximum change of mechanical variables acting in the site. It has something to do with the transfer of earthquake force to structures. Let us remember the simple thing that when one says 'earthquake,' it is thought of as seismic effects in the site, not in the hypocenter.

The said tangent, although the most characteristic of all tangents along the EF function, does not reflect well this energy transfer, particularly in the case of complex EF shape

(Figure 6). This tangent is also a local feature, standing in the middle of the complex shape of the EF, but it is still a peripheral description of force transfer from seismic waves to the structure.

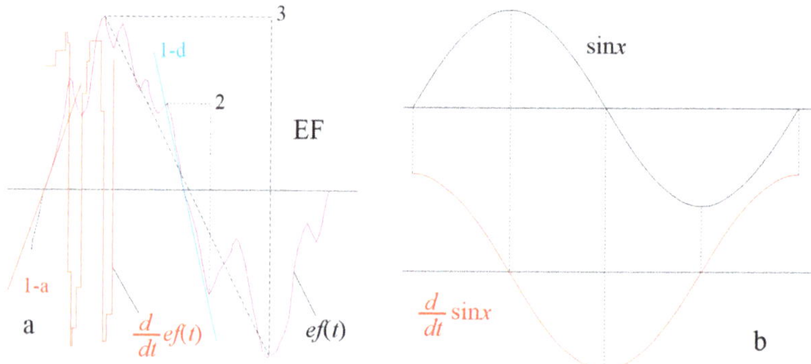

Figure 6. (**a**) The function $ef(t)$ formally describes the EF. The slope of the tangents drawn through the inflection points is depicted by blue solid line (1) and red solid line (1-a). The slope of the EF change is defined by the triangle (2) as a ratio of two increments (amplitude and time), determined by two locally prominent shape details, which define the slope nearest to the case (1). Since the EF shape is complex, the slope (3) defines the change in different ways, which includes maximum amplitudes of SFs. The sharp changes in the first derivative of $ef(t)$ (red solid line) indicate the rate of energy transfer carried by EF waveform, quite opposed to the monotonic change od sin$x$ derivative, given in (**b**).

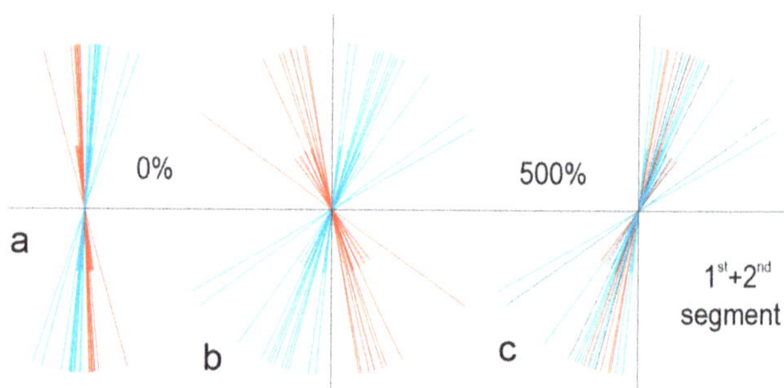

Figure 7. (**a**) Bundle of tangents, taken in inflection points of selected EFs from the 1st and the 2nd segments of seismogram RSN 1616 (Figure 2) and overlapped to coincide their middle points. Blue lines show tangents from parsing zero crossings between subsequent EFs, while blue lines show tangents at zero crossings within the EFs (consult Figure 6a). (**b**) The same as in (**a**) but with the scale on the t-axis stretched out for 500%. (**c**) The same as in (**b**) but with all tangents gathered in one bundle.

In order to take the shape of the EF into consideration, which actually carries the influence, it is more proper to speak of energy transfer since the EF shape itself shows the energy distribution contained in it.

There, one can define the instantaneous energy transfer carried by a single EF. In other words, it is the instantaneous average power, as the energy contained within the EF divided by its duration Δt:

$$p_\Delta = \frac{1}{\Delta t}\int_{\Delta t} EFdt = \frac{1}{\Delta t}\int_{\Delta t} ef(t)dt \quad (25)$$

where the label *EF* symbolically designates the function describing the EF shape (see (9)). Of course, it is a rough measure of energy transfer. It is well known from dynamics that the inertial forces and impulse loads depend on d/dt (Figure 6).

The first derivative of the EF shows the essential difference in the transfer rate, particularly when compared with the sin function, which provides a mild change without discontinuities (Figure 6). The earthquake displays sharp changes, where the flection points are connected with almost linear slopes, quite opposite to sinusoid monotonic change.

Until now, the transfer rate is considered for a single (individual) EF isolated from others. The EFs never act separately but in bursts. The fact that points out the burst influence of EFs on structures are resonant characteristics of structures, having modes of vibration. Thusly, it is necessary to analyze the repetition pattern of EFs.

2.6. Attributes Based on Rearrangement of EFs/SFs

One of a few of the previous attributes was also based on the rearrangement of EFs/SFs, but it was not placed in this subsection because the rearrangement was simple, involving only gathering of EFs at one place, with coinciding their inflection points. Rearrangement as ordering (sorting) is a little 'heavier' operation, which enables a more potent basis for creating new attributes.

The first one is the R-envelope, for which one could even say that it is not only an attribute but also a basis for qualitatively new attributes.

2.6.1. Issue of Seismogram Symmetry

Formally considered, the seismogram fluctuates around the *t*-axis. Physically, the *t*-axis is a reference, around which seismic waves compress or extend the medium. It would be interesting to examine seismogram symmetry with respect to the *t*-axis. The files used for this purpose are recordings of acceleration.

The simplest method to do that is to compare peak envelopes, $A^{+}(t)$ and $A^{-}(t)$, drawn over positive and negative peaks of seismogram waveform, i.e., over peaks of EFs (Figure 2). The difference between these two plain envelopes apparently exists, but they are very indented for a reliable visual assessment.

A more regular picture of symmetry is attained by the R-envelope, both for positive and negative SFs, see Figure 3, and see [10] "Figure 20". As can be seen, there are small differences between the envelopes $R_{+}(\tau)$ and $R_{-}(\tau)$ so that one can speak of an invariant on the population of seismograms [10]. Additionally, one can complain about the small variation range to be useful for qualifying and quantifying the difference between the two envelopes in order to describe the seismogram symmetry.

The cause is simple: rearrangement of SFs equalizes differences, putting similar SFs face-to-face. The seismogram conveys the influence right by its waveform, not by arranged SFs, but by a sequence of EFs. Finally, the chief factor by which the earthquake acts on the environment are transitions, which means that a more proper structural element would be the EF than the SF.

It seems that it would be more proper to create the R-envelope by arranging the EFs, but it is not simple as in the case of SFs, since the EF is ambiguous, having two randomly different maximum amplitudes, and makes the ordering by peak amplitude uncertain.

Then, the better criterion for ordering would be the EF energy: firstly, to calculate the energy of each EF, then to order the EFs by increasing/decreasing energy, and then to draw the R-envelope over maximum amplitudes of respective positive and negative peak amplitudes. This approach is the square-linear combination. Of course, such an envelope would not be labeled by R, but by, e.g., R_E (with E for energy).

A linear-linear version of this energy approach is described in Figure 8. The EFs are ordered twice: once by positive peak amplitudes of EFs (not SFs) and then by negative ones. Each time, the correspondent R-envelope is drawn over the peaks by which the sorting is performed (Figure 9).

The quantized durations Δs make the time variable τ irrelevant, so a new index variable n can be introduced, and all the envelopes from Figure 9 are rewritten as $R_Q^+(n)$, $N_Q^+(n)$, etc.

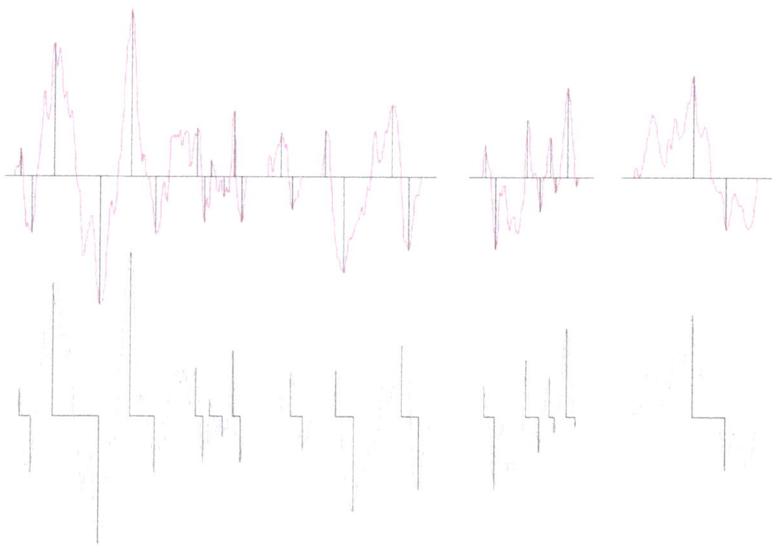

Figure 8. (**up**) Several EFs are taken from Figure 3, and for each one, peak amplitudes of belonging EF^+ and EF^- are drawn. (**down**) The base points of peak amplitudes of each EF are connected by a horizontal line to make a skeleton of EF, in which the horizontal line represents an effective duration of EF. Notice their variability. Dim gray solid lines just indicate the EFs shapes from upper figure.

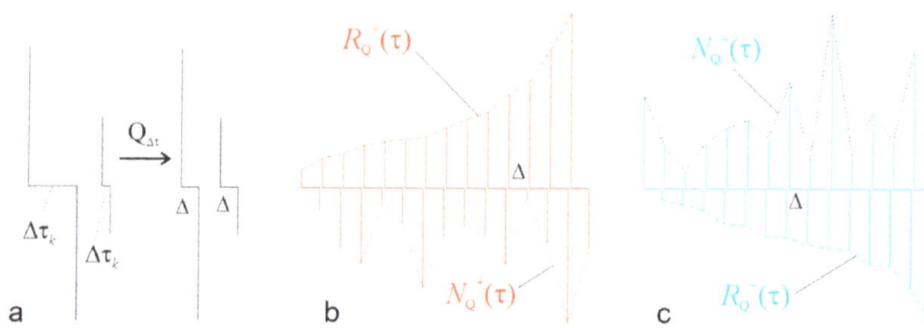

Figure 9. (**a**) The duration of each EF skeleton from Figure 8 is quantized to Δ, keeping peak amplitudes unchanged. (**b**) All EFs skeletons (red solid lines) are arranged by positive peak amplitudes, giving quantized positive R-envelope, labeled as $R_Q^+(\tau)$ (dashed red line). The envelope drawn over the negative peak amplitudes produces random envelope $N_Q^+(\tau)$ (point dashed red line). (**c**) Similar as in (**b**) but with ordering the EFs by negative peak amplitudes.

Two key elements should be noted in Figures 8 and 9. Instead of the structural axis of the EF, already introduced in Figure 5, here, one deals with the skeleton of EF, graphically explained in Figure 8. The other one is the quantizing operation over EF skeleton durations (Figure 9a), and its role is apparent—to minimize the influence of EF shape variability on the result. Consequently, after the quantization, the creation of the R-envelope is meaningless, but not the creation of the quantized envelope labeled by R_Q.

The positive $R_Q^+(\tau)$ is monotonic and topologically identical to the corresponding $R^+(\tau)$ envelope, i.e., the R^+-envelope. The reverse of the $R_Q^+(\tau)$ is the $N_Q^+(\tau)$, which clearly shows the asymmetry of ordered EFs. The similar holds for $R_Q^-(\tau)$ and $N_Q^-(\tau)$.

2.6.2. R-Envelope and New Definition of Seismogram Duration

The duration of the seismogram is a very important attribute. Some of the proposals were quoted in Section 1.

The usual and widespread definition stems from the Husid diagram, for $D_{5\text{-}75}$ and $D_{5\text{-}95}$. Both intervals, determined by respective limits, seem a little arbitrary as a convention. Husid diagram is quite objective functional attribute of a seismogram—it still contains the seismogram's waveform, although in cumulative appearance. Yet, the limits 5%, 75%, and 95% do reflect the subjectivity, where a human, more according to its biased tendency to choose rounded numbers (for example, 5, 10, etc. which has to do with the number of fingers) than in regard to mathematical objectivity, has chosen the rounded numbers. Simply, why exactly 5, but not e.g., 4.4 etc.?

With the R-envelope, one gets a more objective platform to define the seismogram duration. One limit of the seismogram's duration is pretty well defined—it is the highest fluctuation (Figure 3b,c). Another limit of duration is placed somewhere along the monotonic tail of the arranged seismogram (see Figure 3).

The other limit can be defined if one considers peaks of SFs a in rearranged seismogram as the scatter plot. These points are also nodes of the other R-envelope. Linear regression, as the simplest form of regression analysis, is a universal statistical tools that can reveal hidden linear rules behind a scatter plot (diagram).

When a regression line through the "cloud" of the SFs peaks is calculated, one gets a simple geometrical description of the whole rearranged seismogram (Figure 10).

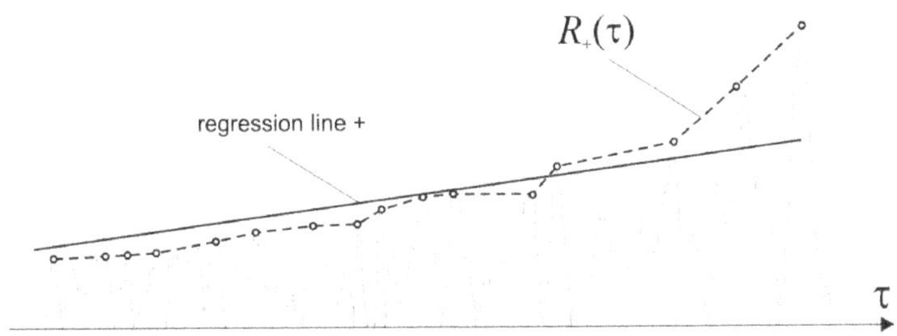

Figure 10. Peak amplitudes of SFs, approximated by triangles, can be simply considered as points of certain scatter diagram, where each of them has its coordinates (x, y). A regression line can then be calculated to show the trend existing in the cloud of points.

3. Results and Analysis

3.1. Attributes from Gathered Fluctuations

Whatever the shape of EF/SF could be, two general categories can be distinguished

- simple, spike-like;
- complex;
- the whole range of shapes between.

To illustrate these typical shapes and for the purpose of this research, a characteristic seismogram is selected, which contains miscellaneous EFs/SFs. It is RSN 1616 (Duzce (Düzce) 1999, Turkey, Lamont 362), with main parameters are: D5-75 14.6 s, D5-95 20.5 s, Mag 7.14, Strike Slip, RJB 23.41 km, Rrup 23.41 km. We used the file DUZCE_362-N.AT2, Horizontal-1 Component time series, taken from the PEER database [33]. Typical examples of said SE/EF categories, taken from RSN 1616, are shown in Figures 1, 3 and 8.

The categorization can be spread over approximating shapes, In this case of triangles, symmetric and asymmetric (Figures 3 and 4). There are also other approximating shapes, but no further elaboration of this issue will be given here.

Displacement of parsed EFs to the same place to coincide with their inflection points creates a system of overlapped EFs, which then show some statistical relations (Figure 11a).

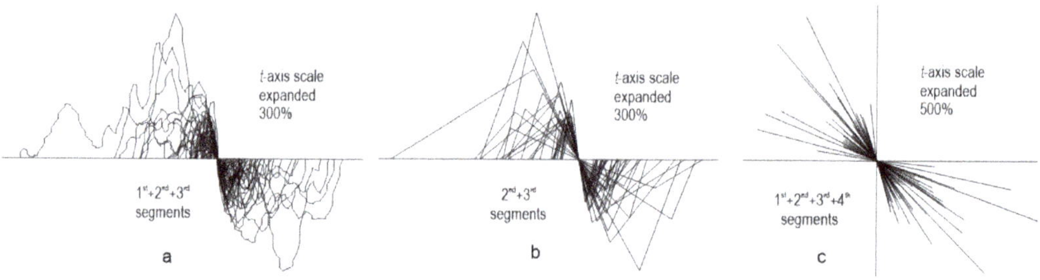

Figure 11. (**a**) The EFs from the first three segments from Figure 2 are overlapped with coinciding inflection points. (**b**) The peaks of SFs within EFs depicted in (**a**), are approximated by triangles, can be simply considered as points of certain scatter diagram, where each of them has its coordinates (x, y). Then a regression line can be calculated to show trend existing in the cloud of points. (**c**) The structural axes of EFs (see Figure 5) make the scatter plot from (**b**) more transparent.

One can think that the mesh is too entangled to show some new information about the content of seismograms, but sometimes even the wood speaks about the trees.

This is for intrinsic EFs. A similar picture can be created from approximated EFs, i.e., $EF^{(A)}$s, which is shown in Figure 11b. Approximation eliminates many unnecessary details of EF shapes, making the whole picture clearer. What one can easily perceive from the 'ordered' mesh of triangles are their peaks and slopes.

One can go further in the direction of mesh clearness by using structural axes of EFs (Figure 11c), whose bundles are completely equivalent to Figure 11b but evidently more transparent.

3.2. Attributes from Rearranged and Ordered Fluctuations

The results of simple manipulations with the EFs/SFs are not only constrained to seismogram waveform but can even reach into the spectral domain.

As can be seen in Figure 2, it is a simple matter to measure durations of SFs. It was performed for SFs within two intervals: for the 1st segment (Figure 12a) and for the 2nd + 3rd segments (Figure 12b).

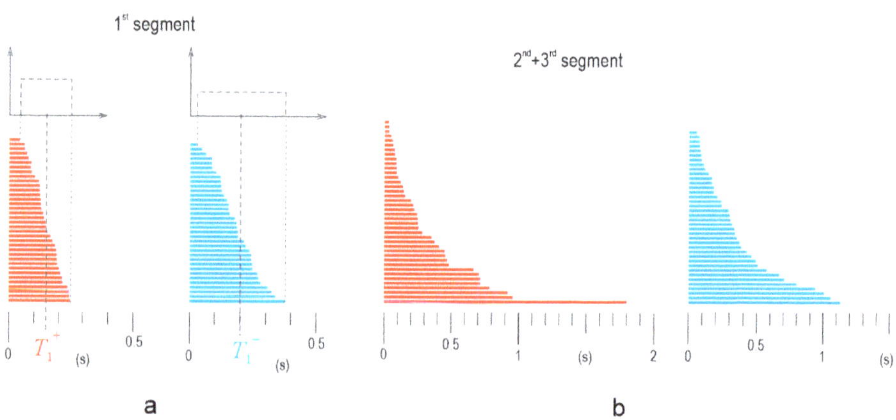

Figure 12. (**a**) The durations of SF^+ (red solid line) and SF^- (blue solid line) within 1st segment in Figure 2 are measured and ordered by length (in seconds). In both cases, they show almost linear change and are approximated by uniform probability distributions (depicted by a dashed black line), with $T_1{}^+$ and $T_1{}^-$ as expectations, respectively. (**b**) The same as in (**a**) but in 2nd and 3rd segments.

In the first case (the 1st segment), the durations of SFs, separately measured for SF^+s and SF^-s, are vertically justified and ordered by the length. Almost linear dependence suggests uniform distribution (of probability), indicated by dash line rectangles in Figure 12a, both for durations of positive and negative SFs. The mean (expectation) of positive SFs (red color) gives $T_1^+ = 0.15$ s. For the negative SFs (blue color), it is $T_1^- = 0.2$ s. Taking into account that the natural form that better corresponds to the sinusoid shape is the EF rather than the SF (see Figure 13), one can figure out that the mean period of fluctuations within the 1st segment is $T_{mean} = 0.15 + 0.2 = 0.35$ s.

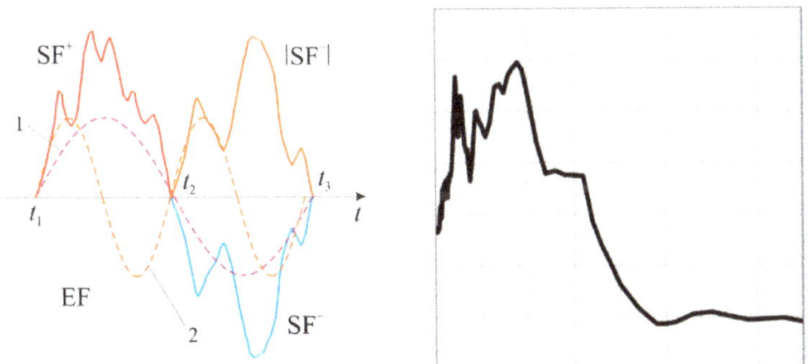

Figure 13. (**left**) The EF consists of SF^+ (red solid line) and SF^- (blue solid line). The period of its basic sinusoid 1, dark red dashed line, is $T_E = t_3 - t_1$. With the SF^- rectified (modulo operation), the two SFs from the EF are one next to the other, having the basic sinusoid 2 (orange dashed line) with the period $T_S = t_2 - t_1$. The T_E and T_S differ in frequency (period) approximately by the factor 2. It is an additional reason to consider EFs in assessments of local features of the seismogram spectrum. (**right**) The spectrum of RSN 1616, taken from the PEER database, shows characteristic peaks. Both axes are with linear scales, the abscissa for the period (in s), and the ordinate is for pSa (in g), with scale increments Dx = 0.5 s and Dy = 0.02 g.

When Figure 11b is concerned, there are more interesting relations. The 1st segment is apparently uniform in regard to fluctuation patterns. In the other interval, which comprises the 2nd and 3rd segments, the SFs are much more versatile, both in amplitude and duration, not even to speak about SF shapes. Otherwise, the other interval is more representative, statistically, than the first one.

3.2.1. R-Envelope—Real Cases

Combining the principle, shown in Figure 3b,c, and a real case, seismogram RSN 1616 shown in Figure 2, the result of the whole procedure consisting of parsing into SFs, approximating the SFs by (asymmetric) triangles, ordering the SFs by peak amplitude, is represented on Figure 14.

The R-envelope is a new characteristic of the seismogram. One could also call it the new waveform or the profile waveform of seismogram. It is a kind of seismogram's transform, although not so strictly defined as, e.g., the Fourier transform. Considered as a "kind of transform", it can also be considered as a "kind of spectrum", relying this claim on its arranged structure. Speaking in terms of attributes, it can be called the functional attribute of the seismogram.

The R-envelope, in the notation $R(\tau)$, is formally a function of (relative) time since it is created over structural elements of the seismogram, the SFs, which brought the time with, either in raw form, the intrinsic SFs, or in approximated form, labeled by $SF^{(A)}$s. If the time is quantized, as in Figures 8 and 9, it becomes irrelevant, and the $R(\tau)$ becomes a discrete function $R(n)$, with a neutral indexed variable, labeled by $R_Q(n)$ and called R_Q-envelope with the letter Q to assign the quantization (see [10] for further details).

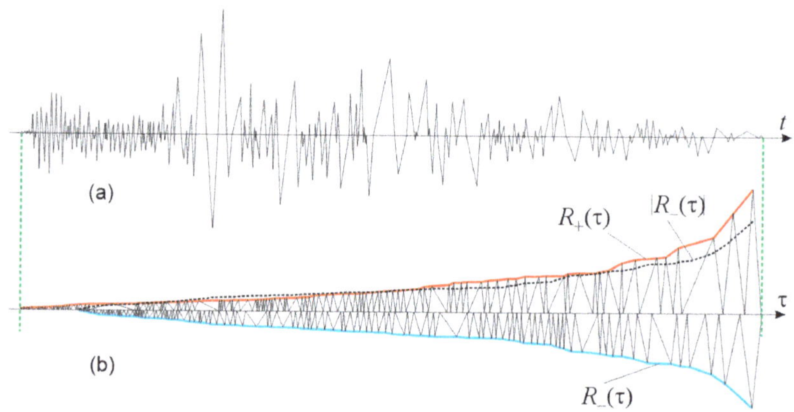

Figure 14. (**a**) The raw seismogram RSN 1616, shown in Figure 2, is segmented into positive and negative SFs, which are approximated by asymmetric triangles (see Figure 4). (**b**) After rearrangement (ordering) of SFs, separately for positive and negative SFs, positive (red solid line) and negative (blue solid line) peak envelopes are drawn labeled by $R^{+}(\tau)$ and $R^{-}(\tau)$, respectively. By the black point line it is depicted the modulo of $R^{-}(\tau)$ for better comparison between the two. The diagram (**b**) is depicted with 200% stretched scale than the (**a**), which is indicated by the green dotted vertical lines.

Following the reference [10], the R_Q-envelopes of several earthquakes are calculated as a family of functions presented on the same diagram (Figure 15).

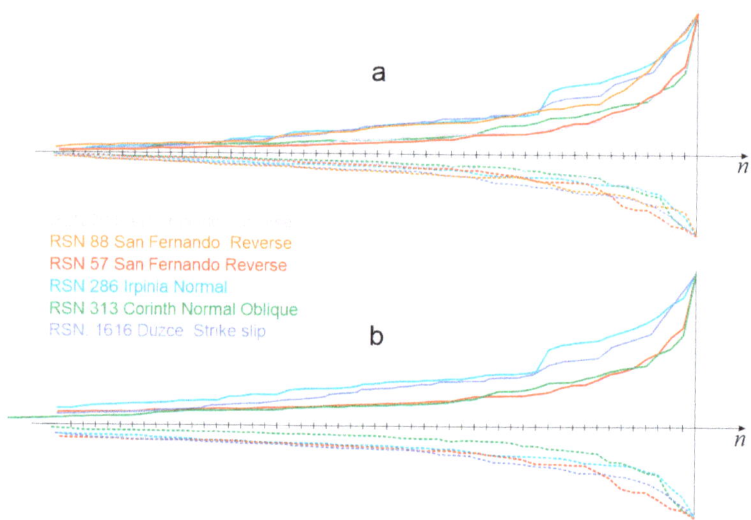

Figure 15. For the list of seismograms given between diagrams (**a**,**b**), an ensemble of R_Q-envelopes, both positive and negative, is represented as the family of functions. Colors of event names are correspondent to lines colors. All seismograms are taken from the PEER seismic database. Individual diagrams are normalized in certain aspects and overlapped. The colors of curves on diagrams correspond with the colors of event names in the list. (**b**) The R_Q-envelopes from indicated seismograms are only overlapped and justified by maximum amplitudes without normalization. (**a**) The R_Qs are normalized by maximum amplitudes individually, where tails are additionally normalized by the smallest tail.

The similarity between R_Qs in Figure 15a strongly suggests the existence of invariant in the population of seismograms (see about this issue in [10]). This similarity does not end with the R_Q-envelope. If one addresses the diagrams in Figure 12, there a striking resemblance to the R_Q-envelopes in Figure 15a can be noticed. The previous claim that the R/R_Q-envelope is an invariant in the population of seismograms can be applied also to distributions in Figure 12b, no matter if there is just one diagram for inferring. It is not difficult to show that these two invariants are actually only the aspects of the fundamental one. In other words, there is a close correlation between these two invariants, i.e., attributes.

Such correlation has a simple explanation from a statistical point of view. The longer the duration of the SF, the bigger is the expected peak amplitude of the SF. Therefore, the diagram of SFs duration is statistically mapped R-envelope.

3.2.2. R-envelope Difference Function and Seismogram Symmetry

A plain envelope is well known from amplitude modulation (AM), with a continuous high-frequency sinusoid carrier, in which positive and negative impulses are symmetrical in respect to abscissa (t-axis) so that it is sufficient to draw only an envelope over positive impulses, knowing that its negative counterpart is symmetrical. Moreover, circuits in AM receivers deal only with positive envelopes.

Randomly alternating processes, such as speech and seismic signals, do not have symmetry in respect to abscissa, so it is natural to define a negative envelope as well. Asymmetry is preserved during seismogram rearrangement hence the envelopes $R_+(\tau)$ and $R_-(\tau)$ are not equal (see Figures 3, 14 and 15), and it is quite logical to define the difference function (Figure 16)

$$\varepsilon(\tau) = R_+(\tau) - R_-(\tau) \tag{26}$$

as a measure of seismogram asymmetry. The $\varepsilon(\tau)$ is a functional attribute and provides an asymmetry profile of the seismogram.

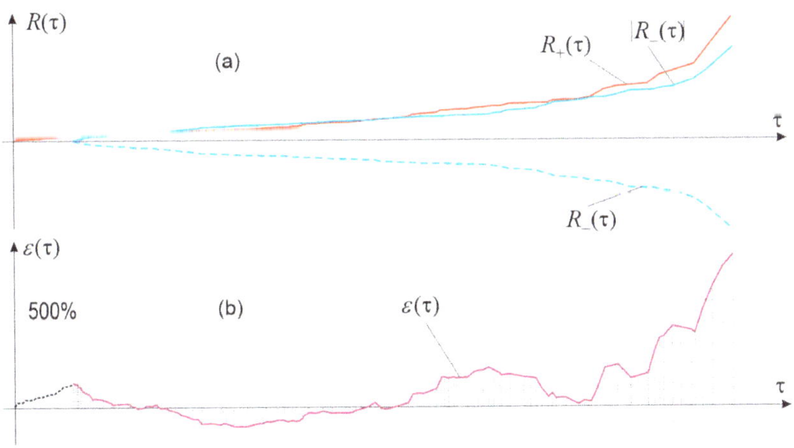

Figure 16. (**a**) The R-envelopes, positive $R^+(\tau)$ (red solid line), and negative $R^-(\tau)$ (blue dashed line), of seismogram RSN 1616 are drawn on the same diagram. For comparison, the modulo of $R^-(\tau)$ is drawn by a blue solid line. (**b**) The difference function of two R-envelopes is drawn with a 500% extended ordinate scale for better insight in details.

One-parametric measure (attribute) of asymmetry is defined simply as

$$A_\varepsilon = \int_0^{T_0} |\varepsilon(\tau)|\, d\tau \tag{27}$$

while cumulative functional attribute, analog to the Husid diagram

$$A_\varepsilon(\tau) = \frac{\int_0^\tau |\varepsilon(\tau)|\, d\tau}{\int_0^{T_0} |\varepsilon(\tau)|\, d\tau} \tag{28}$$

provides a better picture of asymmetry distribution along seismograms. Note that the modulus is used instead of the square to avoid mutual canceling of positive and negative amplitudes in $\varepsilon(\tau)$.

Additional examples of the difference function is represented in Figure 17. In these cases, the R_Q-envelopes are compared, giving ε_Q-functions.

$$\varepsilon_Q(n) = R_Q^+(n) - R_Q^-(n) \tag{29}$$

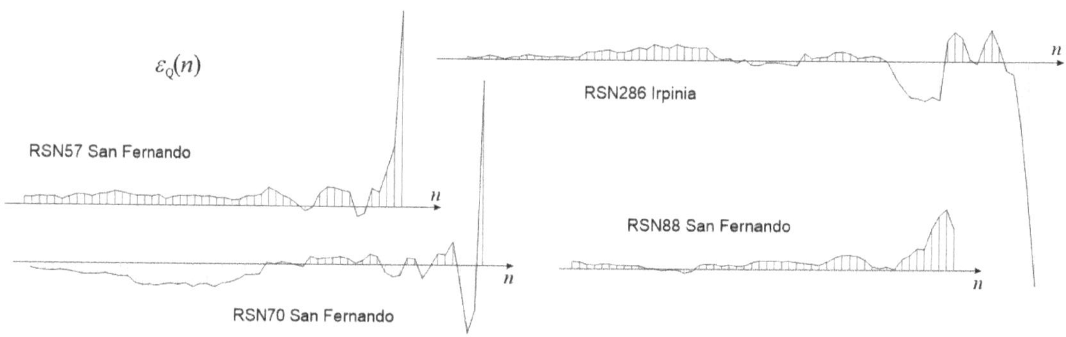

Figure 17. Variations in each difference function $\varepsilon_Q(n)$ indicate the asymmetry profile of the assigned seismogram, given by its RSN. Each difference function is drawn with a 500% extended ordinate scale for better insight. Red lines designate positive differences between corresponding R_Q-envelopes, and blue lines stand for negative differences.

The three cases in Figure 17, RSN57, RSN70 RSN88, belong to the same event, San Fernando 1971, magnitude 6.61, recorded at the sites Castaic—Old Ridge Route, Lake Hughes #1, and Santa Felita Dam, respectively (further details see in [33]), show certain similarities. On the other hand, the similarities between RSN57, RSN79, and RSN286, with polarity excluded, seem to be closer than those between RSN57, RSN70, and RSN88. This suggests the necessity for further research before inferring any more founded claims. Additional cases for study can be found in [10]. In [10], the terms 'head' and 'tail' are used to describe different variability in the diagram of the R/R_Q-envelope. It is much more transferred to the diagram of $\varepsilon_Q(\tau)$.

Another attribute that addresses the issue of seismogram symmetry proposed in Section 2.6.1 (Figures 8 and 9), applied on RSN1616, is presented in Figure 18, and defined by the difference function of random envelopes $N_Q{}^+(n)$ and $N_Q{}^-(n)$, called the random difference function:

$$\eta_Q(n) = N_Q^+(n) - N_Q^-(n) \tag{30}$$

3.2.3. Seismogram Duration Based on Ordered SFs

The Section 2.6.2 explained how the peaks of ordered SFs form a scatter diagram for the calculation of linear regression to find out the seismogram duration (see Figure 10).

A seismogram is separated into two sets of SFs, positive and negative ones which means that one gets two scatter plots from peaks of positive and negative SFs. The elongated distribution of points in these scatter plots simply appeals for the calculation of linear regression. Therefore, one gets two regression lines: regression line + and regression line −. Because of the peak amplitude increase along the sequence of ordered SFs, the regression line + is inclined to t-axis and crosses it at a certain point ($T_1{}^+$ in Figure 19). The similar thing is for $T_1{}^-$.

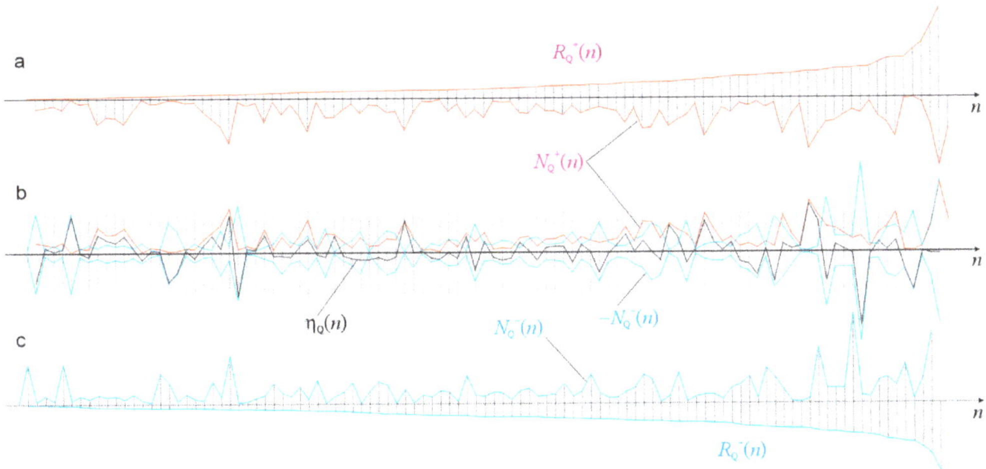

Figure 18. (**a**) The regular R_Q-envelope and the random N_Q-envelope formed on ordered EFs over positive SFs. (**c**) The same as in (**a**) but over negative SFs. (**b**) The random difference function $\eta_Q(n)$, depicted by the black solid line as the geometrical difference between red and blue solid lines, shows the random asymmetry of the seismogram but given by ordered EFs.

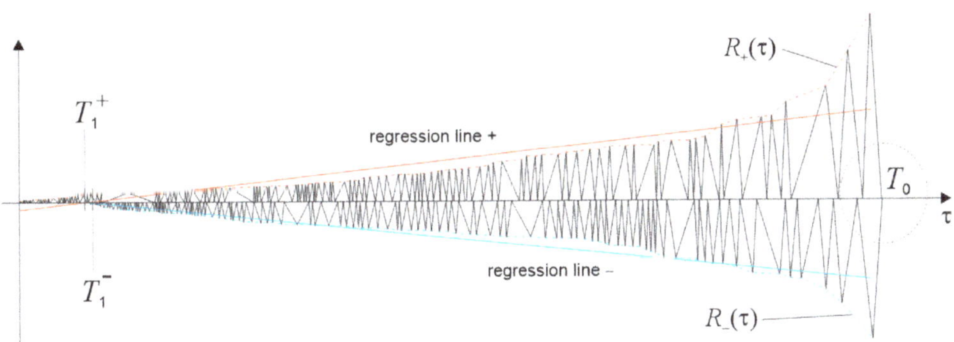

Figure 19. The nodes of R -envelope's polygonal line, i.e., peaks of approximating triangles, can be considered as scatter plots, both for positive and negative SF envelopes. Calculating linear regression for both means putting a line through each of the scatter clouds. These are regression line + and regression line −, which cross the τ-axis at points $T_1{}^+$ and $T_1{}^-$. In this way, we get the effective duration of each R-envelope SF sequence, in combination, yields the duration of the seismogram, which stems from the whole waveform.

The duration, T_D, is determined simply as:

$$T_D = \left(T_0 - T_1^+\right) + \left(T_0 - T_1^-\right) = 2T_0 - \left(T_1^+ + T_1^-\right) \tag{31}$$

A very apparent attribute (general measure) of seismogram's symmetry can be defined directly from Figure 19 as a difference in overall durations of SF$^+$s and SF$^-$s:

$$A_\Delta = \frac{T_1^+}{T_1^-} = \frac{\sum_k \Delta_k^+}{\sum_k \Delta_k^-} \tag{32}$$

4. Discussion

Some issues were partly discussed in previous sections, for the sake of reading continuity. The expression 'seismogram rearrangement from the article title, not ordering or sorting, implies the previous decomposition. The very seismogram indicates elementary repeating pattern, although random, which is named the elementary fluctuation, further divided into a pair of single fluctuations. It is a fundamental concept.

The concept of elementary/single (EF/SF) fluctuation proved itself as fruitful, not only as a base for defining several new attributes, but for the further study of EF forms, particularly of the complex or composite EFs. The decomposition is a prerequisite for recomposition.

4.1. *Mathematical and Physical Meaning of Decomposition*

The seismogram can be formally considered as a mathematical function $s(t)$. Decomposition into EFs/SFs means some mathematical operations, which are actually numerical, since the seismograms are available as numerical files (of some 5000 to 10,000 thousand numerical values).

The proposed approach involves two-stage processing: the prior and the main one. The prior is carried out once, ad forever. Once completed, it reduces the numerical heap of several thousand numbers into a couple of hundreds of number pairs (max amplitude and duration of SFs). This is a high-level numerical compression. Now comes the main processing—it is much easier to manipulate with this small set of numbers, producing various attributes of a seismogram.

The numerical reduction is not the main advantage of the proposed approach. A more important aspect is qualitative: the very fact of parsing (segmentation). Now, one has at disposal the basic structural elements (EFs and SFs) of seismogram and can easily manipulate them, creating unique and specific combinations, i.e., attributes, not found elsewhere, since the other attributes have another definition basis.

An earthquake does not act on a structure at once, as a whole, but sequentially, EF by EF. On the other side, the structure shows its own resonant characteristics, given as vibration modes. These are complex form damped vibrations (composite oscillations of mutually connected structure elements (beams and columns, a real small internal 'earthquakes', as responses to outer excitation. Damped vibrations (from pulse excitation) take some time to dwindle, which can be considered as the memory of structure. The arrival rate of exciting elements in an acting earthquake (at the location of the considered structure) and the resonances of the subjected structure create a complex interference. The concept of the EF, more than the SF, because it is of a more natural shape to reflect the 'push-pull' pattern of excitement, provides an opportunity to compose characteristic local (short) sequences of EFs and to examine structure response instead of applying the whole seismogram.

4.2. *Attributes from Decomposition-Recomposition*

In this paper, the widest definition of seismic attribute is accepted, as any information of interest that can be derived from raw seismic data. They can be roughly classified into parametric and functional attributes. Parametric ones can be single-parametric and group-parametric. As a side note, one could notice some evolution of the notion of attribute in the last two decades.

All attributes proposed in this paper are based on the decomposition approach. Some of them are the result of recomposition (rearrangement). The key value of segmentation is the recomposition, which can yield various results (attributes).

The R/R_Q-envelope is only one of these results, although a significant one. Moreover, there is a new type of random bipartite graph (see [10]), which addresses the issue of seismogram clustering and seismic database organization. The issue of seismogram duration, which is indisputably significant, is treated in a more objective way. As for the R/R_Q -envelope, it revealed an invariant on the population of seismograms, where a small sample was enough for the conclusion. The issue of seismogram symmetry can also be treated in a simple way, without having to engage a huge population of seismograms to infer reliable results (see [57]). Then, applying the principle of recomposition on SFs durations, it is possible to make local assessments of spectrum structure at low numerical cost.

Available attributes could address and describe many issues, but their limitations stem from the basis on which they are defined on. Similarly, the proposed attributes are capable of describing some issues, which the current attributes are not capable of, just because of the definition basis: the segmentation into SFs/EFs, giving the ability of their recomposition. Just to mention some: the ordered sequence of SFs (by SF peak amplitude, to give, e.g., the R/R_Q -envelope, or by SF duration, to give period distribution); the R/R_Q -envelope; the new type of random bipartite graph, as structural description of seismogram; the invariant on seismogram population; etc. There partly belongs the seismogram duration, where some of the available attributes also give very useful definitions. Also for the seismogram symmetry, since there are other ways to describe it, except for the R_Q and N_Q envelopes proposed in this paper, which form the difference functions ε_Q, as a specific description of the seismogram symmetry.

4.3. R-Envelope as Functional Attribute or Transform of Seismogram

It is said that the R/R_Q-envelope is a functional attribute. This claim has the reverse. The R/R_Q-envelope, represented by a polygonal continuous line drawn over SFs peaks, really seems to have a continuous function. However, each segment of a polygonal line is drawn (interpolated) between adjacent peaks. Thus, the segment does not carry any additional information which is not already contained in peak coordinates. It serves only for visual impressions to human vision. In that sense, the R/R_Q-envelope actually represents a group-parametric attribute of the original seismogram, not the functional one, unless polygonal line segments are not taken as a fact of continuity. The same holds for the seismogram: a string of several thousand amplitudes (samples) expressed as real numbers, which peaks the plotting software connects by linear segments, creating a polygonal line just for the needs of the software user (human).

In plain words, a human sees the polygonal line where the computer 'sees' a string of numbers. The R/R_Q-envelope actually does not exist as a function, nor the other 'functions' derived from the R/R_Q-envelope, but it is very useful to treat it formally as a function, not only for the sake of human vision, but for mathematical reasons because in such case it is a single object to manipulate with, and not a string of numbers, i.e., a collection of separate mathematical objects. It is now quite clear that there is no necessity for any R/R_Q-envelope in defining the seismogram duration.

R-Envelope and Spectrum Issues

There is an analogy between the spectrum of the raw seismogram and the rearranged seismogram since both are ordered sequences of elements (SFs and spectral components), although the order is the result of different operations (plain sorting in size and the Fourier transform).

In that sense, the R/R_Q-envelope, as the main product of seismogram rearrangement), can be considered as a kind of its transform. This issue is also considered in [10].

In the case of analytic function, the FT transforms one form into the exact another, and vice versa, showing its cyclic ability. For non-analytic cases, usually expressed as numerical sets, the FT (i.e., DFT, FFT) transforms one set into strictly another one. With digital computers, even analytic cases are processed as numerical sets.

It means that the FT transform operates between two numerical domains, which are actually one and the same domain. This further means that this is not a transformation since it circles within one domain. The same situation then occurs with the DFT: it transfers one set of numbers into another one. The human is the factor that prescribes the meaning to bare numbers prior to and after the transform action. If one would ask the computer, could it tell the difference?

When parsed into SFs (raw or approximated by triangles), the seismogram becomes a finite string of real numbers (peak amplitudes and durations of SFs). When indexed, it becomes a string of integers. The above expression 'vice versa' designates the cyclic ability of a transform, to equally transfer mathematical objects between two domains. The R/R_Q-transform shows limits of such ability. At the integer level, it works completely. At the real-number level (maximum amplitudes or SFs' peaks), it also works, with an addition of another set of real numbers—positions of peaks along the τ-axis. On the SF level, it does not work since it cannot return SFs shapes from R/R_Q-envelope.

There is one more difference between R/R_Q-envelope, as the proclaimed transform of the seismogram, and the FT. Each element (SF) in a recomposed sequence of SFs, which is the basis for creating the R/R_Q-envelope, already existed in the original sequence (seismogram), and it is just moved from the previous to the new position, according to the criterion of increasing/decreasing maximum amplitude of SF. It means that none of the SFs was subject to any transformation during this shifting operation. In other words, it already existed as the object of transformation. On the opposite, each element in the spectrum, the sinusoid component of a certain frequency, is the result of the transformation of the whole original object—seismogram *s*(*t*). This is because the FT (DFT/FFT) is basically defined by an integral dependent on a parameter.

$$I(\alpha) = \int_a^b f(x,\alpha)\,dx \quad S(\omega) = \int_{-\infty}^{\infty} \left[s(t)e^{-i\omega t}\right] dt \tag{33}$$

This further means that one should calculate the whole integral over the variable *t* for each value of parameter ω. It could be called the circle transformation to disambiguate with the cyclic transformation (giving the original after two transformations, $F^{-1}\{F[s(t)]\} = s(t)$. It further means that the whole object of transformation, seismogram *s*(*t*), experiences essential change to yield just a single element (a single sin component with the parameter ω) in the ordered representation of seismogram, i.e., its spectrum. There is also an ordered sequence of SFs, i.e., an ordered picture of a seismogram, but created in a much simple way. Also, since the concept of spectrum is a pretty fundamental one and can attain various forms, one could even name the rearranged sequence of SFs as spectrum since it provides a specific picture of the content of a seismogram, but it is fundamentally different from the FT spectrum, which the frequency spectrum, indicating the main parameter of its products—sinusoid components. Further, the proposed method operates with a far less sized set of numbers, and it does so once, not all over, for each value of the main parameter.

4.4. Is Seismogram Randomness Suitable for Its Statistics?

Seismogram is the product of random processes, and as such, it is formally a random variable.

If the seismogram's complexity is defined by the number of EFs (one attribute more), it does not always provide sufficient statistical "mass" for getting reliable statistical characteristics and conclusions, even with the large number of SFs. Moreover, the waveform complexity is very different in various recordings (acceleration, velocity, displacement) of the same event, even at the same site, and also depends on filtering conditions [7] "article 1.01, p. 22, Figure 13". We considered the usual file sizes of earthquakes in databases [33], with mostly 5000–10,000 amplitudes (numerical values), taken at the rate of 0.01 s.

Although it seems to be a sufficient statistical sample, these 5000 values are not 5000 independent but considerably correlated, yielding effectively a few hundred of practically really independent ones, and, hence, suitable for statistical conclusions.

Another constraint is that we seek for the attributes to be taken from individual seismograms, not so much from a bunch of them, not to speak of a statistical ensemble of seismograms, since the earthquake is an event per se, expressing "individuality", i.e., it is "one-shot" event.

The confirmation of the fact that it is not easy to cluster earthquakes, one can simply find classification schemes in earthquake databases, for example, the PEER database [33].

It is enough to see under which criteria they are grouped, although it is the result of tremendous experience. This issue is considered in more details in [10].

There is a weak homogeneity in the earthquake population. The seismograms of the same event, of the same component, taken at sites with almost equal distances from the hypocenter, show considerable variations in waveform caused by different propagation conditions between fault-site, so that they arrive at a station carrying different stochastics.

Simply, seismograms are too non-stationary events, truly transients, to be suitable for tight clustering. Seismic arrays [8] "Ch. 9" may effectively bridge such gap, but the fact of inhomogeneity remains in power. In this aspect, see the elaboration about the random bipartite graphs as a means for organizing seismic database [10].

4.5. Horizontal Asymptote of R-Envelope?

Since the small fluctuations dominate in seismogram, the arrangement of SFs, by decreasing peak amplitude, makes the R-envelope a monotonic function, in which the R-envelope closely approaches the t-axis, so it might be considered as a horizontal asymptote. Mathematics strictly defines the issue of asymptote.

How close the R-envelope will reach the t-axis, or how much the t-axis is a true horizontal asymptote of the R-envelope, depends on the length of seismogram, i.e., on the moment when recording is interrupted after the basic part of the event and its coda interval. If the recording would be intentionally extended, let us say for the sake of this research, then one would have a long seismogram's tail composed of seismic noise. The one noise component generated by the recorded event will definitely become extinct with time, but not the perpetual seismic noise from other sources all around the never resting Earth. This means that neither recording truncation nor the seismic noise simply allows the R-envelope to infinitesimally reach the t-axis since there always are the smallest but not negligible fluctuations. So, one cannot speak about the asymptotic approaching (reaching) of the R-envelope in the conventional mathematical sense, but just about approaching tendency.

What is more interesting in this issue is the twofold approach the t-axis (better to say, the τ-axis), where both positive and negative R-envelopes approach the same line.

This issue finds its place in the spectrum as well, but with an essential difference: whatever is the truncation of the seismogram, its spectrum will approach the ω-axis infinitesimally close with ω tending to infinity, but the R envelope will not. Also, the spectrum of noisy seismogram will also approach the ω-axis infinitesimally close, though more gradually since the seismic noise has a broader spectrum than the event's seismogram.

4.6. Seismogram Duration—Linear or Non-Linear Regression?

To demonstrate the fertility of the proposed concept it is applied on the issue of seismogram duration using the simplest case of regression analysis—linear regression.

The scatter diagram, hiding certain relationships between two variables, implies that all points are of the same significance. Can one claim that for SFs peaks, which represent maximum amplitudes of SFs? This means that the peaks of the highest amplitudes are equally significant as the smallest ones. In other words, there is no ponderation. It does not make the linear regression questionable for this case since the SFs peaks already involve ponderation simply by being big or being small.

Otherwise, if one claims that the seismogram is not a sample of sufficient size for reliable statistical conclusion, then the duration determined in this way seems to be also unreliable. Here, the regression is applied to a scatter plot, which is not so stochastic one, as it would be if taken from a raw seismogram. The rearrangement of SFs, from an "unkempt" seismogram into an ordered sequence, makes averaging, equalization of original stochastics, and smooths variability, so it equivalently enlarges the sample, enabling more reliable statistical inference.

This does not mean that one cannot apply polynomial regression. A suitable polynomial form can be nothing other than the R-envelope shape fitted with a proper polynomial. It will be an issue for further research.

One of the purposes of ordered SFs, united under the R-envelope to describe them by one mathematical object, was illustrated through the proposed definition of seismogram duration, with the aim to make the definition objective, stemming from the very waveform. Linear regression gave the duration just as an interval, since the interval (duration) is determined from rearranged SFs. Where to place this interval on the raw seismogram? The answer gives the seismogram itself by its beginning, i.e., the position of the first (P-) phase.

The R-envelope and the seismogram duration, as well, are numerically extensive. To estimate its numerical demand, one could simply compare this procedure with integral calculation for, e.g., the Husid diagram.

There is a correlation in each random process. Instead of the correlation function, the usual attribute of stationary signals, the Hurst exponent is more suitable. The rearrangement operation over SFs in a seismogram perturbs the correlation between SFs up to canceling, and more than that, the separation of positive and negative SFs adds to correlation disruption.

$$s(t) = \sum_k^K p_k \cdot i^{(q)}(a_k \cdot t - t_k) + \sum_l^L n_l \cdot i^{(q)}(a_l \cdot t - t_l) \tag{34}$$

4.7. Issue of Zero-Crossing Criterion

Decomposition of raw seismogram into SFs is based on zero crossings. After an in-depth study of the given and other SFs/EFs in the said RSN 1616 time series, one should reconsider the definition of the SF/EF in the part of the t-axis crossing, which sometimes can yield questionable results: such definition detects very small fluctuations, but can overlook considerable ones, especially those within complex SEs. One may also speak of significant and insignificant amplitudes SFs.

That means that the criterion of the t-axis crossing is a point-like one, which does not take into account the context of a random process developing around the crossing point. It also determines the number of SFs.

This criterion is sensitive to the position of the waveform with respect to the t-axis: slight shift of the t-axis (better to say, the reference axis) up or down yields different segmentations of a seismogram into SFs, in respect to the original position of the t-axis. The small SFs are particularly sensitive to such shifts of reference axis (Figure 20), which now can reach internal fluctuation(s) of SF with complex shape and add a new SF in the result of parsing. Then an issue of complex SFs approximation should be modified: complex SF shapes should not be approximated with simple triangles, either symmetric or asymmetric, but with other regular shapes, better suited for actual SF shape, even up to embedding the elements of AI (Artificial Intelligence), e.g., shape recognition in the procedure.

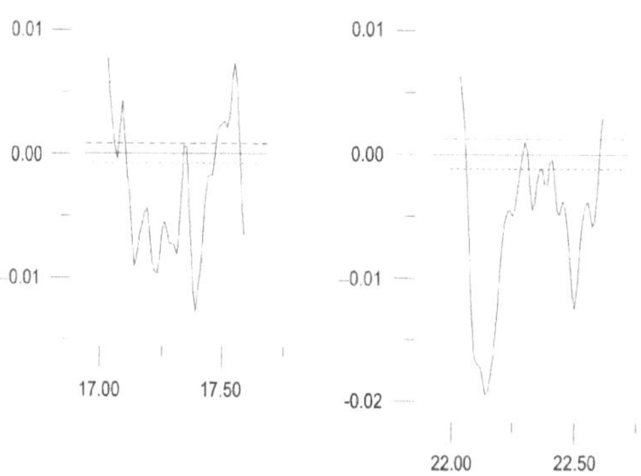

Figure 20. Two examples of random impulses, taken from the seismogram on Figure 2, show ambiguity in the choice of section points between successive SFs, if it is caused by the slight vertical shift of the t-axis (as the referent axis). It is seen that small shifts would result in considerably different SFs. Scale numerals indicate the position of these SFs in the waveform from Figure 2.

The t-axis crossing is an important event in random process development, which generates an actual seismogram. It indicates a transition in the process course, i.e., the pass through the neutral equilibrium, state, which we consider a very important case in random process evolution.

5. Conclusions

The decomposition of seismogram into EFs and SFs turned out to be a fruitful fundamental approach not only for defining new attributes but also for revealing some invariants in the seismogram population. It could be said that it opens the way in a promising direction of further research.

Decomposition enables rearrangement (ordering) of EFs/SFs, which have a fundamentally statistical quality, to produce preliminary mean operation over the raw sample, i.e., seismogram.

The very number of new attributes confirms the effectiveness of said platform, mainly by these presented in the paper, and also by those conceived in this research.

There is a noticeable difference between the currently available and the proposed attributes in addressing certain issues simply because they come from different definition bases. Some of them are disjunct in this sense, and some of them differently treat the same issues, e.g., the seismogram duration.

The arranging of the EFs/SFs makes the seismogram more deterministic than before, since averaging processes take place and give results with new quality, n. Nevertheless contains the same numbers of EFs/SFs before and after. This new quality manifests its effects in new attributes derived from the new arrangements. The main one is the R/R_Q-envelope, showing its potential to derive several new attributes. In that sense, the R-envelope (taken as a continuous curve) can be considered as a kind of seismogram transform.

With this approach, and with the presented results of its application, it is expected to attain a more thorough description of the seismogram, besides the existing attributes, which would find its place in refining and improving regulative in the design and construction of seismic resistant structures, and in other related disciplines as well.

Author Contributions: Conceptualization, R.B.; Resources, R.B. and L.B.; Data curation, R.B. and L.B.; Formal analysis, R.B.; Supervision, R.B.; Investigation, R.B., Visualization, R.B., Methodology, R.B. and L.B.; Writing—original draft, R.B. All authors have read and agreed to the published version of the manuscript.

Funding: This research received no external funding.

Institutional Review Board Statement: Not applicable.

Informed Consent Statement: Not applicable.

Data Availability Statement: Not applicable.

Acknowledgments: Using the seismograms from the PEER Ground Motion Database greatly contributed to our research.

Conflicts of Interest: The authors declare no conflict of interest.

References

1. Dowrick, D. *Earthquake Resistant Design and Risk Reduction*, 2nd ed.; Wiley: Chichester, UK, 2009; ISBN 978-0-470-74701-8.
2. GFZ, Seismic Hazard and Risk Dynamics, Seismic Hazard Assessment for the D-A-CH Countries. Available online: https://www.gfz-potsdam.de/en/section/seismic-hazard-and-risk-dynamics/projects/previous-projects/probabilistic-seismic-hazard-assessments/d-a-ch-seismic-hazard-for-the-d-a-ch-countries-germany-austria-switzerland (accessed on 6 September 2023).
3. A Matter of Development: How to Reduce Vulnerability in the Face of Natural Disasters, UN Economic Commission for Latin America and the Caribbean, March 2000. Available online: https://publications.iadb.org/en/publication/11209/matter-development-how-reduce-vulnerability-face-natural-disasters (accessed on 7 October 2023).
4. Duggal, S.K. *Earthquake-Resistant Design of Structures*, 2nd ed.; Oxford University Press India: Noida, India, 2013. Available online: https://www.technicalbookspdf.com/earthquake-resistant-design-of-structures-2nd-edition/ (accessed on 10 September 2023).

5. *FEMA Earthquake-Resistant Design Concepts, An Introduction to the NEHRP Recommended Seismic Provisions for New Buildings and Other Structures, FEMA P-749/December 2010*; National Institute of Building Sciences Building Seismic Safety Council: Washington, DC, USA, 2010. Available online: https://www.fema.gov/sites/default/files/2020-07/fema_earthquake-resistant-design-concepts_p-749.pdf (accessed on 2 September 2023).
6. Sucuoğlu, H.; Akkar, S. *Basic Earthquake Engineering-From Seismology to Analysis and Design*; Springer: Cham, Switzerland, 2014.
7. Schubert, G. (Ed.) *Treatise on Geophysics*, 1st ed.; Elsevier: Amsterdam, The Netherlands, 2007; Volume 11, p. 6054, ISBN-10: 0444519289. Available online: https://www.geokniga.org/bookfiles/geokniga-treatise-geophysics.pdf (accessed on 10 September 2023).
8. Bormann, P. (Ed.) *IASPEI New Manual of Seismological Observatory Practice*; GFZ German Research Centre for Geosciences: Potsdam, Germany, 2012. [CrossRef]
9. Lindup, G.; Sharpe, R.; Holm, S.; Jarman, S.; Sinclair, R.; Bloomer, A.; Tsamandakis, E.; Smart, C. *Practice Note 19 Seismic Resistance of Pressure Equipment and Its Supports*; Engineering New Zealand: Wellington, New Zealand, 2019; Version 5; ISSN 1176-0907. Available online: https://d2rjvl4n5h2b61.cloudfront.net/media/documents/PN19-SeismicResistancePressureEquipment_V5-2019.pdf (accessed on 10 September 2023).
10. Babić, R.; Babić, L. A New Type of Bipartite Random Graph as a Transform of Seismogram and Its Potential for Organizing Seismic Databases. *Appl. Sci.* **2023**, *13*, 10303. [CrossRef]
11. Turhan Taner, M. Attributes Revisited, Rock Solid Images (RSI), Houston, Texas, May 2003. Available online: https://pdfslide.net/documents/attributes-revisited-rsi.html?page=2 (accessed on 25 September 2023).
12. Subrahmanyam, D.; Rao, P.H. Seismic Attributes—A Review, 7th International Conference & Exposition on Petroleum Geophysics, Hyderabad 2008, P-398. Available online: https://spgindia.org/2008/398.pdf. (accessed on 1 October 2023).
13. Chopra, S.; Marfurt, K.J. Seismic attributes—A historical perspective. *Geophysics* **2005**, *70*, 3SO–28SO. [CrossRef]
14. Koson, S.; Chenrai, P.; Choowong, M. Seismic Attributes and Their Applications in Seismic Geomorphology. *Bull. Earth Sci. Thail.* **2014**, *6*, 1–9. Available online: https://www.geo.sc.chula.ac.th/BEST/volume6/number1/BEST_6_1_001_Sanhasuk_Final.pdf (accessed on 1 October 2023).
15. Ercoli, M.; Forte, E.; Porreca, M.; Carbonell, R.; Pauselli, C.; Minelli, G.; Barchi, M.R. Using seismic attributes in seismotectonic research: An application to the Norcia Mw = 6.5 earthquake (30 October 2016) in central Italy. *Solid Earth* **2020**, *11*, 329–348. [CrossRef]
16. Kushwaha, P.K.; Maurya, S.; Rai, P.; Singh, N. Chapter 15—Prediction of petrophysical parameters using probabilistic neural network technique. In *Basics of Computational Geophysics*; Elsevier: Amsterdam, The Netherlands, 2021; pp. 273–292. [CrossRef]
17. Helal, A.; Farag, K.; Shihata, M. Unconventional Seismic Interpretation Workflow to Enhance Seismic Attributes Results and Extract Geobodiesat Gulf of Mexico Case Study, Project: North Sea f3, January 2015. Available online: https://www.researchgate.net/publication/316343080_UNCONVENTIONAL_SEISMIC_INTERPRETATION_WORKFLOW_TO_ENHANCE_SEISMIC_ATTRIBUTES_RESULTS_AND_EXTRACT_GEOBODIESAT_GULF_OF_MEXICO_CASE_STUDY/figures?lo=1 (accessed on 20 September 2023).
18. Jiao, P.; Alavi, A.H. Artificial intelligence in seismology: Advent, performance and future trends. *Geosci. Front.* **2019**, *11*, 739–744. [CrossRef]
19. Barnes, A. Too Many Seismic Attributes? *CSEG Rec.* **2006**, *31*, 40–45.
20. Lee, J.; Green, T. Predictive Relations for Significant Durations in Stable Continental Regions. In Proceedings of the 14th World Conf. on Earthquake Engineering (14WCEE), Beijing, China, October 2008. Available online: https://www.researchgate.net/publication/266175405_PREDICTIVE_RELATIONS_FOR_SIGNIFICANT_DURATIONS_IN_STABLE_CONTINENTAL_REGIONS/figures?lo=1 (accessed on 15 September 2023).
21. Lee, J. Strong Ground Motion Durations of Directivity Motions. In Proceedings of the 6th International Conference on Earthquake Geotechnical Engineering, Christchurch, New Zealand, 1–4 November 2015. Available online: https://www.issmge.org/uploads/publications/59/60/174.00_Lee.pdf (accessed on 7 September 2023).
22. Anbazhagan, P.; Sheikh, M.N.; Bajaj, K.; Dayana, P.J.M.; Madhura, H.; Reddy, G.R. Empirical models for the prediction of ground motion duration for intraplate earthquakes. *J. Seism.* **2017**, *21*, 1001–1021. [CrossRef]
23. Afshari, K.; Stewart, J.P. Physically Parameterized Prediction Equations for Significant Duration in Active Crustal Regions. *Earthq. Spectra* **2016**, *32*, 2057–2081. [CrossRef]
24. Li, Z.; Kotronis, P.; Wu, H. Simplified approaches for Arias Intensity correction of synthetic accelerograms. *Bull. Earthq. Eng.* **2017**, *15*, 4067–4087. [CrossRef]
25. Mansukhani, S. The Hurst Exponent: Predictability of Time Series. 6 August 2012. Available online: https://pubsonline.informs.org/do/10.1287/LYTX.2012.04.05/full/ (accessed on 9 October 2023).
26. Sychev, V.; Imashev, S. Estimation of Hurst exponent of seismic signal. *Geosyst. Transit. Zones* **2017**, *1*, 50–61. [CrossRef]
27. Galluzzo, D.; Manzo, R.; La Rocca, M.; Nardone, L.; Di Maio, R. Detection of Low Frequency Seismicity at Mt. Vesuvius Based on Coherence and Statistical Moments of Seismic Signals. *Appl. Sci.* **2023**, *13*, 194. [CrossRef]
28. Weisang. Available online: https://www.weisang.com/en/documentation/envelopeanalysis_en/# (accessed on 7 October 2023).
29. Wegler, U.; Lühr, B.G. Scattering Behaviour at Merapi Volcano (Java) Revealed from an Active Seismic Experiment. *Geophys. J. Int.* **2001**, *145*, 579–592. [CrossRef]
30. Sato, H.; Korn, M. Synthesis of Vector-Wave Envelopes in Random Elastic Media on the Basis of the Markov Approximation. *Adv. Geophys.* **2008**, *50*, 43–93. [CrossRef]

31. Sanborn, C.J.; Cormier, V.F.; Fitzpatrick, M. Combined Effects of Deterministic and Statistical Structure on High-Frequency Regional Seismograms. *Geophys. J. Int.* **2017**, *210*, 1143–1159. Available online: https://academic.oup.com/gji/article/210/2/1143/3833065 (accessed on 7 October 2023). [CrossRef]
32. KRSC Kola Regional Seismological Centre May 2018, Generalized Envelopes. Available online: https://www.krsc.ru/?q=en/Envelop (accessed on 7 October 2023).
33. PEER, PEER Ground Motion Database. 2022. Available online: https://ngawest2.berkeley.edu/spectras/new?sourceDb_flag=1 (accessed on 20 September 2023).
34. Havskov, J.; Ottemoller, L. Processing Earthquake Data. 2009. Available online: https://www.geo.uib.no/seismo/SOFTWARE/DOCUMENTATION/processing_earthquake_data.pdf (accessed on 23 September 2023).
35. Upadhyaya, S.; Yeganeh, S.M. *Seasonal Adjustment of National Indices of Industrial Production at International Level*; UNIDO: Vienna, Austria, 2015. [CrossRef]
36. Storchak, D.A.; Schweitzer, J.; Bormann, P. The IASPEI Standard Seismic Phase List. *Seism. Res. Lett.* **2003**, *74*, 761–772. Available online: https://www.isc.ac.uk/standards/phases/ (accessed on 7 October 2023). [CrossRef]
37. Sandwell, D.; Anderson, D.L.; Wessel, P. Global tectonic maps, Special Paper 388, Geological Society of America. 2005. Available online: https://topex.ucsd.edu/sandwell/publications/108.pdf (accessed on 7 October 2023).
38. Anderson Don, L. *New Theory of the Earth*; Cambridge University Press: New York, NY, USA, 2007; ISBN 9780521849593. Available online: https://resolver.caltech.edu/CaltechBOOK:2007.001 (accessed on 7 October 2023).
39. Pischiutta, M.; Rovelli, A.; Salvini, F.; Fletcher, J.B.; Savage, M.K. Directional Amplification at Rock Sites in Fault Damage Zones. *Appl. Sci.* **2023**, *13*, 6060. [CrossRef]
40. Xiao, M.; Xie, J.; Wang, W.; Liu, W.; Sun, J.; Jin, B.; Zhang, T.; Zhao, Y.; Wang, Y. Application of Iterative Virtual Events Internal Multiple Suppression Technique: A Case of Southwest Depression Area of Tarim, China. *Appl. Sci.* **2023**, *13*, 8832. [CrossRef]
41. Kring, D.A.; Boslough, M. Chelyabinsk: Portrait of an asteroid airburst. *Phys. Today* **2014**, *67*, 32–37. [CrossRef]
42. Space-Chelyabinsk. Available online: https://www.space.com/33623-chelyabinsk-meteor-wake-up-call-for-earth.html (accessed on 7 October 2023).
43. Peter Bormann, P.; Wendt, S.; Di Giacomo, D. Chapter 3 Seismic Sources and Source Parameters. In *New Manual of Seismological Observatory Practice (NMSOP-2)*; IASPEI, GFZ German Research Centre for Geosciences: Potsdam, Germany, 2013. [CrossRef]
44. Heimann, S.; González, Á.; Wang, R.; Cesca, S.; Dahm, T. Seismic Characterization of the Chelyabinsk Meteor's Terminal Explosion. *Seismol. Res. Lett.* **2013**, *84*, 1021–1025. [CrossRef]
45. Tauzin, B.; Debayle, E.; Quantin, C.; Coltice, N. Seismoacoustic coupling induced by the breakup of the 15 February 2013 Chelyabinsk meteor. *Geophys. Res. Lett.* **2013**, *40*, 3522–3526. [CrossRef]
46. Malone, S. Seismology in the Air, PNSN. 22 March 2018. Available online: https://pnsn.org/blog/2018/03/22/seismology-in-the-air (accessed on 1 October 2023).
47. Kumar, U.; Chao, B.F.; Hsieh, Y.; Chang, E.T.Y. A meteor shockwave event recorded at seismic and infrasound stations in northern Taiwan. *Geosci. Lett.* **2017**, *4*, 13. [CrossRef]
48. Stevanović, J. Bolide Airbursts as a Seismic Source for the 2018 Mars InSight Mission. *Space Sci. Rev. Oct.* **2017**, *211*, 525–545. [CrossRef]
49. Li, L.; Wang, B.; Peng, Z.; Wang, W. Seismic detections of the 15 February 2013 Chelyabinsk meteor from the dense ChinArray. *Earthq. Sci.* **2016**, *29*, 221–233. [CrossRef]
50. SAGE, Special Event. Chelyabinsk, Russia Bolide (Meteor), Last Updated. 2013-02-28. Available online: https://ds.iris.edu/ds/nodes/dmc/specialevents/2013/02/19/chelyabinsk-russia-bolide-meteor/ (accessed on 1 October 2023).
51. Moshou, A.; Konstantaras, A.; Argyrakis, P.; Petrakis, N.S.; Kapetanakis, T.N.; Vardiambasis, I.O. Data Management and Processing in Seismology: An Application of Big Data Analysis for the Doublet Earthquake of 2021, 03 March, Elassona, Central Greece. *Appl. Sci.* **2022**, *12*, 7446. [CrossRef]
52. Jin, P.; Zhang, C.; Shen, X.; Wang, H.; Pan, C.; Lu, N.; Xu, X. A novel technique for automatic seismic data processing using both integral and local feature of seismograms. *Earthq. Sci.* **2014**, *27*, 337–349. [CrossRef]
53. Gerhard, D. Pitch Extraction and Fundamental Frequency: History and Current Techniques, Technical Report TR-CS 2003-06 November 2003. Available online: https://www.cs.uregina.ca/Research/Techreports/2003-06.pdf (accessed on 7 October 2023).
54. Brad Osgood, B. The Fourier Transform and Its Applications, Lecture Notes for EE 261, Electrical Engineering Department Stanford University. Available online: https://see.stanford.edu/materials/lsoftaee261/book-fall-07.pdf (accessed on 7 October 2023).
55. Babić, R. *Signal Analysis I*; Academic Mind: Belgrade, Serbia, 2000; p. 225. ISBN 86-7466-033-9.
56. Babić, R.; Babić, L.; Jaksić, B. New Aspect of Transition between Line and Continuous Spectrum and Its Relation to Seismic Influence on Structures. *Sci. Iran.* **2018**, *27*, 1515–1524. [CrossRef]
57. Ide, S. Frequent observations of identical onsets of large and small earthquakes. *Nature* **2019**, *573*, 112–116. [CrossRef]

applied sciences

MDPI

Article

A New Type of Bipartite Random Graph as a Transform of Seismogram and Its Potential for Organizing Seismic Databases

Ranko Babić [1,*] and Lidija Babić [2]

[1] Department of Electrical Engineering, Faculty of Technical Sciences, University of Priština in Kosovska Mitrovica, Knjaza Miloša 7, 38220 Kosovska Mitrovica, Serbia

[2] Department of Civil Engineering, Faculty of Technical Sciences, University of Priština in Kosovska Mitrovica, Knjaza Miloša 7, 38220 Kosovska Mitrovica, Serbia; lidija.babic@pr.ac.rs

* Correspondence: ranko.babic@pr.ac.rs; Tel.: +381-63416063

Abstract: This paper proposes a method to reduce seismogram variability as a determining factor in its interpretation, processing, and clustering. By introducing the concept of single fluctuations (SFs), the seismogram can be parsed into a sequence of random impulses with subsequent ordering. This rearrangement of SFs, if they are assigned by positive integers, represents the formal mapping of a regular string of integers into a random one, which can be represented with a bipartite random graph (bigraph). Due to its specific randomness, such a bigraph is considered a new type of random balanced bigraph. The R-envelope and R_Q-envelope, its equidistant version, are defined by tracing the peak envelope over ordered SFs. The equivalence and complementariness of the R_Q and bigraph are considered and discussed, forming a combined characteristic of the seismogram. The R/R_Q provided a considerable reduction in seismogram variability, which was confirmed by creating and analyzing an ensemble of R_Q from several seismograms. In the R_Q domain, distance is defined as a possible basis for metrics and clustering, but the ensemble variability was quite narrow and not as suitable for this purpose. Otherwise, the ensemble shows high redundancy hidden in the seismogram population. As for the bigraph, the mesh of its edges is structuralized in bundles, forming a skeleton, which reflects the internal structural content of the seismogram. The distance over the domain of bigraphs is proposed to show the possibility of clustering. This means that only a combined R_Q and bigraph provides a suitable frame for seismogram representation with reduced variability and, thus, the potential for more effectively organizing seismic databases and a deeper interpretation of seismograms; therefore, R_Qs and bigraphs can be considered as a transform of a seismogram. Many aspects of these concepts are thoroughly discussed. The similarity between concepts of SF and wavelets is briefly considered. This very complex theme is new and promises broad further research. All issues considered in the paper are abundantly illustrated.

Keywords: earthquake; seismogram; earthquake resistance regulative; fluctuation; random bipartite graph; transform; envelope; variability; clustering; database

1. Introduction

Earthquakes can inflict heavy damage and human casualties. Related disciplines have studied earthquakes to widen and deepen understanding of the phenomenon. There is a natural hierarchy in this complex study. At the top is seismology (geoscience), which studies earthquakes as intrinsic processes (see, e.g., [1–3]). The next layer, which relies on the findings and extensive knowledge of the previous discipline, deals with earthquake risk reduction and hazard assessment, creating a basis for the majority of activities in societies that can be affected by earthquakes. There is immense literature related to this area of knowledge, which indicates a high interest of humankind in these effects. This work presents and briefly discusses just a few examples, chosen in a semi-random manner, to help create a representation in this area. For example, the German GFZ (GeoForschungsZentrum, (eng) Research Centre for Geosciences) provides seismic hazard maps [4] as products of the Global Seismic Hazard Assessment Program started in 1993 [5]. The basis for these hazard maps is the seismic database. As for particular research in this area, the references [6–11] are typical and illustrative. There is also a large amount of general literature, including the authors' favorite [12], which covers the transition from hazard to seismic-resistant civil engineering, aimed to alleviate the seismic vulnerability of the built environment [13–15].

The acquired knowledge gradually improves planning, designing, and constructing structures according to the developing standards of earthquake resilience [12]. The perpetual updating of regulations and standards for designing and building structures is derived from the study of earthquakes themselves and the effects they cause on existing structures.

A seismogram contains almost all one can really know about an earthquake, apart from the effects people see, feel, or suffer. It is a recorded waveform of seismic effects at a site, representing fluctuations of related physical (mechanical) quantities as the result of many complex processes taking place in the so-called "information chain" of seismology, with many interrelated sub-systems such as the seismic source, wave propagation through the Earth, distortion by other seismic signals, and the influence of seismic sensors, recorders, and processing techniques on the seismogram [1] "Ch. 1, Figure 1.2, p. 6". An excellent survey of all the phenomena involved can be found in [1] "Ch. 11". The influence of propagation conditions on a waveform's appearance in different seismograms that all originate from the same event can be seen in [16] "Figures 5 and 6".

Therefore, seismogram databases contain all the accumulated intrinsic knowledge of seismicity, where earthquakes are classified according to various criteria, i.e., parameters. Databases are the main sources of seismic information for experts from various branches of knowledge interested in earthquakes and their effects, who then transform these raw data into respective findings, rules, recommendations, and assessments, often incorporated into regulations. These experts include geophysicists, seismologists and observatory personnel, civil engineers, disaster managers, risk planners, urban planners, and governments, not to mention related researchers, students, and training personnel [1] "Ch. 1, pp. 29", [12,17].

Any earthquake in a database is represented by at least one or up to several seismograms and accompanying parameters. Seismic databases can be conditionally categorized as general and specific by content, and as global and local by cover. A typical example of general/global is the PEER Ground Motion Database [18]. The use of the PEER database motivated the authors to conduct this research. Accordingly, this database is the main reference and the main frame for seeking solutions. Also, as the authors have been able to see, other seismic databases are organized on the same principles with the same parametric nomenclature, i.e., on parametric attributes of seismograms (one number for the whole complex waveform). Examples span from specific/local databases [19,20] to general ones [21]. Global databases often redirect the user to local (national) databases [21]. Seismograms are usually given in the form of time series and the spectrum. The spectrum can be considered an ordered picture of the seismogram; nevertheless, both are of random appearance. Considering their shapes, it is apparent that the variability of seismograms is more prominent than that of their spectral counterparts. Therefore, the spectrum is more suitable for practical use, so it appears as the main characteristic of earthquakes in databases.

Difficulties of classification within certain populations rise with the variability of population elements. This is particularly true for seismograms and the system in which they are grouped in databases.

During earlier research on seismograms related to defining seismogram's new attributes, the authors made some interesting findings, which can improve grouping (clustering) criteria. The idea is based on treating the seismogram as composed of random impulses. Recomposing random impulses trained into an ordered sequence can be described as random mapping and represented through a random bipartite graph of a specific kind. Such structuralization of a seismogram significantly reduces its inherent variability and yields certain new characteristics that can be used to improve it by adding to existing grouping criteria (classification) in seismic databases.

1.1. Brief about Bipartite Graphs

Random graphs are very useful and universally applicable mathematical tools. Their offspring, termed complex networks, with more structurality and dynamics, are even more capable of describing the world, which can be considered as being composed of 'systems of systems' and is tremendously versatile and intricate. Among the vast literature, we quote several typical references [22–30].

There are many types of random graphs/networks, such as bipartite random graphs, that have been recognized as particularly insightful representations of many different systems [31,32].

The case presented in this paper, as far as we know, describes a specific application of bipartite graphs not found in the literature until the present day. It involves randomness in such an unusual way, in comparison to existing models, that we were inclined to consider it a new type of random bipartite graph. It falls somewhere between conventional and random bigraphs.

In a bipartite graph (bigraph for short), vertices (nodes) can be divided into two disjoint sets such that no edge connects two vertices of the same set [31].

Generally, two categories of bipartite graphs can be distinguished: conventional and random. In the first, graphs show all existing edges between input and output vertices, while random graphs show possible or potential edges between vertices. Basically, two classical models for random graphs are used: $G(N, L)$, where N vertices are connected with L randomly placed links [24], and $G(N, p)$, where each pair of N vertices is connected with probability p [25].

The core of these models is the fact that random graphs (networks) are statistical ensembles of graphs, where each one has its specific probability of realization [28].

A balanced bipartite graph is one that has an equal number of vertices in both sets, which will be of particular interest in our investigation [31]. Further details can be found in various references [28].

Different terms for basic structural elements of graphs, 'node', 'vertex', 'link', or 'edge', will be used throughout the text.

Regarding nodes, we have to name the two separate sets of nodes of the bipartite graph according to their disposition. The usual graphical representation of these node sets is a linear array of small circles (bold points), where linearity stems from a mapping domain (interval) on an axis and also provides the most transparent outlook of the 'mesh' of edges between input and output nodes. The issue of node disposition contains some interesting aspects.

Linearly disposed nodes are usually arranged in a left–right configuration as two vertical columns. More deeply, it hides the causality orientation of the mapping process: the start of mapping is in the left column, and the end of mapping is in the right one. There is also the vertical disposition, where the beginning set is up, and the end set is down. This defines the implicit direction of the graph.

We will use the terms 'input' and 'output' for sets of vertices (nodes) and adopt vertical disposition as more adequate for our case since the seismogram, taken as a sequence of random impulses, is always presented horizontally.

Despite their seemingly plain geometry, bipartite graphs include many families of different structures [32]. There are many other ways of describing random bipartite graphs using simpler characteristics.

One usually tends to describe any complex entity through simpler properties in order to analyze it in an easier way. Generally, many properties of random graphs are defined and measured, such as diameter, the longest shortest path, average path length, degree distribution, clustering coefficient, etc. More complete and descriptive characteristics are, e.g., the adjacency matrix and the graph spectrum [33,34]. Even the form in which a graph is depicted can be considered a graph property, which is very helpful in its description. The very name 'graph' carries pictorial or geometrical meaning. Also, the adjacency matrix is not just a numerical presentation of the graph but brings a pictorial aspect to the graph's structure.

The random graph proposed in this paper is designated as a seismogram transform, which represents an ordered and simplified seismogram description with greatly reduced original variability, which we expect to be more suitable for organizing seismic databases, which are the fundamental information sources for experts from various knowledge areas.

1.2. Article Goals and Basic Findings

The immense variability of seismograms prevents the effective organization of seismic databases to ease the retrieval of systematic knowledge about seismicity in general, known as data mining or metadata extraction [35].

The issue of proper organization of a seismic database is a very complex matter, but that does not mean that one should not cope with it.

The main idea supporting this research is not to use only the simple parametric attributes of seismograms for grouping (clustering) but to go deeper into the seismogram's structure and propose more complex characteristics suitable for effective clustering.

Identifying the variability as the main factor in solving the problem, the authors will try to find methods of reducing it, thereby hoping to facilitate grouping. Therefore, the term 'potential' in the article's title, indicating conditional relation, means either more or less suitability of the approach.

A few ideas defining the steps of seismogram processing are proposed, which can be assembled into one approach, yielding new characteristics of the seismogram with far less variability than the intrinsic ones found in the seismogram itself. These are the bigraph and its equivalent, better to say its complement, named the R-envelope. Both are based on parsing the seismogram into single random impulses, i.e., elementary/single fluctuations (EF/SF), and the subsequent ordering of SFs to give the said characteristics. The coherence between these two compelled the authors to qualify this approach as the transform.

The presented results confirm the reduced variability. It is more prominent in R_Q. The variability range is narrow, reminiscent of an invariant. This finding deserves thorough further research.

As for the bigraph, the range of variability possesses not only one degree of freedom, as in the case of R_Q but a few more, residing in the system of bundles of edges, recognized in the mesh of edges in the bigraph. The structural information of the seismogram is, thus, preserved in the bigraph. Thus, the meaning of the article's title denotes the new type of bigraph as a transform of seismograms. On the other hand, it does not lessen the importance of R_Q, which found its place in the Graphical Abstract of this paper in the designation $R_Q \oplus BG$.

Because of its new characteristic(s), it was necessary to consider it from various aspects to better describe its properties and applicability.

2. Materials and Methods

2.1. Seismogram as a Sequence of Random Impulses

As stated in the Introduction, a seismogram is the result of many complex processes taking place in the so-called "information chain" of seismology, from the earthquake source to the recording place.

Considered only by its appearance, the waveform, a seismogram is a fluctuational transient of certain mechanical quantities (acceleration, velocity, displacement), where the waveform looks as if it is composed of random impulses [2] "Vol. 1, paper 1.03.1.3, pp. 73, 131", with random amplitudes, durations, and shapes. Although the term "fluctuation" (to fluctuate) refers to the irregular rising and falling (change) of a physical quantity, we will further formalize this term and introduce the concept of elementary fluctuation (EF), consisting of "ups" and "downs", separated by crossings of a certain axis (usually the t-axis). Hence, the seismogram is considered a sequence of EFs since an alternating pattern, although random, occurs in its waveform.

In this sense, the EF is usually composed of two subsequent alternate random impulses, which we will call single fluctuations (SFs).

To put it more clearly:

$$\mathrm{EF}^{(+)} = \mathrm{SF}^{(+)} + \mathrm{SF}^{(-)} \qquad \mathrm{EF}^{(-)} = \mathrm{SF}^{(-)} + \mathrm{SF}^{(+)} \tag{1}$$

Therefore, the EF, or the SF, is the basic structural element of the seismogram (see Figure 1).

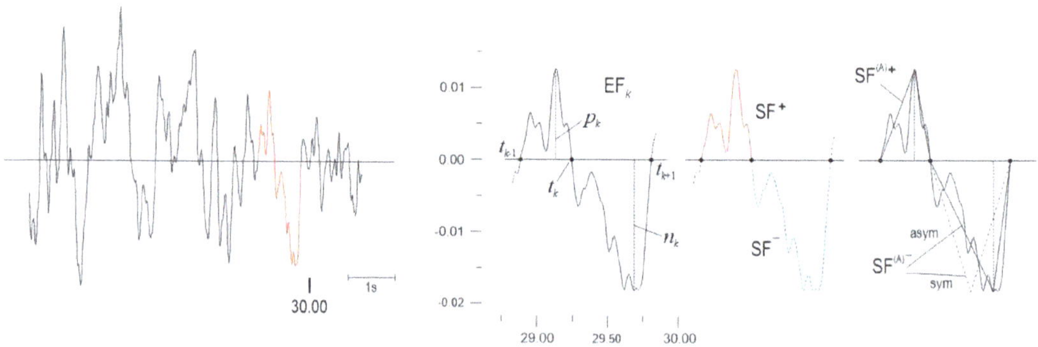

Figure 1. (**left**) A fragment from the seismogram of the RSN 1616 event, Düzce, Turkey, shows the complex random behavior of the seismic process. Units on axes scales are in g and in s. The trace in red color indicates the concept of elementary fluctuation (EF) introduced in this paper. (**right**) The EF can be parsed into two single fluctuations (SFs) as a pair of positive (red line) and negative (blue line) random impulses. (**far right**) Geometrical approximation of SFs by symmetric (sym) and asymmetric (asym) triangles.

For numerical simplification, we express the complex waveform of SFs with two parameters: the maximum amplitude of the SF and its duration (Figure 1), which, geometrically, is equivalent to the approximation of SFs by respective triangles, either symmetrical or asymmetrical (Figure 1). In other words, a seismogram is represented by a non-equidistant sequence of amplitude samples and is a sort of non-uniform sampling.

When we speak of geometry, we will distinguish intrinsic SFs and approximated (by triangles) SFs, which are labeled as SF and $\mathrm{SF}^{(A)}$, respectively.

The concept of the EF or SF enables the parsing of a given seismogram into a sequence of SFs, which is a very important operation. Notice that, from the point of view of this research, it is the same as if the whole seismogram waveform is parsed directly into SFs or, after an interstep of approximation, into respective triangles.

After segmentation, the seismogram is actually divided (separated) into two sequences of SFs (or respective triangles $\mathrm{SF}^{(A)}$s), one comprising only SFs with positive amplitudes and another with negative amplitudes. This can be described in terms of the set theory and is a nice example of set theory application.

The original set of SFs is divided into two disjoint subsets:

$$\{\mathrm{SF}\} = \left\{\mathrm{SF}^{(+)}\right\} \cup \left\{\mathrm{SF}^{(-)}\right\}, \tag{2}$$

$$\left\{\mathrm{SF}^{(+)}\right\} \cap \left\{\mathrm{SF}^{(-)}\right\} = \varnothing, \tag{3}$$

As a next step of the procedure, elements of each of these sequences are assigned with integers:

$$\left\{\mathrm{SF}^{(+)}\right\} = \left\{\mathrm{SF}_i{}^{(+)} : i \in I\right\}, \tag{4}$$

where I is the index set (Figure 2) (see, e.g., [36]).

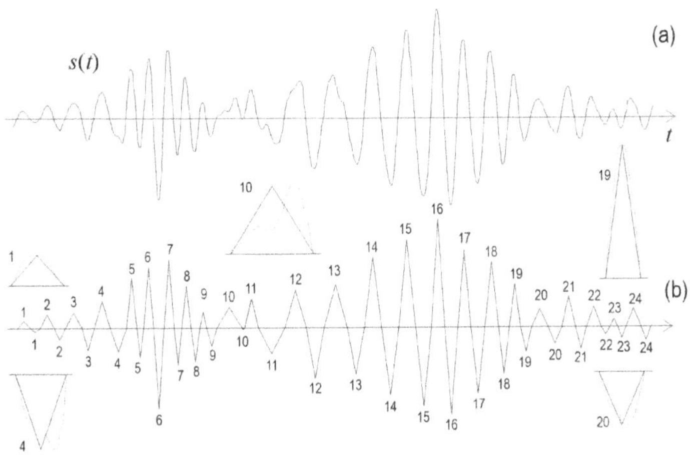

Figure 2. (**a**) Seismogram waveform $s(t)$ is composed of random impulses. (**b**) Approximation by (symmetric) triangles is geometrically equivalent to the numerical reduction of each random impulse to its maximum amplitude and duration. Further reduction involves assigning positive integers to positive and negative triangles. Zoomed-in examples of approximated and indexed SFs (positive: 1, 10, 19, and negative: 4, 20) are inserted around approximated waveform. The original SFs are depicted by dash line.

Similarly, these relations can be easily rewritten for triangles approximating the SFs.

2.2. Ordering of SFs

Having indexed sets of SFs, or $\mathrm{SF}^{(\mathrm{A})}$s, both positive and negative, the next operation is applied: the ordering of each $\mathrm{SF}^{(\mathrm{A})}$ sequence by maximum amplitude. For clarity, the ordering is only applied to positive SFs, which is presented in Figure 3.

A random sequence of SFs is converted into an ordered sequence of SFs, while the initial sequence of naturally ordered integers is turned into randomized indexes. This is obviously a kind of random mapping, which can be written as:

$$\left\{\mathrm{SF}^{(+)}\right\} = \left\{\mathrm{SF}_i{}^{(+)} : i \in I\right\} \overset{F_S}{\rightarrow} \left\{\mathrm{SF}_j{}^{(+)} : j = F_S(i) \in I\right\}, \tag{5}$$

A similar relation can be written for triangles $\{\mathrm{SF}^{(\mathrm{A})}\}$

$$\left\{\mathrm{SF}^{(A+)}\right\} = \left\{\mathrm{SF}_i{}^{(A+)} : i \in I\right\}, \tag{6}$$

It is important to note that the approximation goes from the complex waveform of an SF down to an integer only, which is the ultimate level of approximation (simplification).

Thus, the ordering maps one sequence of integers into another one. This function is another form of mapping. It seems to be a dead end. Reducing either SFs or $\mathrm{SF}^{(\mathrm{A})}$s to integers, one gets simple, bare numbers and loses information for their mapping contained in their max amplitudes. How, then, to obtain an ordered set?

The answer is obvious, but it should be explicated: the amplitude information is hidden in the function that performs the mapping.

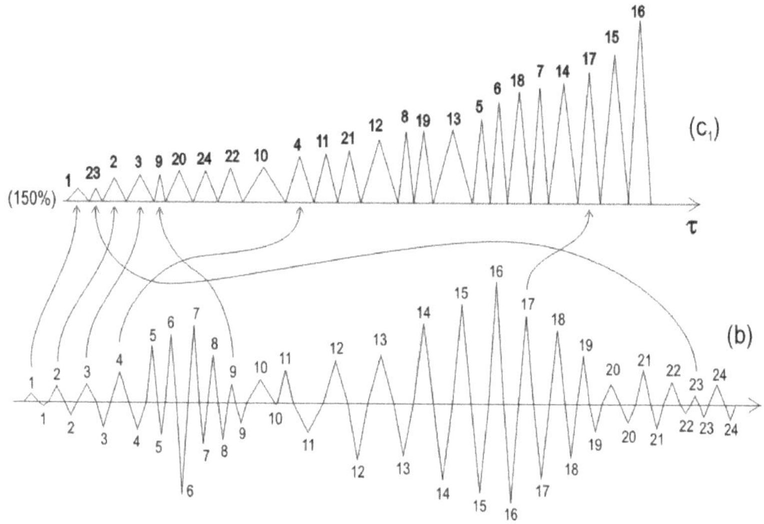

Figure 3. Rearrangement (ordering) of positive $SF^{(A)}$s, from approximated waveform (b), retaken from Figure 2, by maximum amplitude, yields the waveform (c_1). This means the ordered sequence of integers is mapped into a random sequence of integers, or a random sequence of triangles is mapped into an ordered sequence of triangles. The scale on the τ-axis of (c_1) is zoomed in by 150% with respect to diagram (b).

There, one finds the meaning of twofold digitizing of the seismogram: behind the sequence of integers is hidden a corresponding sequence of real numbers (SFs maximum amplitudes), which serve for the ordering operation.

Moreover, the mapping can be considered a sequential operation consisting of a pair of instantaneous acts: taking an input integer, applying the function, and obtaining the result—the output integer. Therefore, it is possible to represent the mapping, which includes input and output and the beginning and ending stages of the mapping process. This is the basis for representing such mapping with the bipartite graph, which shows all acting stages of function.

If reduced to the level of pure integers, the mapping (function) can be written as:

$$\forall i \in I: \ i \overset{F_+}{\rightarrow} F_+(i) = j \in I, \tag{7}$$

for positive SFs (or $SF^{(A)}$s).

This kind of mapping can be represented with bipartite graphs (shown in Figure 4).

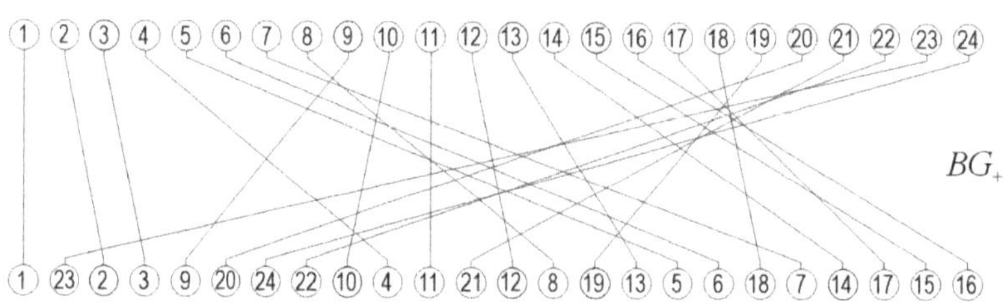

Figure 4. A bipartite graph (vertical configuration), by its edges, connects input and output nodes assigned by integers so that the input ordered sequence shuffles into a random one. The bigraph structure schematically follows the mapping indicated on Figure 3. Such random mapping relies on SFs in the original seismogram and on their random maximum amplitudes.

2.3. New Type of Bipartite Graph as a Transform of Seismogram

2.3.1. Twin Bipartite Graph—Twofold Result of Transform

The previous section considered representation for only positive SFs (or $SF^{(A)}$s) by corresponding bipartite graph (Figures 3 and 4), mainly for the sake of procedure illustration. This means that a bipartite graph is created only for half of the seismogram. The same should be performed for negative SFs. Ordering and mapping (Figure 5) produce an additional bigraph, which means that the (whole) seismogram is represented by a pair of bipartite graphs (Figure 6).

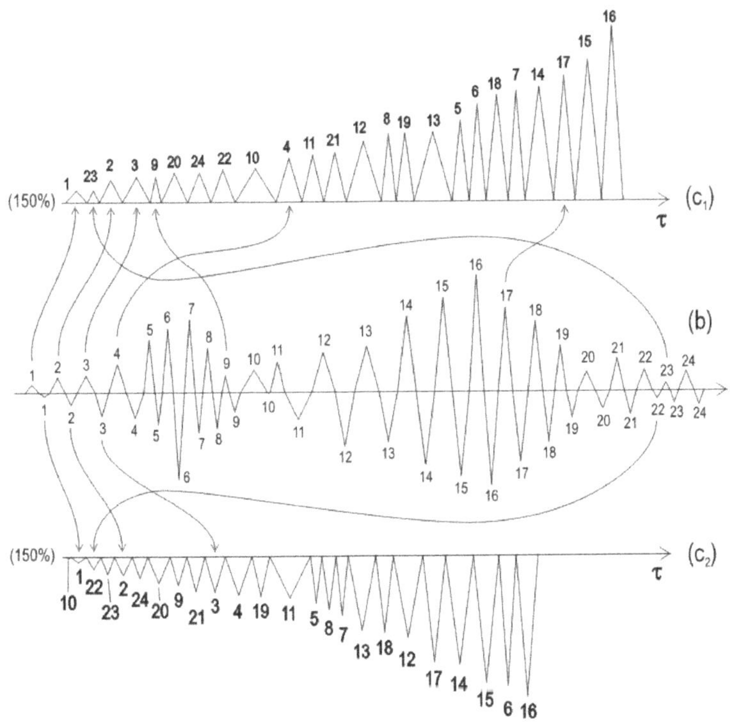

Figure 5. Complete rearrangement of (approximated) seismogram from Figure 2 involves both positive and negative $SF^{(A)}$s, whose mappings are clearly indicated by arrows. Notice that the positive and negative $SF^{(A)}$s are indexed in the same manner, while the respective mappings produce different random sequences of integers. Summed durations of positive and negative $SF^{(A)}$s are different.

Indexing SFs reduces their usually complex waveforms to integers, which means that the entire seismogram's intricate waveform is described by two sequences of integers. In such a reduction of random impulse shapes, only the maximum amplitudes of SFs remain important as a factor for ordering purposes.

The seismogram can be considered as digitized in a twofold manner:

- As a sequence of integers;
- As a sequence of real numbers (maximum amplitudes of SFs, or of approximating triangles $SF^{(A)}$s), whose positions within the sequence are implicitly marked by integers, i.e., indexes.

Splitting the whole seismogram into positive and negative SFs resulted in two separate bigraphs, one unique result of this procedure.

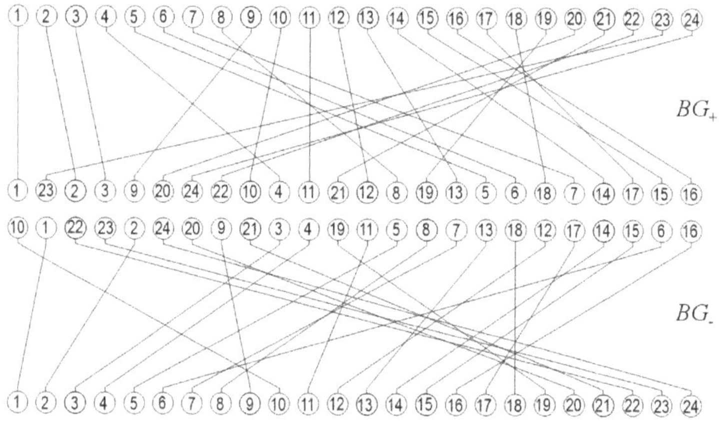

Figure 6. The twin bigraph, composed of two bipartite graphs with two-row geometry, represents mappings (orderings) between sets of indices (integers) for positive and negative SFs, according to mappings on Figure 5. Both bigraphs are positioned in mirror vertical symmetry, where the bigraph BG_- is inverted (input nodes down, output nodes up) to follow the general disposition of SFs in a seismogram.

To overcome this inconsistency, one simply can consider them as united and transpose them into 'the twin bigraph' (Figure 6). There may be the impression that this is only semantic because they remain separated. However, the two bigraphs are not as separated as it may seem from Figure 6. In the background, they are quite correlated since they are parts of a singular seismogram.

Therefore, one can treat the 'double' result of these operations as a whole, as a singular mathematical entity, even though it comprises two actually separated bigraphs. Those two, to some extent, mirror each other so that all operations applied to the seismogram can be treated as a kind of transform.

$$\text{seismogram waveform} = s(t) \overset{\mathscr{T}}{\rightarrow} (BG_+ \oplus BG_-), \tag{8}$$

The operator labeled by $\mathscr{T}$ stands for this transformation. The unity of two bigraphs, i.e., the twin bigraph, is labeled by the symbol $\oplus$, lent from XOR operation (addition modulo 2). Using the symbol $\oplus$ for such a purpose does not mean it executes this operation between two bigraphs strictly, but symbolizes the unity of these two mathematical objects and points out the fact that the two bigraphs should be considered as one. In addition, the operation $\oplus$ brings with it some combinatorial unifying of SF^+s and SF^-s, defined by its truth table, particularly for meaning "one or the other but not both", which additionally vindicates the handy use of strictly defined operation.

Thus, one can write also:

$$s(t) \overset{\mathscr{T}}{\rightarrow} (BG_+ \oplus BG_-),\ \mathscr{T}[s(t)] = (BG_+ \oplus BG_-), \tag{9}$$

Since the BG_- is represented in Figure 6 as geometrically inverted, Equation (9) ought to be modified as:

$$s(t) \overset{\mathscr{T}}{\rightarrow} \left(BG_+ \oplus \overline{BG_-}\right),\ \mathscr{T}[s(t)] = \left(BG_+ \oplus \overline{BG_-}\right), \tag{10}$$

In a nutshell, the transform essentially involves two mappings of integers, both for the positive and negative SFs, which can be designated as functions (bijections) F_+ and F_-

$$\forall i \in I:\ i \overset{F_+}{\rightarrow} F_+(i) = k \in I, \tag{11}$$

$$\forall j \in I: \; j \overset{F_-}{\rightarrow} F_-(j) = l \in I, \quad (12)$$

2.3.2. Coupled Twin Bigraph—Unifying the Twofold

It seems that this transform, as the result of splitting all SFs into positive and negative, produces two separate bigraphs, seemingly separated. They are called twins since there is an invisible unifying factor standing behind them—the seismogram itself. By transferring SFs into integer sequences, the connectedness between the twins fades, becoming less apparent; therefore, it is necessary to conjugate the twins in a more explicit way to make the connections more obvious.

One possibility is to connect the corresponding output nodes of positive and negative bigraphs, thus reestablishing structural relationships that existed in the seismogram at the level of EFs, i.e., EF = $SF^+ + SF^-$. What is the meaning of this proposal?

Actually, in this way, one gets three bigraphs that are overlapped. The two outer bigraphs, formerly the twin bigraphs, are now connected by a new bigraph, where it is not easy to define which nodes are input and which are output. Because of that fact, it deserves a special designation, F_C, and a special name, the connecting bigraph (Figure 7). The whole structure is called the coupled twin bigraph.

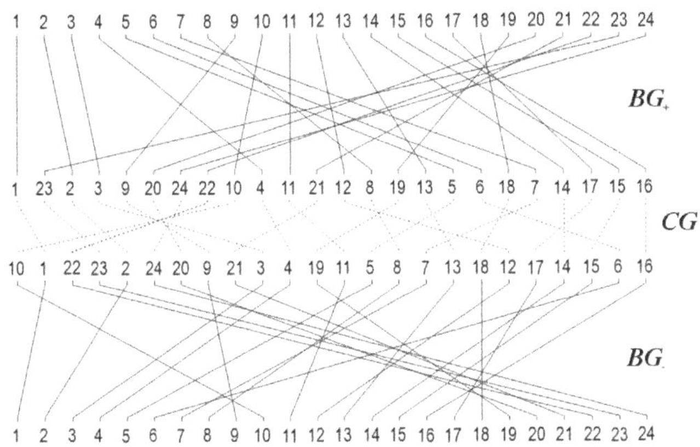

Figure 7. The win bigraph from Figure 6 is unified into a singular entity over the connecting bigraph, which nodes are both output nodes of BG_+ and BG_-, making the coupled twin bigraph. Simultaneously, the connecting bigraph (*CG*) shows symmetry/asymmetry of the twin bigraph, i.e., of the whole seismogram.

This looks like a quadripartite graph: nodes are apparently partitioned into four sets. However, if one recalls the mirror structure of the twin bigraph, based on purposely dividing all SFs into positive and negative subsets where output sets stand opposite each other, there will hardly be any nodes left for a new bigraph, which is called a connecting bigraph.

Nevertheless, there is no paradox. Moreover, it can even be said that such uncertainty confirms the unity of the structure, and now not only in name, as in the case of the twin bigraph, but essentially, reflects the unity of the seismogram's waveform at the integer level.

All previous is said following the bigraph geometry adopted in this paper, as a linear arrangement of nodes, vertically disposed, with input nodes positioned up and output nodes down, where the bigraphs in the twin bigraph are arranged with output nodes face to face. This configuration essentially follows the structure of a seismogram waveform composed of two sequences of SFs on both sides of the t-axis.

2.3.3. Bigraph of the Whole Seismogram

The coupled twin bigraph is proposed to overcome splitting the seismogram into two parts: positive and negative SFs.

There is another, more natural, way to represent the whole seismogram with a single bipartite graph. It is based on a simple operation applied to a certain waveform: rectification. Mathematically, it is expressed by the modulus operation.

$$s(t) \overset{\Downarrow}{\rightarrow} |s(t)| = s_A(t), \tag{13}$$

When negative amplitudes also become positive, then the ordering operation is made over the whole set of SFs, which can be seen in Figure 8. Now, the twin bigraph turns into a conventional balanced bipartite graph, which is presented in Figure 9, and, for distinction, is called an 'absolute bipartite balanced graph'.

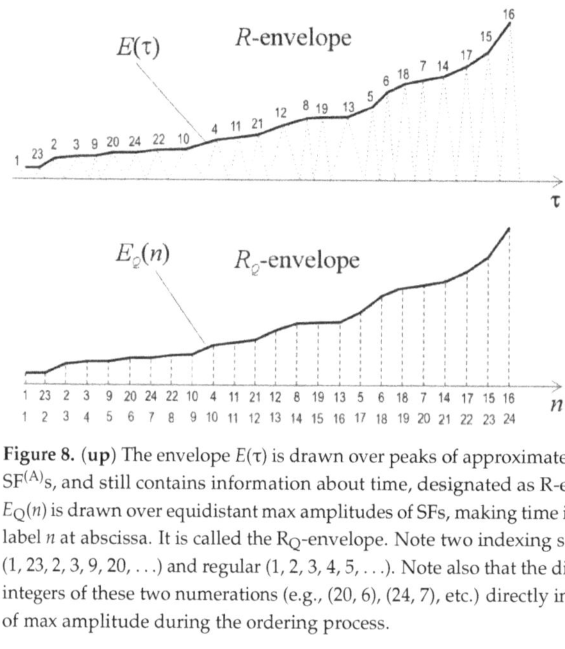

Figure 8. (**up**) The envelope $E(\tau)$ is drawn over peaks of approximated positive SFs, i.e., over positive $SF^{(A)}$s, and still contains information about time, designated as R-envelope. (**down**) The envelope $E_Q(n)$ is drawn over equidistant max amplitudes of SFs, making time information irrelevant, hence the label n at abscissa. It is called the R_Q-envelope. Note two indexing strings along abscissa: positional $(1, 23, 2, 3, 9, 20, \ldots)$ and regular $(1, 2, 3, 4, 5, \ldots)$. Note also that the difference between correspondent integers of these two numerations (e.g., (20, 6), (24, 7), etc.) directly indicates the displacement length of max amplitude during the ordering process.

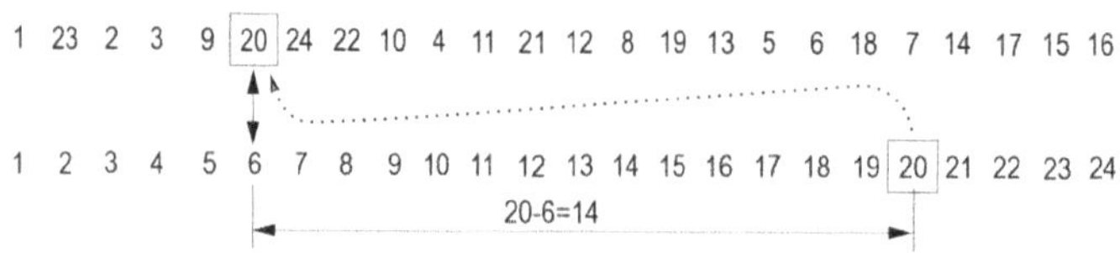

Figure 9. Zoomed-in double indexing from Figure 8, positional and regular, clearly shows the magnitude of displacements of integers during the operation of ordering.

Using operator form, one can write:

$$s(t) \overset{\Downarrow}{\rightarrow} |s(t)| \overset{\mathscr{T}}{\rightarrow} (BG),\ \mathscr{T}[|s(t)|] = (BG), \tag{14}$$

2.3.4. Shuffle of the Twin Bigraf into Singular One

The rectifying operation applied to the seismogram waveform yields a unified (or absolute) bigraph. On the other hand, comparing meshes of edges in the twin bigraph and absolute bigraph, one can notice apparent similarities, which begs a natural question: Is it possible then to "rectify" and unify the twin bigraph to obtain a singular bigraph? How close are these two operations? In other words, can a modulus operation be applied to the twin bigraph itself?

Having on mind Equations (9) and (10), it seems that it is simply necessary to sum graphs A and B to obtain C. Before that, one needs to invert the bigraph BG_- since it was turned in the twin graph. It can be written as:

$$|s(t)| \overset{\mathcal{T}}{\rightarrow} (BG) \overset{?}{\Leftrightarrow} (BG_+ \oplus \overline{BG_-}) \overset{||}{\rightarrow} (BG_+ \oplus |\overline{BG_-}|) = (BG_+ \oplus BG_-) = (BG), \quad (15)$$

The effect is shown in Figure 10.

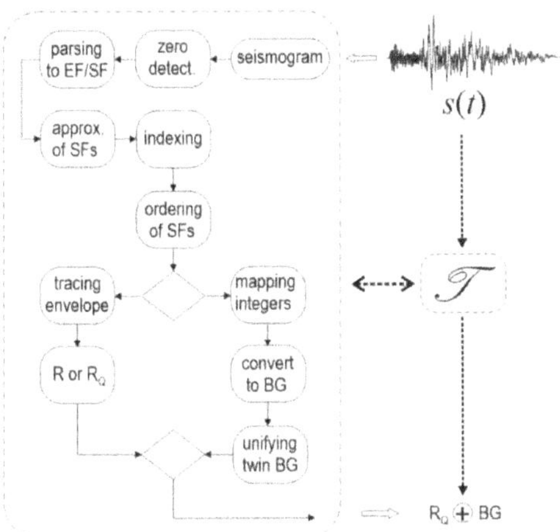

Figure 10. Turning out the seismogram into the sequence of SFs is the basis for creating the R_Q-envelope and the bigraph. All subsequent operations can be gathered into a transform of the seismogram, yielding a result composed of R_Q and BG.

Obviously, if the size of the twin bigraph is N, then the size of the absolute graph is $2N$, assuming an equal number of positive and negative SFs, i.e., the complete alternating structure of each EF = $SF^+ + SF^-$.

2.4. R-Envelope of Ordered Seismogram

In addition to the bigraph as an ordered representation of a seismogram, another characteristic can be introduced, which is also of pictorial appearance. Let its name be the R-envelope (R comes from 'rearranged'). It is also a part of the general frame of this research. Its meaning is graphically explained in Figure 8.

Three forms of the R-envelope can be considered:

- Drawn over the peaks of SFs;
- Drawn over the peaks of $SF^{(A)}$s (over peaks of approximating triangles);
- Drawn over equidistant maximum amplitudes.

In the first two cases, the factor of time also determines the shape of the R-envelope (different durations of SFs). In the third case, the time becomes irrelevant, so the abscissa can be designated as a simple numerical axis, as shown in Figure 8. Because of that, this form of R-envelope is designated as R_Q-envelope (meaning 'quantized').

2.5. R/RQ-Envleope—Bigraph, Equivalence, or Complement?

While the graphical appearances of the bigraph and the R-envelope might seem different, they are deeply equivalent.

The bigraph obviously does not contain any information about time, as opposed to the R-envelope, either drawn over original SFs or over approximated $SF^{(A)}$s: both contain data about durations of SFs. Therefore, the R-envelope is over-equivalent to bigraph and, hence, not truly equivalent.

On the other hand, in depicting the R_Q-envelope, having all $SF^{(A)}$s reduced to their maximum amplitudes (as samples) and equidistantly positioned along the abscissa, one can disregard the time and assign samples by integers, designating earlier positions of SFs in the rearranged seismogram.

Because of that, one simply puts a random sequence on integers along the abscissa.

As can be seen on Figure 9, each vertical pair of integers indicates the mapping (displacement) span, e.g., in the 6th column, one sees numbers 20 and 6, which means that the mapping span is $20 - 6 = 14$, etc. By placing the regular indexing, or just identifying the linear address of a number along the abscissa (i.e., its position), one implicitly depicts the respective bigraph. In other words, the bigraph is contained (hidden) within these two strings of integers. This means that the diagram of the R_Q-envelope contains its (equivalent) bigraph. This is a plain bigraph in which nodes and edges do not have weights.

This still does not mean that the R_Q-envelope and respective diagram are fully equivalent since the R_Q-envelope contains maximum amplitudes of SFs, or $SF^{(A)}$s, the only residual from the seismogram in the R_Q-envelope.

If one considers the weighted instead of the plain bigraph, where the weights of nodes represent maximum amplitudes of SFs, or $SF^{(A)}$s, then a complete equivalence between the R_Q-envelope and respective bigraph is attained.

In brief, the simple bigraph is a 2D structure, while the weighted bigraph is a 3D structure.

The claimed equivalence still seems suspicious since there are two very different mathematical objects to compare: one continuous (R_Q-envelope), and another truly discrete (bigraph).

The point is that the R_Q-envelope is seemingly continuous since it is not a monotonic curve but a polygonal line whose nodes are maximum amplitudes of $SF^{(A)}$s. A polygonal line is composed of linear segments, and each segment is completely determined by its nodes, i.e., the coordinates of each point along a segment are determined by the coordinates of its ends (nodes). In other words, a polygonal line does not contain more information than the nodes that it is drawn through.

Therefore, the R_Q-envelope is actually a string of numbers, the same as those assigned to the nodes of the bigraph, where the polygonal line is drawn only for better visual grasping of the relationships between nodes, which is a representation adapted to the vision of humans.

Variability in the R/R_Q-Envelope and the Bigraph

In general, the variability of an entity (a system) depends on the number of elements it contains and their ability to mutually combine to form various resulting structures. This can be seen in earthquakes, particularly if one considers the waveform as a combination of SFs, which can be combined in different ways to yield different combinations. One of these combinations is the considered seismogram itself, and another one is its ordered version.

Obviously, the inner variability of the considered seismogram is higher in its original form than in its ordered counterpart.

With the nature of the R-envelope in mind, it is expected that it will significantly reduce the inherent seismogram variability. As for the (structural) nature (appearance) of the bigraph, less variability reduction is expected.

These claims ought to be verified by results.

2.6. *Bigraph and/or R_Q-Envelope—Transform of Seismogram*

A (virtual) dilemma 'equivalence or complement', stated in Section 2.6 title, could mean that there are two that are 'the same', and if considered as a result of quoted transform, it makes such transform questionable. What is a transform if it produces the result of the type 'two the same in one'?

If the R_Q-envelope is considered only with regular integers along the abscissa, then all structural information from the seismogram is lost. Let us remember that the structure is not only the collection of elements that form such a structure but rather the relations between the elements. The R_Q-envelope is a truly implicit and abstract description of a seismogram (structure).

Quite the opposite holds for the bigraph—it inherently contains information about the seismogram's structure, although in a certain abstracted form.

Therefore, if one speaks about the favorable result of the transform, in this regard, it will be more appropriate to use the bigraph than the R_Q-envelope, and it seems quite natural. However, by adding to R_Q abscissa positional integers, one obtains the bigraph, now contained in the R_Q-envelope (see Figures 8 and 9).

Because of the crossing between these two, it is decided to consider the pair as a result of quoted transform, but with formal favor to the bigraph as an explicit representation of seismogram structure. Figure 10 attempts to express this in graphical terms.

3. Results

3.1. Seismogram Conversion—Bigraph

Figures 2–7 also belongs to Section 3, although they are already used in Methods to graphically explain the concepts.

When the twin bipartite graph from Figure 6 is unified into one, to give the 'coupled bigraph', as shown in Figure 7, the output nodes from both bigraphs with the same integers are pairwise connected by edges.

Triangle approximation and the modulus operation on the seismogram transform its waveform into a unipolar sequence of spikes, as shown in Figure 11, where rearrangement by maximum amplitude is also seen.

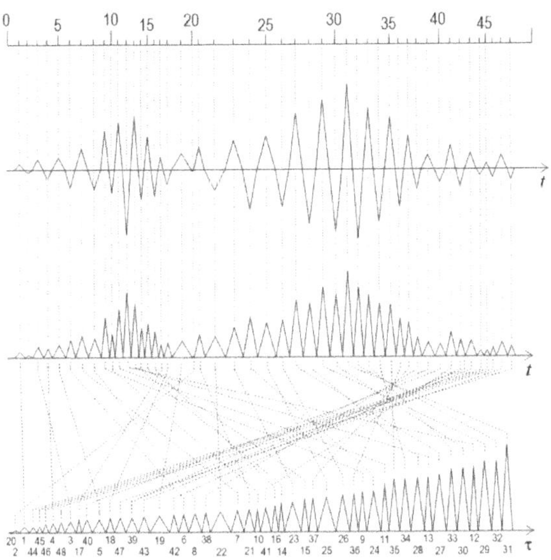

Figure 11. The seismic waveform from Figure 2, approximated by triangles (**up**), is rectified (**middle**) and rearranged by increasing maximum amplitude (**down**). The indexing of spikes goes over all SFs, which is different from that in Figure 2, where separate numeration was applied to positive and negative SFs. The illegibility of the small font of integers around densely distributed spikes is alleviated using a non-linear ruler above the diagrams and vertical reference lines. At the bottom of the down diagram, mapped integers are shown individually.

Now, the ordered sequence of rectified SFs is depicted in a bigraph called the absolute bigraph, shown in Figure 12.

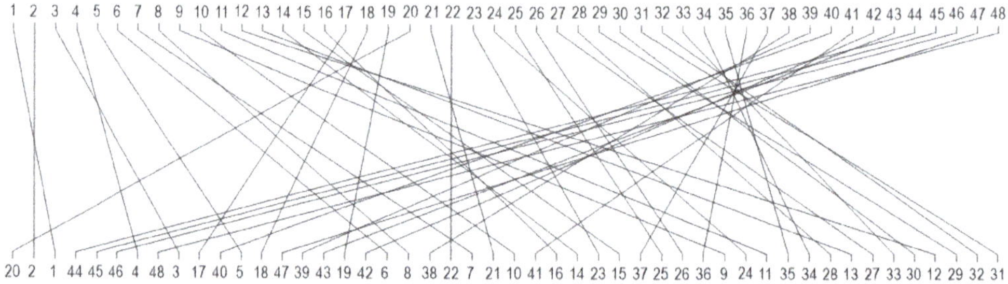

Figure 12. Absolute bigraph of rectified seismogram from Figure 2, shown in Figure 11, which mesh of edges is similar to combined mesh of twin bigraph from Figure 6.

Note the difference in assigning the SFs by integers in twin and absolute bigraphs.

The structure of the corresponding bipartite balanced absolute graph is presented in Figure 9. The mesh of edges reveals the stochastic structure of the sequence of SFs.

As stated in Methods, if the lower bigraph from Figure 6 is inverted (mirrored) to overlap it onto the upper one, to coincide nodes by their numbers, and slightly shifted one over the other to make their nodes visually distinguishable, one obtains a fused bigraph with $2N$ (input/output) nodes, and with $2N$ edges as well, connecting each pair of nodes (Figure 13). Now, this combined bigraph is of the same size (number of edges) as the absolute bigraph. The meshes of edges appear to be quite similar, and it will be interesting for further study.

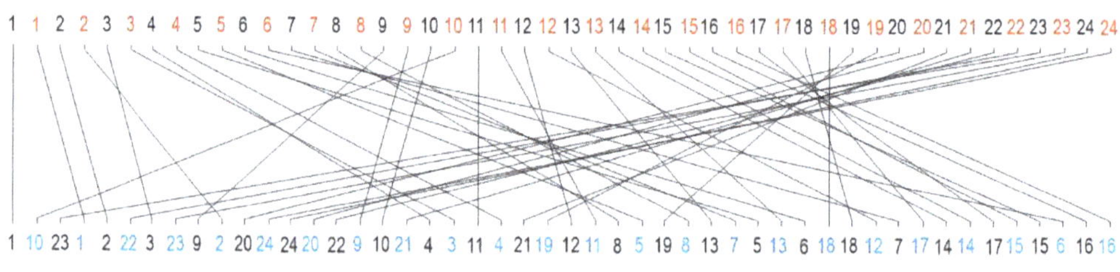

Figure 13. When the down bigraph of the twin bigraph from Figure 6 is rotated (inverted), overlapped, and slightly shifted over the top one to the right (to make both bigraphs from the twin graph visually distinguishable), one gets a fused twin bigraph, very similar to the absolute bigraph, formerly produced by seismogram rectification. Close similarity between meshes of edges in these graphs can be noticed. Indexing of input and output nodes of the down bigraph is given in red and blue integers, respectively, to distinguish each other.

3.2. *Seismogram Conversion—R_Q-Envelope*

In Section 2.4, in Figure 8, one can see a simple example of the R/R_Q-envelope, given for better grasping the concept. Figure 14 presents a real case.

Different variability of the 'head' and the 'tail' is easily seen.

3.2.1. R/R_Q-Envelope and Seismogram Variability

Figure 14 shows a real example of an R_Q-envelope. It is obvious how the R_Q-envelope reduces the intrinsic variability existing in seismograms. Variability is another name for randomness. Rearrangement of random SFs is actually a smoothing, averaging operation.

What one can see in Figure 14 is a gradually decreasing (monotonic) 'tail' and irregular 'head'. This is natural since the tail is (usually) composed of many small SFs (i.e., A_{max}s), while the head is composed of several large SFs (i.e., A_{max}s). Therefore, it is more probable to find an SF among many other SFs whose peak amplitude is close enough to the peak amplitude of currently considered SFs. On the other hand, such probability is low for bigger or big SFs among a small number of big SFs. Because of that, one sees gradual change in the R_Q-envelope along the tail and indented change along the head.

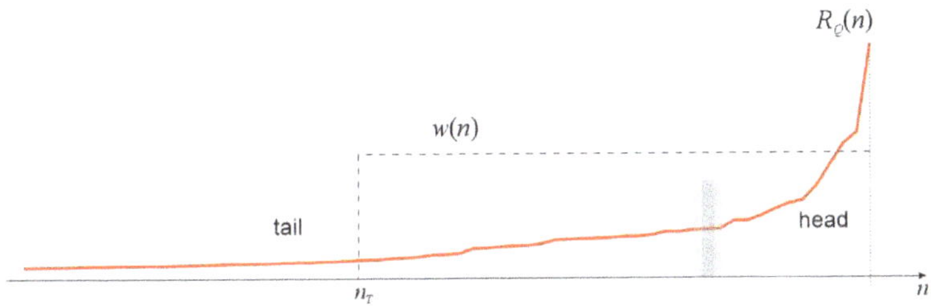

Figure 14. The R_Q-envelope of RSN70 SFERN (file taken from the PEER seismic database), drawn over only positive peaks. Two parts of the envelope are easily distinguished: the 'head', as non-stationary, and the 'tail', as gradually decreasing, almost monotonic. The gray bar indicates an arbitrarily positioned border between the two parts. The weighting function (window) $w(n)$ is depicted by a dashed line to give the idea of tail trimming.

This means that the variability across the seismogram population is expected to be more prominent for heads of their R_Q-envelopes, while the tails 'stick together'.

Taking into account the heavy variability of seismogram duration, either expressed in time units (seconds) or by the number of SFs, namely the slight variability of the R_Q tail corresponding to less significant SFs, gives the basis for applying the weighting function on the R_Q-envelope to trim it to the prescribed length, so that the whole population of R_Q-envelopes, i.e., the whole population of seismograms, can have the same size. The same can be said in terms of bigraph nodes, which will be explained later.

3.2.2. Examples of R_Q-Envelope

In the following, several seismograms are represented by the R_Q-envelope, which provides more complete insight into its main characteristics.

In addition to Figure 14, all Figures 15–19, given below, confirm the previous impression that the R_Q-envelope, and the R-envelope as well, is of similar shape over the whole population of seismograms.

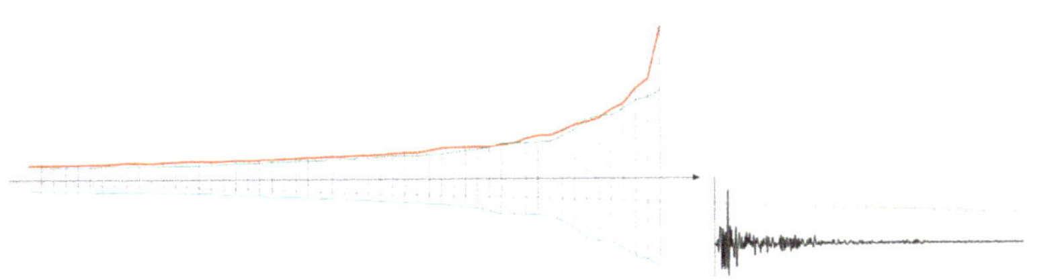

Figure 15. R_Q-envelope of RSN57 (San Fernando), taken from the PEER database, focal mechanism: Reverse, Horizontal-1 Component, contains both positive (red line) and negative (blue line) R_Qs. The modulo of R_Q negative is depicted with a green line. At the right, just for quick visual reference, a small picture of seismogram waveform is inserted, with scale divisions: Dx=10s; Dy=0.2g (acceleration).

Moreover, it can be said that an invariant exists across the seismogram population. The authors expected such a conclusion, but not to that extent. If a certain, large level of seismogram variability remains in the head of the R_Q-envelope, the tail is completely invariant. To further verify this inference, which might be an impression, although it was based on a few examples, more extensive research was carried out on a moderate sample of six seismograms.

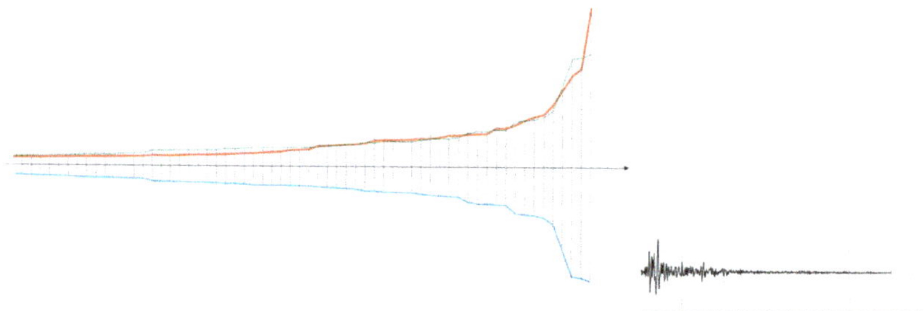

Figure 16. R_Q-envelope of seismogram of RSN70 (San Fernando), also taken from the PEER database, focal mechanism: Reverse, Horizontal-1 Component, already shown in Figure 14. Line colors have the same designation as in Figure 15. At the right, just for visual reference, a small picture of the seismogram waveform is inserted, with scale divisions: Dx = 10 s; Dy = 0.1 g (acceleration).

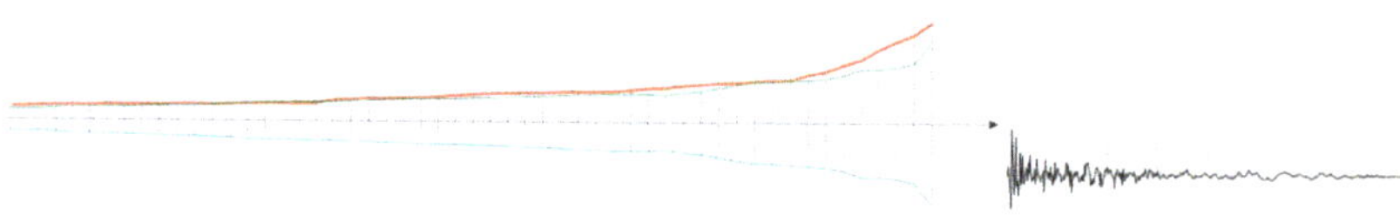

Figure 17. R_Q-envelope of seismogram of RSN88 (San Fernando), also taken from the PEER database, focal mechanism: Reverse, Horizontal-1 Component. Line colors as in Figure 15. At the right, a small picture of the seismogram waveform is inserted, with scale divisions: Dx = 5 s; Dy = 0.1 g.

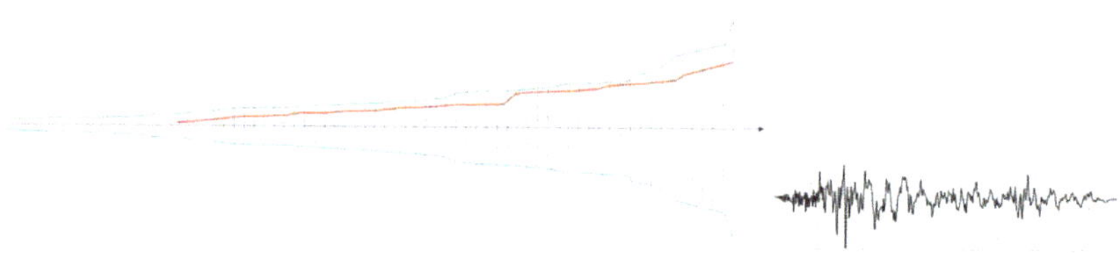

Figure 18. R_Q-envelope of seismogram of RSN286 (Irpinia), also taken from the PEER database, focal mechanism: Normal, Horizontal-1 Component. Line colors as in Figure 15. At the right, a small picture of the seismogram waveform is inserted, with scale divisions: Dx = 5 s; Dy = 0.05 g.

Purposefully, the individual samples were miscellaneous and were not selected from one category defined in the database. The results are given in Figure 19.

The diagrams in Figure 19, particularly on (a), show a narrow variability of R_Q; it seems too narrow to provide proper clustering.

3.2.3. Division Point between Head and Tail of R_Q-Envelope

Figures 15–17 show a very close symmetry of the R_Q-envelope with respect to the n (or τ) axis. In other words, the envelopes R_Q^+ and R_Q^- are similar in shape.

The objective way, although one-parametric, to learn about their closeness is by calculating the difference

$$\varepsilon(\tau) = E_+(\tau) - E_-(\tau), \tag{16}$$

which is shown in Figure 20 for $R(\tau)$ of a typical distant seismogram. Again, one sees almost complete overlapping, closer for tails, less close for heads. Here might be found a criterion for a border separating the head and tail.

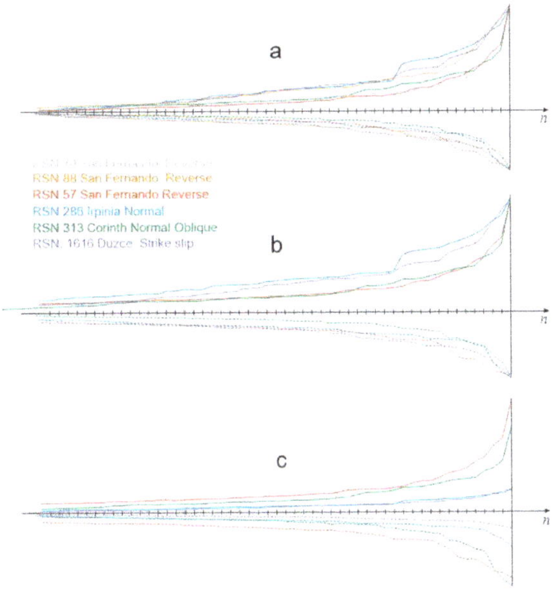

Figure 19. For the list of seismograms, given between diagrams (**a**,**b**), an ensemble of R_Q-envelopes, both positive (solid lines) and negative (dashed lines), is represented as the families of diagrams. All seismograms are taken from the PEER seismic database. Individual diagrams are normalized in certain aspects and overlapped. Colors of curves on diagrams correspond with colors of event names in the list. (**c**) The R_Q-envelopes from indicated seismograms are only overlapped and justified by maximum amplitudes without normalization. (**b**) The four R_Qs from (**c**) are normalized by maximum amplitudes individually. (**a**) As in (**b**), where tails are additionally normalized by the smallest tail.

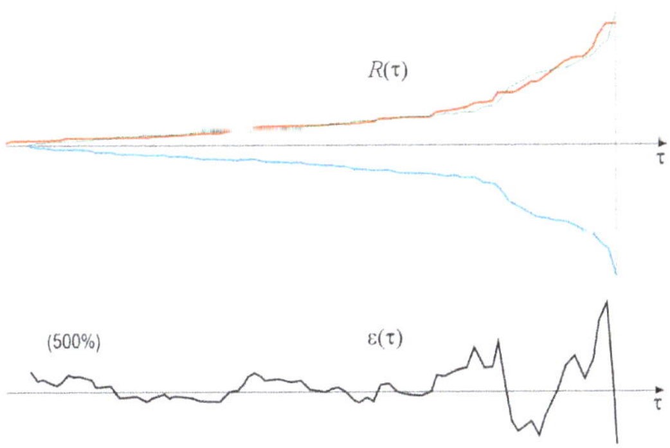

Figure 20. Diagram $\varepsilon(\tau)$, magnified 500% for better clearness, shows the difference between positive (red solid line) and negative (blue solid line) envelopes of a seismogram, where the negative one is overlapped (green solid line) to be compared with the positive one. The change in overlapping measure along the t-axis gives the idea of where to put the separation point between the head and the tail of the R-envelope.

Having in mind all preceding diagrams and in the context of R(τ) symmetry, a general remark, even a recommendation, can be given; it suffices to analyze and process only the positive half of the seismogram, i.e., only positive SFs, thus saving of 50% of all necessary calculations in this approach.

Also, it makes the previous consideration about attaining a singular result in converting a seismogram into its bigraph slightly irrelevant. However, this remark can be regarded only in practical value, while the mathematical significance of seeking the singular product in the transformation process of a seismogram remains in full significance.

3.2.4. Measures of R/R_Q-Envelope Variability

In converting seismograms into an R_Q-envelope or bigraph, the transformation takes place and maps objects from the domain of raw seismograms into the domain of their representations, i.e., images, which are usually simpler than the originals and have less variability. In such cases, it is necessary to define a distance between images to accomplish grouping (clustering) more effectively.

For the case of R-envelope, the distance can be defined as:

$$\Delta_{ij} = \int_{\tau_1}^{\tau_2} \left| \frac{1}{R_i(\tau)|_{\max}} R_i(\tau) - \frac{1}{R_j(\tau)|_{\max}} R_j(\tau) \right| \cdot w(\tau)\, d\tau, \tag{17}$$

which is, actually, the difference between two normalized R-envelopes, where the limits τ_1 and τ_2 are defined by $R(\tau)_{\max}$ and by weighting function $w(\tau)$ (see Figure 14).

A similar equation can be written for $R_Q(n)$ as well.

According to the previous interpretation, which is that the polygonal line defining the shape of R(τ) or of $R_Q(n)$ is essentially not the line but the string of numbers, the distance (17) can be rewritten as:

$$D_{ij} = \sum_{k=1}^{K} \left| \frac{1}{A_{\max}^{(i)}} A_k^{(i)} - \frac{1}{A_{\max}^{(j)}} A_k^{(j)} \right| \cdot w(k), \tag{18}$$

In both cases, one sees the importance of normalization since seismograms can significantly vary in maximum amplitude, while the waveform carries the most significant information and is the main factor in which seismograms differ from each other.

As a general remark, it should be stressed that the convention of the ordering direction of SFs, increasing maximum amplitudes from left to right adopted in this paper, is chosen quite by chance, with no objective reason. This means that the decreasing convention, from left to right, would be equivalent.

The pronounced difference in the shape of the head and the tail in the R_Q-envelope inspired the idea to separately measure the distance for both of them.

3.3. *Bigraph Variability and Issue of Measure*

The R_Q-envelope and the bigraph are considered equivalent representations of seismograms. How, then, can we explain the issue of the variability of the bigraph?

It is easy to notice bundles in the mesh of edges within the bigraph. Moreover, it is easy to identify bundles corresponding to respective groups of SFs just by comparing the output nodes of the bigraph with the shape of the R_Q-envelope. It is easy to grasp this idea by comparing Figures 3 and 4. There, it is easy to select the main bundle of edges, respective to SFs (i.e., $SF^{(A)}$s, or $A_{\max}$s), which form the 'head' of the R_Q-envelope. There are also other bundles that can be identified.

Relying on the claimed equivalence between the bigraph and the R_Q-envelope, it is easy to see that the $R_{Q\max}$ corresponds to the biggest SF/$SF^{(A)}$ or to the rightmost output node n_K in the bigraph (according to the convention of increasing values of SFs).

The edge e_K, connected to the node n_K, designates the mapping of the largest SF/$SF^{(A)}$ and can be named as the structural axis of bigraph (spine, backbone), a reference for considering other edges or bundles of edges.

Obviously, the meaning of these terms depends on the bigraph geometry, which, in this paper, is chosen to be a two-row structure with connecting lines between.

Remembering the previous double indexing of output nodes, positional (mapped) and regular (see Figure 9), it is easy to measure each edge by simply subtracting the correspondent indices, for example:

$$m_\mathrm{d}(n_\mathrm{K}) = m_\mathrm{d}(24) = 24 - 16 = 8$$

This means that the mapping distance of the main node (max SF), which is labeled by index 16 (defining its position in input nodes), is translated (mapped) to position 24 (output nodes), which equals 8 units of displacement.

Similarly, e.g., input node 7, which is mapped to position 20 of the output index string, which makes $m_\mathrm{d}(20) = 20 - 7 = 13$ units of displacement.

For the given bigraph geometry, its height is H_b, and the mapping distance yields the edge slope:

$$\alpha(n_\mathrm{K}) = \alpha(24) = \arcsin(H_\mathrm{b}/m_\mathrm{d}(24))$$

around which are arranged edges with similar slopes, making the primary bundles the main structural element of the bigraph (in its mesh).

$$B_0 = \{e_k : |\alpha(e_k) - \alpha(e_K)| \leq \delta,\ \forall k < K\} \tag{19}$$

In that way, a relationship is established between the head in the diagram of R_Q and the primary bundle in the bigraph mesh. In a similar way, one can define secondary bundles.

Formatting of Bigraps

One dimension of seismogram variability is its duration, which, when transferred into an R_Q or bigraph, means the number of SFs or the bigraph size (number of nodes). Even a group of earthquakes defined by a certain set of grouping criteria shows heavy variability and prevents clustering.

There are two proposals to equalize variable lengths of seismograms.

Extension of bigraph by empty complement—It is well known from number theory and binary coding that, e.g., 011 = 00000011. Applying this simple idea to bipartite graphs, an empty bigraph is added to the current (actual) bigraph to attain the prescribed size. In other words:

$$\#(BG_{E(k)}) = \#(BG_{(k)}) + \#(E_{(k)}) = G, \forall k$$

which one reads out as the size (G) of the extended bigraf is the sum of the sizes of the actual bigraf and its empty complement bigraph to attain a prescribed size (see Figure 21), constant for all bigraphs.

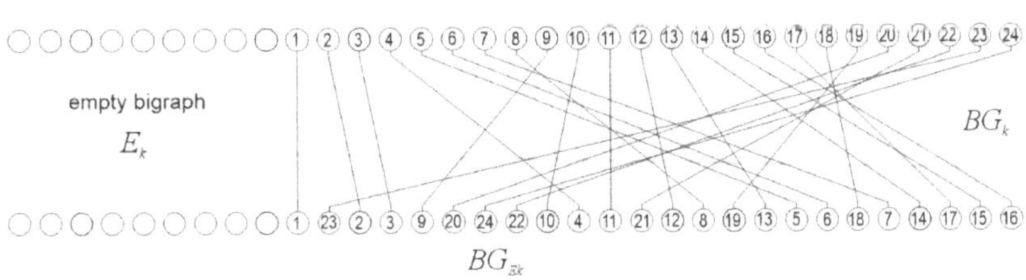

Figure 21. To illustrate the concept of the empty graph, the simple graph from Figure 4 is used, which is extended by an empty graph to the prescribed size. Note that the direction of extension goes along the least significant nodes (SFs).

Another parallel with the empty bigraph can be found in the set theory, with the empty set, but there are significant differences. The empty set has nothing in it but possesses only its existence, pointing to the fact that the set is a frame, and after that, it becomes a real set by containing certain elements.

The empty bigraph is not totally empty: it contains nodes but not the edges between them. Let us note that the Gilbert model of the random graph contains one realization in its statistical ensemble, which is a graph with no edges, similar in the sense of the defined concept of an empty graph [27].

Weighting function—In seismic science, researchers need complete seismograms to study an earthquake's whole background, processes in source, and propagation afterward. In other areas of research and engineering, often there is no need for the whole waveform, but just for certain segments (e.g., seismic civil engineering). By applying the proper weighting function (also known as the window function from digital signal processing), one can extract the most important segment on purpose and reduce the variability across the seismogram sample and, hence, of respective bigraphs.

Note that the simplest, rectangular weighting function is applied since the application of any monotonic one would convert the original bigraph, which is simple (no weighted nodes or edges), into a weighted graph, which would be more complicated for analysis [29].

The weighting function is positioned with respect to the highest SFs (Figure 22) so that it encompasses the most significant SFs and cuts off the tail of less significant ones. Of course, the question of where to place the cut has to be considered.

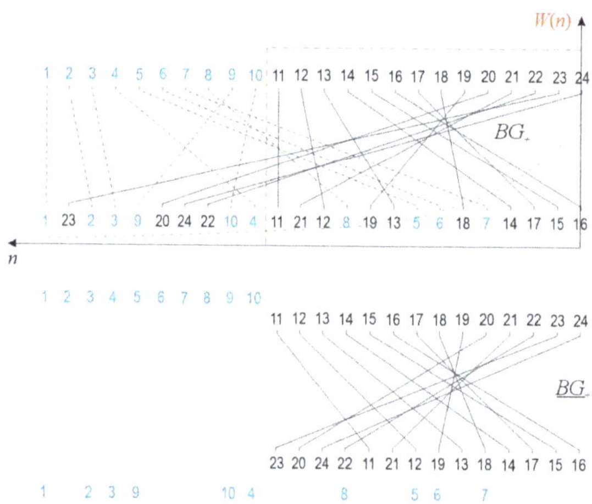

Figure 22. Bigraph BG_+ from Figure 4, is truncated by rectangular weighted function (window, depicted by dashed orange line) $W(n)$ so that less significant SFs are eliminated from describing the positive part of the seismogram. The dashed green line curve on the $W(n)$ diagram indicates the increasing direction of the ordered sequence of positive SFs (the largest SFs are at the right end). Dashed black lines and blue numbers depict discarded edges and nodes, respectively. After that, the remaining nodes and respective edges are rearranged to give a compact bigraph $\underline{BG}_+$.

A very important additional effect of the weighting function application appears. Since the output nodes are randomly interleaved, the edges in the whole mesh of bigraph, which correspond to cut-off input nodes (i.e., insignificant SFs), and form a sub mesh, will disappear, so the remaining bigraph should be compacted, which involves rearrangement of residual bigraphs (Figure 22). The compacting changes the mesh structure. It will be very interesting to compare the edge mesh of original and trimmed bigraphs.

On the contrary, for seismograms of prominently impulse character, i.e., of short duration, in order to avoid such specific cases to determine the trimming length of the predominant rest of the longer seismograms in the population, the concept of the empty bigraph may be of help.

3.4. *Examples of Seismogram Combined Representation by Bigraph and R_Q-Envelope*

The bigraph and the R_Q-envelope create a complementary representation of a seismogram. In the following, a few examples are given and briefly interpreted. All examples are taken from the PEER seismic database except the last one.

3.4.1. R_Q-Envelope and Bigraph of Seismogram RSN1616

This example is interesting because of the fact that complex waveforms of EFs/SFs compose the seismogram, so that the interesting results might be expected. It seems to be obvious from its waveform, given at the top of Figure 23, but the R_Q-envelope shows pretty calm decreasing.

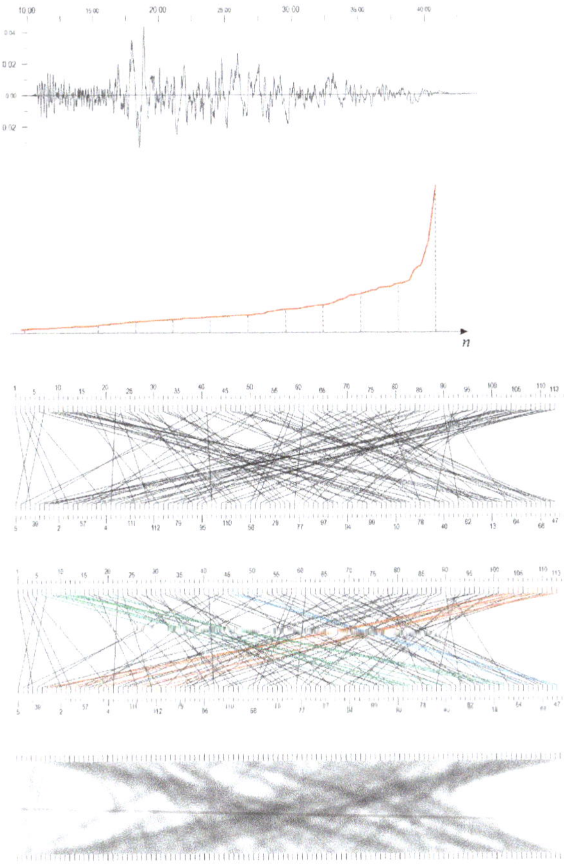

Figure 23. (from (**top**)–(**down**)) Representative seismogram (RSN 1616 Duzce (Düzce) 1999, Turkey, Lamont 362), containing plenty of different SF shapes, with main parameters: D5–75 14.6 s, D5–95 20.5 s, Mag 7.14, Strike-Slip, RJB 23.41 km, Rrup 23.41 km; file used DUZCE_362-N.AT2, Horizontal-1 Component time series). (next) R_Q-envelope for positive SFs (113 points). (next) Bigraph only for positive SFs (black and white). (next) Bigraph only for positive SFs (colored, to indicate main structural bundles): blue edges form the main (primary) bundle of bigraph, the green ones form a secondary bundle, while the red bundle is of the least significance. (next) Bigraph just for positive SFs (black and white, blurred with Gaussian blur to better indicate all structural bundles in the edge mesh). Numeric labeling of input (up) and output (down) nodes are, for visual simplicity, indicated by enumerated scales, where the upper sequence of integers is regular (1, 2, 3, 4, 5, 6, . . . , 120), while the down random one is determined by edges. The size of the bigraph is 113.

The number of nodes in the bigraph, 113 in total, corresponds to the number of significant SFs. The bigraph mesh of edges shows characteristic bundles.

It is important to note that blurred diagram is intentionally blurred to emphasize the bundles of edges, and in such shadowy manner to reveal skeleton of bigraph mesh.

3.4.2. R_Q-Envelope and Bigraph of Seismogram RSN286

This example of seismogram is impulse-like waveform, with asymmetry in negative main amplitudes. Its R_Q-envelope (Figure 24) shows perturbation of the head.

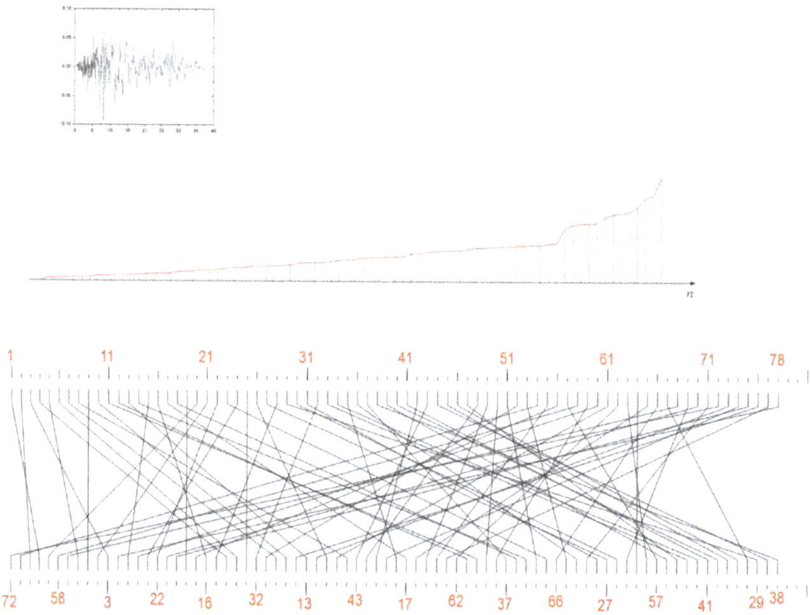

Figure 24. Seismogram RSN286. Insert shows its waveform for reference. R_Q is distinguished by specific head shape. The structural content of the bigraph shows uniform crossing of edges within broad area of mesh.

The bigraph mesh seems to be pretty uniform, without prominent bundles (Figure 24).

3.4.3. R_Q-Envelope and Bigraph of Seismogram RSN313

This example seems pretty similar to the previous one, with still more prominent pulse nature. Again there are prominent negative spikes. The R_Q-envelope is almost regular along its head (Figure 25).

However, the bigraph is of very specific structure, showing two main bundles, almost divided into two separated halves (Figure 24).

3.4.4. R_Q-Envelope and Bigraph of Typical Distant Seismogram

Distant seismograms are typically characterized by resonances, and by quasi periodicity, of the random form. As can be seen on Figure 26, the R_Q-envelope is generally monotonic, with ripples on the head. As for the bigraph, its mesh is almost uniformly intertwined, in a similar way for positive and negative spikes (Figure 26).

The R_Q part of the complement R_Q-BG, as the result of said transform of the seismogram, evidently reduces original variability to a narrow interval. It is so narrow that there seemingly is not enough room for effective clustering of seismogram images. This claim will be verified on a sample of seismograms of sufficient size, which would provide well-grounded statistical inference. This paper had no intention to go that deep into this issue to avoid its unacceptable volume.

Although both the R_Q and bigraph are the result of SFs ordering, the R_Q is a somewhat closed image of the seismogram, while the bigraph is more open, having more degrees of freedom for the clustering process. Otherwise, the bigraph possesses not only a wider interval for clustering with respect to R_Q, but it has the frame for that, provided by several structural elements, bundles, and elements within bundles.

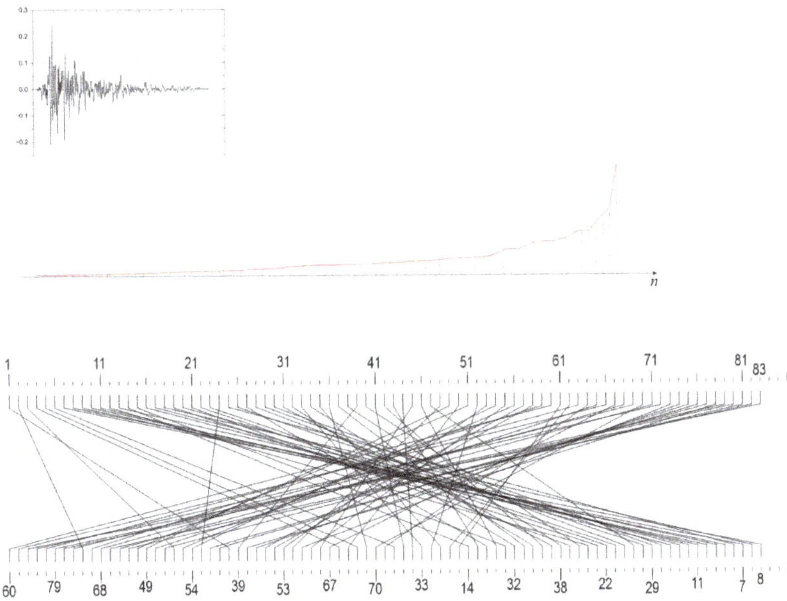

Figure 25. Seismogram RSN313. Insert shows its waveform for reference. Structural content of the bigraph shows almost only two main bundles, sharply covering big and small spikes.

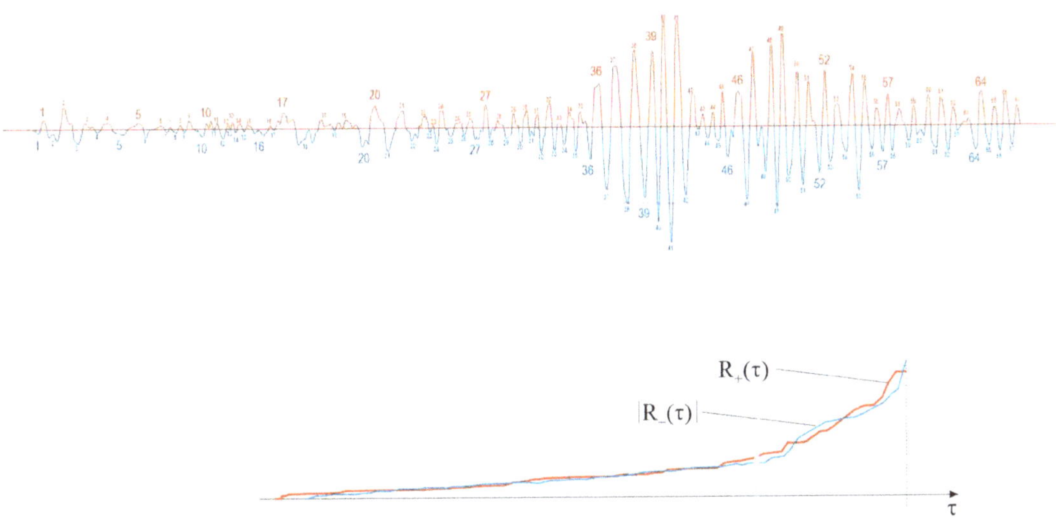

Figure 26. *Cont.*

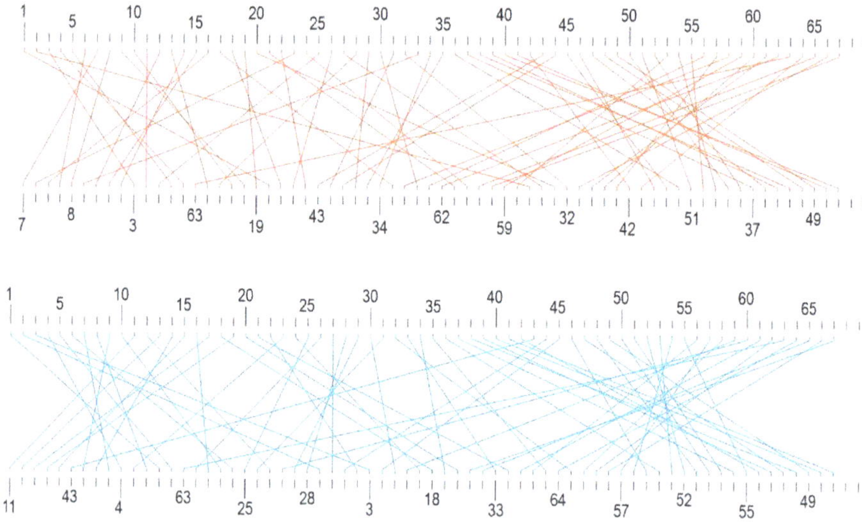

Figure 26. (from (**top**)–(**down**)) An example of a distant seismogram: SFs are labeled by positive integers, separately for positive (red) and negative (blue) SFs. For better legibility of SF indexing, same indices are given in bigger font size, while the others can be easily identified by enumerating 'points' (i.e., indices in small size font) between them. (next) Both R(τ) envelopes, for positive and negative SFs. (next) Bigraph for positive SFs. (next) Bigraph for negative SFs.

4. Discussion

The primary intention of this paper was limited to only introducing the new concept and considering its features in the frame of seismogram interpretation, processing, and, finally, more effective clustering. Some quantitative and qualitative results are given, which contain assessments of issues of reduced variability and clustering. The exact verification of these issues, in addition to a comprehensive presentation of the many items in this paper, would lead to excessive volume; therefore, it is left for the next research.

Seismogram variability and issue of grouping—An earthquake is a very complex random phenomenon, i.e., a very complex random entity with immense variability, which imposes difficulties in grouping (clustering) and database organization.

Typically, grouping (clustering) criteria (parameters) usually include fault (focal mechanism), distance, magnitude, duration (D5–95(s)), Rjb (Joyner–Boore distance), Arias intensity, event name, station name, year, etc. [18]. These are general criteria whose generality confirms said difficulties.

The fault mechanism could serve as a basic classification factor. However, earthquakes affect the surrounding areas, bringing propagation effects into seismograms recorded at certain sites. This is easily seen in, e.g., the PEER database, which is organized according to general criteria despite the fact that it is the result of tremendous accumulated experience and knowledge.

In dealing with a database, one enters parameters for a certain group (class) and can work with any of the presented seismic events (i.e., earthquake), either in the form of its seismogram (time series) or its spectrum. Selecting an individual event from the database by its name is a trivial operation and not concerned with its organization.

It is interesting to notice that the spectrum, as an ordered and more compact image of the seismogram with less variability, is only an entity added to the seismogram, but, as such, not a factor that can be used for database organization, i.e., not a factor but the object of grouping.

The problem of clustering objects that are difficult to cluster is usually solved by converting one object into another using a mathematical operation called transform, mapping it from the original domain into the domain of the object's images, which are now easier to cluster.

This paper proposes an approach to transforming a seismogram into a combination of an R_Q-envelope and bigraph, which, then, deserves the name transform.

More about the variability of R_Q-envelope and bigraph—Previously, R_Q variability was thoroughly considered, and its measure was defined. It can serve as a distance in certain topological spaces, which is the basis for the criterion of clustering (grouping). The variability of the head and the tail in the R_Q-envelope is quite different (see Figure 14).

For the shape of the R_Q-envelope, one can even speak of it as invariant. More precisely, the head of the R_Q is less invariant than the tail of the R_Q, which is inherently invariant, so it is possible to divide the R_Q into two parts and consider their variability in different ways or even measure their variability according to different criteria. In a practical way, the issue of R_Q variability can be brought down to the head variability.

As for bigraph variability, the measure of distance relies on primary and secondary bundles, angle, and number of edges per bundle.

R_Q as invariant and redundancy in seismogram—The form of the R_Q and convergence in its ensemble apparently suggest the existence of a kind of invariant. Moreover, they suggest redundancy in seismograms as well since all seismograms addressed in this paper show very close similarity between their R_Qs. This means that each seismogram contains plenty of redundant information. It is well known that the alternating fluctuation along the waveform is the main cause of redundancy through strong local correlation of alternating amplitudes.

In this regard, an important fact should be underlined: the R_Q is the result of the ordering of SFs by peak amplitudes. This means that in the ordering procedure, all local correlative interrelations that provide seismogram redundancy are disrupted. Then, one would expect redundancy to decrease, but it is increased. The explanation of this paradox is simple: the ordering means putting the most similar SFs one next to the other (which are now closer by peak amplitude), which now means the appearance of additional correlation. This is an interesting effect and deserves further analysis.

Equivalence of R-envelope and bigraph—This issue has been discussed extensively so far. Here, we will add just a few essential remarks.

Instead of the term 'equivalence', it might be more appropriate to qualify the R-envelope and bigraph as complementary characteristics, although their equivalence cannot be disputed.

The R- and R_Q-envelopes contain information on maximum amplitudes of SFs, giving the amplitude content of the seismogram waveform.

If one assumes regular indexing along the abscissa, then the complete information about the mapping structure is lost. However, if one adds positional indexing, then the complete mapping is known, and, actually, the complete bigraph is given below the abscissa (Figures 8 and 9). Thus, the R- or R_Q-envelope with positional indexing contains the complete bigraph. On the other hand, the bigraph, without any additions, directly by its structure, by nodes and mesh of edges, reflects the general structure of the seismogram but without data about maximum amplitudes. This is for the case of a simple bigraph with no weights in nodes and edges. In brief, one can say that the bigraph is more internal, while the R/R_Q-envelope is a more external representation of the seismogram.

Main part of the seismogram—There is an additional distinction in the previous concept. The main part of a seismogram is the one containing significant SFs, i.e., all SFs that are large, which mostly cause significant seismic effects on the built environment. An ordering operation gathers these SFs together, whatever their disposition throughout the seismogram, so that the ordered seismogram is a monotonic object where SFs gradually increase or decrease, depending on the ordering convention (starting from the largest or from the smallest SF). The main part of the rearranged seismogram contains the SFs that are the most important for seismic effects (magnitude, delivered energy, inflicted damage, influence on structures, etc.). The long decreasing tail of the ordered seismogram stems from the tail, which is of secondary influence and significance. Therefore, the ordering makes more explicit the issue of what part of the tail is to be recorded or where to terminate its recording prior to it sinking into seismic noise.

One should be careful when considering the main part of the seismogram. The rearrangement operation gathers all the prominent SFs in one place, sometimes from different areas of the seismogram, so it may produce the impression that they act together. In addition, the SFs are gathered one-sided, positive, negative, or both, with the seismogram rectified. This means they seemingly act in one direction, but the essential capability to influence the built environment and cause damage is its vibratory (alternating) behavior when two maximum opposite SFs, as a part of the most prominent EF, run over the structures.

It is interesting to note that the head of the R_Q-envelope corresponds with the main part.

Does the transform $\mathcal{T}$ produce a simple or weighted bigraph?—This paper is written under the assumption that all forms of bigraphs that result from transform $\mathcal{T}$ are simple, which means that they are not directed nor weighted (the nodes and edges are simple).

Let us remember that the ordering operation rearranges SFs; therefore, the significant ones are separated from less significant or insignificant ones. In other words, the significance of SFs rises along the ordered sequence. Since the SFs are transposed into nodes of a bigraph, it is quite natural to consider input nodes as weighted, different from each other. Similarly, an edge that connects significant (input and output) nodes (read: SF with bigger maximum amplitude) cannot be considered equal to the edge connecting insignificant edges. Therefore, the answer to the title question is obvious. Analysis of weighted bigraphs is more complex than that of simple ones. We followed an implicit assumption in this paper, which was to treat all forms of bigraphs introduced in the paper as simple, as our intention was to present these fundamental concepts in a clear context. In further research, it will be natural to consider weighted bigraphs.

Bipartite graph randomness—In the Introduction, we mentioned deterministic and random bipartite graphs. The first show all existing relations (edges) between input and output nodes. It is seemingly the same in the case of a random graph—one sees the firm structure of edges between nodes in such a graph. Actually, a random graph shows possible (potential) edges between nodes.

This means, in the case of a (deterministic) conventional graph, the actual graph is the only one, while in the second case, the considered graph is just one of many possible realizations, creating a whole statistical ensemble in which any of its elements is a graph with a firm structure, but slightly different from the others. Each 'tossing' (trial) of the potential random graph produces a different realization—a graph with a firm structure.

EFs/SFs and wavelets—There is an apparent similarity between the concept of EF/SF and wavelets. How, otherwise, to call these 'wavelets' when they are small chunks of a complex waveform, a seismogram in this case?

The authors deliberately decided not to use the term wavelet in this case.

The term wavelet can have two associations: as a small wave (impulse) and as a mathematical object. What is more, these associations can be mixed, and one can complain about borrowing that name for a new concept.

The concept of EF/SF truly reminds us of wavelets, which begs for a comparison of these two approaches to seismogram description. Wavelets are particularly suitable for localized signals and random transients; a typical example is the seismic signal represented by a seismogram.

Regarding the comparison between the concept of EF/SF and wavelets, there are essential differences. The wavelets (of a certain type) represent a system of similar elementary waveforms and impulses, a kind of offspring originating from a generative function, the so-called mother function. They form the basis within which a signal is to be decomposed. It is quite similar to the FT, where a sinusoid is a generative function, the function of a parameter, which can create a set of harmonics, a basis for signal decomposition.

As for the concept of EF/SF introduced in this paper, it is impossible to create a basis because they are not only parametrically interrelated but are of random shape, i.e., the SFs do not have shape similarity. In addition, SFs are not 'wavelets' that are applied to the waveform from outside as wavelets do, they come from inside; moreover, they are parts of the waveform they intend to analyze.

Speaking of wavelet transform, spectral descriptions of seismograms are widely used in various areas related to seismic effects and their consequences. The prominent place of spectrum in seismic databases confirms its significance and practical applicability. However, the Fourier transform (FT), with its numerical derivatives (STFT, DFT, and FFT), still dominates the spectral world, while the WT gradually builds its place in it.

5. Conclusions

The theme of the paper proved to be quite interdisciplinary.

The authors' ambition was not to make a complete discovery and a new database but, for the present, only to point out the direction of possible solutions and to propose characteristics that promise solutions.

The proposed characteristics, the R_Q-envelope and bipartite graph, certainly reduced the variability of seismograms.

As for the R_Q, the extent of variability narrowing makes one think that the R_Q is very close to being an invariant of the seismogram population. This finding deserves thorough further research.

On the other hand, the narrow range in which the R_Q-envelope varies makes it a questionable basis for clustering of the seismogram population.

The bigraph, in comparison to the R_Q, also attains narrowed variability but within a domain having a few degrees of freedom. The bigraph goes deeply into the structure of the seismogram, describing its waveform in a more thorough manner through the system of edge bundles. The authors' assessment is that it provides a wider basis for clustering with respect to existing one-parametric criteria.

This transform opens the door for defining various characteristics of seismograms.

The presented approach is obviously not limited only to seismograms but can be applied to all similar stochastic processes.

Author Contributions: Conceptualization, R.B.; Resources, R.B. and L.B.; Data curation, R.B. and L.B.; Formal analysis, R.B.; Supervision, R.B.; Investigation, R.B.; Visualization, R.B., Methodology, R.B. and L.B.; Writing—original draft, R.B. All authors have read and agreed to the published version of the manuscript.

Funding: This research received no external funding.

Institutional Review Board Statement: Not applicable.

Informed Consent Statement: Not applicable.

Data Availability Statement: Not applicable.

Acknowledgments: Using the seismograms from the PEER Ground Motion Database greatly contributed to our research.

Conflicts of Interest: The authors declare no conflict of interest.

References

1. Bormann, P. (Ed.) *IASPEI New Manual of Seismological Observatory Practice*; GFZ German Research Centre for Geosciences: Potsdam, Germany, 2012. Available online: http://nmsop.gfz-potsdam.de (accessed on 10 September 2023). [CrossRef]
2. To, G.F.; Schubert, G. (Eds.) *Treatise on Geophysics*, 1st ed.; 11 Volumes Set; Elsevier: Amsterdam, The Netherlands, 2007; 6054p, ISBN 10:0444519289. Available online: https://www.geokniga.org/bookfiles/geokniga-treatise-geophysics.pdf (accessed on 15 July 2023).
3. Havskov, J.; Ottemöller, L. Processing Earthquake Data, October 2009. Available online: https://www.geo.uib.no/seismo/SOFTWARE/DOCUMENTATION/processing_earthquake_data.pdf (accessed on 22 August 2023).

4. GFZ, Seismic Hazard and Risk Dynamics, Seismic Hazard Assessment for the D-A-CH countries. Available online: https://www.gfz-potsdam.de/en/section/seismic-hazard-and-risk-dynamics/projects/previous-projects/probabilistic-seismic-hazard-assessments/d-a-ch-seismic-hazard-for-the-d-a-ch-countries-germany-austria-switzerland (accessed on 6 September 2023).
5. Grünthal, G.; Mayer-Rosa, D.; Lenhardt, W. Joint strategy for seismic hazard assessment; application for Austria, Germany and Switzerland. In Proceedings of the 21st General Assembly of the International Union of Geodesy and Geophysics (IUGG), Boulder, CO, USA, 2–14 July 1995. B 404.
6. Spence, R.; So, E.; Jenny, S.; Castella, H.; Ewald, M.; Booth, E. The Global Earthquake Vulnerability Estimation System (GEVES): An approach for earthquake risk assessment for insurance applications. *Bull. Earthq. Eng.* **2008**, *6*, 463–483. [CrossRef]
7. Jayaram, N.; Shome, N.; Rahnama, M. Development of earthquake vulnerability functions for tall buildings. *Earthq. Eng. Struct. Dyn.* **2012**, *41*, 1495–1514. [CrossRef]
8. Preciado, A.; Ramirez-Gaytan, A.; Salido-Ruiz, R.A.; Caro-Becerra, J.L.; Lujan-Godinez, R. Earthquake risk assessment methods of unreinforced masonry structures: Hazard and vulnerability. *Earthquakes Struct.* **2015**, *9*, 719–733. [CrossRef]
9. Gaxiola-Camacho, J.R.; Azizsoltani, H.; Villegas-Mercado, F.J.; Haldar, A. A novel reliability technique for implementation of Performance-Based Seismic Design of structures. *Eng. Struct.* **2017**, *142*, 137–147. [CrossRef]
10. Maio, R.; Ferreira, T.M.; Vicente, R. A critical discussion on the earthquake risk mitigation of urban cultural heritage assets. *Int. J. Disaster Risk Reduct.* **2018**, *27*, 239–247. [CrossRef]
11. Shadmaan, S.; Islam, A.I. Estimation of earthquake vulnerability by using analytical hierarchy process. *Nat. Hazards Res.* **2021**, *1*, 153–160. [CrossRef]
12. Dowrick, D. *Earthquake Resistant Design and Risk Reduction*, 2nd ed.; Wiley: Chichester, UK, 2009; ISBN 13:9780470778159.
13. Duggal, S.K. *Earthquake-Resistant Design of Structures*, 2nd ed.; Oxford University Press India: Noida, India, 2013. Available online: https://www.technicalbookspdf.com/earthquake-resistant-design-of-structures-2nd-edition/ (accessed on 10 September 2023).
14. *FEMA Earthquake-Resistant Design Concepts, An Introduction to the NEHRP Recommended Seismic Provisions for New Buildings and Other Structures, FEMA P-749/December 2010*; National Institute of Building Sciences Building Seismic Safety Council: Washington, DC, USA, 2010. Available online: https://www.fema.gov/sites/default/files/2020-07/fema_earthquake-resistant-design-concepts_p-749.pdf (accessed on 2 September 2023).
15. Sucuoğlu, H.; Akkar, S. *Basic Earthquake Engineering-From Seismology to Analysis and Design*; Springer: Cham, Switzerland, 2014. [CrossRef]
16. Moshou, A.; Konstantaras, A.; Argyrakis, P.; Petrakis, N.S.; Kapetanakis, T.N.; Vardiambasis, I.O. Data Management and Processing in Seismology: An Application of Big Data Analysis for the Doublet Earthquake of 2021, 03 March, Elassona, Central Greece. *Appl. Sci.* **2022**, *12*, 7446. [CrossRef]
17. Lindup, G.; Sharpe, R.; Holm, S.; Jarman, S.; Sinclair, R.; Bloomer, A.; Tsamandakis, E.; Smart, C. *Practice Note 19 Seismic Resistance of Pressure Equipment and Its Supports*; engineering New Zealand: Wellington, New Zealand, 2019; ISSN 1176-0907. Version 5. Available online: https://d2rjvl4n5h2b61.cloudfront.net/media/documents/PN19-SeismicResistancePressureEquipment_V5-2019.pdf (accessed on 10 September 2023).
18. PEER. PEER Ground Motion Database. 2022. Available online: https://ngawest2.berkeley.edu/spectras/new?sourceDb_flag=1 (accessed on 9 September 2023).
19. Scasserra, G.; Stewart, J.P.; Kayen, R.E.; Lanzo, G. Database for Earthquake Strong Motion Studies in Italy. *J. Earthq. Eng.* **2009**, *13*, 852–881. [CrossRef]
20. Yuan, X.; Wang, X.; Guo, J.; Dou, A.; Ding, X. Design and realization of RS application system for earthquake emergency based on digital earth. In *IOP Conference Series: Earth and Environmental Science, Proceedings of theDigital Earth Summit, Beijing, China, 7–8 July 2016*; IOP Publishing Ltd.: Bristol, UK, 2016; Volume 46, p. 012037. [CrossRef]
21. USGS National Earthquake Information Center (NEIC), Earthquake Hazards Program. Available online: https://www.usgs.gov/programs/earthquake-hazards/seismogram-displays (accessed on 6 September 2023).
22. Bollobash, B. *Random Graphs*; Cambridge University Press: Cambridge, UK, 2001; M08 30.
23. Barabási, A.-L. Network Science. Available online: http://networksciencebook.com/ (accessed on 10 September 2023).
24. Erdös, P.; Rényi, A. *On the Evolution of Random Graphs*; The Mathematical Institute of the Hungarian Academy of Sciences: Budapest, Hungary, 1960; Volume 5, pp. 17–61.
25. Gilbert, E.N. Random graphs. *Ann. Math. Stat.* **1959**, *30*, 1141–1144. Available online: https://www.stat.cmu.edu/~brian/780/bibliography/09%20Other%20Models/Gilbert%20-%201959%20-%20Random%20graphs.pdf (accessed on 10 September 2023). [CrossRef]
26. Newman, M. *Networks: An Introduction*; OUP: Oxford, UK, 2010; M03 25.
27. Dorogovtsev, S.N. *Lectures on Complex Networks*; Clarendon Press: Oxford, UK, 2010. Available online: https://sweet.ua.pt/sdorogov/dorogovtsev_lectures_first_pages.pdf (accessed on 10 September 2023).
28. Dorogovtsev, S.N.; Mendes, J. The Shortest Part to Complex Networks. July 2004. Available online: https://arxiv.org/abs/cond-mat/0404593 (accessed on 10 September 2023).
29. Clauset, A. Three Lectures on Networks. 2022. Available online: https://aaronclauset.github.io/slides/Clauset_2022_CSSS_IntroNetworks_1.pdf (accessed on 10 September 2023).

30. Sayama, H.; Cramer, C.; Porter, M.A.; Sheetz, L.; Uzzo, S. What are essential concepts about networks? *J. Complex Netw.* **2015**, *4*, 457–474. [CrossRef]
31. Asratian, A.S.; Denley, T.M.J.; Häggkvist, R. *Bipartite Graphs and Their Applications*; Cambridge University Press: Cambridge, UK, 1998.
32. Pavlopoulos, G.A.; Kontou, P.I.; Pavlopoulou, A.; Bouyioukos, C.; Markou, E.; Bagos, P.G. Bipartite graphs in systems biology and medicine: A survey of methods and applications. *GigaScience* **2018**, *7*, 1–31. [CrossRef] [PubMed]
33. Cvetković, D. Applications of Graph Spectra: An Introduction to the Literature. Available online: https://elib.mi.sanu.ac.rs/files/journals/zr/21/n021p007.pdf (accessed on 10 September 2023).
34. Brouwer, A.; Haemers, W. Spectra of Graphs, Monograph, 1 February 2011. Available online: https://www.win.tue.nl/~aeb/2WF02/spectra.pdf (accessed on 10 September 2023).
35. Han, J.; Kamber, M. *Data Mining: Concepts and Techniques*, 2nd ed.; Elsevier: Amsterdam, The Netherlands, 2006.
36. Joshi, K.D. *Introduction to General Topology*; New Age International: New Delhi, India, 1983; reprint 2004.

applied sciences

MDPI

Article

A Fracture Model for Dynamic Sliding Safety Evaluation of a Concrete Dam Subjected to Seismic Excitation

Jeeho Lee [1,*] and Hyeong Seok Lim [2]

[1] Department of Civil & Environmental Engineering, Dongguk University, Seoul 04620, Republic of Korea
[2] Department of Civil & Environmental Engineering, University of Illinois at Urbana-Champaign, Urbana, IL 61801, USA; hl65@illinois.edu
* Correspondence: jeeholee@dgu.edu; Tel.: +82-2-2260-3352

Abstract: The Sliding Safety Factor (SSF) is a crucial criterion for the sliding stability evaluation of concrete dam structures. A concrete gravity dam subjected to strong earthquakes undergoes progressive fractures, in addition to pre-existing fractures, at the dam–foundation interface, which causes a reduction in the shear strength against sliding. In this study, a new SSF is suggested to take account of the progressive fractured area at the dam–foundation interface. A contact and sliding model for the dam–foundation system is also suggested to compute the dynamically varying normal forces and sliding motions for the suggested SSF. To investigate the effect of the progressively fractured area on the sliding safety evaluation, the conventional, improved, and newly suggested SSFs are compared using the dynamic seismic analysis results of a concrete gravity dam. The conventional formulation of the SSF, in which the fractured area is not represented, yields extremely overestimated sliding safety judgements when a dam is subjected to strong earthquakes. On the other hand, the newly suggested SSF with the proposed contact–sliding model provides more realistic and conservative sliding safety evaluation results than the others.

Keywords: Sliding Safety Factor; fracture progress; contact–sliding model; seismic analysis; concrete gravity dam

1. Introduction

Large concrete dams are crucial infrastructures generating backbone energy power and controlling water resource distribution. The earthquake safety of those structures is an important issue, since damaged large dams have an extremely negative impact on social and economic systems across a sizable region, and often those damages are irrecoverable. Especially, the possibility of sliding during an earthquake is important in the safety assessment of a concrete gravity dam, in that sliding can cause the total failure of a dam [1,2].

To quantify the chance of sliding as a whole dam structure, the concept of the Sliding Safety Factor (SSF) has been used widely [3,4]. The SSF of a dam is generally defined as the ratio between the shear strength of the dam–foundation interface and the external shear force applied to the interface. The shear strength consists of the cohesive force, and the multiplication of the vertical force and the coefficient of the internal friction. When the external force exceeds the shear strength, in which the SSF is lower than unity, a dam possibly slides. Since this definition of SSF is straightforward, major dam agents including USACE and USBR have used them to evaluate the sliding stability of their dams [5,6]. However, the conventional SSF formulation cannot explain the sliding stability along the dam–foundation interface, since it does not take into account the fractured area in the interface [7,8]. A concrete gravity dam usually has a pre-existing fractured area due to the hydraulic pressure at the heel area and/or past earthquakes [9,10]. In addition, when a dam is subjected to strong earthquakes, the fracture initiates and gradually propagates at the bottom of a dam [10–12]. These pre-existing and progressive fractures reduce the resistant shear strength against the external shear force [13–16]. Hence, if the fractured area is not considered in the SSF evaluation, the shear strength can be exaggerated against sliding, and this would have an adverse impact on the precision of the sliding safety evaluation. The existing conventional SSF is not able to describe these important situations for the sliding stability of a dam.

To overcome the shortcomings, an improved SSF has been proposed to consider the fractured area in the dam–foundation interface [17,18], where the resistant shear forces at the undamaged area and the fractured area are determined separately. However, in the improved SSF formulation, the total vertical force is decomposed in proportion to the ratio of those areas for simplicity, which produces inaccurate results in dynamic analysis of a dam. In addition, the effect of fracture progression is not considered important in this improved SSF, as well as in the conventional SSF.

In this study, a new SSF is suggested to take account of the progressive fractured area at the dam–foundation interface. In order to use the suggested SSF, a contact and sliding model for the dam–foundation system is required to compute the dynamically varying normal forces and sliding motions, which is presented in the context of the finite element method. To investigate the effect of the fractured area progression on the sliding safety evaluation, the conventional, improved, and newly suggested SSFs were computed using the dynamic earthquake analysis results of a concrete gravity dam subjected to two artificial earthquakes. After comparing the SSF values, there is a discussion about the possibility that the conventional SSF significantly overestimates the resistant capacity of a dam against sliding, and about the importance of an appropriate SSF formulation.

2. Dam Sliding Model

2.1. Conventional and Improved Sliding Safety Factors

To evaluate the dynamic sliding stability of concrete gravity dams subjected to earthquakes, the concept of Sliding Safety Factor (SSF) has been used as the ratio of the shear strength to the shear force on a dam tending to slide along its base [3]. When the SSF is larger than unity during the entire duration of an earthquake, it is supposed that the dam structure is stable against rigid-body sliding. The conventional formulation of the SSF is defined as:

$$\mathrm{SSF} = \frac{cA + N(t)\tan\phi_i}{V(t)} \tag{1}$$

where c is the cohesive strength of the dam–foundation interface, A is the total area of the dam bottom face, N is the total vertical force at the dam base as a function of time t, ϕ_i is the internal friction angle of the dam–foundation interface, and V is the absolute value of the external shear force along the dam base, which is also a function of time t. This sliding criterion represents a simple ratio between the shear strength of the interface at the dam base and the force to slide a dam body.

Although, mainly due to its simplicity, the SSF in Equation (1) is popularly used in static and dynamic stability evaluation, it only accounts for the sliding stability of a dam under the assumption of undamaged dam–foundation interface, and becomes inaccurate or invalid once the fracture propagates along the dam–foundation interface.

To address this problem, the improved SSF has been proposed [17,18]. In the formulation, it is considered that the total shear strength of the interface area is divided into two parts, the shear strength of the undamaged area and one from the fractured (so, damaged) area. Each of them follows Coulomb's friction law with different friction angles, and the SSF is written as:

$$\mathrm{SSF} = \frac{(1-\eta)(cA + N(t)\tan\phi_i) + \eta N(t)\tan\phi_f}{V(t)} \tag{2}$$

where η is the ratio of the damaged area to A, the total area of the dam bottom face, N is the vertical force at the dam–foundation interface, ϕ_i is the internal friction of the undamaged interface region, and ϕ_f is the coulomb friction angle along the fractured surface of the dam–foundation interface. While the internal friction angle is a material property [19], the coulomb friction angle, ϕ_f, represents the surface friction caused by roughness and interlocking along the fractured dam–foundation interface. The ratio η is assumed to be constant to represent a pre-existing fractured region caused by strong uplift hydro-pressure or a preceding earthquake. In Equation (2), it is assumed that the total vertical force at the dam base is divided into the undamaged and the fractured region components in proportion to the ratio of their areas. When a structure is dynamically excited with horizontal and vertical loads, the vertical reaction at its base region cannot be decomposed in proportion to subregional areas. Hence, the vertical force of each area should be evaluated separately for the precise frictional force computation.

2.2. *New Sliding Safety Factor*

In Figure 1, a dam–foundation system having a partially fracture-damaged condition at the dam base interface is depicted. Assuming that the fractured area exists as shown in Figure 1b, the total normal force against the external force is the sum of the normal forces at the undamaged and the fractured areas. However, since the normal force distribution is not uniform, as shown in Figure 1c, the normal force cannot be simply calculated in consideration of the ratio of the fractured area [20,21]. In Figure 1c, it is shown that the magnitude of the vertical force at the toe is larger than the vertical force at the heel, and the distribution varies with time. When a dam is subjected to real earthquakes, the actual distribution of the vertical force will be more irregular.

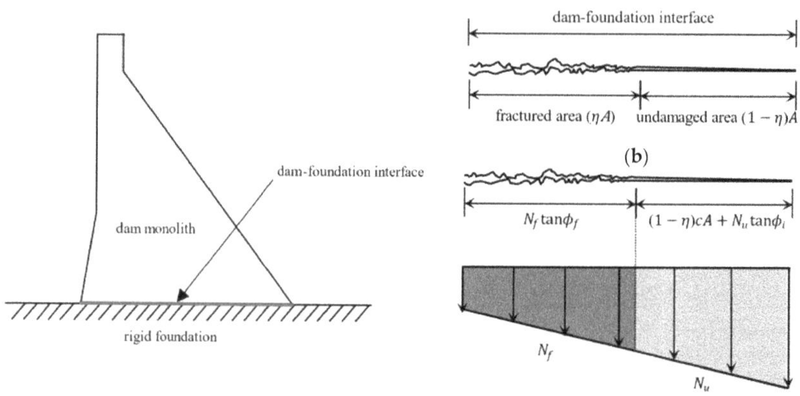

(a) (c)

Figure 1. Dam–foundation interface in the partially fractured condition: (**a**) a schematic dam–foundation model; (**b**) partially fractured dam–foundation interface; (**c**) conceptual normal force distribution along the fractured and undamaged areas.

The engineering guideline of FERC [20] states that the cohesion of the fractured concrete–rock foundation interface should be assumed to be zero. Also, CDA [22] specifies that the cohesion is assumed to be considered within an undamaged concrete structure. Hence, it is reasonable that the cohesive strength in SSF evaluation vanishes to zero at the damaged area where the fracture is severely developed, which is already applied in Equation (2).

While the ratio of the damaged area to the total area, η, is given as a fixed value in Equation (2), for realistic evaluation of the sliding stability of a concrete dam, it is more reasonable that η varies with time as the fracture progresses along the dam–foundation interface. Considering those observations, a modified formulation of SSF is suggested in this study as:

$$\text{SSF} = \frac{(1-\eta(t))cA + N_u\left(t\right)\tan\phi_i + N_d\left(t\right)\tan\phi_f}{V(t)} \tag{3}$$

where N_u is the vertical force at the undamaged area, N_d is the vertical force at the fractured area. In Equation (3), it is noted that η, N_u, and N_d, as well as V, are functions of time. It is also noted that the two vertical forces, N_u and N_d in Equation (3), are computed by a nonlinear finite element analysis considering the nonlinearity of sliding-contact problems, described in Section 2.3, while those vertical forces are inaccurately assumed as the linear proportions of the area, $(1-\eta)N$ and ηN, respectively, in Equation (2). To evaluate the suggested SSF, a contact–sliding model for the dam–foundation interface is necessary for the computation of the fractured area and corresponding vertical forces, which is explained in the following section.

2.3. Interface Contact–Sliding Model

A contact and sliding model for the dam–foundation interface is implemented in the context of the finite element mesh. In this study, it is assumed that the pre-existing or newly generated fracture propagates along the dam–foundation interface plane and the propagation criterion is dominated by the shear mode (mode II) fracture. In addition, the dam base and its corresponding rock-foundation surfaces are assumed to contact and separate vertically along the fractured region.

The dam side of the interface is defined as a slave surface and the foundation side is defined as a master surface. The fracture extension starts from the heel of a dam, since the crack initiation occurs from the heel of a dam due to the hydraulic pressure. The contact surface is defined by the vertical direction and the horizontal direction. As shown in Figure 2, in the vertical direction, mesh penetration between the master and slave surface is controlled [23]. The principle of minimum total potential energy, which is the basis of the finite element method, is applied to the dam–foundation contact model. The basic form of the principle can be represented as:

$$\delta\Pi \equiv \delta U - \delta W = 0 \tag{4}$$

where Π is the total potential energy, U is the strain energy, W is the work done by the external force. Equation (4) is applied to the contact constraint in the vertical direction that controls mesh penetration. The contact constraint is formed by the Lagrange multiplier method and virtual work as:

$$\delta\Pi = \delta p h + p\delta h \tag{5}$$

where h is the clearance between contact interfaces, p is the pressure delivered through the interfaces. Equation (6) shows that the pressure delivered through the interface is zero when there is a positive clearance between the foundation (master) and dam base (slave) surfaces. In case of zero clearance, it is defined that the foundation and dam base surfaces are regarded as contacted and the contact pressure can be delivered through the interface as shown in Equation (7).

$$h > 0,\ p = 0 \tag{6}$$

$$h = 0,\ p \neq 0 \tag{7}$$

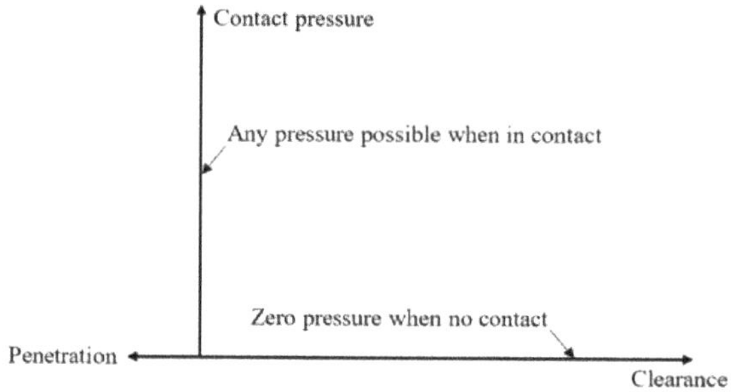

Figure 2. Schematic contact pressure–clearance relationship in contact interface [23].

In the horizontal direction, the fracture progression along the dam–foundation interface is modeled by the stick–slip phenomenon. The Lagrange multiplier method is applied to the minimum total potential energy principle to enforce exact stick–slip conditions [23]. In the i-th iteration, the augmented virtual work can be expressed as:

$$\delta\Pi = \int_S \tau_i \delta\gamma_i + \Delta\gamma_i \lambda dS \tag{8}$$

where τ_i is the friction force at the interface, γ_i is the sliding displacement, λ is the Lagrange multiplier, and S is the interface surface region. In the stick state, when the external shear force τ exceeds the friction force τ_i, the state is changed to sliding as shown in Equation (9). Meanwhile, Equation (10) shows that if the direction of the friction force and the direction of sliding are opposite in the sliding state, the state is changed to stick.

$$\tau > \tau_i \tag{9}$$

$$\Delta\gamma_i \tau_i < 0 \tag{10}$$

By this mechanism, the dam–foundation model follows exact stick–slip conditions. The relationship between stress and strain is depicted in Figure 3. Here, the critical shear force τ_{cri} is the sum of the cohesive forces and the frictional forces of the interface, which is the same as the numerator in Equation (3). There is no additional sliding in the stick state $(\Delta\gamma = 0)$. When the absolute value of an external shear force exceeds the critical shear force, the existing state is changed to sliding.

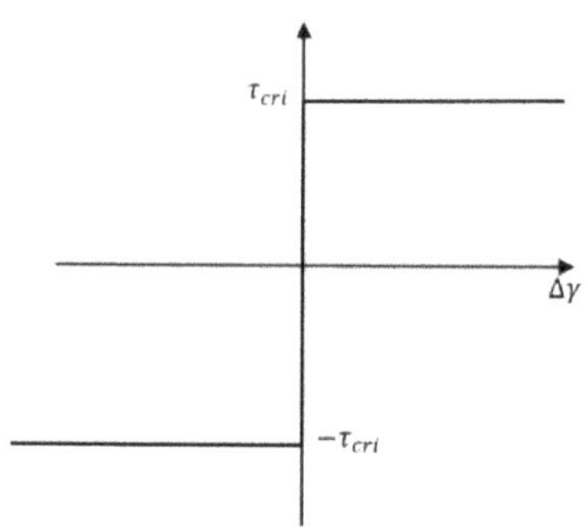

Figure 3. Stick–slip motion with Lagrange multiplier in $\Delta\gamma - \tau$ plane.

3. Numerical Analysis

3.1. Finite Element Model for a Concrete Gravity Dam

To verify the validity of the proposed SSF in Equation (3) and compare it with the other SSF formulations in Equations (1) and (2), the seismic analysis of a 70 m concrete gravity dam is performed as depicted in Figure 4. The dam is modeled by a finite element program, Abaqus. The model consists of 685 eight-node isoparametric plane stress elements. The concrete properties are the elastic modulus E_c = 28.1 GPa, Poisson's ratio ν = 0.12, and the density ρ_c = 2300 kg/m^3. The cohesive force of the concrete is suggested to be 10 percent of its compressive strength. The compressive strength of the concrete is from 20.7 MPa to 34.5 MPa [24]. In this study, the cohesive force of the concrete dam–foundation interface is assumed as 2.07 MPa, which is 10 percent of the minimum compressive strength. The friction coefficients $\tan\phi_i$ and $\tan\phi_f$ are derived from USACE [25] and ACI [26]. According to USACE [25], the coefficient of the internal friction $\tan\phi_i$ is set as 0.7. In ACI [26], the coulomb friction coefficient $\tan\phi_f$ is prescribed as 0.6.

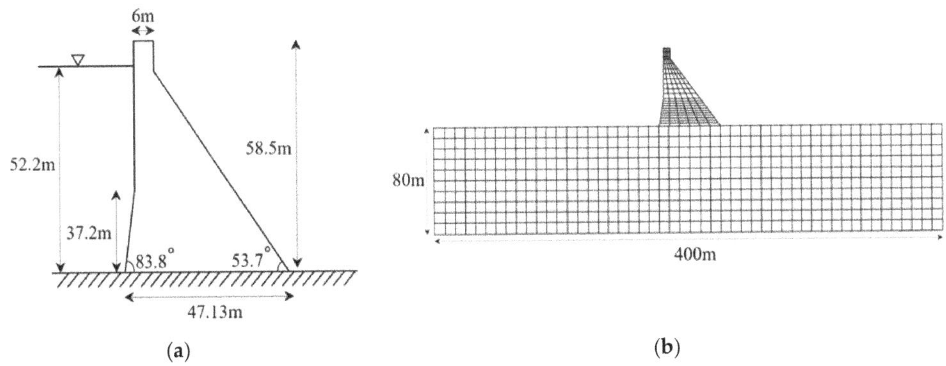

Figure 4. Numerical model for a dam–foundation system: (**a**) non-overflow monolith geometry; (**b**) finite element mesh.

The hydrostatic and hydrodynamic pressures are applied, and the uplift pressure is ignored in this numerical analysis to avoid unnecessary complexity. The hydrodynamic pressure is applied according to the simplified added-mass formulation of Westergaard [27], defined as:

$$m = \frac{7}{8}\rho_w\sqrt{h_w(h_w - y)}A \tag{11}$$

where m is the added mass acting as the hydrodynamic pressure, ρ_w is the density of water, h_w is the depth of the upstream reservoir, y is the vertical height from the surface of water, and A is the area per vertical height.

3.2. Test Cases

There are three different models used for tests, as shown in Table 1: Case 1 is the dam–foundation single-body model, which assumes the undamaged state of the dam. Case 2 is the dam–foundation contact model with a fixed pre-existing fractured area at the upstream dam base, as depicted in Figure 1b. Case 3 is the dam–foundation contact model with a pre-existing fractured area which can be dynamically propagated along its entire dam–foundation interface to the downstream toe. In Cases 2 and 3, pre-existing fractures at the dam base are set to consider the strong hydraulic pressure or earthquakes of the past. The initial ratio of the fractured area to the total area of the dam base is assumed as 50 percent in Cases 2 and 3. Before being fractured, the undamaged area resists against the external force by the shear strength of the dam–foundation interface. After being fractured, the resistant force against the external force follows Coulomb's friction law.

Table 1. SSF equation, initial fractured area ratio, and progress availability for each test case.

Test Case	SSF	Initial A_b (%)	Fracture Progress
1	Equation (1)	100	No
2	Equation (2)	50	No
3	Equation (3)	50	Yes

In Table 2, there is a list of Peak Ground Acceleration (PGA) data from the recent strong earthquakes [28–30]. The highest PGA of an earthquake is 3 g and the lowest one is 0.5 g. In this study, two artificial earthquakes are used, in which the PGAs are 0.64 g and 0.7 g, respectively, as depicted in Figure 5. The effective duration of Artificial earthquake I is 15 s and Artificial earthquake II is 20 s.

Table 2. Significant high-PGA earthquakes.

Name	Country	Year	Magnitude	PGA (g)
Kaikoura	New Zealand	2016	7.8	3
Tohoku	Japan	2011	9.0	2.7
Canterbury	New Zealand	2010	7.1	1.26
Kobe	Japan	1995	6.8	0.8
Chile	Chile	2010	8.8	0.78
Athens	Greece	1999	6.0	0.6
Zarand	Iran	2005	6.4	0.51
Haiti	Haiti	2010	7.0	0.5

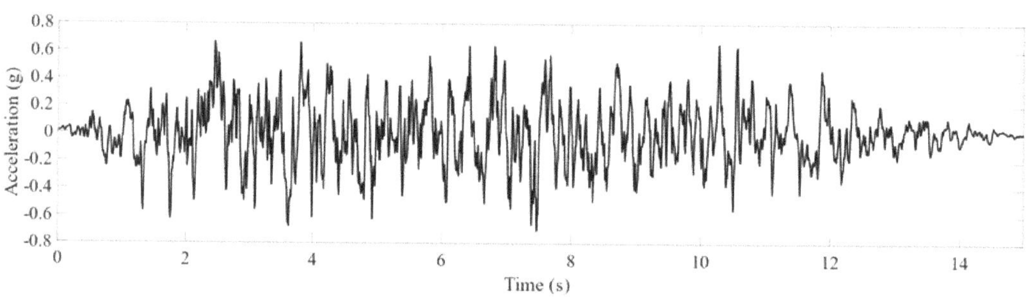

(a)

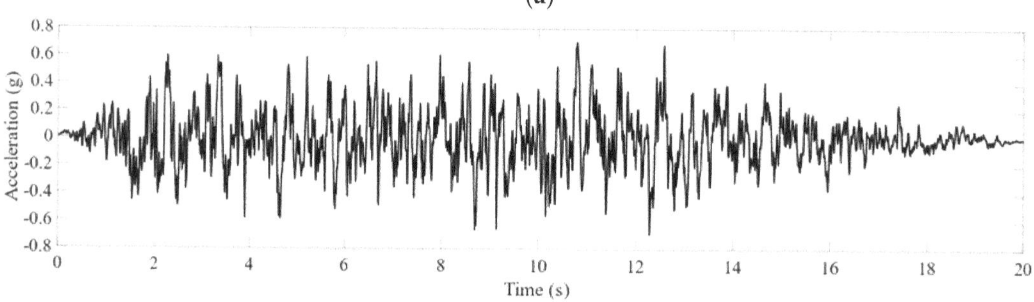

(b)

Figure 5. Artificial earthquake accelerations: (**a**) Artificial earthquake I, PGA = 0.64 g; (**b**) Artificial earthquake II, PGA = 0.7 g.

3.3. Response Results

In Figure 6, horizontal crest displacement of the upstream side from the bottom of the dam and the sliding displacement in Case 3 are shown when the dam is subjected to Artificial earthquakes I and II. Positive values indicate the downstream direction. Since the dam–foundation interfaces are fixed in Cases 1 and 2, the crest displacement and stress distributions of Cases 1 and 2 are mostly identical. However, the SSFs in Cases 1 and 2 are different, and this is described in Section 4.

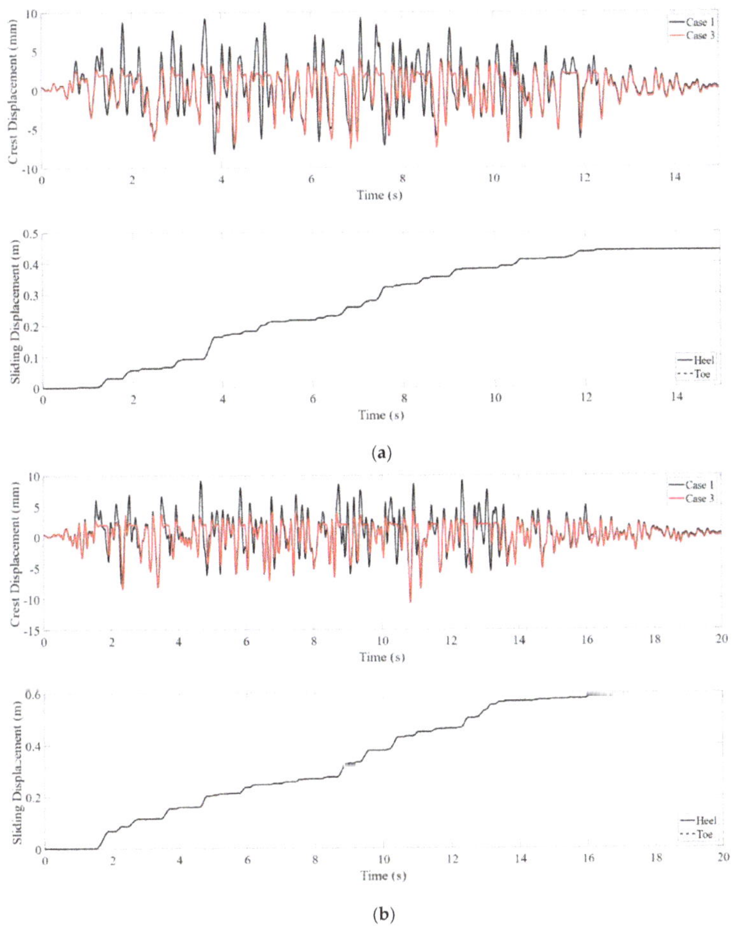

Figure 6. Horizontal crest displacements in Cases 1 and 3, and sliding displacements at heel and toe in Case 3: (**a**) subjected to Artificial earthquake I; (**b**) subjected to Artificial earthquake II.

There is a difference in the crest displacement between Cases 1 and 3 due to the progressive fractured area. There is no sliding in Case 1, whereas the sliding behavior is included in Case 3. When the sliding behavior is observed, the energy dissipation caused by the sliding friction occurs. The crest displacement in Case 3 is affected by the energy dissipation and its amplitude of the crest displacement is relatively smaller than the crest displacement of Case 1 when the sliding behavior is observed. While the sliding behavior is not observed, the crest displacement in Case 3 is mostly identical to the crest displacement in Case 1. Relative displacement contours from the bottom of the dam in Cases 1 and 3 when the largest sliding displacement is observed are shown in Figure 7. According to Figure 7, the sliding behavior affects the overall crest displacement. The overall crest displacement in Case 1 is larger than the crest displacement in Case 3. Also, the difference in the stress distribution can be observed in Figure 8, which depicts the tensile principal stress distribution in Cases 1 and 3. Compared with Case 1, the overall principal stress values in Case 3 are smaller due to energy dissipation from the sliding.

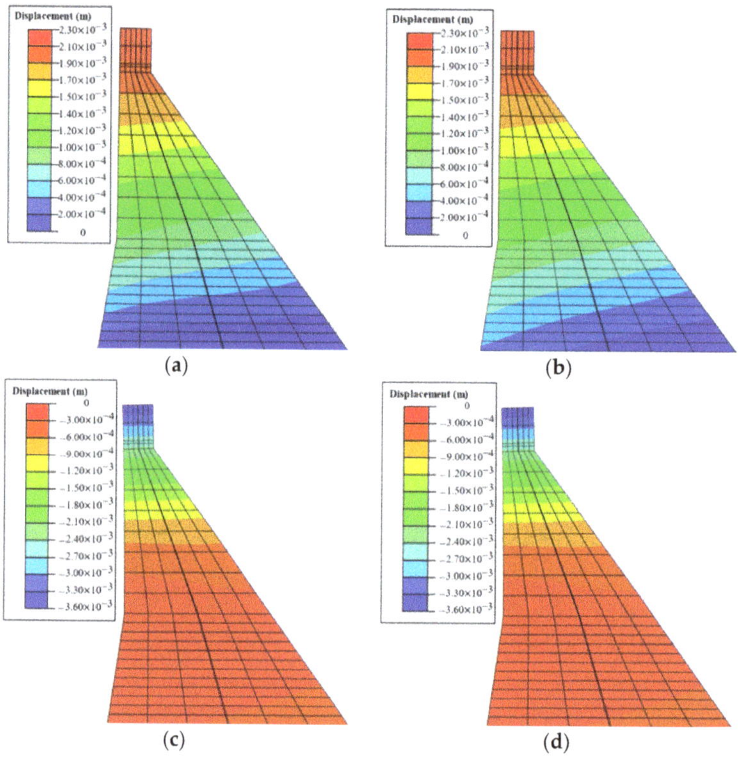

Figure 7. Horizontal displacement contours, relative to the dam base, when the largest sliding displacements are observed: (**a**) Case 1 subjected to Artificial earthquake I (t = 3.76 s); (**b**) Case 3 subjected to Artificial earthquake I (t = 3.76 s); (**c**) Case 1 subjected to Artificial earthquake II (t = 1.86 s); (**d**) Case 3 subjected to Artificial earthquake II (t = 1.86 s).

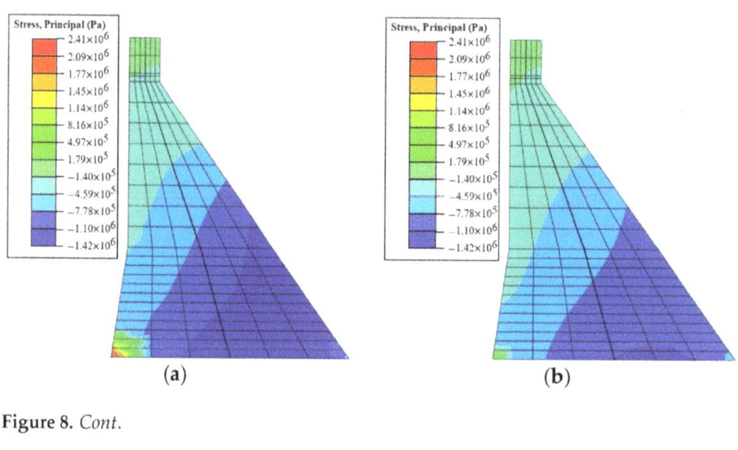

Figure 8. *Cont.*

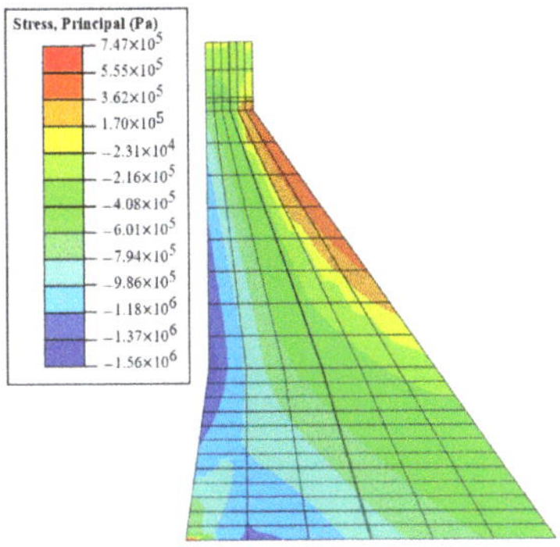

(c)

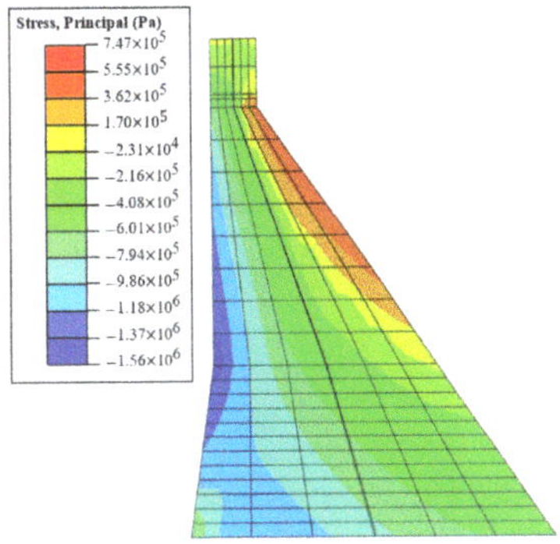

(d)

Figure 8. Tensile principal stresses contours when the largest sliding displacements are observed: (**a**) Case 1 subjected to Artificial earthquake I (t = 3.76 s); (**b**) Case 3 subjected to Artificial earthquake I (t = 3.76 s); (**c**) Case 1 subjected to Artificial earthquake II (t = 1.86 s); (**d**) Case 3 subjected to Artificial earthquake II (t = 1.86 s).

Sliding is the main source of nonlinearity in this study, since the effects of the rocking and the deformation at the bottom of the dam on the sliding behavior of a concrete dam are negligible compared to the effect of the base sliding [31,32]. In Figure 6, it can be observed that sliding displacements at the heel and toe of the dam are nearly identical and coincide with each other in the plot. Hence, it is justified that the rocking and the local deformation of the bottom of a dam can be ignored in sliding analysis.

4. SSF Evaluation

In Figures 9 and 10, reciprocal values of the SSF and the lowest SSF (or highest reciprocal of SSF) point at each test case are shown when the dam model is subjected to Artificial earthquakes I and II. In this study, the reciprocal of the SSF is used in the plot rather than SSF values in order to avoid unimportant high peak values of the SSF generated by the nearly zero external shear force. The red dashed line denotes where the SSF is unity, which indicates the threshold for sliding of a dam structure.

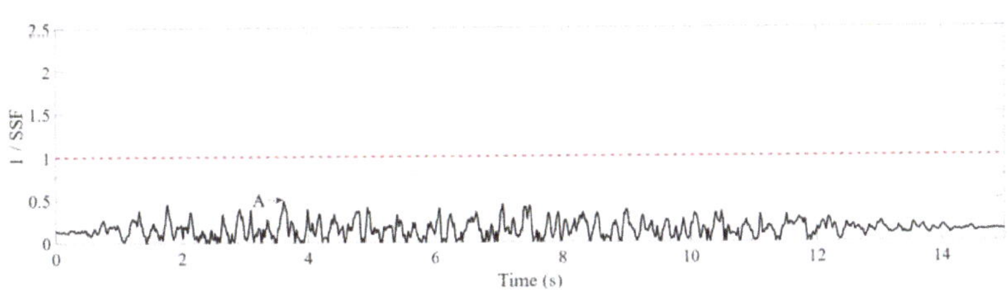

(a)

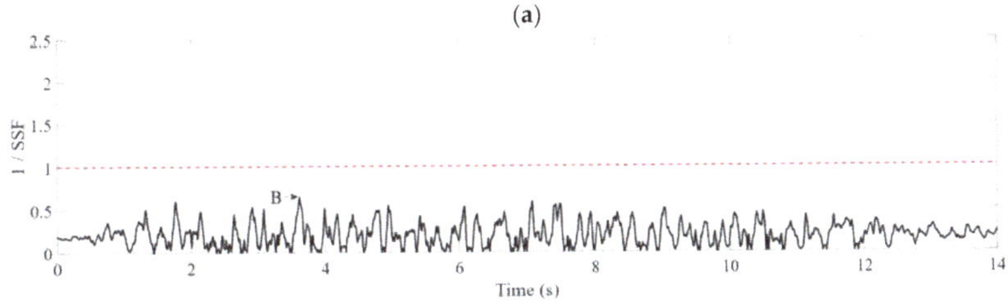

(b)

Figure 9. *Cont.*

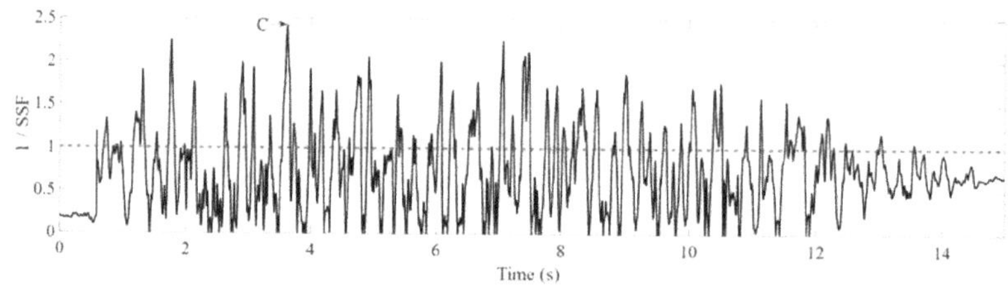

(c)

Figure 9. Reciprocal of SSF subjected to Artificial earthquake I: (**a**) Case 1 (SSF at A = 2.070); (**b**) Case 2 (SSF at B = 1.548); (**c**) Case 3 (SSF at C = 0.413).

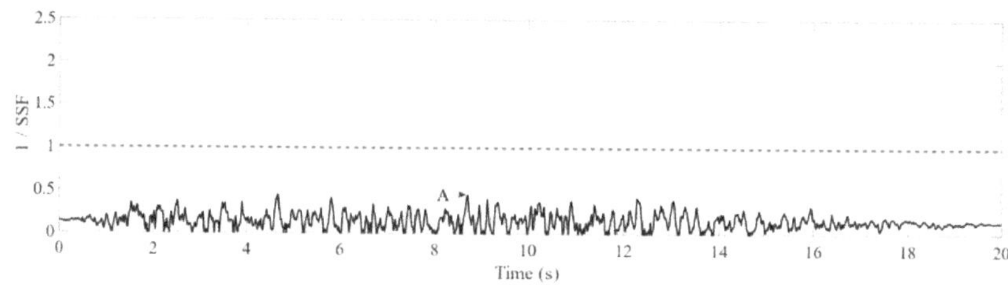

(a)

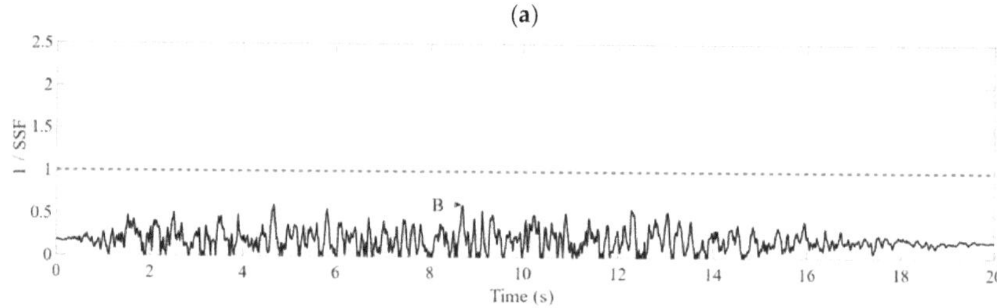

(b)

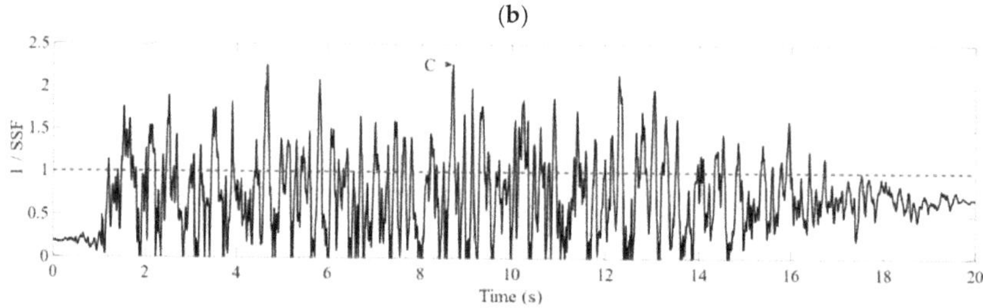

(c)

Figure 10. Reciprocal of SSF subjected to Artificial earthquake II: (**a**) Case 1 (SSF at A = 2.227); (**b**) Case 2 (SSF at B = 1.658); (**c**) Case 3 (SSF at C = 0.442).

In Case 1, the cohesive force is applied to the overall dam–foundation interface. In Case 2, the cohesive force is only applied to the undamaged area. In Case 3, the cohesive force is also applied to the undamaged area, and when the SSF is below unity, it is considered that the dam–foundation interface is fractured, and the cohesive force decreases to zero. In Figure 9, the lowest SSF of Case 1 is 2.070, Case 2 is 1.548, and Case 3 is 0.413, while in Figure 10, the lowest SSF of Case 1 is 2.227, Case 2 is 1.658, and Case 3 is 0.442. The SSF of Case 3 is relatively lower than Cases 1 and 2, since, in the newly fractured region during the fracture propagation, the cohesive force is considered to be zero, the coefficient of the internal friction changes to the coulomb friction coefficient, $\tan\phi_f$, and consequently, the friction force and SSF decrease. Depending on the ratio of the fractured area, the resistant force against the external force highly changes.

According to the results in Table 3, for both artificial earthquakes, the minimums of the SSF in Case 2 are about 75 percent of those in Case 1. Furthermore, the minimums of the SSF in Case 3 show much smaller values, about 20 percent of those in Case 1. Considering that partial fractures at a dam base can exist due to the hydraulic pressure acting at the upstream heel and/or past earthquakes, the SSF in Equation (1), which is widely used as a sliding criterion, significantly overestimates the sliding stability of a dam. It is noted that Cases 1 and 2 show safety satisfaction if unity is used for the safety factor threshold, while only Case 1 passes the sliding safety condition if the safety factor becomes two. The SSF formulation, Equation (3), used in Case 3 produces no satisfaction for the unity safety threshold, which means the other two SSFs possibly result in incorrect judgements on sliding safety.

Table 3. Minimum SSF values for test cases and their ratios.

Artificial Earthquake	SSF				
	Case 1	Case 2	Case 3	Case 2/Case 1 (%)	Case 3/Case 1 (%)
I	2.070	1.548	0.413	74.78	19.95
II	2.227	1.658	0.442	74.50	19.85

5. Discussion

In this study, it is verified that considering the fractured area and its progressive modeling at the dam–foundation interface is crucial in the sliding safety evaluation of a concrete gravity dam. In the numerical analysis study, the conventional formulation of the SSF, in which the fractured area is not represented at all, shows extremely overestimated sliding safety judgements when a dam is subjected to strong earthquakes. This SSF formulation results in about five times safer side values than those evaluated by the newly suggested SSF formulation with initial and progressive fractures, which is based on a more realistic fracture-progress model. The so-called improved SSF formulation, with the fixed fractured area and simplified computation of the vertical force at the fractured area, also shows more than three times safer side values than those by the suggested SSF formulation. Consequently, it can be observed that those existing SSF formulations could provide wrongfully safe judgements on sliding stability checks, which implies that those SSFs possibly exaggerate the sliding safety of dams.

From the numerical simulation, it is shown that the proposed contact–sliding model can reproduce fracture propagation along the dam–foundation interface, which is expected during strong earthquake ground motions, including the two artificial earthquakes used in the present numerical study. In the simulation results, the progressive fracture propagation realistically starts and stops according to the intensity of ground motions. It was observed in the simulation that the concrete gravity dam, subjected to a strong earthquake, slides monotonically toward the downstream due to the reservoir hydrostatic pressure. It is also noted that the fracture rapidly propagates through the entire interface immediately after the dam starts to slide. From these observations, it can reasonably be deduced that it is essential to use an appropriate contact–sliding model for the fracture propagation in an accurate evaluation of the sliding safety for concrete gravity dams. Otherwise, the shear resistance capacity against sliding can become significantly overestimated, as aforementioned in this section. The newly suggested SSF with the proposed contact–sliding model for the dam–foundation interface shows more accurate and conservative sliding safety evaluation results compared to the two existing conventional SSF formulations, since the degradation of the overall shear strength along the dam–foundation interface can be realistically represented in the dam–foundation model and its SSF evaluation.

Although, in this study, the uplift pressure acting along the interface fracture is ignored to avoid complexity, its effects on the SSF evaluation can be significant. The uplift force reduces the vertical forces, which results in weakening the frictional shear resistance forces. Additionally, it can cause mode I fracture propagation with other upward forces acting on the dam body. Since the dynamic distribution of the uplift pressure and its effect on the mode I and/or mixed mode fracture of a concrete dam–foundation interface are quite different from the static one, they remain undergoing research topics, and their impacts on the SSF evaluation should be addressed in a future study.

6. Conclusions

In the present study, it is suggested to use the Sliding Safety Factor (SSF) while considering pre-existing and progressive fractures along the dam foundation to evaluate the sliding safety of a concrete gravity dam subjected to strong earthquakes. In order to use the suggested SSF, it is necessary to implement an appropriate contact and sliding model in the dynamic earthquake analysis, like the one described in this study. This newly suggested SSF with the proposed contact–sliding model provides more realistic and accurate sliding safety factors by considering fracture progression. As a result, it can help avoid suppositional and ambiguous safety factor criteria in dam safety judgements, which will contribute to advances in concrete gravity dam design and evaluation technology. It is also suggested that if the conventional SSF formulation, or its simply improved version, is unavoidably used for a safety analysis, the safety judgement must be made cautiously, taking into account the possibility and impact of fracture progression.

In the current study, it is assumed that the fracture at the dam–foundation interface moves along the interface plane and that the fracture propagation is dominated by the shear force, for simplicity. Although this assumption still gives reasonable and practical results, further studies are needed that include the tensile mode fracture, as well as the shear mode fracture and more general fracture propagation surfaces other than simple planes into the SSF evaluation. Moreover, in order to validate and enhance the suggested model, on-site measurement studies for the sliding displacement of concrete gravity dams subjected to strong ground motions are desirable in future research.

Author Contributions: Conceptualization, J.L.; Data curation, J.L. and H.S.L.; Formal analysis, H.S.L.; Funding acquisition, J.L.; Investigation, J.L. and H.S.L.; Methodology, J.L. and H.S.L.; Project administration, J.L.; Resources, J.L.; Software, H.S.L.; Validation, J.L.; Writing—original draft, J.L. and H.S.L.; Writing—review and editing, J.L. All authors have read and agreed to the published version of the manuscript.

Funding: This research was supported by the Basic Science Research Program through the National Research Foundation of Korea (NRF), funded by the Ministry of Education (NRF-2018R1D1A1B07051141).

Institutional Review Board Statement: Not applicable.

Informed Consent Statement: Not applicable.

Data Availability Statement: Data sharing is not applicable to this article.

Acknowledgments: This work was supported by the Dongguk University Research Fund of 2023.

Conflicts of Interest: The authors declare no conflict of interest.

References

1. Wang, S.; Xu, B.; Zhu, Z.; Li, J.; Lu, J. Reliability Analysis of Concrete Gravity Dams Based on Least Squares Support Vector Machines with an Improved Particle Swarm Optimization Algorithm. *Appl. Sci.* **2022**, *12*, 12315. [CrossRef]
2. Chávez, J.W.; Fenves, G.L. Earthquake response of concrete gravity dams including base sliding. *J. Struct. Eng.* **1995**, *121*, 865–875. [CrossRef]
3. Henry, D.C. Stability of straight Concrete Gravity Dams. *Trans. ASCE* **1933**, *99*, 1041–1061.
4. Nicholson, G.A. *Design of Gravity Dams on Rock Foundations: Sliding Stability Assessment by Limit Equilibrium and Selection of Shear Strength Parameters*; Technical Report GL-83-13; Army Engineer Waterways Experiment Station: Vicksburg, MS, USA, 1983.

5. Headquarters, U.S. *Army Corps of Engineers. Stability Analysis of Concrete Structures*; EM 1110-2-2100; U.S. Army Corps of Engineers: Washington, DC, USA, 2005.
6. Duncan, W.; Huntley, C.; Hokenstrom, J.; Cudworth, A.; McDaniel, T. *Design of Small Dams. A Water Resources Technical Publication*; Final Report, No. PB-95-176368/XAB; U.S. Bureau of Reclamation: Denver, CO, USA, 1987; pp. 328–329.
7. Saichi, T.; Renaud, S.; Bouaanani, N.; Miquel, B. Effects of rock foundation roughness on the sliding stability of concrete gravity dams based on topographic surveys. *J. Eng. Mech.* **2019**, *145*, 04019043. [CrossRef]
8. Wang, X.H.; Zhang, S.R.; Dai, J.G.; Wang, C. Evaluation of base damage and stability of concrete gravity dam subjected to underwater explosion. *Structures* **2022**, *38*, 1502–1514. [CrossRef]
9. Bolzon, G. Collapse mechanisms at the foundation interface of geometrically similar concrete gravity dams. *Eng. Struct.* **2010**, *32*, 1304–1311. [CrossRef]
10. Alliard, P.M.; Léger, P. Earthquake safety evaluation of gravity dams considering aftershocks and reduced drainage efficiency. *J. Eng. Mech.* **2008**, *134*, 12–22. [CrossRef]
11. Krounis, A.; Johansson, F.; Larsson, S. Effects of spatial variation in cohesion over the concrete-rock interface on dam sliding stability. *J. Rock Mech. Geotech. Eng.* **2015**, *7*, 659–667. [CrossRef]
12. Li, J.B.; Gao, X.; Fu, X.A.; Wu, C.; Lin, G. A nonlinear crack model for concrete structure based on an extended scaled boundary finite element method. *Appl. Sci.* **2018**, *8*, 1067. [CrossRef]
13. Sas, G.; Popescu, C.; Bista, D.; Seger, A.; Arntsen, B.; Johansson, F.; Lia, L. Influence of large-scale asperities on the shear strength of concrete-rock interface of small buttress dams. *Eng. Struct.* **2021**, *245*, 112952. [CrossRef]
14. Krounis, A.; Johansson, F.; Spross, J.; Larsson, S. Influence of cohesive strength in probabilistic sliding stability reassessment of concrete dams. *J. Geotech. Geoenviron. Eng.* **2017**, *143*, 04016094. [CrossRef]
15. Javanmardi, F.; Léger, P.; Tinawi, R. Seismic structural stability of concrete gravity dams considering transient uplift pressures in cracks. *Eng. Struct.* **2005**, *27*, 616–628. [CrossRef]
16. Arabshahi, H.; Lotfi, V. Earthquake response of concrete gravity dams including dam–foundation interface nonlinearities. *Eng. Struct.* **2008**, *30*, 3065–3073. [CrossRef]
17. Lo, K.Y.; Ogawa, T.; Lukajic, B.; Tsui, K.K.; Wang, S. Evaluation of strength parameters of concrete-rock interface for dam safety assessment. In Proceedings of the Canadian Dam Safety Conference, Toronto, ON, Canada, 18–19 September 1990; pp. 71–93.
18. Dawson, R.V.; Curtis, D.D.; Donnelly, R.C. *Sliding Resistance of Concrete Gravity Dams*; CEA No. 9331 G 2002; Canadian Electricity Association Technologies Inc.: Montreal, QC, Canada, 1998.
19. Fronteddu, L.; Léger, P.; Tinawi, R. Static and dynamic behavior of concrete lift joints interfaces. *J. Struct. Eng.* **1998**, *124*, 1418–1430. [CrossRef]
20. Federal Energy Regulatory Commission. *Engineering Guidelines for the Evaluation of Hydropower Projects: Chapter 3-Gravity Dams*; Federal Energy Regulatory Commission: Washington, DC, USA, 2016; pp. 3–25.
21. Leclerc, M.; Léger, P.; Tinawi, R. Computer aided stability analysis of gravity dams—CADAM. *Adv. Eng. Softw.* **2003**, *34*, 403–420. [CrossRef]
22. Canadian Dam Association. *Dam Safety Guidelines*; Canadian Dam Association: Toronto, ON, Canada, 2007.
23. Abaqus. *Abaqus Theory Manual*; Dassault Systemes: Johnston, RI, USA, 2023.
24. U.S. Bureau of Reclamation. *Design Criteria for Concrete Arch and Gravity Dams*; U.S Government Printing Office: Washington, DC, USA, 1977; pp. 20–21.
25. Armstrong, R.C. *Engineering and Design: Revision of Thrust Block Criteria in TM 5-813-5/AFM 88-10, Appendix C*; Corps of Engineers: Washington, DC, USA, 1992.
26. ACI Committee 318. *Building Code Requirements for Structural Concrete (ACI 318-08) and Commentary*; ACI: Farmington Hills, MI, USA, 2008.
27. Westergaard, H.M. Water Pressure on Dams during Earthquake. *Trans. ASCE* **1931**, *95*, 418–433. [CrossRef]
28. Anastasiadis, A.; Demosthenous, M.; Karakostas, C.; Klimis, N.; Lekidis, B.; Margaris, B.; Papaioannou, C.; Papazachos, C.; Theodulidis, N. *The Athens (Greece) Earthquake of September 7, 1999: Preliminary Report on Strong Motion Data and Structural Response*; Institute of Engineering Seismology and Earthquake Engineering (ITSAK): Thessaloniki, Greece, 1999.
29. Wood, S.L. Performance of reinforced concrete buildings during the 1985 Chile earthquake: Implications for the design of structural walls. *Earthq. Spectra* **1991**, *7*, 607–638. [CrossRef]
30. DesRoches, R.; Comerio, M.; Eberhard, M.; Mooney, W.; Rix, G.J. Overview of the 2010 Haiti earthquake. *Earthq. Spectra* **2011**, *27* (Suppl. S1), 1–21. [CrossRef]
31. Chopra, A.K.; Zhang, L. *Base Sliding Response of Concrete Gravity Dams to Earthquakes*; Report No. UCB/EERC-91/05; Earthquake Engineering Research Center, University of California at Berkeley: Berkeley, CA, USA, 1991.
32. Chopra, A.K. *Earthquake Engineering for Concrete Dams: Analysis, Design, and Evaluation*; John Wiley & Sons: Hoboken, NJ, USA, 2020; pp. 83–95.

applied sciences

MDPI

Article

Seismic Performance of Frame Structure with Hysteretic Intermediate Discontinuity

Angelo Di Egidio [1], Stefano Pagliaro [1] and Alessandro Contento [2,*]

[1] Department of Civil, Construction-Architectural and Environmental Engineering, University of L'Aquila, 67100 L'Aquila, Italy

[2] College of Engineering, Fuzhou University, Fuzhou 350108, China

* Correspondence: alessandro@fzu.edu.cn

Abstract: The introduction of an intermediate discontinuity in frame structures is commonly named inter-storey isolation. Inter-storey isolation is an effective technique for the seismic protection of new or existing frame structures. The devices that are used to perform the discontinuity, mainly of the structural stiffness, are placed at a higher storey level of the structure and not at the base level. In the latter years, this technique has gained increasing interest because, especially for existing buildings, it is cheaper and technically easier to implement than base isolation. In this paper, the attention is focused on the effects on a frame structure of an intermediate elasto-plastic discontinuity that can be described by a Bouc-Wen model. The frame structure with the intermediate discontinuity is modelled with a 3-DOF reduced model. Its dynamical behaviour is investigated by considering both harmonic and seismic external excitation. The results are summarized in gain maps aimed at finding the parameters that optimize the seismic behaviour of the structure.

Keywords: intermediate discontinuity; hysteretic elasto-plastic devices; archetype system; gain parameters and maps

1. Introduction

The seismic protection of structures is one of the biggest challenges in civil engineering. In recent years, the building technical codes increased the requirements in terms of structural capacity that a structure needs to fulfil, both for new buildings and the seismic retrofit of existing buildings. New buildings can meet the needed requirements by means of a proper design. Instead, improving the seismic response of existing buildings may require invasive interventions that may modify the serviceability of such buildings. There is a wide literature available about the seismic retrofit of structures. Base isolation (BI) and Tuned Mass Damper (TMD) are two widely known techniques. The first studies on base isolation and tuned mass damper systems were presented in [1,2], respectively. Tsai, in [3], and Taniguchi et al., in [4], showed that a TMD placed on the floor immediately above the isolation layer in a base isolated structure can reduce the base displacement. A more complex version of TMD, named Tuned Mass Damper Inerter (TMDI), was presented in [5] and was obtained by coupling a TMD with an inerter. In [5], such a device was intended to reduce the displacement demand of base-isolated structures, but in [6], instead, the same device was used to reduce the vibrations in tall buildings.

Protection systems based on the TMD have been used with limited success in the seismic protection of structures because of the difficulty in tuning the TMD to multiple frequencies, as it is needed for seismic excitations. De Angelis et al. [7] showed that it is possible to decrease a TMD tendency to detuning by designing a TMD with a high ratio between the mass of the TMD and the mass of the structure to be protected. To achieve such a high mass ratio, some authors purposed the combined use of TMD and BI. This combination is commonly named inter-storey isolation: in inter-storey isolation, base-isolated and tuned-mass-damper behaviours coexist.

The inter-storey isolation has become widely used for the seismic protection of structures also because it is usually characterized by a low impact on the aesthetic of the building. Even though there are many works about the use of inter-storey isolation, most of them are related to case studies, and only a few of them are devoted to clarify the general aspects of its mechanical behaviour. For example, Chey et al. added new storeys on the rooftop of an existing 12-storey building [8] and connected the two parts of the structure with an isolation layer, thus making the upper part of the structure work as a TMD to reduce the vibration of the lower part of the structure. Wang et al. [9–11] presented analytical and experimental analyses of the response of building structures with inter-storey isolation. In [12–14], the roof of the building was used as a vibrating damper to protect the structure, whereas in [15,16], entire sections of the studied structures were converted into vibrating masses by means of a uniform disconnection along the height of the building. An analogous technique for the reduction of the seismic displacements is the partial mass isolation (PMI) technique presented in [17], where the performance of the PMI is compared to those of conventional TMD and BI.

In order to highlight the general mechanisms behind inter-storey isolation, a suitable option is to analyse its effects on an archetype model of a structure. In the literature, archetype models are often used to highlight the main characteristics of multi-Degree-Of-Freedom (DOF) models. For instance, in [1,2,18,19], the studied systems are represented by archetype 2-DOF models. A similar approach, aimed at studying the general behaviour of the inter-storey isolation with a 2-DOF archetype model of the structure, has been presented in [20–22]. Specifically, in [20,21] and in this paper, the inter-storey isolation is named intermediate discontinuity, since it represents a sudden change in the stiffness of the structure. A 2-DOF model is suitable to describe systems where the isolation is placed at higher or lower levels, but cannot properly describe systems where the isolation is placed at middle-floors. In [23], a third degree of freedom has been added to the model to improve the describing capacity of the inter-storey isolation. In that paper, the disconnection devices are modelled by linear links with a non-classical Rayleigh damping model. However, the choice of a linear model to describe the behaviour of the disconnection could prevent the correct evaluation of the effectiveness of the introduction of the intermediate discontinuity. Hence, a non-linear modelling of the discontinuity, with constitutive laws closer to the real behaviour of the disconnection devices, could improve the ability of the model to capture the behaviour of real structures.

In this paper, a system with intermediate discontinuity is studied by means of a 3-DOF model, which represents a medium-rise frame structure (i.e., an archetype model), and a Bouc–Wen hysteresis model representing the disconnection devices, which are hypothetically assumed to be lead-rubber-bearing isolators [24,25]. The main purpose of the paper is to define the influence of the hysteretic characteristics of the discontinuity devices on the seismic response of the structure. Since in the paper the numerical simulations refer only to a six-storey frame structure, the results cannot be generalised, but are intended to provide preliminary information about the hysteretic discontinuity. Although the results are not generalizable, the method used to investigate the behaviour of the structure with intermediate discontinuity provides information for most of the medium-rise frame structures with hysteretic inter-storey isolation. Additionally, the paper does not provide technical specifications about the practical realization of an intermediate discontinuity, but it has to be framed as an analytical/numerical study, which precedes any practical implementation.

The first part of the paper is focused on the description of the mechanical behaviour of the system. In particular, Section 2 presents the mechanical model and the equations of motion. The second part is focused on the dynamical response. Specifically, Section 3 studies the response to harmonic excitation, whereas Section 4 presents the response to seismic excitation. The results are summarized in gain maps used to find the optimal parameters for the inter-storey isolation of the archetype model.

2. The Mechanical Model

2.1. Archetype Model

In the following, two archetype systems are used to compare results with and without inter-storey isolation. The Main System archetype (MS) represents the structure with the introduction of an intermediate discontinuity (Figure 1a), whereas the Auxiliary System archetype (AS) represents the same structure without the intermediate discontinuity (Figure 1b). In the case of the MS, the part of the structure below the discontinuity is called substructure, whereas the part above the discontinuity is called superstructure.

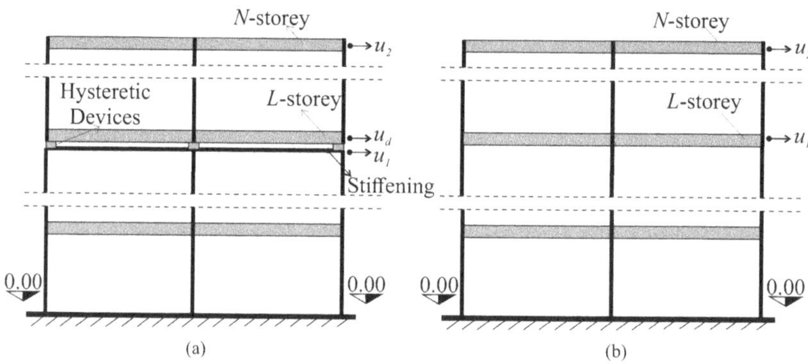

Figure 1. Structural scheme of the (**a**) frame structure with intermediate discontinuity; (**b**) original frame structure (without intermediate discontinuity).

The MS and AS are 3-DOF and 2-DOF equivalent models, respectively (Figure 2). In such models, m_1 and k_1 represent the mass and the stiffness of the substructure, whereas k_2 and m_2 refer to the superstructure. In the MS, the constitutive behaviour of the disconnection devices is modelled by means of the Bouc–Wen model. There are several variations of such a model, for instance, Song and Der Kiureghian [26] proposed a generalized Bouc–Wen model to describe asymmetric hysteresis, whereas Aloisio [27] extended the six-parameters formulation to an eight-parameter formulation to include the cyclical degradation of the material properties. In this paper, the classical Bouc–Wen formulation [28] is deemed fit to accurately describe the constitutive behaviour of the disconnection devices.

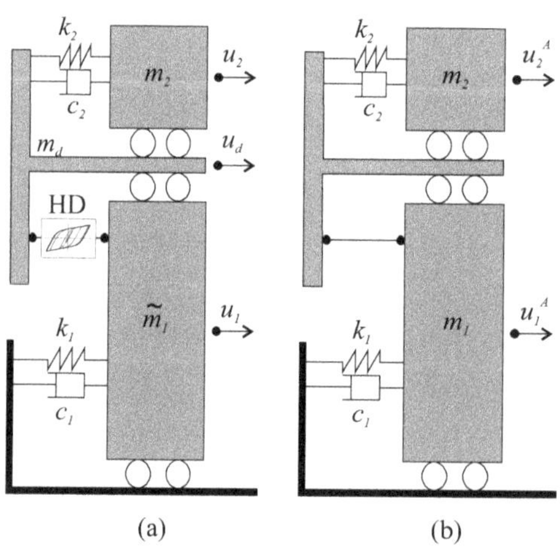

Figure 2. Archetype models: (**a**) MS; (**b**) AS (HD: Elasto-plastic behaviour of hysteretic devices).

It is worth observing that no specific reference is made to any new or existing structures. Moreover, it is assumed that the intermediate discontinuity keeps the structure to be protected in the elastic field. To evaluate the effectiveness of the intermediate discontinuity, the (linear elastic) displacements provided by the MS are compared with those of the AS.

For the MS, the equations of motion, obtained by a direct approach, are written as

$$c_{11}\dot{u}_1(t) + c_{12}\dot{u}_d - F_h(t) + k_1 u_1(t) + m_1\ddot{u}_1(t) = -m_1 a_g(t), \tag{1a}$$

$$c_{21}\dot{u}_1 + c_{22}\dot{u}_d(t) + c_{23}\dot{u}_2(t) + F_h(t) + k_2[u_d(t) - u_2(t)] + m_d\ddot{u}_d(t) = -m_d a_g(t), \tag{1b}$$

$$c_{32}\dot{u}_d(t) + c_{33}\dot{u}_2(t) + k_2 u_2(t) - k_2 u_d(t) + m_2\ddot{u}_2(t) = -m_2 a_g(t), \tag{1c}$$

$$\dot{z}(t) = [\dot{u}_d(t) - \dot{u}_1(t)]\{\tilde{A} - |z(t)|^n\{\beta_1 + \gamma_1 \mathrm{sgn}\{z(t)[\dot{u}_d(t) - \dot{u}_1(t)]\}\}\}u_y^{-1}, \tag{1d}$$

where the restoring force $F_h(t)$ is defined as

$$F_h(t) = \Psi k_d(u_d(t) - u_1(t)) + k_d u_y(1 - \Psi)z(t), \tag{2}$$

$z(t)$ is a hysteretic variable that describes the post-yielding behaviour, whereas $\tilde{A}$, β_1, n, and γ_1 are the Bouc–Wen parameters defined in [28]. The Bouc–Wen parameters are redundant; therefore, both Charalampakis [29] and Ma et al. [30] suggested assigning specific values to some of them. In details, they suggested using $\tilde{A} = 1$, so that $k_d = F_y/u_y$ acquires the meaning of pre-yielding (elastic) stiffness. Additionally, some researchers [29,31] suggested using $\beta_1 + \gamma_1 = 1$, where β_1 and γ_1 are parameters that control the shape and dimension of the hysteresis cycle. The choice to adopt these two constraints, $\tilde{A} = 1$ and $\beta_1 + \gamma_1 = 1$, provides a physical meaning to F_y, u_y, and Ψ. Specifically, F_y is the yielding force, u_y is the yielding displacement, and Ψ is the ratio between post-yielding and pre-yielding (elastic) stiffness. Moreover, the values of the hysteretic dimensionless parameter $z(t)$ remain inside the range $[-1, 1]$. Therefore, in this study, the values adopted for the parameters are $\tilde{A} = 1$, $\beta_1 = \gamma_1 = 1/2$, and $n = 2$.

The damping coefficients c_{ij} in Equations (1a)–(1c) can be grouped in matrix form as

$$\mathbf{C} = \begin{pmatrix} c_{11} & c_{12} & 0 \\ c_{21} & c_{22} & c_{23} \\ 0 & c_{32} & c_{33} \end{pmatrix}, \tag{3}$$

and Matrix $\mathbf{C}$ is derived by means of the classical Rayleigh formulation where $\mathbf{C} = \alpha_R\mathbf{M} + \beta_R\mathbf{K}$. In this equation, α_R and β_R are the Rayleigh coefficients, and it is assumed that the modal damping of the first two modes is $\xi = 0.05$. The elastic stiffness matrix $\mathbf{K}$ and the mass matrix $\mathbf{M}$ are obtained for $\Psi = 1$.

The equations of motion of the AS are described those of a shear-type 2-DOF model. The mechanical behaviour of both MS and AS are considered elastic and linear because the aim of the proposed protection strategy is to maintain the linear elastic behaviour of the structure to be protected. The linear elastic behaviour of the AS is only considered as a reference for the comparison because the AS would conceivably undergo plastic deformations under severe earthquakes. The proposed analytical model does not account for the P-Delta effects that may arise in the columns of the substructure due to the lateral displacement of the superstructure because such displacement is considered negligible, as in [9–11].

2.2. Dynamic Equivalence between Multiple-DOF Systems and Archetype Models

This subsection presents the equivalence procedure used to derive the mechanical characteristics of the AS. Such a procedure was developed mainly for low- and medium-rise frame structures and aims at obtaining the masses m_1 and m_2 and stiffnesses k_1 and k_2 of the AS starting from the geometrical and mechanical characteristics of the multiple-DOF (M-DOF) frame structure without intermediate discontinuity. If the response of the M-DOF frame has a predominant vibration mode, then the AS can correctly represent the behaviour of such a M-DOF system. The equivalence procedure is described in the details in [21]. It is assumed that m_1, m_2, k_1, and k_2 remain the same for the MS. The procedure is based on the following two assumptions:

- The frequency ω_m of the main mode of the M-DOF system without discontinuity and of the first mode of the AS (Figure 2b) are the same;
- The modal displacements of the storey of the discontinuity and top storey (i.e., L-th and N-th storeys in Figure 1a of the M-DOF system without discontinuity are equal to the modal displacements u_1 and u_2 of the AS.

The masses m_1 and m_2 of the AS correspond to the total mass of the substructure and superstructure, respectively. Assuming storeys of equal mass m entails

$$m_1 = mL; \;\; m_2 = m(N-L), \tag{4}$$

where m is the mass of a storey, L is the level where the intermediate discontinuity is introduced, and N is the number of storeys of the frame. The masses of the MS are $\tilde{m}_1$, m_d, and m_2. In particular, the mass of the substructure reads

$$\tilde{m}_1 = m(L-1) + m_0; \;\; m_d = m, \tag{5}$$

where $m_0 = \gamma m$ ($\gamma = 0.5$) is the mass of additional stiffening elements. Such stiffening elements are those that should be added to the level just below the discontinuity to maintain the same lateral stiffness available before the introduction of the discontinuity (Figure 1a). The value $\gamma = 0.5$ is chosen because only horizontal structural elements are needed to restore the stiffness of the disconnected storey, which, consequently, has a mass smaller than that of an entire storey.

The equivalence procedure requires the identification of the main mode of the M-DOF frame structure without discontinuity. It is assumed that the main mode of the structure has the highest modal mass participation factor $\mathcal{L}$. Let Φ be the vector that collects the modal displacements along a reference vertical line and along the direction to which corresponds the highest factor $\mathcal{L}$. It reads

$$\Phi = \{\phi_1, \phi_2, \ldots, \phi_L, \ldots, \phi_{N-1}, \phi_N\}^T \tag{6}$$

The following relationships for equivalent stiffnesses k_1 and k_2 of the AS can be obtained as in [21]:

$$k_1 = \frac{\omega_m^2(m_1\phi_L + m_2\phi_N)}{\phi_L}; \;\; k_2 = \frac{\omega_m^2 m_2 \phi_N}{\phi_N - \phi_L}, \tag{7}$$

where ϕ_L and ϕ_N are the components of the vector Φ that refer to the L and N storeys of the M-DOF system.

It is worth remarking that the stiffness values of the superstructure and substructure are assumed to remain unchanged after the discontinuity is introduced (i.e., k_1 and k_2 obtained by Equation (7) are the same for both AS and MS). The ability of both MS and AS to describe a generic 3D frame structure, without and with intermediate discontinuity, and the limitations of such archetype models were discussed in [23].

2.3. Characteristics of the Frame

The reference building is a regular six-storeys frame structure. The storey mass is $m_p = 301.5$ kg $\times\ 10^3$, that correspond to a storey area of 300 m^2. The viscous damping ratio of the frame structure is $\xi = 5\%$ and the overall storey stiffness is $k_p = 476{,}779$ KN/m. The fundamental period of such a structure is $T = 0.65$ s. For a six-storey building, the intermediate discontinuity can be placed at five different levels. The characteristics of the archetype models depend on the discontinuity level L and are reported in Table 1.

Table 1. Mechanical characteristics of MS and AS archetypes.

L	m_1 [kg × 10^3]	$\tilde{m}_1$ [kg × 10^3]	m_2 [kg × 10^3]	k_1 [kN/m]	k_2 [kN/m]
1	301.5	150.75	1507.50	602,403	182,552
2	603.0	425.25	1206.00	292,175	208,839
3	904.5	735.75	904.50	207,567	250,374
4	1206	1055.25	603.00	177,681	324,132
5	1507.5	1356.75	301.50	167,962	476,779

2.4. Gain Coefficients

The results of the MS are compared with those of the AS to verify if the introduction of the intermediate discontinuity reduces the displacements and drifts of the structure. The comparison between the displacements and drifts of the two models is made through three coefficients that are named gain coefficients and are defined as

$$\alpha_1 = \frac{\max[u_1(t)]}{\max[u_1^A(t)]}; \quad \alpha_2 = \frac{\max[u_2(t) - u_d(t)]}{\max[u_2^A(t) - u_1^A(t)]}; \quad \alpha_3 = \frac{\max[u_d(t) - u_1(t)]}{u_0}, \tag{8}$$

where the displacements u_1, u_2, and u_d refer to the MS (Figure 2a), u_1^A and u_2^A refer to the AS (Figure 2b), and u_0 is the threshold value of the acceptable displacements. The proposed gain coefficients represent a relative evaluation of the effectiveness of the hysteretic intermediate discontinuity. They evaluate the performances of the system with the discontinuity with respect to a reference system, represented by the corresponding linear elastic system without intermediate discontinuity.

The gain coefficient α_1 refers to the substructure and is the ratio between the displacements of the substructures of the MS and AS. Such a coefficient evaluates the performance of the superstructure as TMD for the substructure. The gain coefficient α_2 refers to the superstructure and is the ratio between the drifts of the superstructure of the MS and AS. In this case, α_2 evaluates the behaviour of the intermediate discontinuity as base isolation for the superstructure. Finally, α_3 is the ratio between the displacement of the intermediate discontinuity and the maximum threshold value $u_0 = 15$ cm. Such value was chosen in order to limit excessive displacements on the discontinuity layer that may jeopardize the safety of the structure. Each gain coefficient evaluates a different capability of the intermediate discontinuity. If a gain coefficient is less than unity, the introduction of the discontinuity is beneficial for a specific part of the structure. Therefore, the improvement of the dynamical behaviour of the frame means obtaining gain coefficients lower than unity (i.e., a reduction of displacements and drifts). A simultaneous control on these coefficients leads to a unitary approach to the dynamic enhancement of the structure following from the introduction of the discontinuity.

Two additional gain coefficients are introduced with the purpose to evaluate how the intermediate discontinuity affects the accelerations of the substructure and superstructure. They read

$$\alpha_1^A = \frac{\max[\ddot{u}_1(t)]}{\max[\ddot{u}_1^A(t)]}; \quad \alpha_2^A = \frac{\max[\ddot{u}_2(t) - \ddot{u}_d(t)]}{\max[\ddot{u}_2^A(t) - \ddot{u}_1^A(t)]}, \tag{9}$$

where $\ddot{u}_1(t)$, $\ddot{u}_2(t)$, and $\ddot{u}_d(t)$ are the accelerations of the MS. Similarly, $\ddot{u}_2^A(t)$ and $\ddot{u}_1^A(t)$ are the accelerations of the AS.

2.5. Variable Parameters

The parametric analysis is performed by varying the following parameters:

- Post-yielding to pre-yielding stiffness ratio Ψ.
- Yield displacement ratio $\eta = \frac{u_y}{u_0}$.

- Pre-yielding to storey stiffness ratio $\rho = \frac{k_d}{k_p}$.
- Ratio between m_2 (Equation (4)) and $\tilde{m}_1$ (Equation (5)) $\mu = \frac{m_2}{\tilde{m}_1}$.
- Discontinuity level of the frame structure L.

The results presented in the following sections are obtained by numerically integrating the equations of motion. For this purpose, an original code is developed in the Mathematica software. The equations of motion are integrated in an implicit form and the adopted step size is $\delta t = 0.01$ s.

3. Harmonic Analysis

The harmonic analyses are conducted on the frame whose characteristics are reported in Table 1. Parametric analyses are carried out to identify how the parameters influence the response of the system with the discontinuity. In the analyses, the external excitation is defined as

$$a_g(t) = A\ \mathrm{Sin}(\Omega t), \tag{10}$$

with A amplitude and Ω circular frequency of the excitation. Since the system is non-linear, A is considered as a variable parameter.

3.1. Frequency-Response Curves

A first analysis compares the frequency-response curves of the MS and AS. Such curves represent the dimensionless maximum displacement of the substructure, $\chi_1 = \mathrm{Max}(|u_1(t)|/u_0)$, and the dimensionless drift of the superstructure, $\chi_2 = \mathrm{Max}(|u_2(t) - u_d(t)|/u_0)$, as functions of the frequency ratio $\beta = \Omega/\omega_1$, where $\omega_1 = \sqrt{k_1/m_1}$. In the analyses where the value of η is fixed, such value is assumed as $\eta = 0.2$.

Figure 3 compares the frequency-response curves of the MS obtained for Ψ = 1.0, and the frequency-response curves of the AS. Assuming Ψ = 1.0 entails a linear behaviour of the discontinuity, the comparison shows that the discontinuity introduces a second peak in the frequency-response curves of the MS. Such a peak corresponds to a second resonant frequency, whereas the AS is characterized by a single resonant frequency in the range of the observed frequencies. The existence of a second resonant frequency is a consequence of the presence of a TMD in the system, meaning that, due to the discontinuity, the superstructure works as a TMD for the substructure. The frequency-response curve of χ_2 shows a shift of the main frequency of the MS to lower values and highlights a behaviour as a base isolated structure. Thus, the superstructure acts as a TMD for the substructure and the superstructure behaves as if it was base-isolated. In both cases, it can be asserted that the distance between the two resonant frequencies of the MS is a qualitative measure of the effectiveness of the discontinuity.

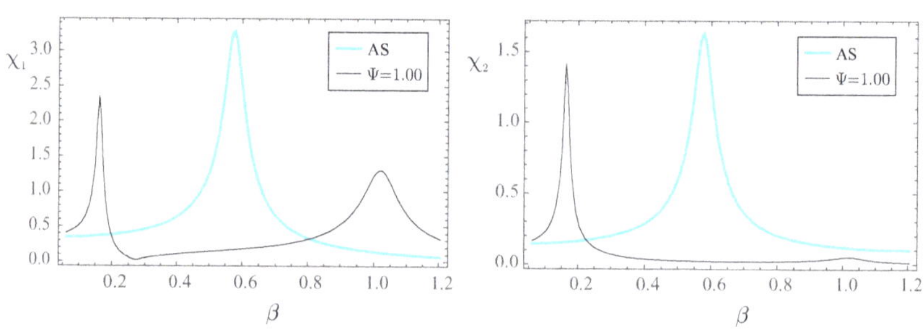

Figure 3. Frequency-response curves: comparison between curves of AS and MS (L = 3 and Ψ = 1.00).

A second analysis investigates the sensitivity of the system to the values of the parameters characterizing the hysteretic behaviour of the disconnection devices. Figure 4 shows the frequency-response curves obtained for several values of the parameters characterizing the Bouc–Wen constitutive law. Specifically, Figure 4a shows the frequency-response curves of the MS for $L = 3$, $A = 0.6$ g, $\rho = 0.2$, and several values of Ψ increasing from $\Psi = 0.10$ to $\Psi = 1.00$. The highest resonant frequency is not significantly sensitive to Ψ; on the contrary, the smallest resonant frequency significantly depends on Ψ. The values of both χ_1 and χ_2 in correspondence of the resonant peaks strongly depend on Ψ. As Ψ decreases, the hysteresis cycles become larger, and the values of the dimensionless displacements in correspondence of the peak decrease. Moreover, the distance between the peaks increases as Ψ decreases. Figure 4b shows the frequency-response curves for $L = 5$, $A = 0.6$ g, $\rho = 0.2$, and the same values of Ψ of the case presented in Figure 4a. When the discontinuity is placed at higher levels, the values of χ_1 and χ_2 at the resonant peak that refers to the lower frequency reduce as Ψ decreases. The value of μ obtained for $L = 5$ is smaller than the value obtained for $L = 3$, and the distance between the two resonant peaks decreases, thus reducing the frequency range where the discontinuity is effective. Figure 4c shows the frequency-response curves obtained for different values of ρ increasing from $\rho = 0.02$ to $\rho = 0.30$, $L = 3$, $A = 0.6$ g, and $\Psi = 0.25$. As ρ increases, both the first and second resonant frequencies shift toward higher values. Moreover, χ_1 manifests higher values at the resonant peaks for smaller values of ρ. The same increase occurs for the values at the left peak of the χ_2 curves. Instead, the value of χ_2 at the higher resonant frequency decreases until it almost vanishes for decreasing values of ρ, thus confirming that, for small values of ρ, the discontinuity acts on the superstructure as a base isolation. The last sub-figure, Figure 4d, shows the frequency-response curves referring to the MS that are obtained for different values of A and assuming $L = 3$, $\Psi = 0.25$, and $\rho = 0.2$. For increasing values of A the position of the peaks slightly shifts leftwards on the β-axis (i.e., the two resonant frequencies are only slightly sensitive to the value of A).

3.2. Gain Maps

This section analyses how Ψ affects the dynamical behaviour of the system with intermediate discontinuity. The effects of increasing values of Ψ are assessed by evaluating the gain coefficient Equations (8) and (9). The results of the parametric analysis are summarised in gain maps that represent the gain coefficients, α_1 and α_2, plotted in the parameter plane $\eta - \rho$. Each map refers to a specific value of Ψ. The maps on the left column of Figure 5 are obtained for $\Psi = 0.25$, whereas the maps on the right column refer to $\Psi = 0.75$. Moreover, the maps of Figure 5a show the α_1 coefficient, whereas the maps of Figure 5b show the α_2 coefficient. The maps are obtained for $\beta = 0.75$ and $A = 0.1$ g; they show a strong dependence on ρ and, up to a threshold value, on η. At higher values of η (where the value of η depends on the considered map), the gain coefficients become independent of η (i.e., in both the α_1 and α_2 maps, the contour levels tend to become horizontal). It is useful to remark that η directly depends on the yielding displacement; hence, high values of η refer to high values of u_y. A frame with intermediate discontinuity and small values of η manifests a stronger non-linear behaviour of the disconnection devices than systems with higher η. For very high value of η, the disconnection devices tend to manifest a linear elastic behaviour since the high-yielding displacement u_y cannot be reached during the motion. The impossibility to reach u_y during the motion is more evident at higher values of the stiffness ratio Ψ, due to the higher post-elastic stiffness of the disconnection devices. For the chosen values of β, A, and Ψ, both α_1 and α_2 are smaller than unity in the whole parameter plane, which means that the intermediate discontinuity reduces the displacement of the substructure and the drift of the superstructure with respect to those of the original frame structure.

Under a harmonic excitation, after a transient time that depends on the damping of the structure, the system undergoes stationary dynamics. To understand the behaviour of the disconnection devices for increasing values of η, the stationary hysteretic cycles are shown for points P_1 and P_3 in Figure 5c, and for points P_2 and P_4 in Figure 5d. The hysteretic cycles show the value of the dimensionless force $\sigma_h = F_h/k_d u_y$ as a function of the dimensionless displacement $\chi_d = u_d(t)/u_y$. By comparing the cycles in the first column of Figure 5c,d, which refer to $\Psi = 0.25$, with the cycles in the second column, which are obtained for $\Psi = 0.75$, it can be observed that a higher value of Ψ corresponds to smaller hysteresis cycles. Similarly, an increase of η also leads to a reduction of the area of the hysteretic cycles (compare the cycles of Figure 5c,d).

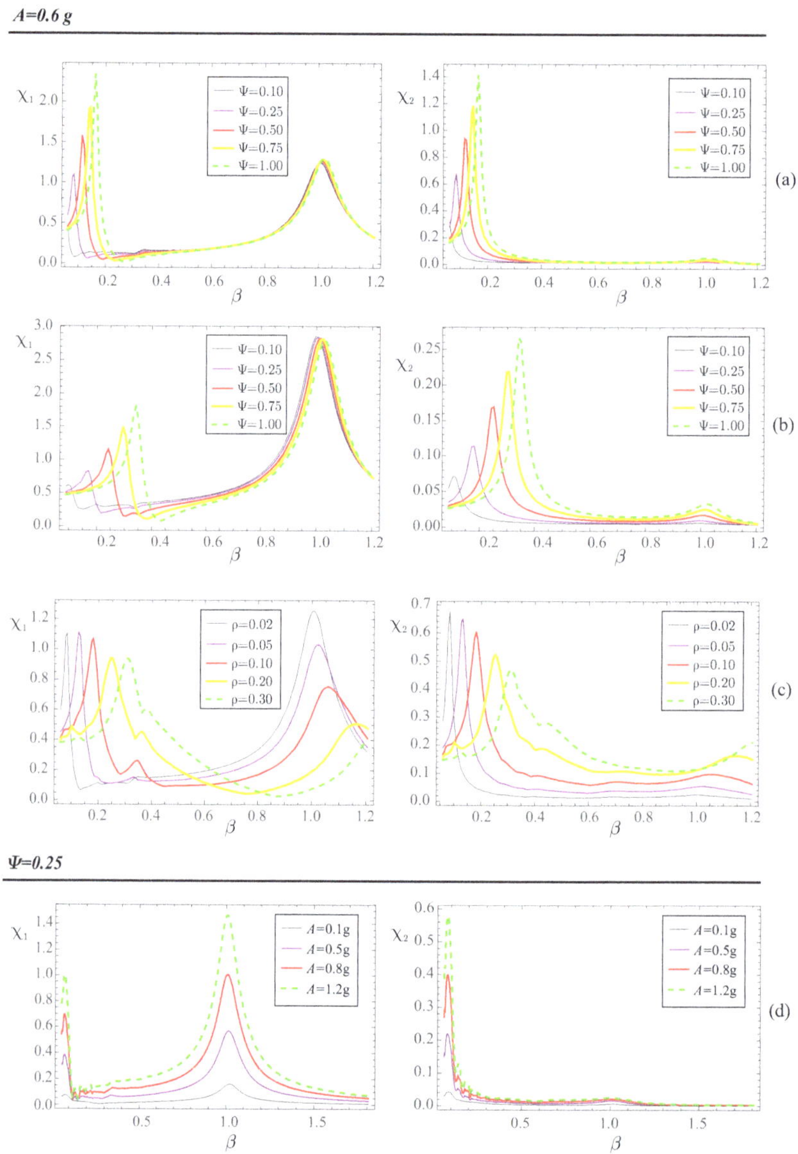

Figure 4. Frequency-response curves: (**a**) $L = 3$, $A = 0.6$ g, $\rho = 0.2$, and different values of the parameter ψ; (**b**) $L = 5$, $A = 0.6$ g, $\rho = 0.2$, and different values of the parameter ψ; (**c**) $L = 3$, $A = 0.6$ g, $\Psi = 0.25$, and different values of ρ; and (**d**) $L = 3$, $\Psi = 0.25$, $\rho = 0.2$, and different values of A.

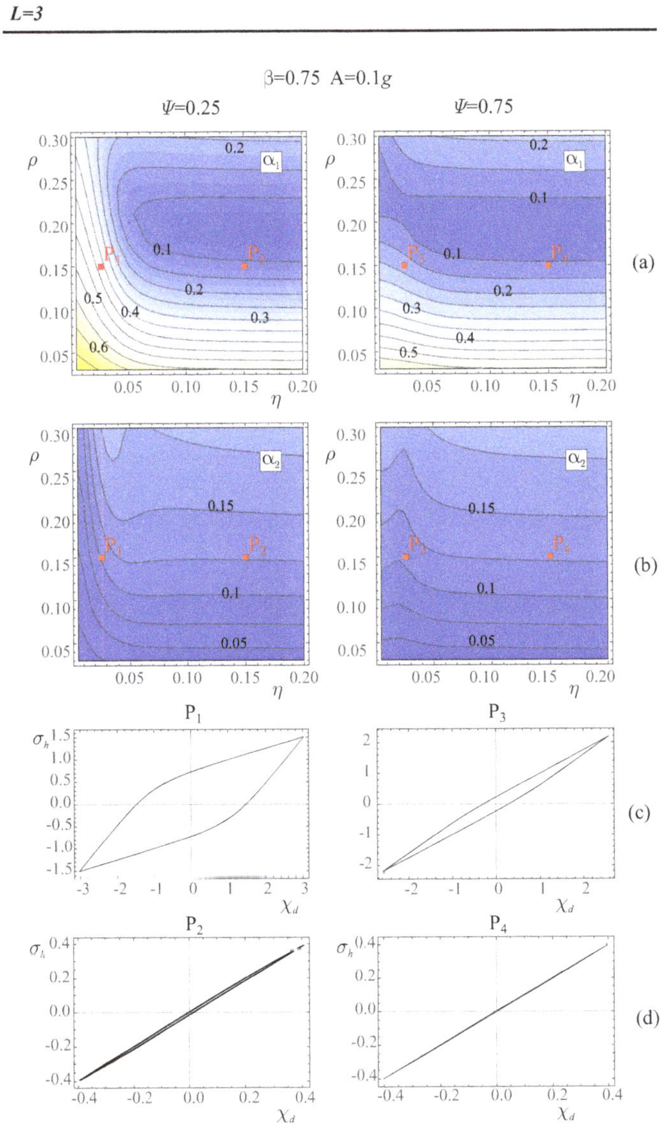

Figure 5. Gain maps for fixed $\beta = 0.75$ and $A = 0.1$ g: (**a**) α_1 maps for two different values of Ψ; (**b**) α_2 maps for two different values of Ψ; (**c**) hysteretic cycles in $P_1 \equiv (0.025, 0.15)$, $\Psi = 0.25$ and $P_3 \equiv (0.025, 0.15)$, $\Psi = 0.75$; (**d**) hysteretic cycles in $P_2 \equiv (0.15, 0.15)$, $\Psi = 0.25$ and $P_4 \equiv (0.15, 0.15)$, $\Psi = 0.75$.

Figure 6 shows the gain maps obtained for $\beta = 1$ and $\Psi = 0.25$. The maps in Figure 6a and Figure 6b are obtained for $A = 0.1$ g and $A = 0.6$ g, respectively, assuming that the discontinuity is located at the third level ($L = 3$). Figure 6c, instead, shows maps obtained for $A = 0.1$ g and a discontinuity located at the fifth level ($L = 5$). The comparison of Figure 6a and Figure 6b highlights the dependence of the gain maps on A. Specifically, in the gain maps of Figure 6b, the contour levels do not tend to a horizontal asymptote because the displacement of the disconnection devices exceeds the yield displacement u_y due to the high value of A, thus assuring wide hysteresis cycles.

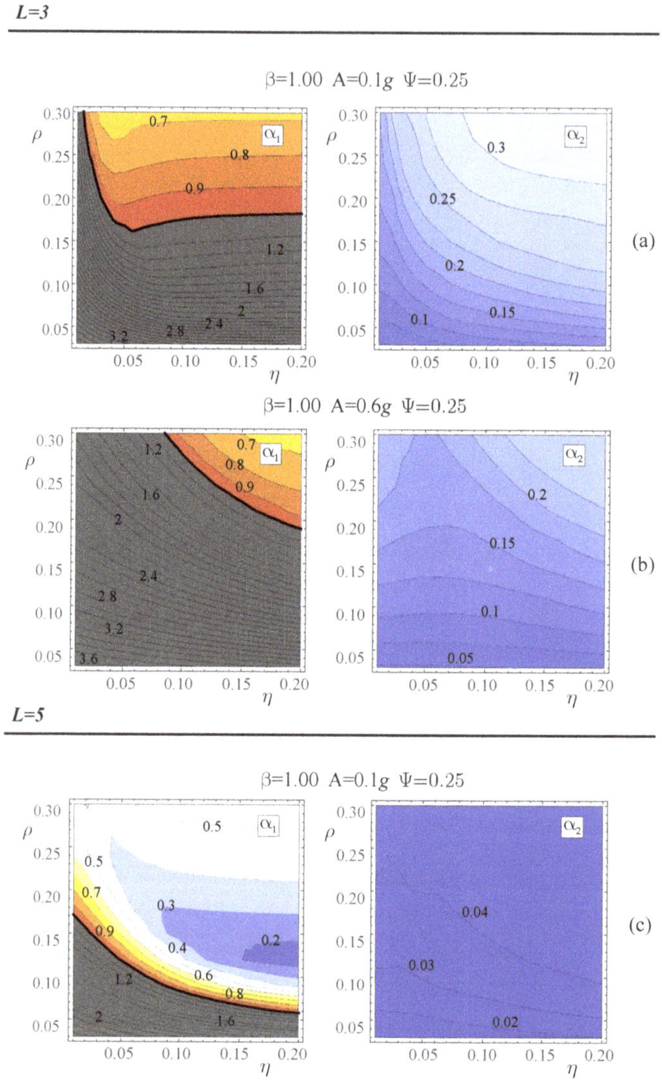

Figure 6. Gain maps for $\beta = 1$: (**a**) $L = 3$, $A = 0.1$ g, and $\Psi = 0.25$; (**b**) $L = 3$, $A = 0.6$ g, and $\Psi = 0.25$; (**c**) $L = 5$, $A = 0.1$ g, and $\Psi = 0.25$.

The α_1 gain maps of Figure 6a,b show that there are large regions of the parameter where the gain coefficients are higher than unity (dark grey regions). In these regions, the discontinuity is not able to reduce the displacement of the substructure. On the contrary, α_2 is smaller than unity in the whole parameter plane, thus assuring a reduction of the superstructure drift for any combination of the parameters. Nevertheless, it is always possible to choose a combination of the design parameters ρ and η inside regions where both the gain coefficients α_1 and α_2 are smaller than unity.

The comparison among the left maps of Figure 5a,b and the maps of Figure 6a (all obtained for $\Psi = 0.25$ and $A = 0.1$ g) shows that the effectiveness of the discontinuity (i.e., the smallness of the gain coefficients) strongly depends on the frequency ratio β.

Finally, Figure 6c shows the gain maps for $L = 5$. In this case, the dark grey regions where α_1 is higher than unity reduce compared to the case with $L = 3$, and, generally, smaller values of both α_1 and α_2 are obtained, thus showing the higher effectiveness of the intermediate discontinuity.

4. Seismic Analysis

This section shows the results of a seismic analysis. The first part of the section aims at providing an initial understanding of the seismic behaviour of the MS. For this purpose, the gain maps show the results of simulations performed using single seismic records as excitation. The two different seismic records considered in this analysis are:

(a) El Centro, CA, Array Station 9, Imperial Valley Irrigation District, component 180;
(b) Parkfield, CO2-065 ground motion recorded during the California earthquake 1966.

Figure 7 shows the time-histories (left graphs) and the pseudo-response spectra (right graphs) of the two seismic records. In the following, each seismic record is called with the corresponding earthquake name (i.e., the names underlined in the list above).

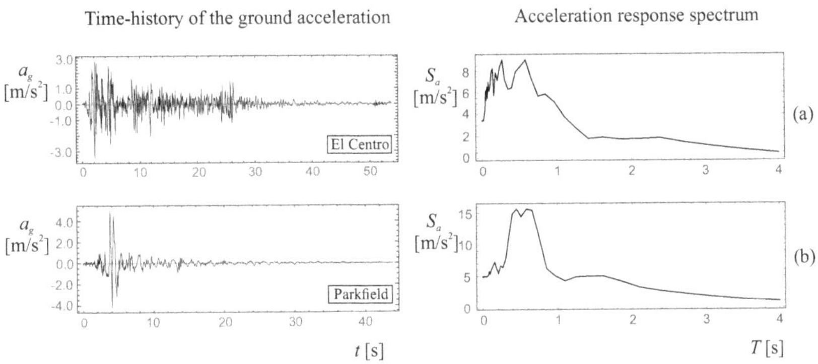

Figure 7. Recorded earthquakes: (**a**) El Centro seismic record; and (**b**) Parkfield seismic record.

Since gain maps referred to a single registration are not useful for a complete assessment of the effectiveness of the intermediate discontinuity, the second part of the section shows the results of an analysis carried out considering $n_e = 305$ seismic records. The complete list of seismic records used in the analysis can be found at http://ing.univaq.it/diegidio/Earthq/Earthquakes_Table.pdf. Additionally, in this case, the results are presented in gain maps. Such gain maps show the mean values of the gain indexes obtained with the different seismic records.

4.1. Gain Maps from a Single Seismic Record

Figures 8–10 show the gain maps obtained with the seismic records of El Centro and Parkfield for $L = 1$, $L = 3$, and $L = 5$, respectively. Specifically Figures 8a–c, 9a–c and 10a–c are obtained for $\Psi = 0.25$, whereas Figures 8b–d, 9b–d, and 10b–d for $\Psi = 0.75$. The regions where the gain coefficients α_1 and α_2 are smaller than unity are coloured, those where these gain coefficients are higher than unity are displayed in dark grey colour. Both the α_1 and α_2 maps show that, as Ψ increases, the gain coefficients increase, thus reducing the effectiveness of the intermediate discontinuity. The α_3 maps show the feasibility of the protection strategy. When α_3 is higher than unity (dark grey zones of the α_3 maps), the displacements of the disconnection devices are higher than the permissible threshold value. The size of the dark grey zones in the α_3 maps increases when Ψ decreases, thus even if lower values of Ψ assure a higher reduction of displacements and drifts, the displacements of the discontinuity level may not be compatible with the capability of the disconnection devices.

In Figures 8–10, the thick black lines on the maps represent the admissible limit of α_3. In each map, only in the region above such lines, the displacement of the discontinuity is below the threshold value u_0. While the α_3 gain maps obtained from the seismic record of El Centro show only small dark grey regions, the α_3 gain maps obtained from the Parkfield seismic record show large regions where $\alpha_3 > 1$ maps. The size of these regions significantly reduces as Ψ increases. Moreover, when the discontinuity level L increases, the size of the dark grey regions tends to reduce.

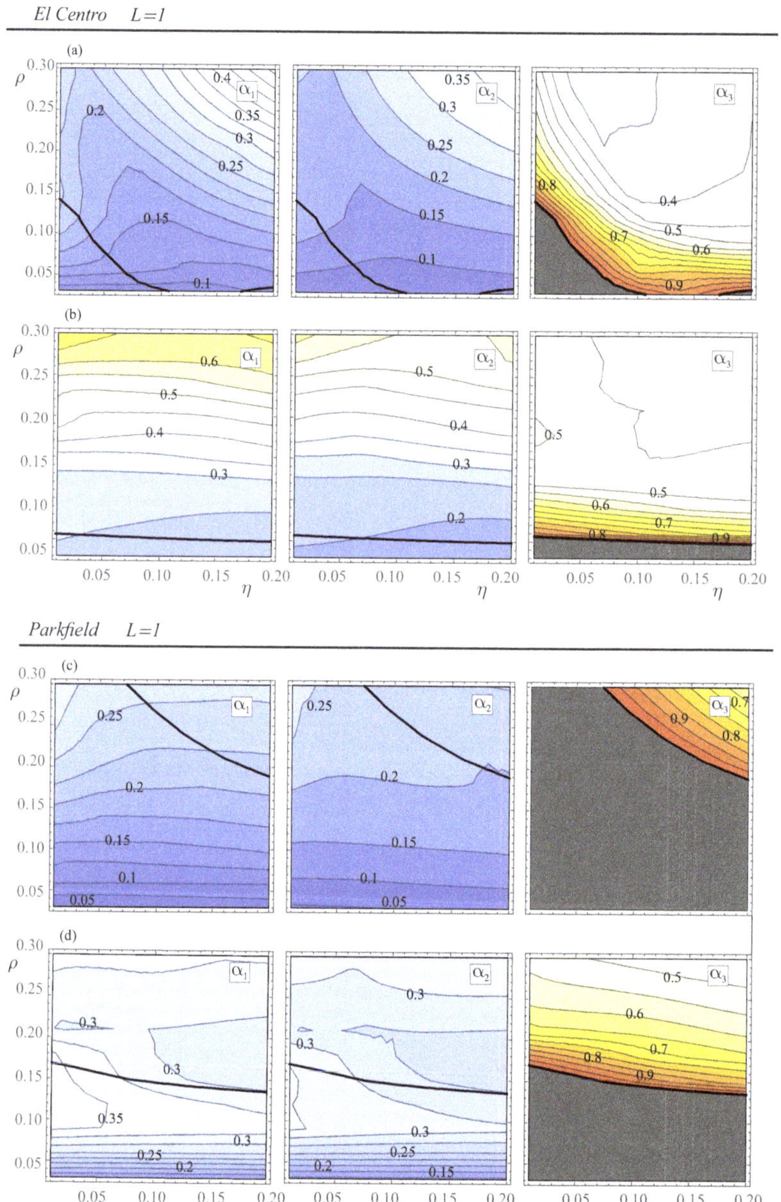

Figure 8. Gain maps of α_1, α_2, and α_3 for $L = 1$: (**a**) El Centro seismic record and $\Psi = 0.25$ (**b**) El Centro seismic record and $\Psi = 0.75$; (**c**) Parkfield Seismic record and $\Psi = 0.25$; and (**d**) Parkfield Seismic record and $\Psi = 0.75$.

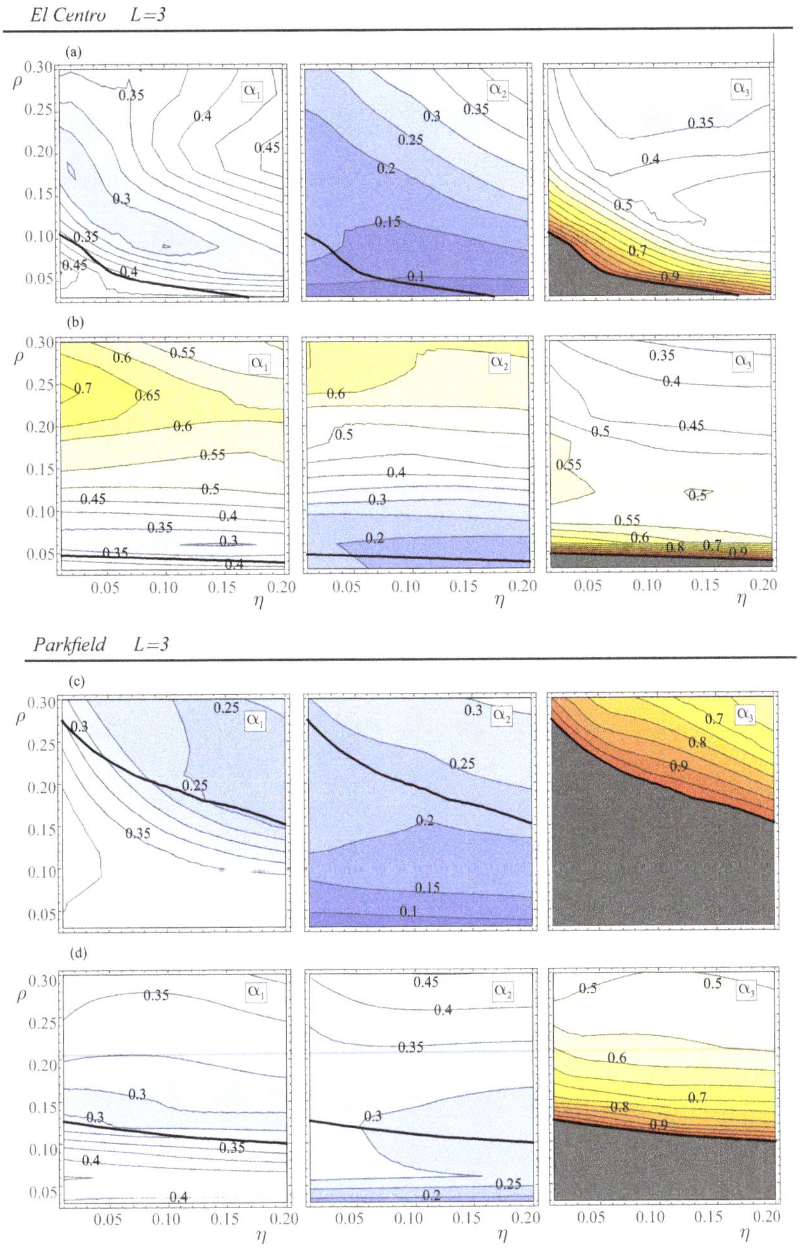

Figure 9. Gain maps of α_1, α_2, and α_3 for $L = 3$: (**a**) El Centro seismic record and $\Psi = 0.25$ (**b**) El Centro seismic record and $\Psi = 0.75$; (**c**) Parkfield Seismic record and $\Psi = 0.25$; and (**d**) Parkfield Seismic record and $\Psi = 0.75$.

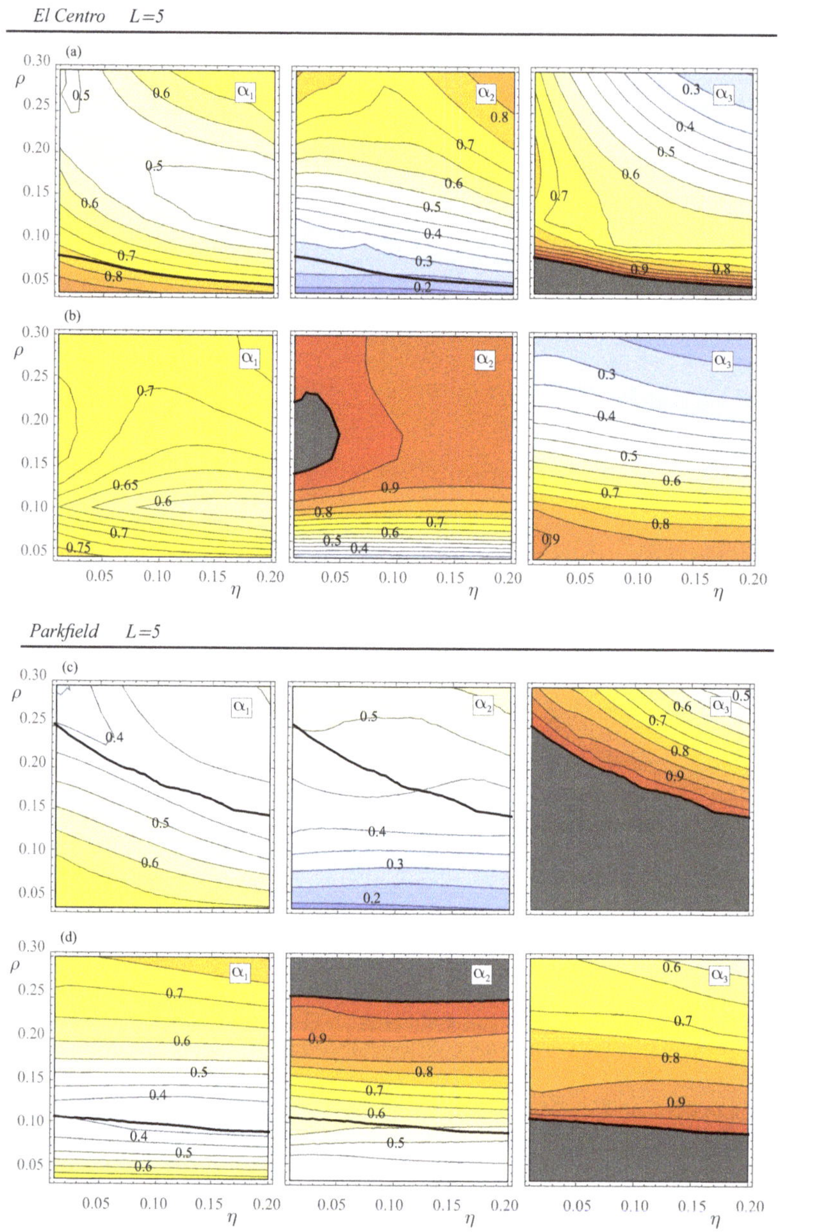

Figure 10. Gain maps of α_1, α_2, and α_3 for $L = 5$: (**a**) El Centro seismic record and $\Psi = 0.25$ (**b**) El Centro seismic record and $\Psi = 0.75$; (**c**) Parkfield Seismic record and $\Psi = 0.25$; and (**d**) Parkfield Seismic record and $\Psi = 0.75$.

Figures 11 and 12 show the α_1^A and α_2^A gain maps under El Centro and Parkfield seismic records, respectively. These maps evaluate the effects of the discontinuity on the accelerations of the structure. The black lines on the maps show the range highlighted in the α_3 maps of Figures 8–10. Only in the regions above such lines are the displacements of the disconnection devices below the admissible threshold. In Figures 11 and 12, the maps in the first two rows refer to α_1^A, whereas the maps in the third and fourth rows refer to α_2^A. The maps in the first and third rows are obtained for $\Psi = 0.25$, the maps in the second and fourth rows refer to $\Psi = 0.75$.

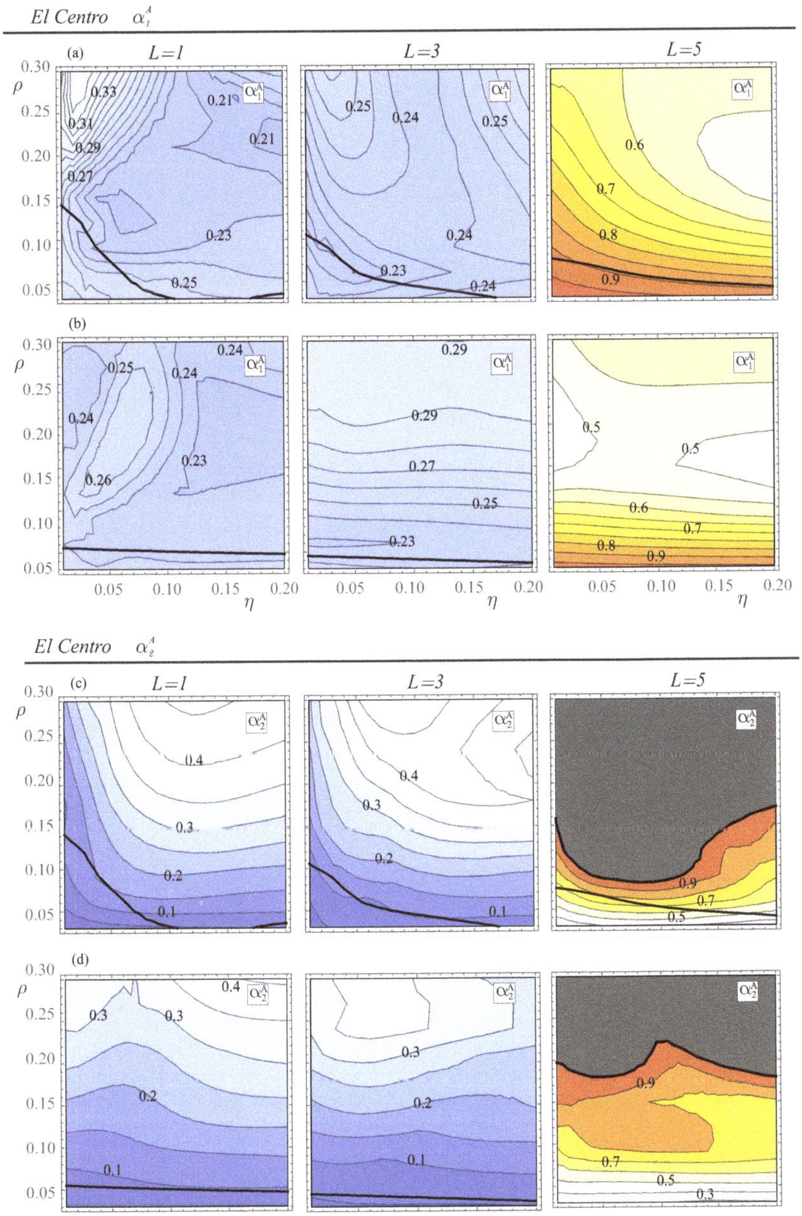

Figure 11. Gain maps of α_1^A and α_2^A for $L = 1$, $L = 3$ and $L = 5$ considering the El Centro seismic record as external excitation: (**a**) maps of α_1^A obtained for $\Psi = 0.25$; (**b**) maps of α_1^A obtained for $\Psi = 0.75$; (**c**) maps of α_2^A obtained for $\Psi = 0.25$; and (**d**) maps of α_2^A obtained for $\Psi = 0.75$.

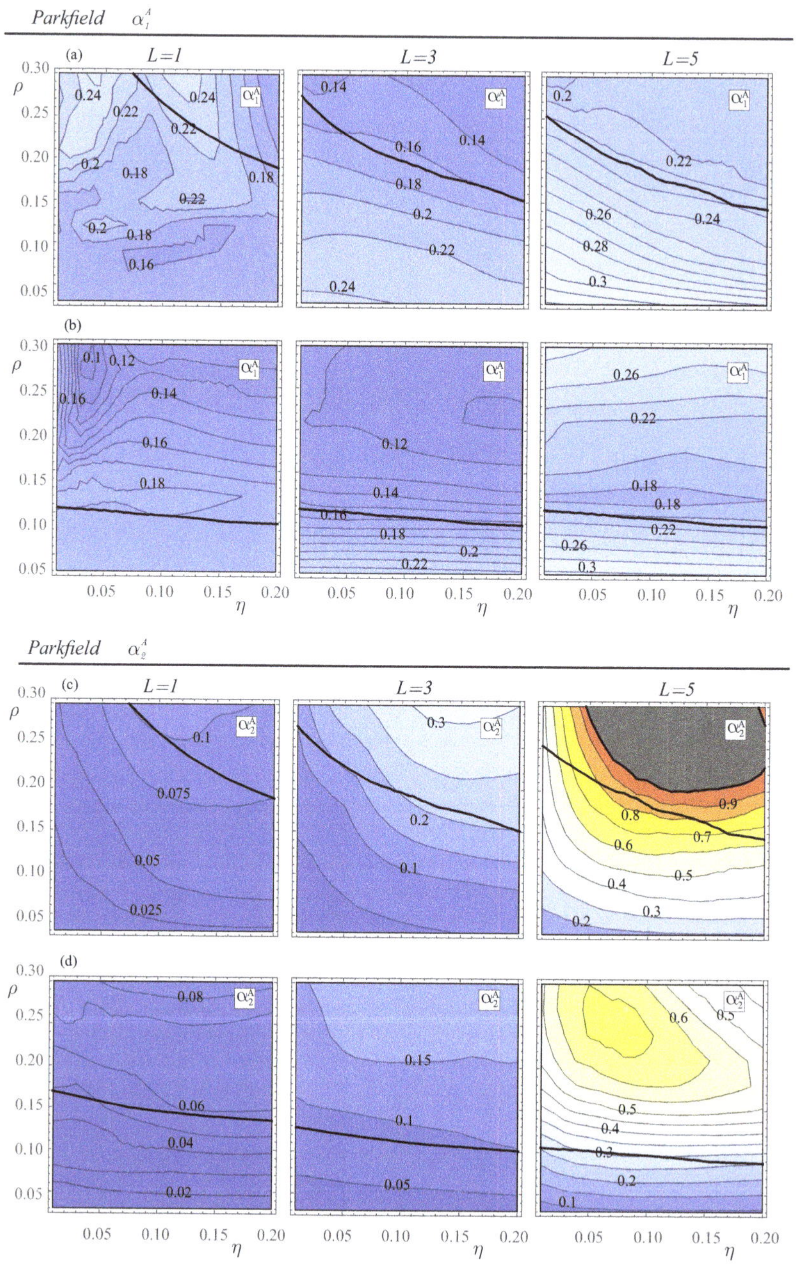

Figure 12. Gain maps of α_1^A and α_2^A for $L = 1$, $L = 3$ and $L = 5$ considering the Parkfield seismic record as external excitation: (**a**) maps of α_1^A obtained for $\Psi = 0.25$; (**b**) maps of α_1^A obtained for $\Psi = 0.75$; (**c**) maps of α_2^A obtained for $\Psi = 0.25$; and (**d**) maps of α_2^A obtained for $\Psi = 0.75$.

The α_1^A maps show that as Ψ increases, α_1^A decreases. Similarly, α_2^A decreases as Ψ increases. Therefore, a lower value of Ψ may increase the accelerations and, at the same time, reduce displacements and drifts. However, both the gain coefficients based on the displacements and drifts (Equation (8)) and those based on the accelerations (Equation (9)) are smaller than unity in a large area of the parameters plane. It is worth observing that α_1^A and α_2^A are lower than α_1 and α_2, thus the use of α_1 and α_2 provides a more severe assessment of the effectiveness of the intermediate discontinuity.

4.2. Gain Maps Obtained from Multiple Seismic Records

The seismic analysis is carried out by using several seismic records obtained from [32]. The magnitude varies from 4 to 8 and the epicentral distance varies from 10 to 30 km. In the analysis, the target parameters are the mean displacements $\overline{u}_i$ that are defined as the mean of the maximum displacements $u_{i,j}$ obtained from the integration of the equations of motion with each seismic record; they read

$$\overline{u}_i = \frac{1}{n_e}\sum_{j=1}^{n_e} \max[u_{i,j}(t)] \tag{11}$$

with $(i = 1, d, 2)$. The displacements $u_{1,j}$, $u_{d,j}$, and $u_{2,j}$ are those of the substructure, discontinuity level, and superstructure obtained from the jth seismic record. The total number of seismic records used in the simulations is n_e. The number of seismic records used is sufficient to reach a coefficient of variation (CoV) of $\overline{u}_i$, $\text{CoV}_{\overline{u}_i} \leq 0.2$. Therefore, the predictability of the model to identify the average displacement is higher than 80%. The $\text{CoV}_{\overline{u}_i}$ is defined as

$$CoV_{\overline{u}_i} = \frac{\sigma(\overline{u}_i)}{\mu(\overline{u}_i)} \tag{12}$$

where $\sigma(\overline{u}_i)$ and $\mu(\overline{u}_i)$ are the standard deviation and mean value of $\overline{u}_i$, respectively. To understand the effectiveness of the disconnection, three coefficients (called mean-gain coefficients in the following) similar to those in Equation (8) are introduced:

$$\overline{\alpha_1} = \frac{\sum_{j=1}^{n_e} \max[u_{1,j}(t)]}{\sum_{j=1}^{n_e} \max[u_{1,j}^A(t)]}; \quad \overline{\alpha_2} = \frac{\sum_{j=1}^{n_e} \max[u_{2,j}(t) - u_{d,j}(t)]}{\sum_{j=1}^{n_e} \max[u_{2,j}^A(t) - u_{1,j}^A(t)]}; \quad \overline{\alpha_3} = \frac{\sum_{j=1}^{n_e} \max[u_{d,j}(t) - u_{1,j}(t)]}{n_e u_0} \tag{13}$$

Figure 13 shows the gain maps of the mean displacements. Such maps represent the mean-gain coefficients in the parameters plane PGA-ρ, where PGA is the peak ground acceleration and ρ is the stiffness ratio previously defined. All the gain maps are obtained for $\eta = 0.2$. Similarly to previous figures, the regions where $\overline{\alpha}_i < 1, (i = 1, 2)$ are named gain regions and are coloured, whereas those where $\overline{\alpha}_i > 1$ are identified by a dark grey colour. The first, second, and third rows of the figure refer to different levels of the discontinuity, $L = 1$, $L = 3$, and $L = 5$, respectively. All maps are obtained for $\Psi = 0.25$.

The maps show wide gain regions. As can be observed, $L = 1$ provides the best performances in the reduction of the displacement of the substructure because the values of $\overline{\alpha}_1$ obtained for $L = 1$ are in general smaller than those obtained for other values of L. On the contrary, $L = 5$ provides the largest reduction of the drift of the superstructure since, in this case, $\overline{\alpha}_2$ is smaller than those in the other cases. The maps of $\overline{\alpha}_3$ show values higher than unity only in the zone with high values of PGA and low values of ρ, specifically in the region where $PGA > 1.1$ g and $\rho < 0.1$. In this region, $\overline{\alpha}_1$ and $\overline{\alpha}_2$ have small values, but the displacement of the discontinuity level exceeds the threshold value u_0 (i.e., $\overline{\alpha}_3$ is higher than unity). Nevertheless, such a region is very small and a slight increase of ρ is sufficient to move in a region of the maps where $\overline{\alpha}_3 \leq 1$ and at the same $\overline{\alpha}_1 \leq 1$ and $\overline{\alpha}_2 \leq 1$. Since all the three coefficients are less than unity the disconnection can be considered a successful protection strategy for the original frame structure. The map obtained for $L = 3$ shows an intermediate behaviour between those obtained for $L = 1$ and $L = 5$, and the map of $\overline{\alpha}_3$ does not show any dark grey regions.

To summarize, the $\overline{\alpha}_1$ maps exhibit a decrease in the performance from $L = 1$ to $L = 5$. This can be explained by assuming that the superstructure acts as a TMD for the substructure and consequently the performance decreases because the mass ratio between the superstructure and substructure decreases [7,22]. Instead, $\overline{\alpha}_2$ decreases from $L = 1$ to $L = 5$. In this case, when $L = 5$, the superstructure is very stiff and the superstructure acts as if it was base-isolated. On the contrary, when $L = 1$, the superstructure has lower stiffness and the performance decreases since the period of the superstructure can approach that of the discontinuity layer. Similar results were already presented in [33] for tower buildings.

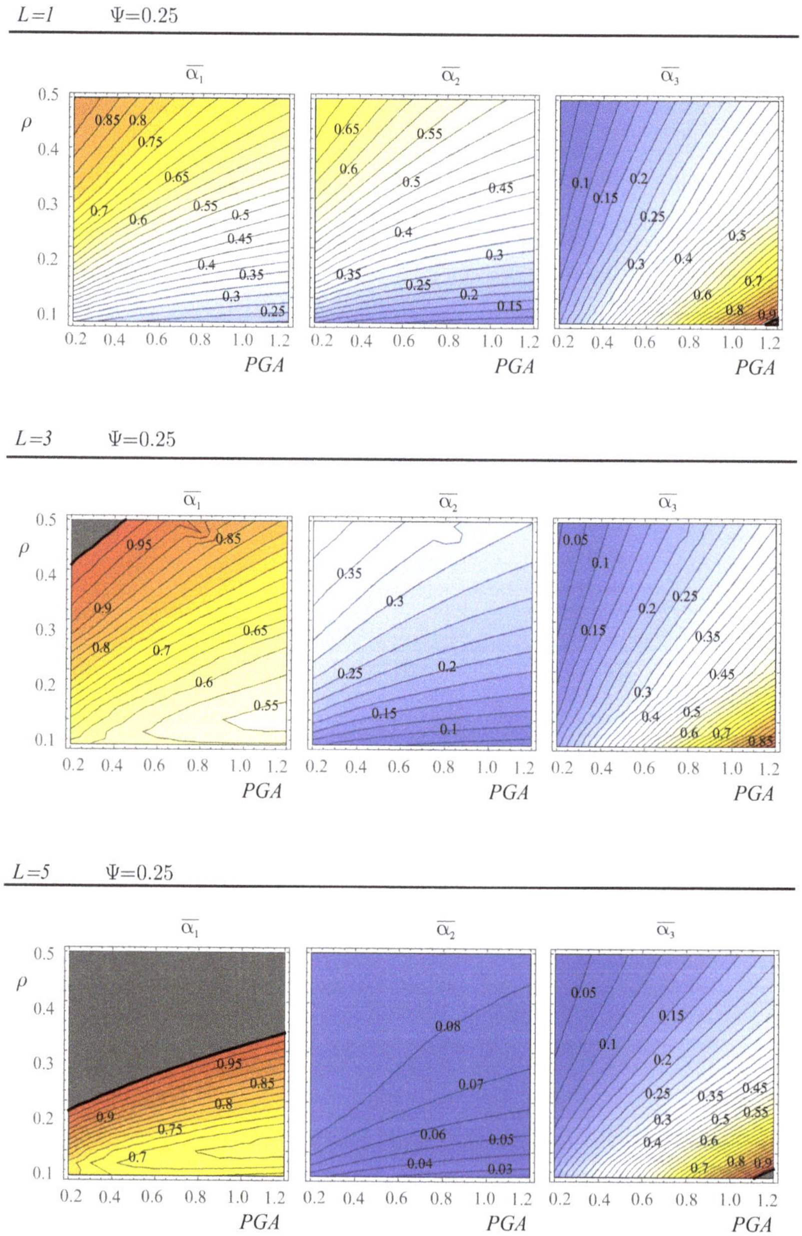

Figure 13. Mean gain maps for different values of $L = 1, 3, 5$ and $\Psi = 0.25$.

Figure 14 shows the mean-gain maps obtained for $L = 3$. The maps in the first row refer to $\Psi = 0.1$, whereas the maps in the second row are obtained for $\Psi = 0.75$. A direct comparison with the maps in the second row of Figure 13 ($L = 3$) allows evaluating the effect of Ψ on the performances of the intermediate discontinuity. The dark grey region of the $\overline{\alpha}_1$ map enlarges with respect to that of Figure 13 for increasing Ψ. Therefore, lower values of Ψ allow a wider range of use. Moreover, when Ψ decreases, the performance of the system increases since there is a higher reduction of the displacement of the substructure (i.e., a decrease of α_1). On the contrary, the coefficients $\overline{\alpha}_2$ and $\overline{\alpha}_3$ are not significantly affected by Ψ. Finally, $\overline{\alpha}_3$ is always smaller than unity when $\Psi < 0.25$, whereas for $\Psi = 0.75$ there is a small region where $\overline{\alpha}_3 \geq 1$.

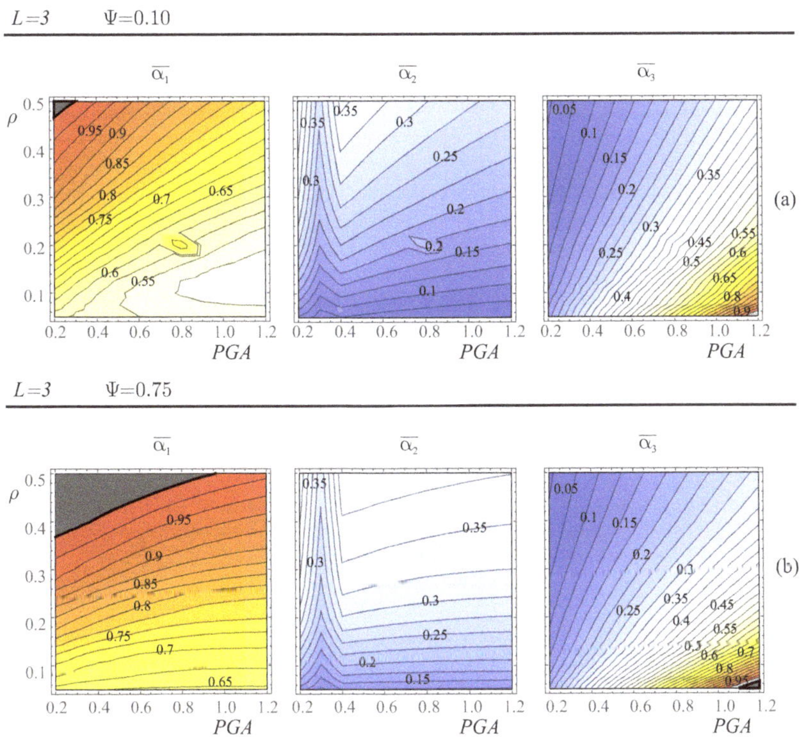

Figure 14. Mean gain maps for $L = 3$ and (**a**) $\Psi = 0.10$; (**b**) $\Psi = 0.75$.

5. Conclusions

This paper proposes a modelling approach for the study of the intermediate discontinuity. The structure with the discontinuity layer has been modelled by a 3-DOF model and the disconnection devices have been described by using the Bouc–Wen model. In order to evaluate the improvement of the dynamical behaviour obtained by the introduction of the discontinuity, two different systems have been analysed. The first system is the frame after the introduction of the discontinuity, called the main system (MS) and the second system is the reference frame, called the auxiliary system (AS).

The first part of the paper is focused on the dynamical behaviour of the MS that is analysed by using harmonic excitation. The results of the analysis have been organized in frequency-response curves and gain maps. It has been found that the value of the post-yielding to pre-yielding stiffness ratio Ψ in the Bouc–Wen model significantly affects the value of the first frequency of the MS. Moreover, by comparing the frequency-response curves of the AS and MS for different values of Ψ, it has been found that lower values of Ψ may extend the range of frequency where the intermediate discontinuity is effective. The harmonic maps highlighted that small values of Ψ may reduce the displacement of the superstructure, but also increase the displacement of the substructure.

The second part of the paper is focused on the seismic response of the models. A preliminary analysis has been carried out considering single seismic records. Five gain coefficients have been defined to evaluate the effects of the introduction of the intermediate discontinuity. Such coefficients have been defined as ratios between the displacements (or the accelerations) of the MS and those of the AS. The results of the analysis have been summarized in gain maps. Such maps show the gain coefficients in a specific parameters plane. The results also show that an increase of Ψ reduces the displacement of the substructure and increases that of the superstructure. Moreover, higher values of Ψ reduce the displacement of the discontinuity layer, thus reducing the likelihood of such displacement exceeding a threshold value. It has been found that another possibility to reduce the displacement of the discontinuity layer is to change the level of the discontinuity L. As L increases, the displacement of the discontinuity layer decreases due to the filtering effect of the substructure. Additionally, the effects of the intermediate discontinuity on the accelerations acting on the structure have been evaluated, and it has been found that when Ψ decreases, the accelerations may increase.

Finally, a seismic analysis is carried out considering several seismic records within definite ranges of magnitude and epicentral distance. The results have been organized in gain maps that show the gain coefficients in different parameters planes. Such maps may be a useful auxiliary tool in the understanding of the behaviour resulting from the introduction of an intermediate discontinuity in a frame structure.

Author Contributions: Conceptualization, A.D.E., S.P. and A.C.; Methodology, A.D.E. and A.C.; Software, S.P.; Formal analysis, S.P.; Investigation, A.D.E., S.P. and A.C.; Writing—original draft, S.P.; Writing—review & editing, A.D.E. and A.C.; Supervision, A.D.E. and A.C. All authors have read and agreed to the published version of the manuscript.

Funding: This research was partially funded by CARISPAQ Foundation, year 2022.

Data Availability Statement: Data, models, and code used during the study are available from the corresponding author by request. Specifically, the following data are available: (*i*) data of the results, and (*ii*) the source code for the numerical integration procedure.

Conflicts of Interest: The authors declare no conflict of interest.

References

1. Kelly, J. Base isolation: Linear theory and design. *Earthq. Spectra* **1995**, *6*, 223–244. [CrossRef]
2. Den Hartog, J.P. *Mechanical Vibrations*, 4th ed.; McGraw-Hill: New York, NY, USA, 1956.
3. Tsai, H.H. The effect of tuned-mass dampers on the seismic response of base-isolated structures. *Int. J. Solids Struct.* **1995**, *32*, 1195–1210. [CrossRef]
4. Taniguchi, T.; Kiureghian, A.D.; Melkumyan, M. Effect of tuned mass damper on displacement demand of base-isolated structures. *Eng. Struct.* **2008**, *30*, 3478–3488. [CrossRef]
5. De Domenico, D.; Ricciardi, G. An enhanced base isolation system equipped with optimal Tuned Mass Damper Inerter (TMDI). *Earthq. Eng. Struct. Dyn.* **2018**, *47*, 1169–1192. [CrossRef]
6. Giaralis, A.; Taflanidis, A. Optimal tuned mass-damper-inerter (TMDI) design for seismically excited MDOF structures with model uncertainties based on reliability criteria. *Struct. Control Health Monit.* **2018**, *25*, e2082. [CrossRef]
7. De Angelis, M.; Perno, S.; Reggio, A. Dynamic response and optimal design of structures with large mass ratio TMD. *Earthq. Eng. Struct. Dyn.* **2012**, *41*, 41–60. [CrossRef]
8. Chey, M.; Chase, J.; Mander, J.B.; Carr, A. Innovative seismic retrofitting strategy of added stories isolation system. *Front. Struct. Civ. Eng.* **2013**, *7*, 13–23. [CrossRef]
9. Wang, S.J.; Chang, K.C.; Hwang, J.S.; Lee, B.H. Simplified analysis of mid-story seismically isolated buildings. *Earthq. Eng. Struct. Dyn.* **2011**, *40*, 119–133. [CrossRef]
10. Wang, S.J.; Chang, K.C.; Hwang, J.S.; Hsiao, J.Y.; Lee, B.H.; Hung, Y.C. Dynamic behavior of a building structure tested with base and mid-story isolation systems. *Eng. Struct.* **2012**, *42*, 420–433. [CrossRef]
11. Wang, S.J.; Hwang, J.S.; Chang, K.C.; Lin, M.H.; Lee, B.H. Analytical and experimental studies on midstory isolated buildings with modal coupling effect. *Earthq. Eng. Struct. Dyn.* **2013**, *42*, 201–219. [CrossRef]

12. Villaverde, R.; Mosqueda, G. Aseismic roof isolation system: Analytic and shake table studies. *Earthq. Eng. Struct. Dyn.* **1999**, *28*, 217–234. [CrossRef]
13. Villaverde, R.; Aguirre, M.; Hamilton, C. Aseismic Roof Isolation System Built with Steel Oval Elements. *Earthq. Spectra* **2005**, *21*, 225–241. [CrossRef]
14. Matta, E.; De Stefano, A. Robust design of mass-uncertain rolling-pendulum TMDs for the seismic protection of buildings. *Mech. Syst. Signal Process.* **2009**, *23*, 127–147. [CrossRef]
15. Feng, M.; Mita, A. Vibration Control of Tall Buildings Using Mega SubConfiguration. *J. Eng. Mech.* **1995**, *121*, 1082–1088. [CrossRef]
16. Sadek, F.; Mohraz, B.; Taylor, A.W.; Chung, R.M. A method of estimating the parameters of tuned mass dampers for seismic applications. *Earthq. Eng. Struct. Dyn.* **1997**, *26*, 617–635. [CrossRef]
17. Anajafi, H.; Medina, R. Comparison of the seismic performance of a partial mass isolation technique with conventional TMD and base-isolation systems under broad-band and narrow-band excitations. *Eng. Struct.* **2018**, *158*, 110–123. [CrossRef]
18. Rana, R.; Soong, T. Parametric analysis and simplified design of tuned mass damper. *Eng. Struct.* **1998**, *20*, 193–204. [CrossRef]
19. Miranda, J.C. On tuned mass dampers for reducing the seismic response of structures. *Earthq. Eng. Struct. Dyn.* **2005**, *34*, 847–865. [CrossRef]
20. Fabrizio, C.; Di Egidio, A.; de Leo, A. Top disconnection versus base disconnection in structures subjected to harmonic base excitation. *Eng. Struct.* **2017**, *152*, 660–670. [CrossRef]
21. Fabrizio, C.; de Leo, A.; Di Egidio, A. Tuned mass damper and base isolation: A unitary approach for the seismic protection of conventional frame structures. *J. Eng. Mech.* **2019**, *145*, 04019011. [CrossRef]
22. Reggio, A.; De Angelis, M. Optimal energy-based seismic design of non-conventional Tuned Mass Damper (TMD) implemented via inter-story isolation. *Earthq. Eng. Struct. Dyn.* **2015**, *44*, 1623–1642. [CrossRef]
23. Pagliaro, S.; Di Egidio, A. Archetype Dynamically Equivalent 3-DOF Model to Evaluate Seismic Performances of Intermediate Discontinuity in Frame Structures. *J. Eng. Mech.* **2022**, *148*, 04022004. [CrossRef]
24. De Domenico, D.; Ricciardi, G. Optimal design and seismic performance of tuned mass damper inerter (TMDI) for structures with nonlinear base isolation systems. *Earthq. Eng. Struct. Dyn.* **2018**, *47*, 2539–2560.
25. Wang, S.J.; Lin, M.H.; Chiang, Y.S.; Hwang, J.S. Mechanical behavior of lead rubber bearings under and after nonproportional plane loading. *Earthq. Eng. Struct. Dyn.* **2019**, *48*, 1508–1531. [CrossRef]
26. Song, J.; Der Kiureghian, A. Generalized Bouc-Wen model for highly asymmetric hysteresis. *J. Eng. Mech.* **2006**, *132*, 610–618. [CrossRef]
27. Aloisio, A.; Alaggio, R.; Kohler, J.; Fragiacomo, M. Extension of Generalized Bouc-Wen Hysteresis Modeling of Wood Joints and Structural Systems. *J. Eng. Mech.* **2020**, *146*, 04020001. [CrossRef]
28. Wen, Y.K. Method for random vibration of hysteretic systems. *J. Eng. Mech.* **1976**, *102*, 249–263. [CrossRef]
29. Charalampakis, A. Parameters of Bouc-Wen hysteretic model revisited. In Proceedings of the 9th HSTAM International Congress on Mechanics, Limassol, Cyprus, 12–14 July 2010.
30. Ma, F.; Zhang, H.; Bockstedte, A.; Foliente, G.; Paevere, P. Parameter Analysis of the Differential Model of Hysteresis. *J. Eng. Mech.* **2004**, *71*, 342–349. [CrossRef]
31. Constantinou, M.; Adnane, M. *Dynamics of Soil-Base-Isolated Structure Systems: Evaluation of Two Models for Yielding Systems*; Report to NSAF; Department of Civil Engineering, Drexel University: Philadelphia, PA, USA, 1987.
32. Pacific Earthquake Engineering Research Center (PEER). *Strong Ground Motion Database*. Available online: https://peer.berkeley.edu. (accessed on 6 November 2021).
33. Ziyaeifar, M.; Noguchi, H. Partial Mass Isolation in Tall Buildings. *Earthq. Eng. Struct. Dyn.* **1998**, *27*, 49–65. [CrossRef]

applied sciences

MDPI

Review

Pushover Analysis in Seismic Engineering: A Detailed Chronology and Review of Techniques for Structural Assessment

Kevin Karanja Kuria * and Orsolya Katalin Kegyes-Brassai

Department of Structural Engineering and Geotechnics, Széchenyi István University, 9026 Győr, Hungary; kegyesbo@sze.hu
* Correspondence: kevin.karanja.kuria@sze.hu

Abstract: This study analyzes the progression, utilization, and inherent challenges of traditional nonlinear static procedures (NSPs) such as the capacity spectrum method, the displacement coefficient method, and the N2 method for evaluating seismic performance in structures. These methods, along with advanced versions such as multi-mode, modal, adaptive, and energy-based pushover analysis, help determine seismic demands, enriching our grasp on structural behaviors and guiding design choices. While these methods have improved accuracy by considering major vibration modes, they often fall short in addressing intricate aspects such as bidirectional responses, torsional effects, soil-structure interplay, and variations in displacement coefficients. Nevertheless, NSPs offer a more comprehensive and detailed analysis compared to rapid visual screening methods, providing a deeper understanding of potential vulnerabilities and more accurate predictions of structural performance. Their efficiency and reduced computational demands, compared to the comprehensive nonlinear response history analysis (NLRHA), make NSPs a favored tool for engineers aiming for swift seismic performance checks. Their accuracy and application become crucial when gauging seismic risks and potential damage across multiple structures. This paper underscores the ongoing refinements to these methods, reflecting the sustained attention they receive from both industry professionals and researchers.

Keywords: nonlinear static analysis; detailed vulnerability assessment; existing buildings; seismic performance evaluation

1. Introduction

The assumption of linear elastic behavior is fundamental in seismic response evaluations, but it may not accurately estimate actual structural damage [1,2]. More precise analysis methods are needed. Dynamic time history analysis is rigorous but impractical for regular use due to its complexity and the requirement for detailed MDOF models [3]. Nonlinear static pushover analysis, a simpler alternative, is gaining traction for predicting lateral strength and inelastic deformation, although it requires detailed MDOF models [4]. Newer nonlinear algorithms using pushover analysis data have shown promise for more realistic seismic demand estimates and are being integrated into performance-based engineering guidelines [5–9]. Seismic analysis procedures are categorized into linear and nonlinear methods, each further divided into static and dynamic procedures. The linear static technique is widely used due to its simplicity [10,11], while linear dynamic techniques offer diverse approaches for studying time-varying structural reactions during seismic events [12,13]. Figure 1 illustrates the combination of these seismic analysis procedures.

Despite the practicality of linear procedures, their accuracy can be limited, especially for complex structures or significant seismic events, necessitating the development of nonlinear procedures [14]. Nonlinear time history (NTH) analyses are the most precise for estimating seismic demands, but their complexity and high computational requirements limit their use in routine engineering applications [15]. Nonlinear static procedures (NSPs) have been adopted as alternatives and involve monotonously increasing predefined load patterns until a target displacement is reached [16], as recommended in [5]. However, these simplified NSPs can be inaccurate in predicting inelastic seismic demands when higher mode and inelastic effects are significant [16,17]. To address these limitations, enhanced procedures, such as modal pushover analysis (MPA), modified modal pushover analysis (MMPA) [18], and upper-bound pushover analysis (UBPA) [19], have been developed. In addition, nonlinear procedures, including nonlinear modal response history analysis [20] and nonlinear response history analysis [21], are available. These methods consider the nonlinear behavior of materials and large deformations, but they require significant computational resources [22,23].

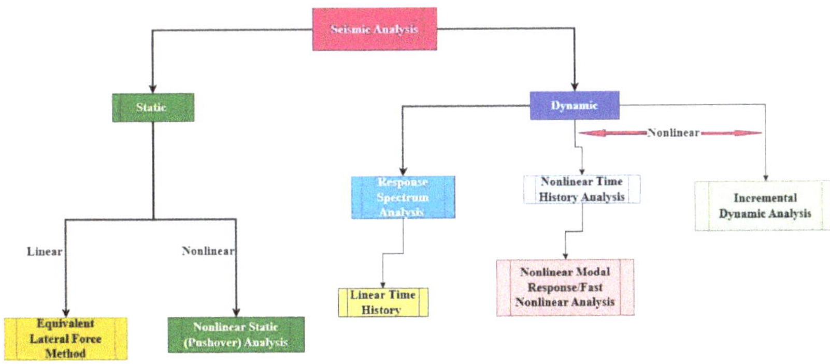

Figure 1. Flowchart highlighting the various seismic analysis methods.

To further improve the prediction of inelastic seismic demands, adaptive pushover procedures were developed. These procedures involve progressively updating load vectors to reflect changes in system modal attributes during the inelastic phase [24]. Additional approaches, such as the adaptive modal combination (AMC) procedure, use adaptive load patterns [16,25,26]. These developments represent significant strides towards enhancing the accuracy and practicality of seismic demand estimation.

The novelty of this review lies in its comprehensive integration of various pushover methods into a single document, including their procedures, and highlighting their respective advancements. As we navigate through the following sections, we aim to provide an extensive understanding of each method, its advantages and disadvantages, and its applicability in seismic analysis. The subsequent sections are structured as follows:

- Section 2: an examination of widely adopted nonlinear static analysis techniques. This section represents a comprehensive review of the most utilized nonlinear static analysis procedures, tracing their development and application in seismic studies.
- Section 3: an exploration of pushover analysis methods using first modal inertial/invariant single load vectors. This section represents an in-depth examination of pushover analysis methods that employ first modal inertial or invariant single load vectors, detailing their methodology and significance.
- Section 4: a study of pushover analysis with invariant multi-mode vectors. This section investigates the intricacies of pushover analysis techniques that utilize invariant multi-mode vectors, highlighting their unique features and applications.
- Section 5: an examination of adaptive load vector techniques. This section explores adaptive load vector methods, examining their adaptability and how they differ from traditional techniques.
- Section 6: an overview of research efforts assessing various pushover analysis approaches. This section is a summary of the extensive research initiatives undertaken to evaluate the effectiveness and applicability of different pushover analysis techniques.

- Section 6.1: a comparison between nonlinear static pushover analysis and nonlinear dynamic analysis techniques. This section details a comparative study highlighting the differences and similarities between nonlinear static pushover analysis methods and their dynamic counterparts.
- Section 7: an analytical discussion. This section engages in a critical discussion, exploring the methodologies, their strengths, potential limitations, and the implications of their findings within the broader context of seismic analysis.
- Section 8: final remarks and takeaways.

This section outlines concluding thoughts on the presented methodologies, encapsulating the key findings, and offering insights into potential future directions in the field. This comprehensive examination aims to provide valuable insights into the wide array of pushover methods, aiding engineers and researchers in selecting the most suitable approach for their specific seismic analysis requirements.

2. A Review of Commonly Used Nonlinear Static Analysis Procedures

Nonlinear static (pushover) analysis is an essential simplified static nonlinear procedure used for estimating structural deformations. It is employed to achieve various set objectives, including assessing and devising seismic retrofit solutions for existing buildings, as well as designing novel structures that incorporate non-conventional structural materials, systems, or features not compliant with current building code requirements. Additionally, evaluating the performance of buildings is necessary to tailor them to the specific demands and preferences of owners and stakeholders. These endeavors play a crucial role in enhancing the resilience and safety of buildings, both old and new, while ensuring they align with the unique needs and objectives of the stakeholders involved [27]. A host of techniques have been developed over time, and several, such as the N2 method, the capacity spectrum method (CSM), and the displacement coefficient method (DCM) [14], are widely adopted in the field [2,4].

Nonlinear static procedures (NSPs), also called pushover analyses, are widely used in seismic engineering to assess structures' seismic vulnerability. These procedures are more computationally efficient than dynamic analyses while still providing valuable insights into a structure's behavior under lateral loads. NSPs categorize the lateral load patterns applied to the structural model during the analysis into three primary groups: first-modal inertia, multi-mode, and adaptive mode vectors. An explanation of each category follows.

1. First modal inertia/invariant single load vectors (FEMA 356) pushover analysis:

In a first modal inertia pushover analysis, the lateral load pattern is based on a single fundamental mode of vibration of the structure. The applied lateral load is proportional to the mass participating in the first mode of vibration. This approach considers primarily the inertia forces of the first mode and neglects higher modes' contributions. The analysis is often more straightforward than multimodal pushover, but it may not capture the full response behavior of complex structures that significantly engage higher modes during seismic events [8]. This method proposed three lateral load patterns, as shown below [16,28]:

- Distribution according to the fundamental mode shape:

$$F_i = W_i \varnothing_i, \tag{1}$$

- Inverted triangle distribution:

$$F_i = \frac{W_i h_i}{\sum_{i=1}^{n} W_l h_l} * V_b, \tag{2}$$

- The value of V_b is given by:

$$V_b = S_d(T_n)W, \tag{3}$$

- Load distribution as stipulated in FEMA 356 [8]:

$$F_i = \frac{W_i h_i}{\sum_{i=1}^{n} W_l h_l^k} * V_b, \tag{4}$$

- Uniform distribution of load pattern:

$$F_i = W_i, \tag{5}$$

2. Invariant multi-mode vectors pushover analysis: In this procedure, different modes are combined [29]. Appropriate modes are identified and included in the analysis in a manner that aligns with the analysis undertaken. The following expression is generally applied in determining the distribution of the loads:

$$F_i = \sum_{j=1}^{n} \alpha_{mr} \Gamma_j M_i \varphi_{ij} S_a(\zeta_j, T_j), \tag{6}$$

3. Adaptive load vectors: This is a method where the load pattern is not constant but changes and updates instantaneously during the analysis, based on the evolving response of the structure. The following expression is generally applied in determining the distribution of the loads.

Under an assumed initial loading pattern, [30] determined the incremental loading by applying the following expression:

$$\Delta F_i^{k+1} = V_b^k \left(\frac{F_i^k}{V_b^k} - \frac{F_i^{k-1}}{V_b^{k-1}} \right) + \Delta V_b^{k+1} \left(\frac{F_i^k}{V_b^k} \right) \tag{7}$$

To account for higher modes, this was later modified as suggested by [31], and the following equation was applied:

$$F_i = \frac{W_i \sqrt{\sum_{j=1}^{n} (\varphi_{i,j} \Gamma_j)^2}}{\sum_{L=1}^{N} W_L \sqrt{\sum_{j=1}^{n} (\varphi_{L,j} \Gamma_j)^2}} * \Delta V_b + F_i^{old}, \tag{8}$$

In Equation (1) to (8), various symbols represent key aspects of seismic load distribution in structures. F_i is the force distribution at each level, W_i denotes the weight at each level, and Φ_i indicates displacement in the fundamental mode. h_i is the height of each level, V_b represents the base shear, and other symbols, such as Γ_j, M_i, Φ_{ij}, $S_a(\zeta_j, T_j)$, ΔF_i, and ΔV_b are used for multi-modal analysis and adaptive load vectors in seismic analysis.

Choosing a nonlinear static procedure (NSP) for seismic vulnerability assessment involves balancing accuracy needs with structural complexity. Simple pushover analysis is convenient but may miss multimodal effects, while multimodal analysis demands substantial computational resources, especially for complex structures. Figure 2 illustrates the evolution of the pushover analysis methods, emphasizing the ongoing drive to enhance seismic demand estimation methods and the continuous need for research in structural engineering. As computational technology advances, more sophisticated nonlinear analysis procedures, such as nonlinear response history analysis (NLRHA), have become feasible but remain computationally intense and time-consuming [32]. In response, pushover-based methodologies, such as the N2 method, the capacity spectrum method (CSM), and the displacement coefficient method (DCM), have emerged as efficient, simplified alternatives that use empirical approaches to directly estimate seismic demand and capacity, thereby reducing computational burdens [2,33,34]. The N2 method provides a simplified estimation of seismic demands based on a plastic hinge analysis approach, the CSM uses a capacity spectrum to assess structural response, and the displacement coefficient method estimates seismic demands based on displacement coefficients derived from experimental data [2–4]. In conclusion, nonlinear static (pushover) analysis techniques continue to evolve due to advancements in computational capabilities, demands of complex and unsymmetrical 3D geometries, the requirement for improved understanding of complex nonlinear phenomena, and the ongoing need for accurate seismic vulnerability assessment methodologies [4].

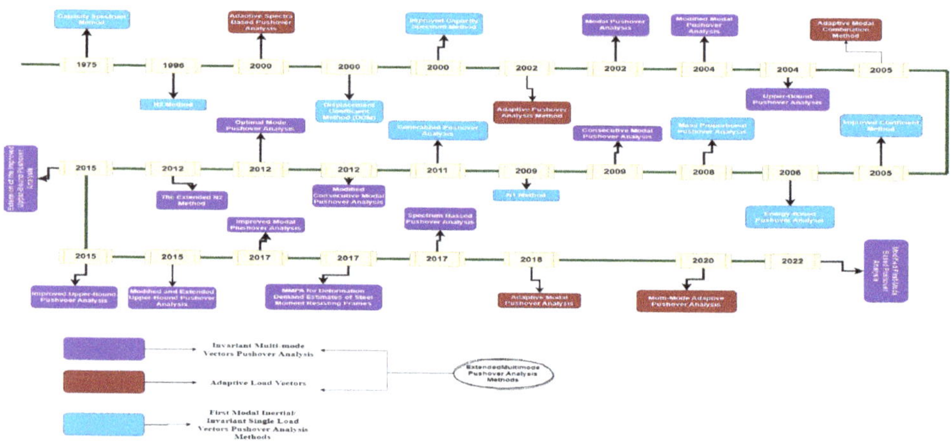

Figure 2. Graphical timeline of the development and evolution of pushover analysis methods.

These developments are changing the way we understand and respond to seismic activities, ensuring the safety and sustainability of our buildings [1–4]. These methods, developed over the past forty years, are described in the next section.

3. The First Modal Inertial/Invariant Single Load Vectors Pushover Analysis Methods

This section provides a comprehensive exploration of modal inertial pushover analysis methods, which are fundamental techniques used to understand and predict how structures respond to seismic forces. Several of these methods have been integrated into established seismic design codes, reflecting their importance and widespread acceptance in the field of structural engineering.

3.1. The Capacity Spectrum Method (CSM) ATC 40 [5]

The capacity spectrum method, initially proposed by Freeman et al. in 1975 [35], was introduced as an alternative tool for rapid assessment of buildings. [36] It compares a structure's ability to resist lateral loads with seismic demand (response spectrum). The capacity/pushover curve represents the structure's capacity. The primary pushover recommended in ATC 40 (1996) [37] is based on the equivalent linearization technique [5], applying a linearization approach that uses a capacity spectrum. The following steps describe the procedure involved in this methodology [38]:

1. Pushover analysis and conversion: execute a pushover analysis to chart the structure's capacity curve, showcasing base shear, V_b, against the building's roof displacement, D_t. Adapt this curve into acceleration-displacement (AD) metrics using the equivalent single degree of freedom (ESDOF) model. Utilize the first mode participation factor, C_0, and the modal mass for the conversion: $D^* = D/C_0$ and $A = V_b/M$.
2. Graphical representation: superimpose the derived capacity curve with the 5%-damped elastic response spectrum, ensuring both are in AD format.
3. Deformation demand and pseudo-acceleration: opt for an initial peak deformation demand '*dt*' and pinpoint its linked pseudo-acceleration *A* from the capacity curve, starting with a damping ratio $\zeta = 5$.
4. Ductility and damping calculation: Measure ductility using $m = D^*/u_y$ and compute the hysteretic damping with $\zeta_h = 2(m-1)/pm$. Determine the equivalent damping

ratio with $\zeta_{eq} = \zeta_{el} + k\zeta_h$, where k is a factor adjusting for the system's hysteretic behavior. Refine the '*dt*' estimate using the elastic demand curve for ζ_{eq}.

5. Convergence check: Monitor the displacement '*dt*' for stabilization. Once consistent, the target displacement for the multi-degree-of-freedom (MDOF) system is set as '$dt = C_0 dt$'.
6. The equivalent linearization process necessitates an understanding of the displacement ductility ratio. The ATC-40 document outlines three iterative methods: procedures A, B, and C. While procedures A and B are clear and easily programmable, procedure C is a visual method not apt for coding. Specifics of these methods can be found in the ATC-40 [5] document and are omitted here for conciseness. The procedures are clearly defined in the [5] document. In his work [39], he also applied this methodology.

3.2. *The Improved Capacity Spectrum Method*

The Improved Capacity Spectrum Method (ICSM) updates the original Capacity Spectrum Method (CSM) for seismic analysis by introducing constant-ductility inelastic design spectra, as per FEMA-440 and [40–42]. Key advancements include new calculations for effective period and damping. ICSM's main feature is its technique for estimating seismic demands using the equivalent Single Degree of Freedom (SDOF) system, where the capacity spectrum is mapped against the inelastic demand spectrum at various ductility levels, as shown in Figure 3 and represented by the various color lines. The intersection of one of these spectra determines the deformation demand, requiring iterative analysis. The selection is made when the ductility factor μ from the capacity diagram aligns with the ductility at the point where it intersects with the demand curve. ICSM also imposes strength constraints and iterative methods for target displacement calculation, enhancing the accuracy and robustness of seismic design frameworks.

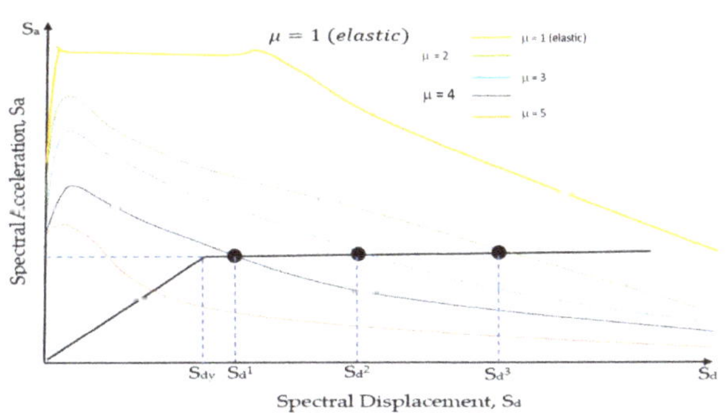

Figure 3. Illustration of the application of the ICSM method [43].

3.3. *The N2 Method, Eurocode 8*

The N2 method, initially developed by [43] and later improved by [44], is a widely used nonlinear static analysis method recommended for use by EC8 [42]. It combines pushover analysis of an MDOF model and response spectrum analysis of a corresponding SDOF model to achieve a balance between reliability and applicability for everyday design use. The N2 method differs from the capacity spectrum method in that it has adopted an $R - \mu - T$ approach rather than the over damped spectra applied in the CSM method. The detailed procedure for conducting the N2 method is outlined below,

1. Pushover Analysis: perform the analysis to obtain the capacity curve represented in V_b–D terms.
2. Conversion and approximation: Convert the pushover curve of the MDOF system into the capacity diagram of an ESDOF system. Approximate this curve with an idealized elastic-perfect plastic relationship to determine the period T_e of the ESDOF.

3. Target displacement calculation: the target displacement is computed as follows:

$$d_{et}^{*} = S_a(T_e)\left[\frac{T_e}{2\Pi}\right]^2 \tag{9}$$

where $S_a(T_e)$ represents the elastic acceleration response spectrum at the period T_e. To ascertain the target displacement d_t^{*}, different formulas are proposed for varying period ranges.

In the short period range ($T* \leq T_C$), if the condition $F_y'/m' \leq S_a(T_e)$ holds, the response remains elastic, leading to $dt' = de't$ and $dt = C_0 dt'$. If this condition is not met, the response turns nonlinear, and the ESDOF maximum displacement is given by:

$$d_t' = \frac{d_{et}' t}{\sqrt{1 + (qu-1) * \frac{TC}{Te} \leq d_{et}' t}} \tag{10}$$

where q_u is the acceleration ratio of the structure with unlimited elastic capacity, times the modal mass m* over its yield force, defined as:

$$qu = \frac{S_a(T_e) * m'}{F_y'} \tag{11}$$

In the medium and long period range ($T^* \geq T_C$), the target displacement for the inelastic system matches that of an elastic structure, so $dt' = de't$. The MDOF system's displacement is consistently determined as $dt = C_0 dt'$.

3.4. The Displacement Coefficient Method FEMA 356 [8], FEMA 440 [40], FEMA 273 [45] and ASCE 41 [7]

The displacement coefficient method, initially used in FEMA 273 [45] and later incorporated in FEMA 356 [8], aims at improving the accuracy of nonlinear static procedures (NSPs) by directly estimating the values of the peak inelastic displacement using modifying factors. The displacement coefficient method, a simplified approach, is less accurate for irregular or nonlinear structures. The displacement coefficient method calculates displacement coefficients for each mode of vibration, allowing for better prediction of inelastic behavior and determination of the control-node/target displacement [38]. The DCM involves the calculation of displacement coefficients for each mode of vibration of the structure to estimate the displacement response [14].

1. Carry out a pushover analysis of the multi-degree-of-freedom system (MDOF) similar to 3.1(1). At this step, the eigenvalue analysis of the system is conducted to establish the modal properties of the system.
2. The following relationship is then used to determine the target displacement.

$$\delta_t = C_0 C_1 C_2 C_3 S_a \frac{T_e^2}{4\Pi^2} g, \tag{12}$$

In their work, Asıkoğlu et al. [38] clearly defined all the associated parameters. Figure 4 depicts the displacement coefficient method in a graphical manner.

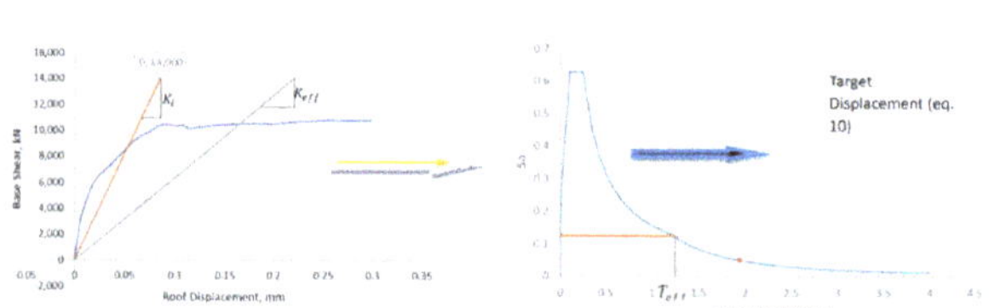

Figure 4. Graphical representation of the DCM procedure [8]. Image based on [40].

K_i is the initial or elastic lateral stiffness of a building, reflecting its stiffness under small, elastic deformations. K_{eff}, the effective lateral stiffness, is calculated at a point where the base shear force is 60% of the building's effective yield strength, derived from the bilinear representation of the capacity curve, indicating the building's behavior under larger forces.

3.5. The Improved Coefficient Method

The FEMA-440 document [40] suggests several enhancements over the FEMA-356 coefficient method [8] for target displacement estimation. Firstly, it introduces a revised equation for C_1 as follows:

$$C_1 = 1.0 + \frac{R-1}{\alpha T_e^2} \tag{13}$$

Here, α is set to 130, 90, and 60 for site classes B, C, and D, respectively (refer to FEMA-440 [44] for a detailed breakdown of site classes).

The next enhancement pertains to the coefficient C_2, expressed as follows:

$$C_2 = 1.0 + \frac{1}{800}\left(\frac{R-1}{T}\right)^2 \tag{14}$$

Lastly, FEMA-440 suggests omitting the coefficient C_3 and introduces a strength constraint to prevent dynamic instability. This strength constraint is defined by setting a maximum value for R, as illustrated below:

$$R_{max} = \frac{\Delta_d}{\Delta_y} + \frac{\alpha_2^{-t}}{4}; t = 1.0 + 0.5\ln(T) \tag{15}$$

In this equation, Δ_d represents the deformation at peak strength, Δ_y denotes the yield deformation, and α_2 is the negative slope of the force-deformation curve's strength-degradation segment.

3.6. The Energy-Based Pushover Analysis Method

Energy-based pushover analysis [46–48] is a method that enhances nonlinear static analysis by integrating inertial properties and kinetic energy into force or displacement patterns. This method is particularly relevant for seismic input as it utilizes the fundamental mode shape Φ, effective damping, and a velocity profile to represent a structure's dynamic behavior and to scale the response spectrum. The method is insightful for understanding kinetic energy dissipation through plastic behavior but has limitations, especially for braced structures [31,49,50].

The procedure of energy-based pushover analysis can be summarized as follows:

1. Model selection: choose a structural model based on the building's characteristics and identify the dominant mode shape (Φ) through modal analysis.
2. Force application: apply lateral forces (F_k) according to the mode shape and mass (m_i) distribution.
3. Incremental loading: incrementally apply loads and measure displacements (d_{jk}) at each step.
4. Equivalent SDOF system: calculate both the equivalent mass ($m^* = \Sigma m_i$) and the base shear ($F^* = \Sigma F_i$).
5. Displacement calculation: sum incremental displacements to find total equivalent displacement ($\delta^* = \Sigma(\Delta\delta)$).
6. Capacity curve construction: plot the capacity curve as base shear (F^*) versus total equivalent displacement (δ^*).
7. Energy analysis: calculate the ductility ratio ($\mu^* = \delta^*_{max}/\delta^*_y$) and compute the energy term ($\Pi_x = 1/2F^*_y\delta^*_y(2\mu^* - 1)$).

8. Assessment of limitations: review the obtained results, being aware of potential constraints, particularly when dealing with braced structures. The cumulative equivalent displacement, denoted as δ^*, is determined by summing up all the incremental values of $\Delta\delta^*$. Figure 5 depicts the overall concept of this method

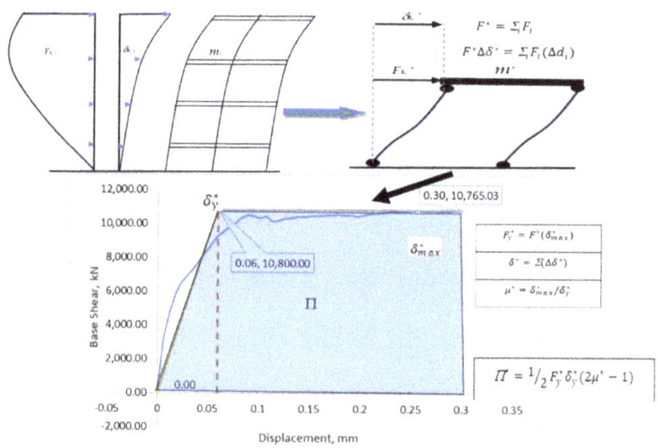

Figure 5. Graphical representation of the energy-based pushover analysis method [28].

This method was applied on a three-level steel structure conceptualized for Los Angeles within the framework of the SAC initiative.

3.7. The N1 Method

The N1 pushover analysis method offers a practical alternative to the N2 method, assessing seismic response using nonlinear static analysis without a defined equivalent SDOF system.

The title 'N1' indicates its characteristics: 'N' refers to the method being nonlinear and '1' denotes that it employs a single multi-degree-of-freedom (MDOF) structural model. The N1 method aims to evaluate displacement demand by adjusting the results obtained from response spectrum analysis (RSA) to reflect nonlinear structural behavior. This method borrows from FEMA 368 and FEMA 369 [51] methodologies, enhancing them by allowing for stiffness reduction and thereby lengthening the time. This is a departure from the traditional N2 method, which typically acquires this by the bilinearization of the capacity curve. The N1 method has been shown to match the accuracy of the N2 method, but only when a lateral load distribution proportional to the first mode shape is included. The following relationship illustrates the comparison between the N1 and N2 methods:

$$D_{el} = \Phi_n \Gamma_1 S_{de}(T_1), \tag{16}$$

In the equation, $\Delta_{e\ell}$ represents a displacement or deformation parameter. Φ_n is the *nth* mode shape of the structure, indicating how it deforms in its *nth* mode of vibration. Γ_1 is the modal participation factor for the first mode, quantifying the contribution of this mode to the overall response. $S_{de}(T_1)$ denotes the spectral displacement for the first mode, which depends on the fundamental period T_1 of the structure. This equation is used in seismic analysis to estimate displacements under certain loading conditions.

The N1 method for pushover analysis is used in irregular buildings with vertical irregularity, primarily for comparison or seismic demands. Researchers have refined it into the "N1 corrected method," aligning more closely with the N2 method.

3.8. The Mass Proportional Pushover Analysis

The mass proportional pushover (MPP) procedure is a pushover analysis approach introduced by Kim and Kurama [52] to estimate the peak seismic lateral displacement demands for structures. It offers a simplified method that eliminates the need for conducting a modal analysis or modal combination, even when higher mode effects are significant. The main advantage of the MPP procedure is that it uses a single pushover analysis for the structure, with the lateral force distribution being proportional to the total seismic masses assigned to the structure at different levels. This approach consolidates the effects of higher modes into a single invariant lateral force distribution, making it a more efficient and practical method for estimating displacement demands.

1. Perform single nonlinear pushover analysis: carry out a single nonlinear pushover analysis of the structure with the lateral force distribution given by $[m]g\{1\} = [w]\{1\}$.
2. Determine MDOF base shear force vs. roof displacement relationship: establish the relationship between the MDOF base shear force Vb and the roof displacement, u_r, from the pushover analysis of the structure.
3. Determine equivalent SDOF pseudo-acceleration vs. displacement relationship: calculate the equivalent SDOF pseudo-acceleration A versus displacement D relationship using the following relationships:

$$A = \frac{V_b}{M}, \tag{17}$$

$$A = \frac{u_r}{\Gamma}, \tag{18}$$

4. Find maximum SDOF displacement Dmax: solve the equation $\ddot{D}_n + 2\xi_n\omega_n\dot{D}_n + \frac{F_{sn}}{L_n} = -\ddot{u}_g(t)$ with $Fs/L = A$ to find the maximum SDOF displacement Dmax.
5. Calculate maximum MDOF roof and floor lateral displacements: compute the maximum MDOF roof and floor lateral displacements of the structure using the following relationship:

$$\{u_{max}\} = D_{max}\Gamma\{u_e\}$$

3.9. The Modified First-Mode-Based Pushover Analysis

The modified first-mode-based pushover analysis (MFPA) was proposed by Aman Mola Worku et al. [53] as an advanced assessment method. It represents the global force–displacement capacity graphically, which enables a comparison with the representative response spectrum. MFPA assumes that the maximum inelastic deformation of a nonlinear single-degree-of-freedom (SDOF) system can be estimated from the maximum deformation of a linear elastic SDOF system which has a higher period and a higher damping ratio than the initial values of a nonlinear system. In brief, the procedure involves the following steps:

1. Definition of the structure's nonlinear model.
2. Conducting an eigenvalue analysis on the structural model to determine the period and normalized mode shape.
3. Apply the following relationship to determine the load pattern for the MFPA:

$$S_j^* = \beta_j m_j \varnothing_{1j} S_a(T_1, \xi_1) \tag{19}$$

4. Conduct a static pushover analysis to get the V_b–U_r, which is later transferred to the idealized capacity curve.
5. Determine the target displacement value from the idealized pushover curve.
6. Establish the building's performance estimate at the Maximum Considered Earthquake.

4. Invariant Multi-Mode Vectors Pushover Analysis

The procedure involves combining different modes, as cited in [29]. Relevant modes are identified and incorporated into the analysis. A particular expression is used to determine the distribution of the loads in this process.

4.1. The Modal Pushover Analysis

The modal pushover analysis (MPA), introduced by Chopra et al. [54], is a seismic performance evaluation method designed to enhance the accuracy of predicting how structures respond to earthquakes. Unlike traditional pushover analyses, MPA uses a modal-based approach that distributes inertia forces for each mode, which are then combined to approximate the total seismic demand on inelastic structures. For elastic structures, MPA is equivalent to response spectrum analysis (RSA). This method aims to provide a more accurate and efficient way to assess the seismic performance of structures, offering comparable accuracy to RSA for inelastic systems [54]. A step-by-step summary of the MPA procedure to estimate the seismic demands for building is presented as a sequence of steps:

1. Compute the natural frequencies and modes of the building for linear elastic vibration.
2. Perform nonlinear static analysis to develop the base shear-roof displacement pushover curve for each mode.
3. Idealize the pushover curve as a bilinear curve, considering any negative post-yielding stiffness.
4. Convert the pushover curve to the force-displacement relationship for the n-th mode inelastic SDF system.
5. Calculate the peak deformation of the n-th mode SDF system using the force–deformation relationship and damping ratio, using the following second order differential equation.

$$\ddot{D}_n + 2\zeta_n\omega_n\dot{D}_n + \frac{F_{sn}\left(D_n, \dot{D}_n\right)}{L_n} = -\ddot{u}_g(t), \tag{20}$$

6. Determine the peak roof displacement associated with the n-th mode SDF system. The following relationship applies:

$$u_{rno} = \Gamma_n \Phi_{rn} D_n, \tag{21}$$

7. Extract the desired responses from the pushover database at the roof displacement equal to the peak displacement.
8. Repeat Steps 3–7 for all modes.
9. Compute the dynamic response due to each mode by subtracting the contribution of gravity loads alone.
10. Combine the gravity response and peak modal responses using the complete quadratic combination (CQC) rule to obtain the total response. The following relationship applies:

$$r_{MPA} = \sqrt{\sum\nolimits_{n=1}^{j} r_{no}^2}, \tag{22}$$

4.2. Multi-Mode Pushover for Deformation Demand Estimates of Steel Moment-Resisting Frames

The procedure proposed by [55] is a method for estimating the deformation demands of steel moment-resisting frames during seismic events [56]. The multi-mode pushover procedure accurately predicts seismic demands of steel moment-resisting frames by calculating the target interstory drift and using modal combinations. It considers frequency content, higher mode effects, and the interaction between modes. It offers superior response predictions with minimal computational effort. This approach allows for more accurate prediction of the deformation demands of the structure. The multi-mode pushover procedure involves the following steps:

1. Modal analysis: Perform a modal analysis to determine the natural frequencies, mode shapes, and modal participation factors of the structure. This can be done using a general-purpose nonlinear static analysis tool.
2. Response spectrum: Obtain the response spectrum for the design earthquake. This spectrum represents the maximum response of the structure at different periods.
3. Mode combination: Combine the modal responses to obtain the total response of the structure. The modal responses are combined using a modal combination rule, such as the complete quadratic combination (CQC) rule or the square root of the sum of squares (SRSS) rule.
4. Target interstory drift: Estimate the target interstory drift, which is the maximum interstory drift that the structure is expected to experience during the earthquake. This can be done using one of three procedures: the modal pushover procedure (MPP), the adaptive capacity spectrum method (ACSM), or the multi-mode pushover procedure (MP3).
5. Generalized force vectors: Calculate a set of generalized force vectors using the modal analysis results. These force vectors represent the instantaneous force distribution acting on the structure when the interstory drift at one story reaches its maximum value during dynamic response to the seismic excitation.
6. Maximum response: determine the maximum value of each response parameter by taking the envelope of the results obtained from the generalized force vectors.

4.3. Improved Modal Pushover Analysis

In contrast to the modal pushover analysis (MPA) procedure which utilizes invariant multi-mode lateral load pattern vectors for obtaining responses, the improved modal pushover analysis (IMPA) [57] incorporates the redistribution of inertia forces after the yielding of the structure. The key enhancement in IMPA is its use of post-yielding structural deflection shape as an invariant lateral load pattern. However, to circumvent extensive computation, it suggests a two-phase lateral load distribution for the first mode, while the force patterns for the higher modes follow the MPA approach. The steps encompassed in the IMPA procedure are as follows:

1. Initiate with first mode MPA: Begin with the initial steps (1–3) of the MPA method for the primary 'mode'. The initial load pattern is determined by the lateral force distribution, represented as $S_1{}^* = m\Phi_1$.
2. Determine displacement at yield: from the pushover analysis in the first step, ascertain the structure's displacement vector, $\psi 1y$, at its yielding point.
3. Progress with an updated load pattern: From the point of structural yielding, extend the pushover analysis using a new load distribution $S1y* = m\Phi 1y$. This is the second-phase lateral load pattern. Utilize this updated curve and the subsequent Steps 4–7 of the MPA method to assess the structure's behavior.
4. Compute overall response (IMPA): Merge the primary 'mode' response from Step 3 with the responses of higher modes derived from the conventional MPA method. Combine using methods like the square root of the sum of squares (SRSS) or complete quadratic combination (CQC).

4.4. The Modified Modal Pushover Analysis

Chopra et al. [18] developed the modified modal pushover analysis (MMPA) to estimate seismic demands in frame buildings, with a key simplification that assumes higher vibration modes are elastic. This reduces computational effort compared to the traditional modal pushover analysis (MPA). The study found that MMPA generally yields larger and potentially more accurate demand estimates, especially in cases where MPA underestimates demands. However, for cases where MPA already overestimates demand, MMPA may increase conservatism. MMPA's main limitation is in lightly damped systems (with damping less than 5%), where its linear elastic assumption for higher modes may not accurately represent inelastic behavior, potentially leading to an overestimation of demand. The accuracy of the MMPA procedure is also found to be sensitive to specific building and ground motion characteristics. The following phases in the modified modal pushover analysis (MMPA) approach to calculate the peak inelastic response of a multistory structure with a plan symmetric about two orthogonal axes subjected to earthquake ground motion along an axis of symmetry are summarized below:

1. Modal pushover analysis application: implement Steps 1–4 like those outlined in the standard modal pushover analysis.
2. Determine peak deformation $D1$: For the primary mode's inelastic single degree of freedom (SDF) system, compute $D1$ using parameters $T1$ and $\zeta 1$ [20]. This can be achieved via nonlinear response history analysis (RHA), an inelastic design spectrum, or specific empirical formulas.
3. Ascertain peak roof displacement $ur1$: using the equation $ur1 = \Gamma 1 \Phi r1 D1$, calculate $ur1$ linked to the primary mode's inelastic SDF system.
4. Retrieve relevant responses: from the pushover data (from Step 2), extract response values $r1 + g$ for the combined effects of gravity and lateral forces at a displacement of $ur1 + urg$.
5. Calculate primary mode dynamic response $r1$: determine $r1$ using the relation $r1 = r1 + g - rg$, where rg represents the contribution from gravity loads.
6. Evaluate higher-mode dynamic responses: For modes beyond the first, assume the system retains its elasticity and compute the dynamic responses. This mirrors the traditional modal analysis for a linear multi-degree-of-freedom (MDF) system, eliminating the need for extra pushover analyses and thus saving computational time. The number of modes to consider for accurate results varies based on the building's height.
7. Compute overall response (demand): integrate the gravity response with the peak 'modal' responses using the square root of the sum of squares (SRSS) rule: $r \approx max[rg \pm \Sigma n rn2]$, where $r1$ is derived from Step 5 and $rn(n > 1)$ from Step 6.

4.5. The Generalized Pushover Analysis (GPA)

Sucuoglu and Gunay [58] introduced a new method called generalized pushover analysis (GPA). The GPA method uses generalized force vectors to predict building sway during earthquakes. Pushover analyses create an envelope of results, allowing for better understanding of building reactions under different conditions.

The GPA algorithm is composed of the following five basic steps:

1. Eigenvalue analysis: calculate natural frequencies ω_n, modal vectors u_n, and modal participation factors Γ_n from eigenvalue analysis.
2. Response spectrum analysis: obtain modal spectral amplitudes A_n, D_n from the corresponding linear elastic spectra.
3. Calculate modal interstory drift ratios at the j-th story, $\Delta_{j,max}$, and the maximum interstory drift ratio at the j-th story, $\Delta_{j,max}$, from RSA.
4. Generalized force vectors. Compute generalized force vectors fj that produce the maximum response Δ_j, as calculated from the expression below:

$$f_j(t_{max}) = \sum\nolimits_n \left(\Gamma_n m \varphi_n A_n \frac{\Delta_{j,n}}{\Delta_{j,max}} \right), \quad (23)$$

Target interstory drift demands:
calculate the target interstory drift, Δ_{jt}, at the j-th story using Equation (24):

$$\Delta_{jt} = \sum\nolimits_n \Gamma_n \frac{\Delta_{j,n}}{\Delta_{j,max}} D_n \left(\varphi_{n,j} - \varphi_{n,j-1} \right), \quad (24)$$

Note: If the first mode spectral displacement demand D_1 in Equation (24) is replaced with its inelastic counterpart, D_1^*, then a first mode pushover analysis is conducted. D_1^* is obtained from either NRHA or the inelastic response spectrum of the bi-linear SDOF system. For higher modes $n = 2 - N$, D_n values are obtained from the linear elastic response spectrum.

5. Generalized pushover analysis (GPA): Conduct N GPAs. In the j-th GPA ($j = 1 - N$), incrementally push the structure in the lateral direction with a force distribution proportional to f_j.

At the end of each loading increment, i, compare the interstory drift Δ_{ji} obtained at the j-th story with the target interstory drift Δ_{jt} calculated from Equation (24).

Continue displacement-controlled incremental loading ($i = 1, 2, \ldots$) at the j-th GPA until Δ_{ji} reaches Δ_{jt}.

4.6. The Consecutive Modal Pushover Analysis (CMP)

The consecutive modal pushover (CMP) procedure, introduced by [59], is an innovative pushover method designed to address the limitations of the conventional nonlinear static procedure (NSP) when analyzing higher-mode responses in tall buildings. Unlike the traditional NSP, the CMP procedure integrates both multi-stage and single-stage pushover analyses, with final structural responses determined by combining the results of these separate analyses. The research demonstrated that the CMP procedure significantly improves the accuracy of predicting seismic demands of tall buildings, especially in estimating hinge plastic rotations at mid and upper floor levels, when compared to results from nonlinear response history analysis (NL-RHA) and modal pushover analysis (MPA). Thus, the CMP procedure represents a substantial advancement in seismic demand prediction for tall buildings, effectively incorporating higher-mode effects [59]. The steps explained below illustrate the details of this procedure.

1. Frequency and mode shape analysis: Conduct an eigen analysis on the linear elastic structure, focusing on the initial three modes. Ensure mode shapes are normalized such that the roof component, denoted as Φn, is set to one.
2. Determine lateral force distribution: calculate $Sn* = m\Phi n$, where $Sn*$ represents the distribution of incremental lateral forces throughout the structure's height for the nth stage.
3. Total roof displacement calculation: ascertain the structure's total target roof displacement, represented as δt.
4. CMP analysis steps:
   - For medium-rise structures, execute a single-stage pushover analysis using an inverted triangular load pattern. For taller structures, use a uniform force distribution. Continue until the roof's control node reaches the predefined δt.
   - Implement a two-stage pushover analysis, where the second stage starts from the end state of the first stage.
   - For structures with a fundamental period of 2.2 s or more, conduct a three-stage pushover analysis. Each stage begins from the end state of the preceding stage.
5. Determine the peak responses: From the earlier pushover analyses, calculate the peak values for targeted responses, such as displacements, drifts, and hinge rotations. Represent these peak values from the one-, two-, and three-stage analyses as $r2$, and $r3$ respectively.
6. Calculate the overall peak responses:
   - If $T < 2.2$ s, then $r = max\{r1,r2\}$.
   - If $T \geq 2.2$ s, then $r = max\{r1,r2,r3\}$.
7. Sequential modal analysis note: The structure's nonlinear behavior is intertwined with its response and is influenced by the loading path. In the CMP approach, modal pushover analyses should be sequentially executed, starting with the primary mode, and advancing to the subsequent ones, as detailed in [60].

4.7. Modified Consecutive Modal Pushover Analysis

The study by [61] aimed to refine seismic demand estimation in asymmetric-plan tall buildings, accounting for higher modes and torsion. It introduced the modified consecutive modal pushover (MCMP) procedure, which was applied to 10, 15, and 20-story structures. The results, compared with modal pushover analysis (MPA) [54] and nonlinear time history analysis (NLTHA [42], showed that the MCMP procedure provided accurate seismic demand estimates, as validated by NLTHA. Notably, the MCMP excelled in predicting specific structural responses, such as plastic hinge rotations, and outperformed both FEMA load distribution patterns and MPA, especially regarding displacements and story drifts. These benefits were most evident in mid- and upper-story buildings, marking the MCMP as a valuable tool for assessing seismic demands in taller structures.

1. Carry out the initial steps, 1 and 2, as per the procedure explained in Section 4.6 above.
2. Determine the total target displacement and increment. Calculate the values for the total target displacement and target displacement increment using the following relationships:

$$U_{ri} = \beta_i \delta_i \tag{25}$$

$$\beta_i = 1 - \sum_{j=1}^{N_s-1} \alpha_j, \tag{26}$$

3. Apply gravity loads. Consider gravity loads as an initial condition for each pushover analysis.
4. Conduct a single-stage pushover analysis. Perform a pushover study employing a uniform lateral load distribution until the target displacement at the roof is reached.
5. Conduct a multi-stage pushover analysis. Determine the number of analysis stages, ns, using the following relationship:

$$\sum_{i=1}^{N_s} M_i^* \geq 0.9 \sum_{i=0}^{N} m_j, \tag{27}$$

- For the first stage, apply the initial distribution $\varphi 1s1^* = m\varphi 1$ to the building to reach the displacement increment $ur1 = \beta 1\delta t$.
- Continue the excitation with the incremental forces $S_2^* = m\varphi_2$, until the displacement increments at the roof equals $u_{r2} = \beta_2\delta_2$ in the second stage.
- Perform the pushover analysis until the number of stages equals Ns and the displacement at the roof reaches the total target displacement δt.
- At each stage of the multi-stage analysis, the initial condition is the same as the state at the end of the previous stage.

6. Ascertain the peak values of seismic reactions. For both single-stage and multi-stage pushover analyses, determine the peak values of seismic reactions, such as displacements, story drifts, rotations of plastic hinges, and axial force of the braces.
7. Determine peak responses' envelope. Calculate the envelope of peak responses, r, as follows:

$$r = max\{r_s, r_M\}, \tag{28}$$

4.8. Upper-Bound (UBPA) Pushover Analysis

Engineers often use simplified nonlinear static analytical procedures, also known as pushover analyses, to assess the seismic demands of tall buildings, as these methods are simpler than nonlinear response history analysis. However, this conventional approach has limitations in accurately predicting the inelastic seismic demands of high-rise buildings. To address this, a study [19] introduced an improved simplified pushover analysis method, which incorporates higher-mode effects into the analysis. This method consists of three key components: (1) response spectrum-based higher-mode displacement contribution ratios; (2) an alternative formula for determining the lateral load pattern; and (3) an upper-bound modal combination rule for establishing the target roof displacement. This proposed method was validated using five buildings of varying heights and compared with three other types of analysis [19]. Results show that the proposed method is more accurate in estimating important response attributes of tall buildings, including roof displacements, story drift ratios, and plastic hinge rotations.

The principal steps of the proposed pushover analysis procedure are as follows:

1. Build the selection and elastic properties. Select a building and determine its elastic structural properties, including natural frequencies and mode shapes. Normalize mode shape Φ_n such that the roof degree of freedom Φ_{rn} = 1 for all modes.
2. Calculate the upper-bound contribution of the second mode. Use the elastic response spectrum of the selected earthquake to determine the upper-bound of the contribution of the second mode $(q_2/q_1)UB$, as given by:

$$\left(\frac{q_2}{q_1}\right)_{UB} = \left|\frac{\Gamma_2 D_2}{\Gamma_1 D_1}\right|, \tag{29}$$

3. Determine the lateral load distribution vector. Compute the distribution vector of the lateral loads over the height of the building using:

$$f_s = F_1 + F_2 = \omega_1^2 m\varphi_1 + \omega_2^2 m\varphi_2 \left(\frac{q_2}{q_1}\right)_{UB}, \tag{30}$$

4. Compute the target roof displacement Ur. Determine the target roof displacement Ur as per the expression:

$$u_{r,UB} = u_{r,TLP}\left[1 + \left(\frac{q_2}{q_1}\right)_{UB}\right], \tag{31}$$

5. Perform a pushover analysis. Conduct a pushover analysis using the lateral load calculated in Step 3 until the target displacement computed in Step 4 is reached (single-run analysis).
6. Extract the seismic responses. Determine the desired seismic responses (e.g., story drifts, element internal forces, etc.) at the step corresponding to the target displacement obtained from the single run in the previous step.

4.9. Modified and Extended Upper-Bound (MUB) Pushover Analysis

The upper-bound (UB) pushover analysis method, initially developed for 2D tall building frames, tends to underestimate seismic demands at lower levels of the structures [19]. To address this issue, a study [62] suggested combining the seismic demands from the UB pushover method with those from a conventional pushover analysis, aiming to better regulate seismic demands at lower building levels. Additionally, the study proposed an extension of the UB pushover method to accommodate tall buildings with unsymmetrical plans. This extension, termed the extended upper-bound (EUB) method, incorporates upper-bound lateral forces with patterns tailored to various unsymmetrical-plan tall buildings and considers higher-mode effects in calculating these forces. The results indicate that the EUB method can accurately estimate inelastic seismic demands of unsymmetrical-plan tall buildings, considering both higher modes and torsional effects.

1. Initial calculations: perform calculations as outlined in Steps 1 to 4 of Section 4.8 above.
2. Pushover analyses: After applying gravity loads, conduct both the upper-bound pushover and conventional pushover analyses. Continue these analyses until the control node, located at the roof's center of mass, reaches the target displacement determined in the previous step.
    - For medium-rise unsymmetric-plan buildings, perform a conventional pushover analysis using an inverted triangular load pattern, and use a uniform force

distribution for high-rise buildings until the target displacement is reached. For torsionally flexible systems, perform the conventional pushover analysis using the lateral force *fs*1, which is derived from Equation (30).

- Perform the upper-bound pushover analysis using the upper-bound lateral forces *fsi* derived from Equation (30) until the target displacement at the roof is reached.
- Determine the peak seismic demands. Calculate the peak values of the seismic demands individually for the upper-bound pushover and conventional pushover analyses. Denote the peak values obtained from these analyses by r_{fsi} and r_1, respectively.

3. Determine the envelope of results. Calculate the envelope (r) of the results computed in the previous step as follows:

$$r = \begin{cases} max\left\{r_1, r_{fs2}\right\} & for\ the\ TS\ and\ TF\ systems \\ max\left\{r_1, r_{fs2}, r_{fs3}, r_{fs4}\right\} & for\ the\ RSS\ Systems \end{cases}, \tag{32}$$

where

$r_1 = r_{TLP}$ for the TS and TSS medium-rise systems
$r_1 = r_{ULP}$ for the TS and TSS high-rise systems
$r_1 = r_{fs1}$ for the TF systems

4.10. The Improved Upper-Bound (IUB) Pushover Analysis

To enhance the accuracy of seismic behavior predictions in high-rise buildings, an advanced and improved upper-bound pushover procedure was proposed. This method, suggested by [63], refines the lateral load pattern and has undergone testing on nine, twelve, fifteen, and twenty-story steel buildings. The findings demonstrate that this new approach is generally more accurate and aligns more closely with nonlinear time history analysis (NTHA), showing its effectiveness and potential for improved seismic behavior prediction. The improved upper-bound (IUB) pushover procedure involves the following steps:

1. Initial calculations: perform Steps 1 to 3 in a similar way as outlined in Section 4.8 above.
2. Define the lateral load distribution vector. Use the following relationship to define the distribution vector of the lateral forces over the height of the structural system:

$$f_s = \omega_1^2 m \varphi_1 + \left(\omega_2^2 m \varphi_2 \left(\frac{q_2}{q_1}\right)_{UB}\right) * 0.5, \tag{33}$$

3. Evaluate the target lateral displacement at roof. Evaluate the roof's target lateral displacement, U_r, by applying the following relationship:

$$U_r = U_{rM1}\left(1 + \left(\frac{q_2}{q_1}\right)_{UB} * 0.5\right), \tag{34}$$

4. Perform pushover analysis. Use the lateral load determined in Step 2 to perform a pushover analysis of the structure until the target displacement, determined in Step 3, is reached (single-run analysis).
5. Identify seismic responses. Determine the relevant seismic responses at the step corresponding to the target displacement from the single run in the preceding phase. This includes parameters such as story drifts, element internal forces, etc.

4.11. The Extension of the Improved Upper-Bound Pushover Analysis Procedure (EIUB)

This approach was developed by [64] to accurately predict the behavior of both ordinary and tall structures. The extended IUB examines the seismic response of irregular structures with setbacks. This procedure aims at expanding the improved upper-bound (IUB) analysis to account for the influence of setbacks on the seismic response of mid-rise structures. In addition to the first two modes of vibration utilized in the IUB approach [63], a third mode is used to construct the applied load vector, which may portray the irregularity characteristic. The proposed expanded IUB version was applied to fifteen 10-story structures with varying setback configurations, as well as a reference structure. This new version's performance was assessed in terms of target displacement, story displacements, story drift, and plastic hinge rotations, which were compared to those produced from prior pushover processes using nonlinear time history analysis (NTHA) as a benchmark.

The extension of the improved upper-bound pushover analysis procedure involves the following steps, summarized:

1. Conduct an eigenvalue analysis. Conduct an eigenvalue analysis to compute the natural frequencies of the structure, ωn, and the mode shapes, Φn, such that the lateral component of Φn at the roof equals unity.
2. Determine the upper-bound of mode contributions. Using the elastic response spectrum of the specified earthquake data, determine the upper-bound of the contribution of the 2nd and 3rd modes q_i/q_1, as described by the following relationship:

$$f' = 0.48q_1\left[\left(\omega_1^2 M\varphi_1\right) \pm \frac{0.26}{0.48}\left(\omega_1^2 M\varphi_2\frac{q_2}{q_1}\right) \pm \frac{0.26}{0.48}\left(\omega_3^2 M\varphi_3\frac{q_3}{q_1}\right)\right], \tag{35}$$

3. Calculate the lateral load distribution vector. Calculate the lateral load distribution vector throughout the building's height using the following relationship:

$$f_{i,EIUB} = \max\left(f''_{i,UB}, f_{i,Unif}\right), \tag{36}$$

4. Calculate the target roof displacement. Calculate the desired roof displacement Ur by applying either one of the following relationships:

$$U_r = U_{rM1}\left(1 + \frac{q_2}{q_1}C_r\right), \tag{37}$$

$$U_r = U_{rM1}\left(1 + \frac{q_2}{q_1}C_r + \frac{q_3}{q_1}C_r\right), \tag{38}$$

5. Perform a pushover analysis. Perform a pushover analysis using the lateral load estimated in Step 3 until the target displacement calculated in Step 4 is reached (single-run analysis).
6. Determine the seismic responses. Determine the maximum values of the seismic responses from the single-run analysis conducted in step 5.

4.12. The Extended N2 Method

The extended N2 method was developed by Maja Kreslin and Peter Fajfar at the University of Ljubljana [65]. The extended N2 method improves the estimation of seismic demand compared to the basic N2 method by considering higher-mode effects both in plan and in elevation [65]. The extended N2 method provides a more accurate distribution of seismic demand throughout the structure, particularly in the upper part of the building and at the flexible edges. This improvement is achieved by combining the results of basic pushover analysis and those of standard elastic modal analysis. The extended N2 method yields fair, conservative estimates of response and greatly improves the prediction of seismic demand in the specified areas. The extended N2 method involves the following detailed procedure:

1. Perform the basic N2 analysis [42,65], which includes a pushover analysis to determine the target displacement at the roof level. This analysis is done separately for each horizontal direction of the building.
2. Perform an elastic modal analysis of the building using a 3D mathematical model [66]. This analysis considers all relevant modes of vibration in each horizontal direction. The results of the modal analysis are normalized so that the roof displacement

at the center of mass (CM) is equal to the target displacement obtained from the pushover analysis.

3. Determine the seismic demand by applying correction factors to the results of the pushover analysis. Two sets of correction factors are used: one for displacements in plan and the other for story drifts along the elevation. The correction factors are determined separately for each horizontal direction and depend on the location in plan.

4.13. The Spectrum-Based Pushover Analysis (SPA) Method

The proposed spectrum-based pushover analysis (SPA) method was developed by Zhang et al. [67] to establish a quick and effective seismic assessment method for tall buildings. The spectrum-based pushover analysis (SPA) procedure consists of several steps, as follows:

1. Modelling of the building and applying the seismic gravity loads.
2. Eigenvalue analysis of the structure to compute the (ω_n), $T_{i,}$ (αi, βi, λi), values.
3. Compute the force vector of different modes: calculate $s_n* = m\Phi_n$, where s_n* represents the distribution of incremental lateral forces over the height of the structure for the nth stage of the multi-stage pushover analysis.
4. Obtain the design spectrum and compute the target displacements. Obtain the design spectrum for the sites of interest or the mean spectrum of a series of ground motion data. Compute the target roof displacements of the structure for different modes u_{ir0}, the total roof displacement u_{r0}, and contributions using Equations (26) and (27):

$$u_{ir} = \alpha_i * u_{r0}, \tag{39}$$

$$u_{ir0} \approx \left(\sum_{i=1}^{N} (u_{ir0})^2\right)^{0.5}, \tag{40}$$

5. The spectrum-based pushover analysis procedure (SPA):
6. In the first stage, perform a nonlinear static pushover analysis using incremental lateral forces $S_1{}^* = m\Phi 1$ until the roof displacement reaches $u_{1r} = \alpha_1 u_{r0}$. Obtain the peak response for this stage, noted by r_1. In the second stage, apply incremental lateral forces $S_2{}^* = m\Phi_2$ until the roof displacement equals $u_{2r} = \alpha_2 u_{r0}$. Note that the initial condition in this stage is the same as the state at the last step of analysis in the last stage. Obtain the peak response $r2$ for this stage. Repeat Step 5.2 for the remaining modes considered. For each ith step, the target displacement is $uir = \alpha iu_{r0}$, and the initial condition remains consistent with the state at the end of the previous stage.
7. Envelop the peak response for each step: determine the maximum response from all stages using:

$$r = max\{r_1, r_2, \ldots.., r_n\}, \tag{41}$$

The value of r represents the seismic responses estimated by the SPA procedure.

5. Adaptive Load Vectors

This method involves a dynamic load pattern that updates in real time during the analysis, reflecting the ongoing response of the structure.

5.1. The Adaptive Pushover Analysis Method

The adaptive pushover method represents the most advanced nonlinear static method [38,68], as it updates the load pattern uniformly in the inelastic range through each analysis step. The method consolidates instantaneous mode shapes, enables spectral amplifications, and determines modal shapes, using complex calculations and flexible options for modal forces scaling. The basic steps of this method can thus be summarized as follows:

1. Eigenvalue analysis: perform an eigenvalue analysis using the stiffness matrix at the previous step.
2. Displacement calculation: calculate the displacement profile for the current step based on the modal response. The following relationship is applied:

$$D_{ij} = \Gamma_J \Phi_{ij} S_D(j), \tag{42}$$

3. Displacement normalization: normalize the displacements obtained in the previous step to keep the top displacement proportional to the load factor by applying the following formula:

$$\overline{D_i} = \frac{D_i}{maxD_i}, \tag{43}$$

4. Load factor (λ) update: update the load factor and calculate the displacement vector based on the updated load factor.
5. Displacement application: apply the updated displacement to the structure and solve the system of equations.
6. Stiffness matrix update: calculate the updated stiffness matrix after the loading is applied.
7. Loop: return to the first step of the loop to proceed with the next step.
8. The flow chart in Figure 6 illustrates this procedure.

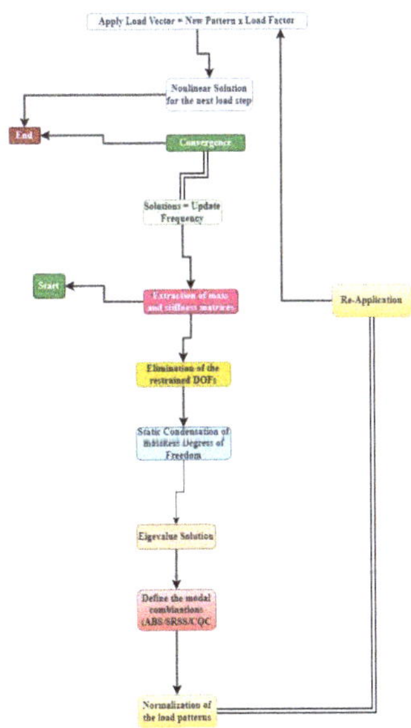

Figure 6. Illustration of the adaptive pushover procedure [38,69].

5.2. The Adaptive Modal Pushover Analysis (AMPA)

The Adaptive modal pushover analysis (AMPA) method was introduced by [70] in their paper titled "Adaptive Spectra-Based Pushover Procedure for Seismic Evaluation of Structures". Gupta and Kunnath's adaptive modal pushover analysis addresses the limitations of the traditional modal pushover analysis by iteratively updating the modal load pattern based on the deformed shape of the structure at each step of the analysis. This adaptive process allows for better consideration of higher-mode effects, nonlinear behavior, and changes in structural characteristics under seismic loads, leading to a more accurate assessment of the structure's performance.

The adaptive modal pushover analysis (AMPA) procedure can be summarized as follows:

1. Create a mathematical model: develop a comprehensive mathematical model of the structure.
2. Define nonlinear relationships: establish the nonlinear force-deformation relationships of the structural components. This includes the initial stiffness, yield moment, and post-yield stiffness, or an encompassing force-deformation envelope.
3. Compute the damped elastic spectrum: calculate the damped elastic response spectrum for site-specific ground motions, considering appropriate damping constants, such as 5% for reinforced concrete structures.
4. Perform an eigenvalue analysis: Conduct an eigenvalue analysis to determine periods and eigenvalues. Compute the modal participation factors using:

$$\Gamma_j = \frac{1}{g}\sum_{i=1}^{i=N} W_i \varnothing_{ij}, \tag{44}$$

5. Calculate the story forces for modes: compute the forces at each story level for each mode using:

$$\Gamma_j = \frac{1}{g}\sum_{i=1}^{i=N} W_i \varnothing_{ij}, \tag{45}$$

6. Compute and combine the base shears: calculate the modal base shears V_j and combine them using the SRSS method to compute the building base shear V:

$$V_j = \sum_{i=1}^{i=N} F_{ij}, \tag{46}$$

$$V = \sqrt{\sum_{j=1}^{N} V_j^2}, \tag{47}$$

7. Scale story forces: uniformly scale the story forces by a scale factor S_n, expressed as:

$$V_j = S_n V_j, \tag{48}$$

where

$$S_n = \frac{V_B}{N_S V}, \tag{49}$$

8. Perform a static analysis for each mode: Conduct a static analysis using scaled incremental story forces for each mode independently. Ensure a bidirectional analysis for modes other than the fundamental mode.
9. Compute element responses: calculate the element forces, displacements, story drifts, and member rotations using the SRSS combination, updating the results of the previous step accordingly.
10. Check and update for yielding: At the end of each step, compare the member forces. If yielding is detected, recompute the member and global stiffness matrices and return to Step 4.
11. Iterative analysis: repeat the process until the maximum base shear or global drift exceeds a predetermined limit.

5.3. The Adaptive Modal Combination Method

The adaptive modal combination (AMC) approach, proposed by [71], is an innovative method for evaluating the earthquake resilience of building structures. It integrates elements of the ATC-40's capacity spectrum approach, Gupta and Kunnath's adaptive pushover technique, and Chopra and Goel's modal pushover analysis [54]. The AMC method uniquely employs lateral forces to mimic the instantaneous inertia force distributions during seismic events and dynamically updates target displacements using energy-based modal capacity curves and constant-ductility capacity spectra. This feature eliminates the need for pre-estimating target displacement before conducting the pushover analysis. The method has been rigorously tested on steel moment-frame buildings, demonstrating accurate estimations of critical seismic performance parameters, such as roof displacement and interstory drift [71]. The procedure can be summarized as follows:

1. Modal property assessment: analyze the structure to determine its natural frequencies, mode shapes, and the modal participation factors based on its current state.
2. Adaptive lateral load pattern construction: for each mode, devise an adaptive lateral load pattern using:

$$S_n^{(i)} = m\Phi_n^{(i)}, \tag{50}$$

where $S_n^{(i)}$ is the load distribution, m is the mass matrix, and $\Phi_n^{(i)}$ is the mode shape at step i. This pattern can be recalculated at specific intervals, balancing computational demands with accuracy.

3. Energy-based incremental evaluation: determine the next step of the capacity curve for each ESDOF system using:

$$\Delta D_n^{(i)} = \Delta E_n^{(i)} / V_{b,n}^{(i)}, \tag{51}$$

where $\Delta E_n^{(i)}$ is the increment of work done by the lateral force pattern and $V_{b,n}^{(i)}$ is the base shear at step i.

4. Inelastic response assessment: If the system exhibits inelastic behavior, approximate the global system ductility and post-yield stiffness ratio. A preliminary pushover analysis might be required to establish these parameters.
5. Generate capacity spectra: for the specific ground motion under consideration, produce capacity spectra in an acceleration-displacement response spectrum (ADRS) format.
6. Overlay the modal capacity curve with demand spectra: Plot the modal capacity curve alongside inelastic demand spectra. The intersection point represents the dynamic target for the pushover analysis, given by:

$$S_{a,n}^{(i)} = V_{b,n} / \left(\alpha_n^{(i)} W\right), \tag{52}$$

7. Extract the response parameters: at the step where the dynamic target point is achieved, retrieve the desired response parameters, such as displacements or member rotations.
8. Repeat the above steps for the necessary number of modes, typically the first few, for low to medium-rise buildings. Finally, combine the peak modal responses using an appropriate scheme, such as the square root of the sum of squares (SRSS), to determine the total response:

$$r\left(\left(\sum_n \left(r_n^{(ip)}\right)^2\right)^{0.5}\right) \tag{53}$$

5.4. Adaptive Upper-Bound (AUB) Pushover Analysis

The adaptive upper-bound (AUB) pushover analysis procedure was developed by A.Y. Rahmani et al. [72] to assess the performance of tall buildings. It considers the effects of higher modes and the dynamic changes in structural characteristics during inelastic responses. The accuracy of the proposed procedure is tested using four steel moment-resisting frame buildings. The procedure can be summarized as follows:

1. Initialization: Begin with the model. Set the step to $k = 0$. Initialize the load factor $\Delta\lambda$ and conduct an eigenvalue analysis. The upper-bound (UB) load vector is determined using the given Equation (54):

$$f_{UB} = \omega_1^2 M\varphi_1 + \omega_2^2 M\varphi_2 \left(\frac{q_2}{q_1}\right)_{UB}, \tag{54}$$

2. Target displacement calculation: using Equation (55), compute the predefined target displacement:

$$u_r = u_{rTLP}\left[1 + \left(\frac{q_2}{q_1}\right)_{UB}\right], \tag{55}$$

3. Load vector application: implement the newly determined load vector.
4. Displacement check: Examine if the roof displacement has reached the predefined target. If it has, the procedure ends; if not, proceed to the next step.
5. Frequency update: Check if there is a need to update the frequency due to yielding. If required, apply the UB lateral load pattern as per Equation (54).
6. Eigenvalue analysis: conduct an eigenvalue analysis using the updated stiffness (K) and mass (M) matrices.
7. Inelastic response spectrum construction: construct the inelastic response spectrum considering the current ductility ratio using Equation (56):

$$\mu_k = \frac{d_k}{d_y}, \tag{56}$$

8. Determine values: Identify the values of S_{d1k} and S_{d2k}. Then, calculate the (q_2/q_1) (k) ratio using Equation (57):

$$\left(\frac{q_2}{q_1}\right)_{UB} = \left|\frac{\Gamma_2 S_{d2}}{\Gamma_1 S_{d1}}\right|, \tag{57}$$

9. Lateral load pattern calculation: determine the new lateral load pattern using Equation (58):

$$f_i^{(k)} = \max\left(f_{i,UB}^{(k)}, F_{I,Unif}^{(k)}\right), \tag{58}$$

10. AUB lateral load pattern application: apply the AUB lateral load pattern as per Equation (59):

$$P^{(k)} = P^{(k-1)} + \Delta P^{(k)} = P^{(k-1)} + \Delta\lambda^{(k)}\overline{f}^{(k)} \circ P_0, \tag{59}$$

11. Iteration: return to Step 3 and repeat the process until the roof displacement meets the predefined target.

5.5. A Multi-Mode Adaptive Displacement-Based Pushover Analysis Procedure for Estimating the Seismic Demands of RC Moment-Resisting Frames (MADP)

The multi-mode adaptive displacement-based pushover (MADP) method, cited in [56], combines 'multi-modal' and 'adaptive' pushover approaches. It starts with a load pattern based on the structure's elastic mode shape and updates it as the analysis progresses. The method uses the structure's 'first-mode' inelastic mode shape to estimate target displacement accurately and is employed to assess the seismic performance of RC moment-resisting frames. Key advantages include overcoming traditional NSP limitations by considering higher-mode effects and structural changes [73], enabling adaptive and detailed seismic demand evaluation, and utilizing a simple, computationally efficient target-displacement calculation method [44].

The procedure is summarized in the following steps:

1. Create a structural mathematical model: develop a mathematical model of the structure that integrates the monotonic nonlinear behavior and degradation features of the structural components.

2. Eigenvalue analysis and mode shapes: Conduct an eigenvalue analysis of the elastic building model to compute the natural frequencies ω_n and the associated mode shapes Φn. Normalize the mode shapes so that the roof component of Φn equals unity ($\Phi nr = 1$).
3. Compute the lateral force distribution: calculate $Sn = m\Phi n$, where Sn represents the distribution of lateral forces across the structure's height for the first stage of the multistage pushover analysis.
4. Perform a nonlinear static analysis: Conduct a standard nonlinear static analysis using the lateral load pattern $Sn = m\Phi n$. From this analysis, determine the base shear-roof displacement ($Sn - m\Phi U_n$) or pushover curve of the structure for the n-th mode. (See Figure 7a.)
5. Idealize the pushover curve as bilinear: Draw a straight line ('Line 1' in Figure 7b) through the origin to match the elastic segment of the pushover curve. Draw a flat line ('Line 2' in Figure 7b) through the pushover curve at the maximum base-shear capacity, $V_{bn,\,\max}$. The ultimate roof displacement, u_{rnu}, is assumed as a point of the pushover curve where a loss of 20% in base-shear capacity is observed (see Figure 7b).
6. Determine the force–deformation relationship: calculate the force–deformation relationship (see Figure 7d) of the 'nth-mode' inelastic SDOF system based on the bilinear idealized pushover curve using the following relationships:

$$\frac{F_{sn}}{L_n} = \frac{V_{bn}}{M_n}, \tag{60}$$

$$D_n = \frac{u_{bn}}{\Gamma_n \Phi_{nr}}, \tag{61}$$

$$L_n = \frac{m_n}{\Gamma_n}, \tag{62}$$

7. Construct the inelastic SDOF system: Create the 'nth-mode' inelastic SDOF system (see Figure 7c) with unit mass and a force-deformation relationship defined in Step 2. Assign a damping ratio equal to the value for the nth mode of the original MDOF structure (ζn).
8. Compute the peak deformation of the SDOF model:
9. Calculate the peak deformation of the SDOF model, D_n, using Equation (62), where D_i is the absolute peak deformation for the ith ground motion record.
10. Calculate the target displacement for nth mode: determine the target displacement of the structure at the roof level for the nth mode, u_{rnt}, using Equation (63):

$$u_{rnt} = \Phi_{nr} \Gamma_n \hat{D}_n \tag{63}$$

11. Perform adaptive multi-stage pushover analyses (the MADP procedure): Apply gravity loads and perform adaptive displacement–control pushover analyses. This involves multiple sub-steps, including calculating new lateral load patterns based on the formation of plastic hinges and updating the target displacement values (see Steps 10.1–10.5.4) [64].
12. Repeat for the essential modes: Repeat Steps 3–11 for as many modes as deemed necessary. Typically the first three modes of vibration are adequate for most common buildings.
13. Determine the final structural responses: combine the modal responses obtained from Step 12 using an appropriate modal combination rule, such as square root of the sum of squares (SRSS) or complete quadratic combination (CQC), as guided by references [37].

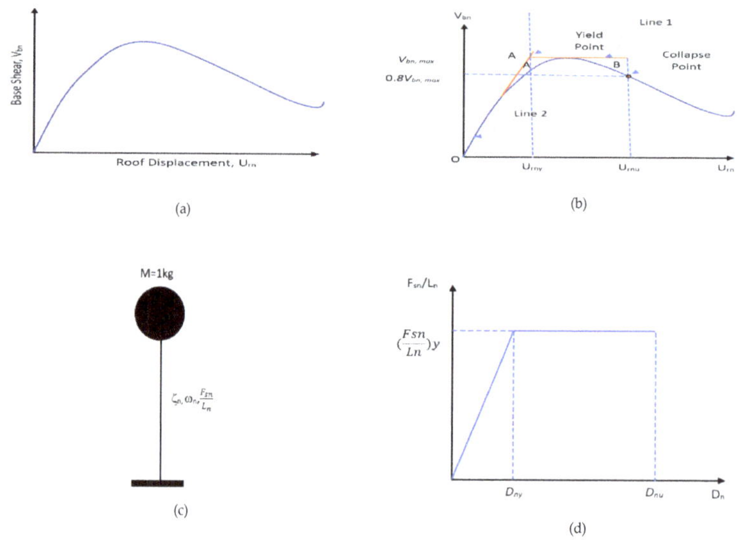

Figure 7. A schematic representation of (**a**) the structure's 'nth-mode' pushover curve, (**b**) the proposed bilinear idealization technique, (**c**) the 'nth-mode' inelastic SDOF model with unit mass (m = 1 kg), n damping ratio, n natural frequency, and (**d**) the force–deformation relationship as defined by a bilinear idealized pushover curve and Equations (60) and (61).

5.6. Optimal Mode Pushover Analysis

The optimal multi-mode pushover analysis method, developed by Attard and Fafitis in 2005 [74], is a dynamic seismic analysis approach that adjusts continually based on a building's evolving vibration properties during a simulation. In each step of the analysis, an "optimal" vibration shape for the building is calculated, forming the basis for a simplified vibration model. Lateral forces, based on the building's previous vibration characteristics, are then applied until the building yields. After each yielding event, the building's stiffness and vibration properties are recalculated and the simplified model is updated, informing the next step of the analysis. This iterative method is designed to provide valuable insights for creating buildings that are more resistant to earthquakes.

1. Define the objective function. Define the objective function using specified constraints and variables.
2. Compute a variant load pattern. Calculate a variant load pattern and optimally determine a representative single-DOF system (R-SDOF).
3. Determine the building's capacity curve. Establish the building's capacity curve and calculate a story force distribution based on the building's vibration properties. Use this distribution to push the building until a plastic hinge develops [5].
4. Capacity curve degradation: As the capacity curve degrades, new vibration properties of the degraded building are determined. Each time a yield stage is reached, this is treated as a 'new' building [5].
5. Analysis at different stages of degradation. Perform building analysis at various stages of degradation. This considers modes of vibration of the yielded building, which may have multiple yield mechanisms, in the final target displacement prediction.
6. Implement an optimization algorithm. Develop an optimization algorithm to minimize the target displacement error between the pushover and nonlinear time history analyses of a few buildings for a specific ground record [8].
7. Determine optimal parameters. Determine optimal capacity and demand parameters. Use these parameters to combine the individual mode shapes from each stage of

yielding into a single overall response spectrum for the structure. This integrated response spectrum can then be used to assess the performance and safety of the structure under various loading conditions.

6. A Summary of Research Initiatives on Evaluating Different Pushover Analysis Techniques

This section evaluates the efficacy of specific pushover analysis methods by comparing them to popular techniques, such as nonlinear time history analysis and incremental dynamic analysis. Their reliability and relevance are emphasized using post-earthquake data. Moreover, their adequacy is ascertained by contrasting them with outcomes from comprehensive seismic risk assessment methodologies.

6.1. A Comparison of Nonlinear Static Pushover Analysis Methods and Nonlinear Dynamic Analysis Methods

In seismic engineering, the comparison between nonlinear static pushover analysis and dynamic methods, including nonlinear time history and incremental dynamic analysis, has been a significant area of research. Rahmani et al. [72] introduced an adaptive upper-bound pushover analysis for tall buildings, showing its results align closely with nonlinear time history analysis. Concurrently, Rossetto et al. highlighted the efficiency of combining earthquake nonlinear response history analysis with other methods as an alternative to full dynamic analyses. Numerous studies have explored the use of pushover analysis methods alongside incremental dynamic and nonlinear time history analysis. The following sections present insights from research that employed various pushover techniques to assess seismic structural behavior.

The capacity spectrum method (CSM), a traditional tool in seismic engineering, has been found to sometimes underestimate the response of structures due to incorrect assumptions about the equivalent damping ratio. However, an improved version of the CSM [75] was introduced, which considers damage effects. This improved method has shown efficacy when compared with nonlinear dynamic procedures. Furthermore, the results from the study by Gentile and Galasso [76] indicate that nonlinear time history analysis of equivalent SDoF systems is not substantially superior to a nonlinear static analysis coupled with the capacity spectrum method.

The N2 method has been applied to assess the seismic behavior of asymmetric buildings. Research by Bosco et al. [77] found that improved nonlinear static methods provide more accurate estimates of maximum dynamic response than the original N2 method.

The extended N2 method has been applied to structures like an 8-story reinforced concrete building [67]. The results obtained showed that this method reproduces the real torsional behavior of the analyzed buildings effectively.

In the study by Yang Liu et al., the reinforced concrete (RC) shear wall structure's significance in medium- and high-rise buildings for seismic resistance was discussed. They introduced the spectrum-based pushover analysis (SPA) as an effective method for predicting seismic demand and compared its accuracy with the nonlinear response time history analysis (NLRHA) and other methods [67]. K. Shakeri and Samaneh Ghorbani further explored a pushover procedure tailored for buildings with bi-axial eccentricity under bi-directional seismic excitation [78]. Additionally, Phaiboon Panyakapo [79] validated the cyclic pushover analysis's efficacy in gauging seismic demands.

The seismic performance assessment of structures has been a focal point of research, leading to the development of various methodologies over the years. Among these, the modal pushover analysis (MPA) and its advanced variants have emerged as pivotal tools to address the inherent limitations of traditional pushover analysis, especially when dealing with structures that exhibit intricate modal interactions. K. Shakeri and Samaneh Ghorbani introduced a modal pushover procedure tailored for asymmetric-plan buildings subjected to bi-directional ground motions, demonstrating its accuracy in predicting peak seismic responses [78,79]. Jing-Zhou Zhang and colleagues developed an improved consecutive modal pushover (ICMP) procedure, which factored in the inelastic properties of structures and the interplay between vibration modes. Their findings suggested that this method surpassed conventional procedures in estimating seismic demands for tall edifices [80]. Additionally, F. Khoshnoudian and M. Kiani proposed a modified consecutive modal pushover (MCMP) procedure designed for one-way asymmetric-plan tall structures with dual systems. This method was found to outperform the FEMA load distribution and the standard MPA procedure in terms of accuracy [61].

When juxtaposed with the nonlinear response time history analysis (NLRHA), Yang Liu and J. Kuang's research indicated that the spectrum-based pushover analysis (SPA) method was adept at predicting seismic demands, especially in structures where the seismic demand and subsequent damage were pronounced [67]. H. R. Khoshnoud and Kadir Marsono further underscored that the MPA procedure was more precise in estimating the seismic demands of buildings compared to the FEMA356 procedure [81]. Notably, R. Goel and A. Chopra [54] have been instrumental in evaluating the efficacy of pushover analysis methodologies, emphasizing the importance of MPA in the seismic performance-based design of structures [54].

Exploring the realm of upper-bound pushover analysis, A. Y. Rahmani and colleagues introduced an adaptive upper-bound pushover analysis, tailored for the seismic performance assessment of tall buildings. This method considers both the effects of higher modes and the progressive changes in the dynamic characteristics of structures during inelastic response. The results of this approach were found to be closer to those obtained by NTHA [74]. An enhanced version of the upper-bound pushover procedure was proposed for the seismic assessment of high-rise moment-resisting steel frames, showcasing better results and alignment with outcomes from NTHA [63].

To conclude, the realm of seismic engineering has witnessed rigorous exploration and comparison between nonlinear static pushover analysis with dynamic methodologies. Pioneering works by researchers such as Rahmani et al. [64], Rossetto et al., Yang Liu et al. [82], and the notable contributions of Goel and Chopra [54] have underscored the significance of advanced pushover techniques. These methodologies, especially the modal pushover analysis and its variants, have emerged as robust tools, often aligning closely with the results from nonlinear time history analysis, thereby enhancing our understanding and prediction of seismic demands in structures.

7. Discussion

The study looks at the numerous nonlinear static analysis methods applied to evaluate the seismic vulnerabilities and dangers of various structures. Numerous studies have undoubtedly been conducted to address these concerns because of the shortcomings of traditional pushover methods. The pushover analysis method, a performance-based design approach, was developed to identify and evaluate the degree of deformation induced by lateral loads. The research methods, their uniqueness, and their limitations/benefits are clearly defined in the discussion below.

7.1. The First Modal Inertial Pushover Analysis Methods

- The capacity spectrum method (CSM). It is a graphical tool in seismic analysis that visually represents the supply-demand equation, accommodating multiple limit states on a structure's load-displacement curve. It is simpler and faster than other methods, making it efficient for evaluating a structure's earthquake resilience. While it can enhance result accuracy by providing larger seismic demand estimates, it has drawbacks: it uses highly damped elastic spectra, which may not accurately reflect inelastic systems' responses, and it may overestimate seismic demands in irregular, multi-story concrete buildings.
- The improved capacity spectrum method (ICSM) represents an advancement over the traditional capacity spectrum method by introducing new formulas for effective period and damping, as well as iterative procedures for determining target displacement and setting strength limitations. A notable feature of ICSM is its integration of constant-

ductility inelastic design spectra, which aligns a structure's capacity more closely with its inelastic spectrum. This method enhances the accuracy and realism in seismic performance evaluations but also adds complexity due to its iterative nature and the need for precise input parameters. While it offers a more accurate representation of seismic behavior, the increased complexity and reliance on accurate data may pose challenges in practical applications.

- The N2 method uses time-independent lateral displacement shapes and an inelastic response spectrum approximating structural and ground motion characteristics. While simple and useful for a range of buildings, including those with irregularities, it faces limitations in predicting seismic demands for structures influenced by higher modes and is less effective for torsionally flexible or high-rise buildings. The method's reliance on idealized spectra can also lead to inaccuracies in representing real structural behavior.
- The displacement coefficient method (DCM), introduced in FEMA 273 and later incorporated into FEMA 356, is unique in nonlinear static procedures (NSPs) for its direct estimation of peak inelastic displacements using modifying factors and its approach of calculating displacement coefficients for each mode of vibration. This method enhances the accuracy in predicting inelastic behavior and determining control-node/target displacements in NSPs. However, DCM faces challenges with less accuracy when applied to irregular or highly nonlinear structures and the complexity of calculating displacement coefficients for each vibration mode, which can limit its applicability in certain seismic scenarios or structural types.
- The improved coefficient method, as an advanced version of the coefficient method, uniquely estimates maximum forces and displacements in nonlinear static analysis, focusing on roof displacement to assess structural response. ICM offers more precise estimates of frame roof displacement and base shear than other methods, with a user-friendly approach that enhances design and evaluation processes and enables easy parametric studies. However, its accuracy declines for local response measures like story drifts, and it may introduce errors due to partial plastic mechanisms. Additionally, its reliance on highly damped elastic spectra for seismic demand estimation might not accurately represent the dynamic response of inelastic systems.
- The energy-based pushover analysis enhances nonlinear static analysis by integrating inertial properties and kinetic energy, making it particularly apt for seismic scenarios. It utilizes the fundamental mode shape, effective damping, and a velocity profile to represent a structure's dynamic behavior, offering deep insights into kinetic energy dissipation through plastic behavior. However, its effectiveness is limited for certain structural types, particularly braced structures, indicating a constraint in its applicability across different architectural designs. Despite this, the method stands out for its ability to merge dynamic aspects into static analysis, providing a more nuanced understanding of seismic impact on structures.
- The N1 method stands out for not requiring a well-defined equivalent single-degree-of-freedom (SDOF) system, facilitating its use in practical structures. It evaluates displacement demand using response spectrum analysis and adjusts for nonlinear behavior, offering accuracy comparable to the N2 method for certain load distributions. While effective for irregular or high-rise buildings by considering higher modes, the N1 method is primarily suitable for low-rise buildings and less effective for high-rise or torsionally flexible plan-asymmetric structures. A significant limitation is its inability to accurately predict seismic demands in structures heavily influenced by higher modes, owing to its limited capture of these effects.
- The mass proportional pushover (MPP) procedure, introduced by Kim and Kurama, simplifies pushover analysis for estimating peak seismic lateral displacement demands. This method stands out by removing the need for complex modal analysis or combination, even when higher-mode effects are significant. Its main advantage lies in conducting a single pushover analysis with lateral forces distributed in proportion

to the seismic masses at different levels of the structure, effectively consolidating higher-mode effects into one invariant distribution. This unique approach not only streamlines the estimation of displacement demands but also enhances practicality and efficiency in seismic analysis, particularly for structures where higher-mode influences are crucial.

7.2. The Invariant Multi-Mode Vectors Pushover Analysis

- The modal pushover analysis (MPA) is distinguished by its inclusion of higher modes in estimating seismic demands, offering more accuracy than traditional pushover methods and a precision akin to nonlinear response history analysis for inelastic range responses. Particularly effective for irregular structures, MPA outperforms adaptive pushover analysis in drift profiles and capacity determination, and it accurately captures phenomena in structures with strength irregularities. While simpler than adaptive pushover methods, it is especially adept at estimating demands for unsymmetrical-plan buildings. However, MPA may fall short in accuracy compared to dynamic analysis for buildings with a strong first story or lower half and might inaccurately estimate seismic demands in vertically irregular frames. It also requires incorporating enough modes to accurately estimate story drifts in such structures, highlighting a balance between its advanced capabilities and inherent limitations.
- The multi-mode pushover procedure, designed for steel moment-resisting frames, excels in estimating deformation demands during seismic events. This method distinguishes itself by accurately predicting seismic demands through the calculation of the target interstory drift and the use of modal combinations, considering the frequency content, higher-mode effects, and the interactions between modes. Its ability to provide superior response predictions with minimal computational effort marks a significant advantage, especially in practical engineering applications where efficiency is key. The focus on a comprehensive range of seismic influences ensures more precise deformation demand predictions, enhancing the structural resilience against seismic events. While the specific limitations of this method are not detailed, general challenges could include accurately capturing complex seismic interactions in various structural forms and the need for precise parameter considerations to maintain its effectiveness in diverse seismic scenarios.
- The improved modal pushover analysis (IMPA) method advances beyond the standard modal pushover analysis (MPA) by incorporating the redistribution of inertia forces following the yielding of structures, a significant deviation from MPA's invariant multi-mode lateral load patterns. Its primary enhancement lies in using the post-yielding structural deflection shape as an invariant lateral load pattern, offering a more accurate reflection of dynamic behavior during seismic events. To balance computational efficiency with accuracy, IMPA employs a two-phase lateral load distribution for the first mode and retains the MPA approach for higher modes. This method represents a sophisticated approach in seismic analysis, aiming to achieve a practical yet precise evaluation of structural responses under seismic loading. However, it might entail complexities in execution due to its advanced analytical requirements.
- The modified modal pushover analysis (MMPA), developed by simplifies seismic demand estimation in frame buildings, assumes that higher vibration modes remain elastic, thereby reducing computational efforts compared to traditional modal pushover analysis (MPA). MMPA often yields larger, more accurate demand estimates, particularly where MPA underestimates, but it can lead to increased conservatism if MPA already overestimates demand. Its primary limitation lies in lightly damped systems (less than 5% damping), where its linear elastic assumption for higher modes might not accurately represent inelastic behavior, potentially resulting in demand overestimation. The method's accuracy is also notably sensitive to the specific characteristics of buildings and ground motions, making its applicability variable depending on structural details and seismic conditions.

- Generalized pushover analysis (GPA) offers a balance between simplicity and accuracy. GPA utilizes generalized force vectors target deformation demands and generates an envelope of results. It is suitable for preliminary design and design review but limited to elastic behavior and higher-mode consideration.
- The upper-bound pushover analysis method, integrating aspects of the capacity spectrum method, energy balance, and modified modal pushover analysis, uniquely considers higher-mode effects in the nonlinear seismic evaluation of planar building frames, using a single load vector combining the first and a factored second mode shape. This method excels in predicting seismic demands in high-rise buildings, where higher-mode effects are significant, and offers a simplified procedure that yields more accurate nonlinear response estimates, applicable to both new and existing medium high-rise buildings. However, it tends to underestimate nonlinear responses at lower levels while overestimating them at higher levels in medium high-rise structures. Its accuracy may falter, particularly in irregular structures, and it is unsuitable for torsional flexible plan-asymmetric irregular structures. Additionally, the method overlooks variations in inertial load patterns and dynamic responses due to degrading strength and stiffness and does not account for other energy dissipation sources such as duration effects, viscous damping, and kinetic energy.
- The modified and extended upper-bound (MUB) pushover analysis method emerges as an advanced and versatile approach to estimating seismic demands for tall buildings. It seamlessly combines the advantages of the UB pushover method (higher-mode consideration) and the conventional pushover method (accurate lower-level demands) to provide a more balanced and accurate assessment. Its adaptability to various building configurations, including symmetric and unsymmetrical plans, makes it a valuable tool for a broad spectrum of structures. However, it is important to be mindful of its simplifying assumptions and ensure adequate data availability for a comprehensive analysis.
- The improved upper-bound (IUB) pushover analysis refines the lateral load pattern, resulting in more accurate predictions of roof displacements, story drift ratios, and plastic hinge rotations for tall buildings. It is more complex to implement than conventional pushover analysis but provides more reliable results. Its effectiveness and potential for improving seismic behavior prediction have been demonstrated through testing on various steel buildings.
- The extension of the improved upper-bound pushover analysis procedure (EIUB) extends the capabilities of conventional IUB to accurately predict the seismic response of irregular structures with setbacks. It incorporates a third mode of vibration to account for the irregularity characteristic, resulting in improved predictions of roof displacements, story drift ratios, and plastic hinge rotations for irregular buildings. However, its complexity and data requirements make it more suitable for advanced engineering analyses.
- The extended N2 method stands out as a refined version of the conventional N2 method, leveraging its strengths to enhance the accuracy of seismic demand estimation. Its unique feature lies in the integration of results from both basic pushover analysis and standard elastic modal analysis, providing a more reliable assessment of seismic demand distribution throughout the structure. This approach particularly excels in predicting seismic demand at the upper part of the building and at flexible edges, where the conventional N2 method may fall short. However, its complexity and data requirements make it more suitable for advanced engineering analyses compared to the conventional N2 method.
- Spectrum-based pushover analysis (SPA) utilizes response spectra to estimate seismic demand in tall buildings. It offers a balance between simplicity and accuracy, making it suitable for preliminary design and design review purposes. SPA's key features include response spectra integration, simplified pushover analysis, and the generation

of an envelope of results. However, its simplifying assumptions, limited consideration of higher modes, and overestimation of lower-level demands limit its applicability.

7.3. Adaptive Load Vectors

- Adaptive pushover analysis allows for the application of exact force profiles calculated by modal analysis at every step, resulting in a more realistic response that is consistent with the expected system behavior under seismic loading. Adaptive pushover analysis is often more accurate than conventional methods in determining the drift profiles and capacity of irregular structures. However, it may require greater computational efforts for calculating the response at discrete times.
- Adaptive modal pushover analysis (AMPA) is an advanced pushover analysis method that enhances seismic demand estimation through iterative modal load pattern updates based on the structure's deformed shape. This approach improves accuracy, particularly for taller buildings with complex geometries, but requires higher computational efforts and detailed structural information.
- The adaptive modal combination method (AMC) enhances the precision of seismic demand estimation by dynamically updating the target displacements using energy-based modal capacity curves and constant-ductility capacity spectra. AMC eliminates the need for pre-estimating the target displacement, a limitation of conventional pushover analysis. It accounts for instantaneous inertia force distributions during seismic events, making it more accurate for tall buildings with complex geometries and irregularities. However, its complexity and data requirements make it more suitable for advanced engineering analyses compared to conventional pushover analysis.
- Adaptive upper-bound (AUB) pushover analysis refines the accuracy of seismic demand estimation for tall buildings by considering higher-mode effects and dynamically updating the load pattern based on the structure's deformed shape. This approach improves the accuracy of predictions, particularly for taller buildings with complex geometries, but requires more detailed structural information and computational resources compared to conventional pushover analysis.
- Multi-mode adaptive displacement-based pushover (MADP) combines the strengths of both multi-modal and adaptive pushover approaches. It considers higher-mode effects and structural changes to provide a more comprehensive and realistic assessment of seismic demand. MADP utilizes the structure's first-mode inelastic mode shape to accurately estimate the target displacement, making it particularly useful for assessing the seismic performance of RC moment-resisting frames. However, its complexity and data requirements make it more suitable for advanced engineering analyses.
- Optimal mode pushover analysis (OMPA) is a dynamic seismic analysis method that continuously adjusts based on the building's evolving vibration properties during a simulation. It utilizes an "optimal" vibration shape for the building in each step, forming the basis for a simplified vibration model. Lateral forces are applied until yielding events occur, and the building's stiffness and vibration properties are recalculated after each yielding event. This iterative method provides more comprehensive and accurate seismic demand estimates, particularly for structures with complex geometries. However, OMPC's complexity and data requirements make it more suitable for advanced engineering analyses.

8. Conclusions

This paper primarily focuses on outlining methodologies and the evaluation steps for assessing the seismic performance of structures. Emphasizing the critical importance of this performance, especially in earthquake-sensitive areas, the paper underscores the direct impact on inhabitant safety and the economic viability of infrastructure. It presents a comprehensive review of various pushover analysis methods, detailing the unique characteristics and historical development of each technique. The paper also includes in-depth evaluations and comparisons of these methods, anchored in the analysis of previous research, with a specific focus on approaches such as the adaptive modal combination (AMC) method.

The main goal of this research was to offer a comprehensive overview of pushover analysis methods, targeting a varied audience that includes engineers, architects, and urban planners. The study acknowledges the challenge of comparing different pushover analysis techniques due to their complex nature. It focuses on assessing various methodologies regarding their computational efficiency and practical application, aiming to elucidate their effectiveness and suitability in specific contexts.

One of the salient advantages of the methods discussed is their adaptability across different geographical regions, with necessary modifications. These techniques can be instrumental in formulating regional seismic risk scenarios and strategies.

However, it is imperative to note that different pushover analysis methods might yield varying results when gauging seismic performance. These variations arise from the distinct parameters each method considers, with the influence of each parameter differing across methodologies. As a result, there is a pressing need to calibrate existing pushover analysis techniques or to develop new methods. Understanding the current pushover analysis methods in-depth is crucial to make informed modifications or developments, addressing challenges like site-specific characteristics and data uncertainties.

To enhance the existing pushover analysis methods:

- A diverse range of buildings could be examined using refined seismic assessment techniques.
- The outcomes of pre-seismic evaluations, possibly using digital tools, can be juxtaposed with post-seismic data to ascertain the accuracy of the pushover analysis methods.
- Advanced computational techniques, such as machine learning algorithms, can be integrated with post-seismic building evaluation data to rectify the limitations of traditional pushover analysis methods.

By employing these strategies, refinements can be introduced to the parameters used in pushover analysis, leading to significant advancements in the methodology. A profound understanding of each parameter's impact on structural performance is essential, and their relative significance should be determined to make the pushover analysis more robust and reliable.

Author Contributions: Conceptualization, K.K.K. and O.K.K.-B.; investigation, K.K.K.; writing—original draft preparation, K.K.K.; writing—review and editing, O.K.K.-B.; supervision, O.K.K.-B. All authors have read and agreed to the published version of the manuscript.

Funding: This research received no external funding.

Institutional Review Board Statement: Not applicable.

Informed Consent Statement: Not applicable.

Data Availability Statement: Data are contained withing the article.

Conflicts of Interest: The authors declare no conflicts of interest.

References

1. Singh, R.; Awadhiya, G.P. Pushover Analysis of Multistorey Building. Master's Thesis, Delhi College of Engineering, Delhi, India, 2008.
2. Shehu, R. Implementation of Pushover Analysis for Seismic Assessment of Masonry Towers: Issues and Practical Recommendations. *Buildings* **2021**, *11*, 71. [CrossRef]
3. Bektaş, N.; Kegyes-Brassai, O. Conventional RVS Methods for Seismic Risk Assessment for Estimating the Current Situation of Existing Buildings: A State-of-the-Art Review. *Sustainability* **2022**, *14*, 2583. [CrossRef]
4. Zameeruddin, M.; Sangle, K.K. Performance-based Seismic Assessment of Reinforced Concrete Moment Resisting Frame. *J. King Saud Univ. Eng. Sci.* **2021**, *33*, 153–165. [CrossRef]
5. Applied Technology Council. *Seismic Evaluation and Retrofit of Concrete Buildings*; aTe Applied Technology Council Proposition 122 Seismic Retrofit Practices Improvement Program; California Seismic Safety Commission: Redwood City, CA, USA, 1996; Volume 1, Available online: www.4downloader.ir (accessed on 14 August 2023).

6. British Standards Institution. *Eurocode 8: Design of Structures for Earthquake Resistance*; British Standards Institution: Milton Keynes, UK, 2005.
7. American Society of Civil Engineers; American Society of Civil Engineers and Structural Engineering Institute. *Seismic Evaluation and Retrofit of Existing Buildings: ASCE/SEI, 41-17*; American Society of Civil Engineers: Reston, VA, USA, 2017.
8. Federal Emergency Management Agency. *FEMA356/November 2000 Pre-Standard and Commentary for the Seismic Rehabilitation of Buildings*; Federal Emergency Management Agency: Washington, DC, USA, 2000.
9. Ministry of Land, Infrastructure, Transport and Tourism. *White Paper on Land, Infrastructure, Transport and Tourism, Japan*; Ministry of Land, Infrastructure, Transport and Tourism: Tokyo, Japan, 2019.
10. Ferraioli, M.; Lavino, A. Irregularity Effects of Masonry Infills on Nonlinear Seismic Behaviour of RC Buildings. *Math. Probl. Eng.* **2020**, *2020*, 4086320. [CrossRef]
11. Bisch, P.; Carvalho, E.; Degee, H.; Fajfar, P.; Fardis, M.; Franchin, P.; Kreslin, M.; Pecker, A.; Pinto, P.; Plumier, A.; et al. *Eurocode 8: Seismic Design of Buildings Worked Examples Support to the Implementation, Harmonization and Further Development of the Eurocodes*; Publications Office of the European Union: Luxembourg, 2012. [CrossRef]
12. Gupta, A.K. *Response Spectrum Method in Seismic Analysis and Design of Structures*; Routledge: New York, NY, USA, 1990; pp. 1–170. [CrossRef]
13. Chopra, A.K.; Goel, R.K. Evaluation of NSP to Estimate Seismic Deformation: SDF Systems. *J. Struct. Eng.* **2000**, *126*, 482–490. [CrossRef]
14. Abass, H.A.; Jarallah, H.K. Comparative Study of the Seismic Assessment according to ATC-40, FEMA-356 and FEMA-440 for Existing Hospital Building Located at Baghdad City. Available online: https://www.acapublishing.com/dosyalar/baski/PACE_2021_342.pdf (accessed on 17 November 2022).
15. Benavent-Climent, A.; Escolano-Margarit, D. Shaking table tests of structures with hysteretic dampers: Experimental results versus prediction using non-linear static methods. *Bull. Earthq. Eng.* **2012**, *10*, 1857–1883. [CrossRef]
16. Kalkan, E.; Kunnath, S.K. Assessment of current nonlinear static procedures for seismic evaluation of buildings. *Eng. Struct.* **2007**, *29*, 305–316. [CrossRef]
17. Paret, T.F.; Sasaki, K.K.; Eilbeck, D.H.; Freeman, S.A. Approximate Inelastic Procedures to Identify Failure Mechanisms From Higher Mode Effects. Paper No. 966; In *Eleventh World Conference on Earthquake Engineering*; Elseviser: Amsterdam, The Netherlands, 1996.
18. Chopra, A.K.; Goel, R.K.; Chintanapakdee, C. Evaluation of a Modified MPA Procedure Assuming Higher Modes as Elastic to Estimate Seismic Demands. *Earthq. Spectra* **2004**, *20*, 757–778. [CrossRef]
19. Jan, T.S.; Liu, M.W.; Kao, Y.C. An upper-bound pushover analysis procedure for estimating the seismic demands of high-rise buildings. *Eng. Struct.* **2004**, *26*, 117–128. [CrossRef]
20. Lin, J.-L.; Tsaur, C.-C.; Tsai, K.-C. Two-degree-of-freedom modal response history analysis of buildings with specific vertical irregularities. *Eng. Struct.* **2019**, *184*, 505–523. [CrossRef]
21. Jun, D.-H.; Kang, H.-G. Nonlinear Response Analysis of Multistory Buildings Subjected to Synthetic Motions Compatible with Design Spectrum. In Proceedings of the 15th World Conference on Earthquake Engineering, Lisbon, Portugal, 24–28 September 2012.
22. Lima, C.; Martinelli, E.; Faella, C. Simplified Nonlinear Static Procedures for Seismic Assessment of Masonry Infilled RC Structures Confinement of Masonry Columns with FRP Materials View Project International Journal of Structural Glass and Advanced Materials Research View Project. Available online: https://www.researchgate.net/publication/320034814 (accessed on 3 July 2023).
23. Jalayer, F.; Cornell, C.A. Alternative non-linear demand estimation methods for probability-based seismic assessments. *Earthq. Eng. Struct. Dyn.* **2009**, *38*, 951–972. [CrossRef]
24. Gupta, B.; Kunnath, S.K. Adaptive Spectra-Based Pushover Procedure for Seismic Evaluation of Structures. *Earthq. Spectra* **2000**, *16*, 367–391. [CrossRef]
25. Papanikolaou, V.K.; Elnashai, A.S.; Pareja, J.F. Evaluation of Conventional and Adaptive Pushover Analysis Ii: Comparative Results. *J. Earthq. Eng.* **2006**, *10*, 127–151. [CrossRef]
26. Antoniou, S.; Pinho, R. Development and Verification of a Displacement-Based Adaptive Pushover Procedure. *J. Earthq. Eng.* **2004**, *8*, 643–661. [CrossRef]
27. Deierlein, G.G.; Reinhorn, A.M.; Willford, M.R. Nonlinear Structural Analysis for Seismic Design a Guide for Practicing Engineers. Available online: www.curee.org (accessed on 22 October 2022).
28. Themelis, S. Pushover Analysis for Seismic Assessment and Design of Structures. 2008. Available online: http://www.ros.hw.ac.uk/handle/10399/2170 (accessed on 22 October 2022).
29. Fajfar, P. Analysis in seismic provisions for buildings: Past, present and future. The fifth Prof. Nicholas Ambraseys lecture. *Bull. Earthq. Eng.* **2018**, *16*, 2567–2608. [CrossRef]
30. Bracci, J.M.; Kunnath, S.K.; Reinhorn, A.M. Seismic Performance and Retrofit Evaluation of Reinforced Concrete Structures. *J. Struct. Eng.* **1997**, *123*, 3–10. [CrossRef]
31. Antoniou, S. Advanced Inelastic Static Analysis for Seismic Assessment of Structures. Ph.D. Thesis, Imperial College London, London, UK, 2002.
32. Khan, M.A. Performance-Based Design and Retrofit Methods for Buildings. In *Earthquake-Resistant Structures*; Elsevier: Amsterdam, The Netherlands, 2013; pp. 317–351. [CrossRef]

33. Bektaş, N.; Kegyes-Brassai, O. A comparative study of rapid visual screening methods to detailed seismic assessment of a reinforced concrete residential building. In Proceedings of the 3rd European Conference on Earthquake Engineering & Seismology, Bucharest, Romania, 4–9 September 2022.
34. Ghobarah, A. Performance-based design in earthquake engineering: State of development. *Eng. Struct.* **2001**, *23*, 878–884. [CrossRef]
35. Freeman, S.A.; Nicoletti, J.P.; Tyrell, J.V. Evaluations of existing buildings for seismic risk—a case study of Puget Sound Naval Shipyard. In Proceedings of the U.S. National Conference on Earthquake Engineering, Oakland, CA, USA, 18–20 June 1975; pp. 113–122.
36. Freeman, S.A. Review of the Development of the Capacity Spectrum Method. *ISET J. Earthq.Technol.* **2004**, *41*, 1–13.
37. Comartin, C.D.; Niewiarowski, R.W.; Freeman, S.A.; Turner, F.M.; Comartin, R.W.N.C.D.; Gajan, S.; Raychowdhury, P.; Hutchinson, T.C.; Kutter, B.L.; Stewart, J.P. Seismic Evaluation and Retrofit of Concrete Buildings: A Practical Overview of the ATC 40 Document. *Earthq. Spectra* **2000**, *16*, 241–261. [CrossRef]
38. Aşıkoğlu, A.; Vasconcelos, G.; Lourenço, P.B. Overview on the Nonlinear Static Procedures and Performance-Based Approach on Modern Unreinforced Masonry Buildings with Structural Irregularity. *Buildings* **2021**, *11*, 147. [CrossRef]
39. Kuria, K.K.; Kegyes-Brassai, O.K. Nonlinear Static Analysis for Seismic Evaluation of Existing RC Hospital Building. *Appl. Sci.* **2023**, *13*, 11626. [CrossRef]
40. FEMA. *Improvement of Nonlinear Static Seismic Analysis Procedures FEMA 440*; Applied Technology Council: Redwood City, CA, USA, 2005.
41. Goel, R.K. Evaluation of Current Nonlinear Static Procedures Using Strong Motion Records. In *SMIP07 Seminar on Utilization of Strong-Motion Data*; 2007; pp. 57–80.
42. *EN 1998-1*; Eurocode 8: Design of Structures for Earthquake Resistance—Part 1: General Rules, Seismic Actions and Rules for Buildings. British Standard Institution: Milton Keynes, UK, 2004.
43. Peter, F. N2-A Method for Nonlinear Seismic Analysis of Regular Buildings. In Proceedings of the Ninth World Conference on Earthquake Engineering, Tokyo, Japan, 2–9 August 1988.
44. Fajfar, P. A Nonlinear Analysis Method for Performance-Based Seismic Design. *Earthq. Spectra* **2000**, *16*, 573–592. [CrossRef]
45. Seminar, C.H. *Nehrp Guidelines for the Seismic Rehabilitation of Buildings (FEMA 273)*; Federal Emergency Management Agency: Washington, DC, USA, 1997.
46. Hernández-Montes, E.; Kwon, O.-S.; Aschheim, M.A. An Energy-Based Formulation for First-and Multiple-Mode Nonlinear Static (Pushover) Analyses. *J. Earthq. Eng.* **2004**, *8*, 69–88. [CrossRef]
47. Mendes Pontes, A.R. *The Forests of Maraca, Northern Brazilian Amazonia: Their Structure, Phenology, Fruit Assessment and an Illustrative Guide to Their Fruit*; Editora da Universidade Federal de Pernambuco: Vitória de Santo Antão, Brazil, 2013.
48. Kotanidis, C.; Doudoumis, I.N. Energy-Based Approach of Static Pushover Analysis. In Proceedings of the 14th World Conference on Earthquake Engineering, Beijing, China, 12–17 October 2008.
49. Leelataviwat, S.; Saewon, W.; Goel, S.C. An Energy Based Method for Seismic Evaluation of Structures. In Proceedings of the 14th World Conference on Earthquake Engineering, Beijing, China, 12–17 October 2008.
50. Coleman, J.; Spacone, E. Localization Issues in Force-Based Frame Elements. *J. Struct. Eng.* **2001**, *127*, 1257–1265. [CrossRef]
51. Building Seismic Safety Council. *NEHRP Recommended Provisions for Seismic Regulations for new Buildings and Other Structures Part 2: Commentary (FEMA 369)*; Building Seismic Safety Council: Washington, DC, USA, 2000.
52. Kim, S.-P.; Kurama, Y.C. An alternative pushover analysis procedure to estimate seismic displacement demands. *Eng. Struct.* **2008**, *30*, 3793–3807. [CrossRef]
53. Worku, A.M.; Hsiao, P.-C. An improved first-mode-based pushover analytical procedure for assessing seismic performance of special moment resisting frame building structures. *Eng. Struct.* **2021**, *252*, 113587. [CrossRef]
54. Chopra, A.K.; Goel, R.K. A modal pushover analysis procedure for estimating seismic demands for buildings. *Earthq. Eng. Struct. Dyn.* **2001**, *31*, 561–582. [CrossRef]
55. Ferraioli, M. Multi-mode pushover procedure for deformation demand estimates of steel moment-resisting frames. *Int. J. Steel Struct.* **2017**, *17*, 653–676. [CrossRef]
56. Jalilkhani, M.; Ghasemi, S.H.; Danesh, M. A multi-mode adaptive pushover analysis procedure for estimating the seismic demands of RC moment-resisting frames. *Eng. Struct.* **2020**, *213*, 110528. [CrossRef]
57. Mao, J.; Zhai, C.; Xie, L. An improved modal pushover analysis procedure for estimating seismic demands of structures. *Earthq. Eng. Eng. Vib.* **2008**, *7*, 25–31. [CrossRef]
58. Sucuoğlu, H.; Günay, M.S. Generalized force vectors for multi-mode pushover analysis. *Earthq. Eng. Struct. Dyn.* **2011**, *40*, 55–74. [CrossRef]
59. Poursha, M.; Khoshnoudian, F.; Moghadam, A. A consecutive modal pushover procedure for estimating the seismic demands of tall buildings. *Eng. Struct.* **2009**, *31*, 591–599. [CrossRef]
60. Kim, S.; Enzo, D. Pushover Analysis Procedure in Earthquake Engineering. *Earthq. Spectra* **1999**, *15*, 417–434. [CrossRef]
61. Khoshnoudian, F.; Kiani, M. Modified consecutive modal pushover procedure for seismic investigation of one-way asymmetric-plan tall buildings. *Earthq. Eng. Eng. Vib.* **2012**, *11*, 221–232. [CrossRef]
62. Poursha, M.; Samarin, E.T. The modified and extended upper-bound (UB) pushover method for the multi-mode pushover analysis of unsymmetric-plan tall buildings. *Soil Dyn. Earthq. Eng.* **2015**, *71*, 114–127. [CrossRef]

63. Rahmani, A.Y.; Bourahla, N.; Bento, R.; Badaoui, M. An improved upper-bound pushover procedure for seismic assessment of high-rise moment resisting steel frames. *Bull. Earthq. Eng.* **2018**, *16*, 315–339. [CrossRef]
64. Rahmani, A.Y.; Badaoui, M.; Bourahla, N.; Bento, R. Extension of the improved upper-bound pushover analysis for seismic assessment of steel moment resisting frames with setbacks. *Bull. Earthq. Eng.* **2022**, *20*, 7609–7640. [CrossRef]
65. Kreslin, M.; Fajfar, P. The extended N2 method considering higher mode effects in both plan and elevation. *Bull. Earthq. Eng.* **2012**, *10*, 695–715. [CrossRef]
66. Chopra, A.K. *Dynamics of Structures: Theory and Applications to Earthquake Engineering*, 4th ed.; Prentice Hall: Haboken, NJ, USA, 2011.
67. Liu, Y.; Kuang, J.S. Spectrum-based pushover analysis for estimating seismic demand of tall buildings. *Bull. Earthq. Eng.* **2017**, *15*, 4193–4214. [CrossRef]
68. Rofooei, F.; Rofooei, F.R.; Attari, N.K.; Rasekh, A.; Shodja, A.H. Adaptive pushover analysis. *Asian J. Civ. Eng. (Build. Hous.)* **2007**, *8*, 343–358.
69. Papanikolaou, V.K.; Elnashai, A.S. Evaluation of Conventional and Adaptive Pushover Analysis I: Methodology. *J. Earthq. Eng.* **2005**, *9*, 923–941. [CrossRef]
70. Sürmeli, M.; Yüksel, E. An adaptive modal pushover analysis procedure (VMPA-A) for buildings subjected to bi-directional ground motions. *Bull. Earthq. Eng.* **2018**, *16*, 5257–5277. [CrossRef]
71. Kalkan, E.; Kunnath, S.K. Adaptive Modal Combination Procedure for Nonlinear Static Analysis of Building Structures. *J. Struct. Eng.* **2006**, *132*, 1721–1731. [CrossRef]
72. Rahmani, A.; Bourahla, N.; Bento, R.; Badaoui, M. Adaptive upper-bound pushover analysis for high-rise moment steel frames. *Structures* **2019**, *20*, 912–923. [CrossRef]
73. Chopra, A.K.; Goel, R.K. *A Modal Pushover Analysis Procedure to Estimate Seismic Demands for Buildings*; Pacific Earthquake Engineering Research Center: Berkeley, CA, USA, 2001.
74. Attard, T.; Fafitis, A. Modeling of higher-mode effects using an optimal multi-modal pushover analysis. *WIT Trans. Built Environ.* **2005**, *81*, 405–414.
75. Barbat, A.H.; Pujades, L.G.; Lantada, N. Seismic damage evaluation in urban areas using the capacity spectrum method: Application to Barcelona. *Soil Dyn. Earthq. Eng.* **2008**, *28*, 851–865. [CrossRef]
76. Gentile, R.; Galasso, C. Simplicity versus accuracy trade-off in estimating seismic fragility of existing reinforced concrete buildings. *Soil Dyn. Earthq. Eng.* **2021**, *144*, 106678. [CrossRef]
77. Bosco, M.; Ghersi, A.; Marino, E.M.; Rossi, P.P. Comparison of nonlinear static methods for the assessment of asymmetric buildings. *Bull. Earthq. Eng.* **2013**, *11*, 2287–2308. [CrossRef]
78. Shakeri, K.; Ghorbani, S. A pushover procedure for seismic assessment of buildings with bi-axial eccentricity under bi-directional seismic excitation. *Soil Dyn. Earthq. Eng.* **2015**, *69*, 1–15. [CrossRef]
79. Panyakapo, P. Cyclic Pushover Analysis procedure to estimate seismic demands for buildings. *Eng. Struct.* **2014**, *66*, 10–23. [CrossRef]
80. Zhang, J.-Z.; Jiang, J.; Li, G.-Q. An improved consecutive modal pushover procedure for estimating seismic demands of multi-storey framed buildings. *Struct. Des. Tall Spéc. Build.* **2017**, *26*, e1336. [CrossRef]
81. Khoshnoud, H.R.; Marsono, K. Assessment of *FEMA*356 nonlinear static procedure and modal pushover analysis for seismic evaluation of buildings. *Struct. Eng. Mech.* **2012**, *41*, 243–262. [CrossRef]
82. Yang, P.; Wang, Y. A Study on Improvement of Pushover Analysis. In Proceedings of the12th World Conference on Earthquake Engineering, Auckland, New Zealand, 30 January–4 February 2000.

MDPI AG
Grosspeteranlage 5
4052 Basel
Switzerland
Tel.: +41 61 683 77 34

Applied Sciences Editorial Office
E-mail: applsci@mdpi.com
www.mdpi.com/journal/applsci

www.ingramcontent.com/pod-product-compliance
Lightning Source LLC
Chambersburg PA
CBHW061557220926
51850CB00005B/358
9783725851638